# Valuable Metals Recovery by Mineral Processing and Hydrometallurgy

# Valuable Metals Recovery by Mineral Processing and Hydrometallurgy

Editors

**Marinela Panayotova**
**Vladko Panayotov**

MDPI

Basel • Beijing • Wuhan • Barcelona • Belgrade • Novi Sad • Cluj • Manchester

*Editors*
Marinela Panayotova
University of Mining and Geology
Sofia
Bulgaria

Vladko Panayotov
Bulgarian Academy of Sciences
Sofia
Bulgaria

*Editorial Office*
MDPI AG
Grosspeteranlage 5
4052 Basel, Switzerland

This is a reprint of articles from the Special Issue published online in the open access journal *Minerals* (ISSN 2075-163X) (available at: https://www.mdpi.com/journal/minerals/special_issues/VMRMPH).

For citation purposes, cite each article independently as indicated on the article page online and as indicated below:

| Lastname, A.A.; Lastname, B.B. Article Title. *Journal Name* **Year**, *Volume Number*, Page Range. |
|---|

**ISBN 978-3-7258-2157-0 (Hbk)**
**ISBN 978-3-7258-2158-7 (PDF)**
**doi.org/10.3390/books978-3-7258-2158-7**

# Contents

# About the Editors

**Marinela Panayotova**

Marinela Panayotova is a Full Professor in Mineral Processing and Recycling at the University of Mining and Geology, Sofia, Bulgaria. She has an M.Sc. in Inorganic and Analytical Chemistry and in Environmental Engineering, a PhD in Physical Chemistry (Corrosion Prevention) and is an Assoc. Prof. in Physical Chemistry. According to Elsevier (2019 and 2020), she was ranked in the World's Top 2% Scientists. Her professional and research interests are focused on the processing of raw materials (ores, concentrates, liquid and solid waste), covering hydrometallurgical, flotational, physicochemical, chemical and electrochemical methods for recycling and processing; environment protection, involving processes of mobilizing and immobilizing water, soil and air pollutants; technologies for wastewater and gases purification; EIA; the chemistry of RES; corrosion and the corrosion protection of metals and concrete; and nanomaterials. She is an author/coauthor of 21 books (monographs, textbooks, and teaching aids), seven of them written in English; eleven invited chapters of monographs written by an international team in English; and over 200 publications. She is the co-owner of 24 issued patents and utility models and the leader/a participant in 47 projects and 8 EIAs. She won Inventor of the Year in 2014, Bulgaria. She is a member of the EGRC, ECE, and UN (5th, 6th and 8th sessions). She is the Editor in Chief of the *Sustainable Extraction and Processing of Raw Materials (SEPRM)* Journal and a member of the Editorial Board of four other international journals.

**Vladko Panayotov**

Vladko Panayotov is a Corresponding Member of the Bulgarian academy of science and a part-time Professor of Flotation, the University of Mining and Geology, Sofia. He is an engineer in the field of chemical technology and a Ph.D. and a Doctor of Technical Sciences (Dr. Habil.) in Mineral Processing. His fields of interest include new technologies in mineral processing (lead–zinc, copper–pyrite, and copper–molybdenum selections), wastewater and washing acid purification, technogenic waste utilization, and electrochemical technologies. Teams led by Mr. Panayotov have implemented 7 technologies and 4 systems for flotation improvement. He has managed co-managed more than 20 research and industry implementation projects. The technologies he and his team have developed have been presented in many countries, including Germany, France, Belgium, Italy, Turkey, Russia, Chile, India, Peru, Bolivia, etc. He is the owner and co-owner of 42 inventions, patents and utility models. He is the author/co-author of 24 books (monographs, textbooks, and teaching aids); 7 invited chapters of monographs by an international team (written in English); and over 210 publications. He was a two-term member of the Scientific Council of the International Mineral Processing Congress. He is the recipient of more than 20 international and national awards for contributions to the development of science and technologies, receiving among them the highest Bulgarian orders "Stara Planina - first degree" and "St. Cyril and Methodius" with a necklace, for his contribution to the development of mining and metallurgy (at the industrial and education level) in Bulgaria; a record of his name in the Golden Book of Bulgarian Inventors; Inventor of the Year, 2014; and the Bernardo O'Higgins Order of Chile.

# Preface

This reprint is based on the papers published in the Special Issue of Minerals on "Valuable Metals Recovery by Mineral Processing and Hydrometallurgy".

Metals will be increasingly needed in future to meet the demand for the sustainable development of our society. Use of lower-grade and more complex for processing ores and metal extraction waste will be needed. This requires novel approaches and technologies for mineral processing and the hydrometallurgical extraction of metals.

This Special Issue presents excellent examples of recent relevant research on the sustainable and environmentally friendly recovery of metals from difficult-to-process or low-grade ores, low-grade concentrates and metal extraction waste.

This reprint can serve as a valuable knowledge resource for researchers, engineers, and students interested in delving into the latest advancements in the recovery of metals.

We are thankful to all authors for their contributions and the Minerals Editorial Office for their professional guidance and management of this Special Issue. It would not have been possible to create this reprint without their support.

**Marinela Panayotova and Vladko Panayotov**
*Editors*

minerals

MDPI

*Editorial*

# Editorial for the Special Issue: "Valuable Metals Recovery by Mineral Processing and Hydrometallurgy"

**Marinela Panayotova [1,*] and Vladko Panayotov [2]**

1 Department of Chemistry, University of Mining and Geology St. Ivan Rilski, 1700 Sofia, Bulgaria
2 Engineering Sciences Unit, Bulgarian Academy of Sciences, 1000 Sofia, Bulgaria
* Correspondence: marichim@mgu.bg

Achieving the sustainable development of our society and facilitating breakthrough innovations necessitates metals. All metals (not only those classified as precious and critical or strategic for the present and near future), required both for advanced technologies and our everyday life, are valuable, and their demand will increase with time.

Humankind has been driven to use more efficient and complex techniques for processing ores, which requires novel approaches and technologies for mineral processing and the extraction of metals.

In order to obtain enough metals for future development, the recovery of metals from secondary sources, such as solid and liquid waste from mining, mineral processing, and extractive metallurgy, has to be considered.

Potential challenges to future metal extraction technologies also include soaring energy prices, the necessity to secure safe and clean water, and the need to use resources efficiently and comprehensively, while protecting the environment.

The solutions to all the aforementioned problems require new theoretical backgrounds and technical developments, including reagents and operational innovations of processes and technologies, when the problems associated with both primary ores and secondary technogenic resources are tackled.

Papers discussing all the above mentioned aspects and challenges and presenting solutions to metal recovery through mineral processing and hydrometallurgy are gathered in this Special Issue.

All metals, after being mined, are processed through a mineral beneficiation stage, followed by metallurgy/hydrometallurgy, or sometimes enter directly to hydrometallurgy.

The flotation concentration of metal-bearing ores is probably the most widely used mineral beneficiation process. Its efficiency depends on the minerals' preparation, i.e., grinding, flotation reagent types and concentrations, and other parameters of the flotation process, such as pulp pH values, oxidation–reduction conditions, temperature, applied external energy impacts, etc.

The first part of this collection is devoted to flotation.

The effects of different grinding media (mild steel, MS, and stainless steel, SS) and milling conditions (wet or dry, the latter with a potential to decrease the freshwater use) on the results from copper pyrite (Cu-Py) flotation are studied [1]. Galvanic interaction between sulphide minerals and the grinding media is eliminated by dry grinding. This results in higher redox potential, irrespective of type of milling media compared to wet grinding, thus leading to a high flotation rate of Cu and obtaining Cu concentrate with a higher grade, especially when SS is used. However, in this case, the Cu recovery is lower due to the instability of the froth structure and the coalescence of air bubbles in flotation. Flotation froth is the most stable structure in the case where MS balls are used in milling. The highest separation efficiency, and hence, the best flotation performance, is achieved under wet grinding conditions with MS balls.

**Citation:** Panayotova, M.; Panayotov, V. Editorial for the Special Issue: "Valuable Metals Recovery by Mineral Processing and Hydrometallurgy". *Minerals* **2024**, *14*, 653. https://doi.org/10.3390/min14070653

Received: 18 June 2024
Accepted: 24 June 2024
Published: 26 June 2024

Furthermore, studies on the dependence of the technological results of non-ferrous sulfide ores (copper–arsenic-bearing and non-bearing arsenic, lead–zinc, and polymetallic) flotation on the pulp oxidation–reduction potential (presented with respect to the standard hydrogen electrode, SHE, i.e., Eh) are reviewed [2]. Findings on the relation of Eh and collector-less flotation are presented. Changes in the pulp potential due to different gases application and various reagents addition are considered. The influence of the grinding medium on the pulp Eh, and hence, on the flotation results, is presented through various examples. The relation between the oxidation–reduction potential and reagent effects is exhibited and explained. pH–Eh ranges of different minerals' flotation, as recorded in various studies, are summarized and visualized jointly for all mentioned ores. It is concluded that the pulp Eh and pH values govern the reactions occurring between the various minerals in the pulp and collectors, and thus, the flotation results. It is proposed to use jointly pulp Eh and pH values and the pH–Eh diagrams (that include the main species of the elements forming the minerals in a given flotation pulp and the main reagent species) to control flotation process.

As mentioned, use of the most suitable flotation reagents is decisive for the overall process results. This is especially important when it comes to critical high-tech and strategic metals such as molybdenum (Mo), rhenium (Re), and Cu. It is known that 25% to 30% of these metals end up in rough flotation tailings when conventional reagents are used. A new reagent—dithiopyrylmethane (DTM) composite—is proposed [3]. It forms a complex DTM–Re compound and chemically adsorbs on rhenium-containing molybdenite, ensuring an increase in Re recovery into the bulk Cu–Mo–Re concentrate by 17%, reducing the loss of Re with flotation tailings twofold, and subsequently providing 97.6% Re extraction from molybdenum concentrate by autoclave leaching.

Finally, in this part of the Special Issue, the results of theoretical and experimental studies are presents on (i) the substantiation and development of criteria that characterize the efficiency of overcoming the energy barrier of internal cohesion forces during the grinding of complex ores; and (ii) the substantiation of comprehensive criteria for assessing the effectiveness of physical and energy impacts during ore beneficiation and a complex criterion that can be used for the intensification of the entire mineral beneficiation process [4].

Hydrometallurgical technologies are a powerful means of extracting metals both from concentrates and directly from poor copper ores and ores containing precious metals or rare earth elements (REEs). The second part of this Special Issue is devoted to hydrometallurgical methods.

Bioleaching is an environmentally friendly technique that uses microorganisms to remove metals from ore in cases when the traditional extraction methods (most often applying strong acids or alkaline solutions) are not economically viable. In addition to the most well-known and widely used microorganisms, new ones are being explored. A new strain of *Leptospirillum* sp., namely, *L. ferrodiazotropum* Ksh-L, was isolated from a dump-bioleaching system of the Kashen copper–molybdenum mine and studied for its ability to leach Cu and iron (Fe) from ore [5]. Ksh-L can efficiently oxidize chalcopyrite, thus mobilizing Cu and Fe in leachate. When *L. ferrodiazotropum* Ksh-L is used in association with *At. thiooxidans* ATCC 19377, the leaching efficiency is nearly doubled. It is proposed to use this association to enhance the efficiency of copper extraction from chalcopyrite.

Gold (Au) recovery from refractory ores is a long-standing problem that still attracts scientists' attention. Guner et al. demonstrated that diagnostic leaching, in combination with automated mineralogy, can be used as a fast and practical method to examine the processability of run-of-mine ore and flotation concentrate [6]. Diagnostic leaching provides direct indicators of the material's processability, while the automated mineralogy gives very detailed information about the gold deportment and minerals that are present in the gangue and can affect the further processing of Au-bearing concentrate. Based on the findings of the proposed combination of methods for mineral study, the consecutive steps are proposed for the material treatment and Au recovery.

Most gold deposits are associated with sulphides such as pyrite and arsenopyrite. When the minerals exhibit refractory properties due to Au occlusion in the minerals' matrix, conventional cyanidation usually extracts less than 50% of the Au that is present in ores. In such cases, simultaneous pressure oxidation and cyanidation is often applied in order to dissolve the precious metal. The effectiveness of this extraction process is influenced by the cyanide (NaCN) concentration, oxygen pressure, and temperature. Optimal ranges of those leaching conditions have been proposed by Soto-Uribe et al. [7].

Carlin-type gold ore is typical refractory ore where gold ore grains are finely distributed and present as invisible or submicroscopic gold, encapsulated in arsenopyrite and pyrite. The leaching of these ores is difficult because the fine gold is wrapped in pyrite and arsenopyrite. The oxidation roasting pre-treatment technique is proposed to treat such gold ore [8]. A two-stage roasting process is proposed: in the first stage, arsenic is removed; in the second stage, sulfur is removed. The roasting almost completely oxidizes pyrite and arsenopyrite to hematite. The share of exposed gold and oxide-encapsulated gold in the ore increases considerably (over 88% of the total Au) and the share of sulfide-encapsulated gold is reduced almost to 0%. Exposed Au and the Au encapsulated in the hematite porous structure is easily leached by cyanide solution, ensuring total Au extraction of over 83% from this ore type.

Further study revealed the oxidation and dissolution behavior of pyrite in an alkaline environment, with the aim of facilitating Au exposure in the subsequent leaching [9]. The conclusion from the applied electrochemical testing techniques (cyclic voltammetry, linear scanning voltammetry, current–time, and electrochemical impedance spectroscopy), complemented with analytical methods such as XPS and SEM–EDS, is that the oxidative dissolution of pyrite occurs preferably in an alkaline environment with a pH of around 12, and an oxidation potential above 0.9 V (vs. SHE). As a result of pyrite oxidation, stable species (polysulfides, $S_x^{2-}$, and thiosulfates, $S_2O_3^{2-}$) are generated, which possess oxidizing and coordinating abilities in the alkaline system. The idea has been posited to leach gold by making full use of the sulfur in gold-bearing sulfide minerals.

Hydrometallurgy is also used in obtaining other metals. One possibility is intensifying the leaching process through the preliminary energy impact (a combination of thermal, electrochemical, and ultrasonic treatment) to ensure the intensive breakdown of mineral complexes and grains in eudialyte concentrate, with the aim of enhancing zirconium (Zr) and REE recovery [10]. The proposed treatment ensures 87.0–91.7% recovery of Zr and 76.0–81.1% recovery of REE in the pregnant leach solution (PLS). A process for the selective recovery of Zr and REEs from the PLS of eudialyte concentrate is proposed, based on a combination of chemical precipitation, sorption, and elution methods.

Currently, when rich ore deposits are practically exhausted all over the world and the development of depleted and/or complex ores is required, more and more attention is being paid to extraction from old mining operations and metallurgical waste, which are often richer sources of metals than currently developed deposits. Another unconventional source of metals is liquid waste from mining, mineral processing, and extractive metallurgy. The third part of this Special Issue is dedicated to metal extraction from solid and liquid mining waste.

For example, Kiprono et al. pointed out that a considerable degree of mining and mineral processing tailings that hold strategic metals are available in Kenya [11]. Modern flotation and hydrometallurgical methods are outlined as an important solution to efficiently extract strategic and critical metals from these secondary sources while reducing environmental effects.

Brown corundum dust is a solid waste generated during the preparation of brown corundum using bauxite as raw material. The dust has a relatively high gallium (Ga) content. However, Ga cannot be leached with satisfactory recovery through the application of conventional water and/or acid leaching. The reason is that Ga is dispersed in the potassium-rich phase, wrapped by amorphous silicate and the corundum phase. A process route involving roasting activation (with sodium carbonate) and sulfuric acid ($H_2SO_4$) leaching is proposed to extract Ga from brown corundum dust [12]. Due to roasting, the corundum and amorphous silicate are transformed into sodium silicate and aluminosilicate salts that enable Ga to pass to the pregnant leach solution when $H_2SO_4$ leaching is applied.

Boiler ash is a solid waste formed by the combustion of heavy fuel oil in power plants. It contains valuable metals, such as vanadium, nickel (Ni), zinc (Zn), iron (Fe), and magnesium (Mg). Salt roasting and water/acid leaching of boiler ash are conventional approaches to extract high-purity vanadium and concentrates of other metals in the residue. Hydrometallurgical processing of the water leaching solid residue aiming for the high-purity and cost-effective extraction of valuable metals such as Ni, Zn, Fe, and Mg is proposed [13]. It is based on $H_2SO_4$ leaching at elevated temperature. After Fe and Mg removal as $Fe(OH)_3$ and $MgC_2O_4{\cdot}2H_2O$, high-purity nickel and zinc hydroxides are precipitated, which are transformed to the corresponding pure oxides by calcination.

Valuable metals can be recovered from wastewater from mining, metallurgy, and other industries through their adsorption on natural zeolites as safe and abundant materials. Tsitsishvili et al. [14] identified the ability of Georgian heulandite, treated with dilute hydrochloric acid solutions, to uptake relatively high amounts of silver, copper, and zinc.

Overall, the papers gathered in this Special Issue show that valuable metals can be recovered from difficult-to-process or impoverished ores, as well as from nontraditional sources by applying up-to-date beneficiation methods and hydrometallurgical approaches, based on a solid theoretical background, supported by modern analytical methods and software modeling.

**Conflicts of Interest:** The author declares no conflicts of interest.

## References


1. Can, N.M.; Başaran, Ç. Effects of Different Grinding Media and Milling Conditions on the Flotation Performance of a Copper-Pyrite Ore. *Minerals* **2023**, *13*, 85. [CrossRef]
2. Panayotova, M. Control of Non-Ferrous Metal-Sulfide Minerals' Flotation via Pulp Potential. *Minerals* **2023**, *13*, 1512. [CrossRef]
3. Chanturiya, V.A.; Matveeva, T.N.; Getman, V.V.; Karkeshkina, A.Y.; Gromova, N.K. Substantiation of New Reagent Compositions for the Effective Extraction of Rhenium in the Processing of Complex Molybdenum Ores. *Minerals* **2023**, *13*, 372. [CrossRef]
4. Aleksandrova, T.; Nikolaeva, N.; Afanasova, A.; Romashev, A.; Kuznetsov, V. Justification for Criteria for Evaluating Activation and Destruction Processes of Complex Ores. *Minerals* **2023**, *13*, 684. [CrossRef]
5. Khachatryan, A.; Vardanyan, N.; Willscher, S.; Sevoyan, G.; Zhang, R.; Vardanyan, A. Bioleaching of Chalcopyrite by a New Strain Leptospirillum ferrodiazotrophum Ksh-L Isolated from a Dump-Bioleaching System of Kashen Copper-Molybdenum Mine. *Minerals* **2024**, *14*, 26. [CrossRef]
6. Guner, M.K.; Bulut, G.; Hassanzadeh, A.; Lode, S.; Aasly, K. Automated Mineralogy and Diagnostic Leaching Studies on Bulk Sulfide Flotation Concentrate of a Refractory Gold Ore. *Minerals* **2023**, *13*, 1243. [CrossRef]
7. Soto-Uribe, J.C.; Valenzuela-Garcia, J.L.; Salazar-Campoy, M.M.; Parga-Torres, J.R.; Tiburcio-Munive, G.; Encinas-Romero, M.A.; Vazquez-Vazquez, V.M. Gold Extraction from a Refractory Sulfide Concentrate by Simultaneous Pressure Leaching/Oxidation. *Minerals* **2023**, *13*, 116. [CrossRef]
8. Ma, L.; Liu, X.; Wang, L.; Qi, J. Mineral Phase Evolution during Oxidation Roasting Pretreatment of Typical Carlin Gold Ore and Effects on Gold Leaching Efficiency. *Minerals* **2023**, *13*, 558. [CrossRef]
9. Huang, Y.; Jia, Z.; Wang, W.; Yao, J.; Gao, R.; Xu, L.; Zhang, H.; Zhang, Y.; Song, X. Study on Electrochemical Behavior of Oxidized Pyrite in Alkaline Electrolyte. *Minerals* **2023**, *13*, 1070. [CrossRef]
10. Chanturiya, V.A.; Minenko, V.G.; Samusev, A.L.; Ryazantseva, M.V.; Koporulina, E.V. Physicochemical and Energy Methods to Improve the Recovery of Rare Earth Elements from Eudialyte Concentrate. *Minerals* **2023**, *13*, 414. [CrossRef]
11. Kiprono, N.R.; Smoliński, T.; Rogowski, M.; Herdzik-Koniecko, I.; Sudlitz, M.; Chmielewski, A.G. Kenya's Mineral Landscape: A Review of the Mining Status and Potential Recovery of Strategic and Critical Metals through Hydrometallurgical and Flotation Techniques. *Minerals* **2024**, *14*, 21. [CrossRef]
12. Zhang, J.; Chang, Y.; Gao, C.; Hui, X.; Jokilaakso, A. Extraction of Gallium from Brown Corundum Dust by Roasting—Acid Leaching Process. *Minerals* **2023**, *13*, 900. [CrossRef]
13. Ibrahim, A.H.; Lyu, X.; Atia, B.M.; Gado, M.A.; El Deeb, A.B. Cost-Effective and High Purity Valuable Metals Extraction from Water Leaching Solid Residues Obtained as a By-Product from Processing the Egyptian Boiler Ash. *Minerals* **2022**, *12*, 1084. [CrossRef]
14. Tsitsishvili, V.; Panayotova, M.; Mirdzveli, N.; Dzhakipbekova, N.; Panayotov, V.; Dolaberidze, N.; Nijaradze, M. Acid Resistance and Ion-Exchange Capacity of Natural Mixtures of Heulandite and Chabazite. *Minerals* **2023**, *13*, 364. [CrossRef]

minerals

Review

# Kenya's Mineral Landscape: A Review of the Mining Status and Potential Recovery of Strategic and Critical Metals through Hydrometallurgical and Flotation Techniques

**Nelson R. Kiprono *, Tomasz Smoliński, Marcin Rogowski, Irena Herdzik-Koniecko, Marcin Sudlitz and Andrzej G. Chmielewski ***

Institute of Nuclear Chemistry and Technology, Dorodna 16, 03-195 Warsaw, Poland; t.smolinski@ichtj.waw.pl (T.S.); m.rogowski@ichtj.waw.pl (M.R.); i.herdzik@ichtj.waw.pl (I.H.-K.); m.sudlitz@ichtj.waw.pl (M.S.)
* Correspondence: n.kiprono@ichtj.waw.pl (N.R.K.); a.chmielewski@ichtj.waw.pl (A.G.C.); Tel.: +48-50-398-3882 (N.R.K.); +48-22-504-1205 (A.G.C.)

**Abstract:** Kenya is an East African country with the third-largest economy in sub-Saharan Africa. The demand for metals and minerals continues to increase due to urbanization, population rise, and new infrastructure growth in different countries. Kenya formally confirmed the discovery of oil and various minerals in April 2013, launching itself as a new player in Africa's rapidly expanding extractive sector. This review paper highlights the mining status in Kenya and the role of hydrometallurgical and flotation processes in the recovery of deficit metals from ores and mine wastes. The nation's 2030 Vision is anticipated to benefit greatly from the proceeds from the sale of oil, gas, and valuable metals. Because Kenya was originally mapped as an agricultural region, less mineral prospecting was done in earlier times. The country's mining industry is now dominated by the manufacture of non-metallic goods, and it is largely neglected for minerals. One of the most serious problems for the mining industry in Kenya is the production of tailings that hold strategic metals. The material is already ground, which means the most energy-consuming process has been already applied, and chemical engineering processes like leaching are more feasible at this point. Hydrometallurgical and flotation recovery of valuable metals from wastes, high and low-grade ores, or tailings is essential. The resources will be preserved, which ensures sustainability in the growth of the mining industry.

**Keywords:** mining; ores; tailings; hydrometallurgy; recovery; strategic metals; critical metals; flotation

**Citation:** Kiprono, N.R.; Smoliński, T.; Rogowski, M.; Herdzik-Koniecko, I.; Sudlitz, M.; Chmielewski, A.G. Kenya's Mineral Landscape: A Review of the Mining Status and Potential Recovery of Strategic and Critical Metals through Hydrometallurgical and Flotation Techniques. *Minerals* **2024**, *14*, 21. https://doi.org/10.3390/min14010021

Academic Editors: Marinela Ivanova Panayotova and Vladko Panayotov

Received: 27 October 2023
Revised: 18 December 2023
Accepted: 22 December 2023
Published: 24 December 2023



## 1. Introduction

Mining is the practice of extracting valuable minerals or other geological materials from the Earth's crust, for economic development. These materials can be found in seams, reefs, lodes, orebodies, etc. Sustainable growth satisfies existing demands without jeopardizing the capacity of future generations to meet their needs. Justifiable development must be implemented in all human pursuits, and this requires responsible mining. It is no longer possible to extract raw materials just based on economic considerations; social and environmental factors must also be taken into account. As a result, minimal environmental effects, human health safety, full resource utilization, and waste-free mines should all be goals of good practice standards for mining and processing mineral resources [1]. Mineral-rich nations make use of these resources to shift their economy in the direction of sustainable development. Because of their abundant mineral resources, countries like South Africa, Australia, the United States, China, and Ghana, among others, are now industrialized [2]. The mining industry is anticipated to be the main engine of industrial expansion in low-income countries, particularly in Africa. These African nations include, among others, Kenya, Ghana, South Africa, Mali, and the D.R.C. (Democratic Republic of the Congo). Initiatives such as the Africa Mining Vision (AMV), a "pathway" developed by

African states themselves, place the continent's long-term and wide development goals at the center of all policy formulation related to mineral exploitation. It was started to aid in the economic transformation of Africa's mineral-rich regions [3].

Since the Industrial Revolution, the use of energy resources (primarily fossil fuels) has increased dramatically on a global scale [4]. In the coming decades, it is projected that the global energy sector will experience a continuous shift towards renewable energy alternatives [5]. With an expected rise of electricity demand of 150% from the year 2010 to 2050, renewable energy solutions, including photovoltaic power, hydroelectric power, and wind power, are going to play a significant part in satisfying future global demands [6]. The need for key metals will increase substantially as a result of the broad acceptance of technologies with low carbon footprints [5]. As millions of humans adopt contemporary ways of life, the demand for vital metals is, in fact, on the rise [7]. Still, the loss of minerals required for renewable energy technologies could make it more difficult to move from a fossil fuel-based to eco-friendly substitutes. Kenya is one of the ideal places that can contribute strategic and critical metals to the global economic chain. Therefore, this study reviews Kenya's present situation concerning strategic and crucial metals and discusses techniques for recovering them through hydrometallurgical processes. The article also, in its last section, introduces the concept of strategic and critical metal recovery using flotation techniques. Proposals and prospects for the future application of the described technologies in Kenya have also been given.

*Strategic and Critical Metals*

Strategic metals are resources required for vital use in times of crisis, but their acquisition is uncertain in terms of quality, quantity, and timing and any justification for their supply would require advanced planning [8]. When a metal is essential to the state's economic strategy, which includes its defense, energy, and security strategies, it is also said to be strategic. However, metal could be seen as strategic for a certain business or set of industries, including those in the automobile, renewable energy, aerospace, nuclear, ICT (Information and Communication Technologies), and electronics sectors, and so on. Each region has specific types of metal that have been determined to be crucial and strategically important for their economies. Kenya classified Ti, gas, oil, Cu, diamond, Au, Nb, gypsum, sand, and fluorspar, amongst others, as strategic and critical materials for Vision 2030 [8]. However, a region like the United Kingdom outlined Sb, Be, Cr, Au, Hf, In, Co, Ga, Ge, platinum group metals (Pt, Ir, Pd, Os, Rh, Ru, Re), REEs (Rare Earth Elements), Ta, Ti, W, Li, Mg, Ni, Nb, and V as strategic and critical materials.

Critical metals are those that are expensive, difficult to substitute for, geologically rare, prone to potential supply constraints, and required for an economically significant drive. These metals are significant due to their unique features. The possibility of supply restrictions seems to be the most important aspect of these characteristics. Unlike in the 1950s, almost every metal in the periodic table is needed currently to produce different products for society, Figure 1. A good percentage of these crucial metals were disregarded and treated as mine waste fifty to sixty years ago [9].

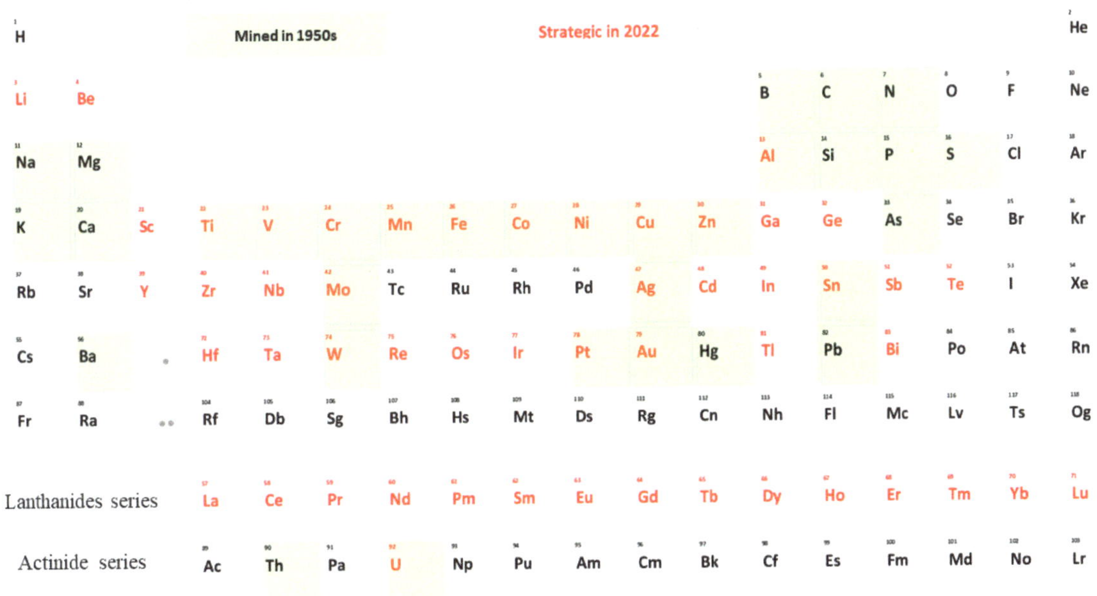


**Figure 1.** A periodic table displaying the aspects of various types of metals and their economic status [10].

Metals such as REEs, Ge, Ta, Se, Sn, In, Ga, and Te are geologically sparse, though crucial [11]. Relatively low atomic weight La, Ce, Pr, Nd, and Pm are categorized as light REEs (LREEs) while Sm, Tm, Yb, Eu, Dy, Ho, Gd, Tb, Er, and Lu are classified as heavy REEs (HREEs) because of their high atomic weight. Since they are widespread, even though it is rare to locate them in quantities large enough to permit the production of profitable minerals, these elements are referred to as REEs [12]. Due to their occurrence in ore deposits comparable to those of lanthanides and exhibiting the same chemical properties, Y and Sc are likewise classified as REEs [13,14].

When contrasted to the other elements in the periodic table, REEs exhibit distinct magnetic, electrical, optical, luminescent, and chemical properties [15]. Considering that the majority of REEs have similar oxidation states and atomic radii, they can substitute one another in a variety of crystal lattices. Due to this substitutional capacity, numerous REE occurrences can be found in a single mineral, which leads to a widespread distribution of REEs in the Earth's crust. They can be found in a variety of mineral forms, including oxides, carbonates, phosphates, silicates, and halides. Generally, REEs are high on the priority list of essential metals in many nations where these metals are in short supply [13].

## 2. Mining Situation in Kenya

Kenya, located in the eastern part of Africa, is richly endowed with different types of mineral resources. The Kenyan government assessed its mineral potential and produced a working document in 1999, and over 400 mineral occurrences were identified. The government introduced regulations to drive this sector [16]. Vision 2030 recognizes the such industry as one of the key drivers of the country's economic development. Growth in this sphere is expected to increase the number of jobs, the income of the workers, and the country's GDP (Gross Domestic Product). The 2010 constitution created Counties and gave them the capacity to establish and carry out their development strategies. This new dispensation is forcing Counties to assess their resources and exploit them for their local development. The resources vary from County to County and are highly dependent on the topography, drainage, ecology, and climatic conditions of the County. The worldwide demand for minerals is projected to progressively rise at a mean rate of approximately

3% annually [17]. That is likely to trigger more exploration and mining activities in the resource-endowed regions of the country.

Because Kenya was originally mapped as an agricultural region, less mineral prospecting was done in earlier times. The country's mining industry is now dominated by the manufacture of non-metallic goods, and it is largely neglected for minerals. For instance, Kenya is the world's third-largest supplier of soda ash and ranks seventh in terms of fluorspar production. Additionally, Kenya formally confirmed the discovery of oil and various minerals in April 2013, launching itself as a new player in Africa's rapidly expanding extractive sector [18]. The non-renewable resources mentioned above are, regrettably, mainly found in places with high rates of poverty, ongoing droughts, instability, and long-term marginalization by the government. Conflicts between local communities, international corporations, and governmental organizations are consequently high [19].

Fe ore, Ti, and Au are amongst the metallic minerals now produced in the country. Kenya's export data show that this industry is constantly expanding. Along with industrial minerals like talc, gypsum, dolomite, and gemstones, Kenya's mineral suite also contains resources like Ag, Cu, and Zn, REEs, coal, limestone, soda ash, Mn ore, fluorspar ($CaF_2$), diatomite, chromite, Nb, and silica sand. Natural fluorspar ($CaF_2$), largely mined in Kenya, is typically associated with other minerals such as phosphates ($PO_4^{3-}$), quartz ($SiO_2$), celestite ($SrSO_4$), galena (PbS), calcite ($CaCO_3$), barite ($BaSO_4$), and chalcopyrite ($CuFeS_2$) [20]. The most valuable REE resources in the world are also located in the coastal region (Jombo Complex) at Mrima Hills, where Cortec Mining Kenya Limited estimated that they could have an in-ground value of up to USD 62.4 billion [19,21,22]. Mrima Hill is home to a carbonatite cluster. The laterite capping resulting from the weathering of the hill's igneous constituents has economic grades of REE and Nb mineralization [23].

In terms of Counties, Turkana County has oil, gypsum, graphite, Zr, Nb, Ba, Pb, Fe, Te, Ag, Mn, Cu, and Ni ores, whereas Taita Taveta principally possesses Fe, Mn, Cu, and quartz. While Kwale County is abundant in Zn, Pb, Cu, U, Ti, REEs, Nb, and oil, Kitui County has significant amounts of Fe, coal, Cu, limestone, amethyst, and sapphire. Lamu County is rich in Ti, gas, and oil, whereas Migori County has commercially exploitable deposits of Zn, diamond, Au, Al, Fe, and Cu. Au generation in Migori and other regions is a significant driver of the nation's economic development and growth, contributing approximately USD 10.3 million/year to the country's economy [24]. Transmara and Migori districts began to mine Au in the year 1930. However, industrial mining became significant recently in the country due to the discovery of Au-bearing reefs in the Lolgorien region. The local population and various enterprises that perform mining operations in the area now rely heavily on these activities as one of their main sources of income [25]. Figure 2 shows some of the minerals that have been confirmed to be available for economic purposes in Kenya.

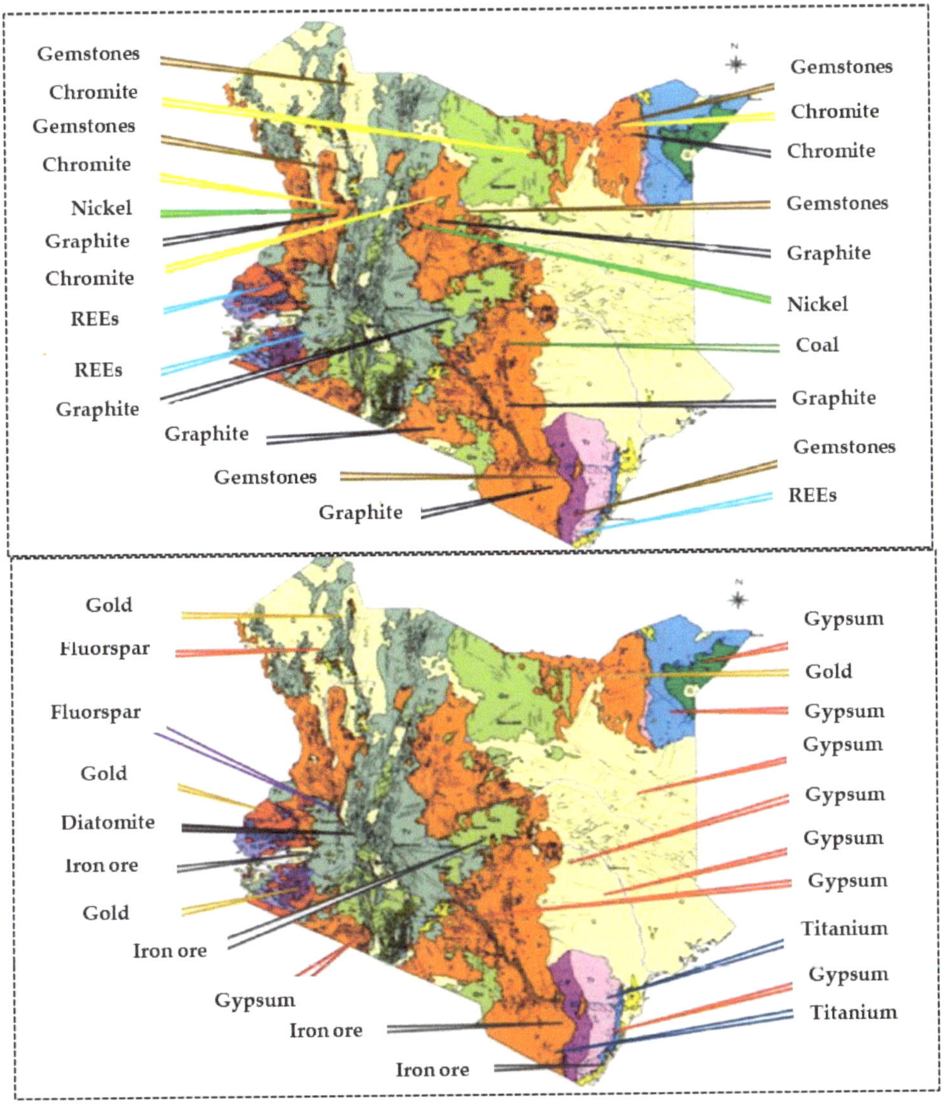


**Figure 2.** Maps showing mineral occurrence in Kenya [26].

For the nation to achieve the objectives of Vision 2030, the Fe and steel industry will play a significant role in the industrialization process [27]. Fe minerals are widely present in western Kenya, especially in Nyanza. Furthermore, there are untapped reserves of Fe ore in various Kenyan Counties, including Taita Taveta (Manyatta), Siaya (Samia), and Taraka (Marimanti). Although complicated, the process of making steel essentially involves melting Fe ore in a blast furnace along with coke. In some rural areas of Kenya and the neighbouring nations, the manufacturing of sponge Fe has been accomplished by smelting Fe ore using charcoal and a forced air draught from bellows made of goatskin. The majority of the ores were first crushed and sent straight to a blast furnace, especially those with 50% or more Fe without beneficiation. To increase the Fe content and get the concentrate ready for the blast furnace, the majority of extracted ores nowadays go through beneficiation [27].

Some of the platinum group metals have also been reported in Kenya. In 14 petrologically distinct spinel peridotite xenoliths found in the Marsabit volcanic zone (Kenya Rift), platinum group metals were identified [28]. However, their concentrations are very low. Marsabit's xenoliths have a total platinum group metals concentration that ranges from 14 to 32 ppb. Unfortunately, such concentrations are too low to be recovered for economic purposes.

Understanding the principles influencing element distribution in the Earth's crust is important. In particular, the role played by rocks and minerals in the uptake of certain metals from groundwaters requires comprehension of the movement and fixation of elements. Scientists have researched the compositions of several hot water springs, related minerals, rocks, and fossil bones from Kanjera and Kanam near Lake Victoria in Kenya. The location selected for the study was ideal due to the presence of volcanic hot springs and sedimentary deposits connected to Homa Mountain's carbonatite complex, which is comparatively rich in REEs [29]. The findings provided evidence of the absorption and mobility of REEs. Bicarbonate and fluoride were found to be abundant in the spring waters of both locations, and Kanjera's Zr concentration was exceedingly high. All examined rock samples had higher REE contents than chondrites. The rocks further showed higher concentrations of light REEs (Nd, Sm, La, Ce, Eu, and Gd) than heavy REEs (Tb, Yb, Lu, Ho, and Tm). The travertine specimen seemed to be moderately discoloured by Mn and/or $Fe_2O_3$ or hydroxide deposits. The acid-insoluble component of the travertine (5.08% by weight) included less than 11% of each REE but more than 60% of the aggregate concentration of some elements, including Cr, Hf, and Ta. The REEs, Th, and U were generally present in fairly large quantities in the carbonate (acid-soluble) fraction. It was notable because such levels were precipitated from percolating ground water. Contrarily, the trona sample only contained detectable levels of the following elements: Ce, Sm, Sc, Cr Eu, Co, and Th. It was concluded that groundwater–rock interaction has a significant impact on rock and mineral composition [29].

### 2.1. *Environmental Impacts of Mining Activities*

Mining in Kenya and Africa at large is considered one of the main pillars of socio-economic benefits. However, the same industry is known to be amongst the main sources of pollutants. The influence of mining on environmental contamination can be both geogenic (related to parent rocks) and anthropogenic, encompassing man-made activities such as chemical use [30]. The discovery of mines and their exploitation was solely determined by the economic aspect when the exploitation of natural resources began many years ago. However, tailings production is one of the major issues facing the mining industry. Tailings encompass the remains of the fine-grained (1 to 600 μm) ground rock after the metals of value have been removed from the mined ore [31,32]. Such materials are very reactive, attributable to their minute particle size, with the variation of their chemical and physical properties depending on the ore type. Most of the tailings result during the beneficiation process.

Nearly all the heavy metals in the ores exist as compounds with sulphide in their structures. That contributes to the acid mine drainage (AMD) in the region, responsible for the death of several fish in places like Lake Victoria and the surrounding rivers. AMD is produced when sulphide-containing minerals such as pyrite are exposed to water and oxygen, where oxidation of the minerals results in the generation of hazardous $H_2SO_4$ acid. AMD is one of the serious environmental challenges associated with the discharge of acidic, metal, and sulphate-containing industrial wastewater into the surroundings [33].

Additionally, wastes produced by mining activities contain elevated concentrations of metalloids and metals. The elements can be mobilized, infiltrating into the surface water and groundwater [34]. A significant fraction of such elements in industrial wastes are non-biodegradable and highly toxic. Chemical contamination occurs when the reagents added by the mineral processing industries are released. For instance, cyanide or Hg for extracting Au or $H_2SO_4$ acids used in leaching Cu oxides can be supplemented under controllable conditions but may leak into the environment if not properly handled. Alternatively, chemical pollution can occur when naturally occurring minerals are oxidized [35]. This is mainly predominant when sulphide minerals are involved while extracting Au. Other metals released during artisanal Au mining in regions like Migori and related industries include Cu, Cd, Ni, Zn, Pb, and As, all discharged into soil and water bodies [24].

Kenya is one of the countries in Africa that aims to intensify the extraction of REEs and other metals from various sources. The intricate process of mining resources such as REEs may result in negative effects on the environment and human health. The effects of mining on the environment, nevertheless, rarely receive adequate attention in the majority of developing nations. Deforestation, water contamination, soil erosion, changes in landscape structure, and the destruction of wildlife habitats are a few of the environmental effects of recovering REEs from ore. The effects of REEs are influenced by the type of rock around its deposit, the availability of additional metals or substances within the rocks, the climate, and the distance from streams and lakes to the mine.

Since agricultural land is lost due to the expansion of the mining sites, vegetation in mining areas is also destroyed during ore extraction [36]. Moreover, mine closure results in numerous wastes and the degradation of the land once minerals have been extracted to a considerable extent and the remaining ones are no longer commercially viable for extraction. Information for mitigating potentially hazardous environmental effects following the closure of mining activities and preventing particulate matter emissions from mines should be made available. This is critical, along with the responsible use and disposal of metal-bearing technology devices [36]. Furthermore, treatment of tailings and the repurposing of them will be crucial.

### 2.2. *The Presence of Critical and Strategic Metals in the Mine Wastes*

Valuable metals can be economically recovered from the tailings produced duringmining operations. When selecting the tailings for use in an alternate application, consideration must be granted to the tailings' physical and chemical characteristics [37]. In the dominant geological circumstances within which they are found, the minerals that make up economically valuable ores are largely stable. Their chemical stability decreases when they are exposed to the atmosphere. That explains the high reactivity of tailings. Despite containing 94% of the processed ore, these wastes still contain valuable metals [38]. Thus, from the perspective of protecting the environment and preserving raw materials, continued treatment of these wastes is possible. Additionally, the material has already been pulverized, consuming the high amount of energy in the process; hence, chemical engineering procedures such as leaching become subsequently more practical at this stage [39].

Mine waste deposits are increasingly becoming dependable sources of critical and strategic metals, though some of them could be regarded as low-grade ores based on the current standards [40]. Closing the consumer cycles through the liberation of these metals from complex and dilute waste streams will become significantly viable economically [41]. Access to mine deposits is easy because they are localized at the surface without considerable overburden. The mineralogical composition of the tailings can vary depending on the ores containing the metals of interest. Discarded tailings have been found to contain significant amounts of critical metals such as Ga in bauxite residues or Ge or In in the residues of Zn ores, among others. The processing of the tailings from the ponds and waste-rock heaps of the Penouta mine in Spain led to the recovery of Ta and Nb [42]. However, the crucial influencing factors include the quantities or volumes of the critical raw materials, their concentrations, quality of tailings, and the composition of the minerals in the extractive wastes.

There have been reports of tailings in Kenyan mining industries, including the short-lived Au mine in the Migori district. Both open-cut and subsurface mining techniques were used. After that, the ore was pulverized and panned, leading to huge heaps of tailings and waste rocks. Several elements like Ti, Cr, Rb, Sr, Y, Mn, Zn, As, Au, Pb, Fe, Co, Cu, Zr, and Nb exist in various quantities in the Migori region [25,43]. Such minerals are crucial, and they can be reprocessed for economic gain. For instance, through the application of magnetic separation methods, valuable $TiO_2$ can be extracted from the waste [44].

Western Kenya is another region endowed with minerals such as Au, which was exploited by the British firm known as Rosterman Au Mine industries in the 1930s before they exited the site in 1952. Since the firm's exit, artisan-based miners have been extracting Au deposits from the tailings of the abandoned mines. Alluvial mining is one of the mining techniques used by artisan miners, which encompasses the excavation of the ore from the slow-moving parts of River Isiukhu [45]. The tailings and the gravel of the ore obtained are transported to appropriate locations for processing through panning or sluicing. In the majority of the mining sites, the sluiced tailings are scavenged numerous times before they are ultimately disposed of, to achieve recovery of the Au deposits lost in the previous operations [32].

## 3. Recovery of the Strategic and Critical Metals from Various Sources

The numerous minerals that are already present in Kenya can be recovered using a variety of techniques devised by academics from around the world. For the benefit of the nation, Kenyan researchers and the mining sector can adopt and improve such techniques. Compared to pyrometallurgical methods (roasting, smelting, calcining, and thermal refining of the minerals), hydrometallurgical technologies (leaching, solvent extraction, precipitation, and ion exchange amongst others) offer a greater chance of efficiently recovering very low concentrations of metals with less energy use. Pyrometallurgy has been linked to low metal recoveries, increased air pollution, and huge consumption of energy. Consequently, our review article places a strong emphasis on hydrometallurgy because it is a better alternative.

### 3.1. Leaching

Leaching is one of the crucial steps in the recovery of metals from different materials. However, the process is non-selective; hence, there is a necessity to recover metals from the pregnant solutions using other selective methods like solvent extraction, ion exchange precipitation, etc. The reagents used in the leaching process vary. For extracting base metals such as Cu, chemicals like inorganic acids can be used [46]. These agents are commonly used with oxidants, which enhances the efficiency of the process. For precious metals such as Au, the reagents used are halogens, cyanide, and thiocyanate. Cyanide is the most excellent in the reclamation of precious metals from both secondary and mineral sources [47].

There are various types of leaching approaches, including bioleaching and chemical leaching. Bioleaching is the biological solubilization of metals from wastes and minerals. That means micro-organisms (for example, *Acidithiobacillus ferrooxidans* and *Acidithiobacillus thiooxidans*) converting solid, insoluble metals and their compounds into extractable forms. Nonetheless, chemical leaching involves the use of bases/acids, chelating agents, surfactants, redox agents, or salts to liberate the metal from the solid phase to the aqueous phase. Both organic and inorganic acids can be utilized for chemical leaching. EDTA (Ethylenediaminetetraacetic acid) is one of the common chelating agents for various metals.However, NaClO and $H_2O_2$ are the prominent oxidizing agents employed to improve the recovery of the metals from the solid phases [47].

Several technological methods use chemical leaching to process ores (in situ, heap, and dump leaching). The in situ approach involves the application of a lixiviant ($H_2SO_4$, thiosulphates, thiourea, and cyanides) which is pumped directly into the channels in the rocks or inactive mines. After some time, the resulting pregnant leach solution is pumped to the surface and then taken for post-processing for metal recovery. The technique can be used in the recovery of U and Cu [48]. Contrarily, in heap leaching, the piles are placed in the open air on a previously prepared impermeable pad (plastic, clay). The largest grains are at the base of the pile and the most pulverized at the top.The particle sizes of typical ores range from 10 to 100 mm. The leaching agent is sprinkled on the upper surface of the layer and percolates through it. An aqueous solution with dissolved metals is received through the collecting channel [48]. Similar to this method is dump leaching, except that the ore is taken directly from the mine and stacked without crushing. The method is used for recovering Cu and precious metals and can be applied in extracting low-grade ores [49].

### *3.2. Solvent Extraction of the Common Transition Metals and Uranium*

Given its capacity to cope with enormous quantities of diluted pregnant liquor of aqueous solution, solvent extraction is a widely recognized hydrometallurgical procedure. The technique is frequently used in the recovery of leached d-block, REEs, and precious metals using extractants. The extractant is an organic ligand that is intended to coordinate with the cations in a targeted manner [50]. This procedure is mostly accomplished by vigorously agitating the two immiscible media, enabling the solute(s) to be transported from one phase to the other in a controlled manner. Normally, there are two phases involved: an organic phase, having the extractant dissolved in a suitable hydrocarbon diluent; and an acidic aqueous medium, having the target metal ion(s) [51]. The pure extractant is occasionally employed as an organic solvent. However, because of its high viscosity, pure extractants are typically not used. For high product purity and large-scale manufacturing, solvent extraction is a more practical technology that has gained widespread acceptance. The above technique is incredibly practical in hydrometallurgy since it does not require specialized equipment, and it uses inexpensive chemicals to extract metals from their secondary and primary resources [52].

In recent years, there has been an evident expansion in the industrial application of Ni and Co. This may be seen in two ways: the rise of the manufacture of such metals globally, and on the other side, the presence of progressively greater quantities of different wastes containing Co and Ni [51,53–55]. These metals' unique characteristics, such as their exceptionally high melting and boiling temperatures, hardness, thermal conductivity, electrical and changing oxidation states, magnetism, and propensity for forming alloys with additional elements, make their recovery and recycling intriguing at the moment. The isolation of Ni and Co requires high precision due to their comparable physicochemical characteristics. The most important step in the process is selecting an appropriate extractant, which is typically dependent on the acidity and constitution of the leaching mixture [56]. By utilizing diverse techniques, researchers have attempted to enhance the solvent extraction approach. An efficient extractant for the separation of Ni and Co was reported to be Cyanex 272 [bis(2,4,4-trimethylpentyl) phosphinic acid]. Cyanex 272 and D2EHPA [Di-(2-Ethyl Hexyl) phosphoric acid] can be used to separate Zn and Mn from Ni in the leach solution. According to the observations, Cyanex 272 and D2EHPA produced a superior separation of Ni and Co [56].

Kenya was ranked as the 10th highest exporter of Nb, Ta, V, and Zr ore worldwide in 2021, with USD 44.1M in exports of these metals [57]. Nb, Ta, V, and Zr ore was rated as Kenya's 29th most exported good that year. Such critical metals can be recovered through hydrometallurgical procedures. In recognition of its exceptional physicochemical properties, V—often referred to as the "vitamin of modern industry"—is used extensively in various industries. The two primary categories of V recovery methods are roasting extraction and direct extraction [58]. The extraction of $V^{4+}$ and $Mn^{2+}$ was explored using new technology for selectively separating and extracting them from a co-leaching mixture of roasted stone coal and pyrolusite via solvent extraction. According to the findings, 99.13% of $V^{4+}$ was recovered using three counter-current extraction steps with 5% ($v/v$) EHEHPA (2-ethylhexyl phosphonic acid mono-2-ethylhexyl ester) and an organic/aqueous (O/A) ratio of 1:1 under a starting aqueous pH of 2.0. With three counter-current extraction stages, 99.74% of $Mn^{2+}$ was extracted by applying 40% ($v/v$) D2EHPA with a 60 mol% saponification effectiveness at an O/A phase ratio of 2:1 and a preliminary aqueous pH of 3.5. Following a two-stage counter-current experiment using 1 M $H_2SO_4$ at a 2:3 O/A phase ratio, $Mn^{2+}$ was found to be completely scrubbed off [59].

A synergistic recovery method for $Cu^{2+}$ and $Zn^{2+}$ ions from chloride aqueous solutions has also been developed. The selective extractant was obtained through a combination of trioctylphosphine oxide (TOPO) and Aliquat 336. Employing 0.06 M Aliquat 336 and 0.025 M TOPO in kerosene, $Zn^{2+}$ ions were effectively separated from $Cu^{2+}$ ions [60]. Equivalent amounts of both phases (10 mL each) were physically mixed for 20 min during the process at a fixed temperature (25 ± 2 °C). The mass balance was then used to determine the distribution ratio (*D*) of the cations measured by an atomic absorption spectrometer, Equation (1).

$$D = \frac{[Me]_{org}}{[Me]_{aq}} \tag{1}$$

Equation (2) gives the extraction percentage (% *E*),

$$\% \, E = \frac{D}{D+1} \times 100 \tag{2}$$

where the concentrations of the metal ions in the aqueous and organic media are $[Me]_{aq}$ and $[Me]_{org}$, respectively.

$$S = \frac{D_{M1}}{D_{M2}} \tag{3}$$

Equation (3) was used to obtain the selectivity coefficient of metal *M1* over metal *M2*.

Low recovery of $Cu^{2+}$ with TOPO was observed in the 0.1 to 2 M range. However, the extraction efficiency of $Zn^{2+}$ and $Cu^{2+}$ from HCl solutions with Aliquat 336 was reported to be quite high. In HCl solutions containing 0.05 M Aliquat 336, selectivity coefficients of $Zn^{2+}$ over $Cu^{2+}$ were very low. The extraction of $Cu^{2+}$ and $Zn^{2+}$ to the organic phase can be linked to results in the formation of their anionic complexes at greater chloride concentrations. The $Cl^-$ ions' impact increases the development of recoverable anionic compounds such as $ZnCl_3{}^-$, $ZnCl_4{}^{2-}$, and $CuCl_4{}^{2-}$ species in the aqueous phase. That increases the extraction of Zn and Cu. As a result, Aliquat 336 is highly effective at removing these metal ions from the solution. This chemical functions as an anion exchanger. The reaction between the Aliquat 336 extractant ($R_4N^+Cl$) and the metal can be represented as Equation (4) [60]:

$$(MCl_4)^{2-}_{(aq)} + 2R_4NCl_{(org)} \Rightarrow (R_4N)_2(MCl_4)_{(org)} + 2Cl^- \tag{4}$$

The Institute of Nuclear Chemistry and Technology (INCT) in Warsaw, Poland, has also created several technological schemes for recovering strategic metals from different waste materials. The institute examined if Zn, Mn, and Mg could be separated from used Zn electrolytes selectively, Figure 3. D2EHPA diluted in n-heptane was used to extract these elements. The separation of the metallic compounds from the $SO_4{}^{2-}$ mixture having 0.89 g/dm$^3$ of Mn, 15.9 g/dm$^3$ of Zn, and 24.3 g/dm$^3$ of Mg was evaluated. According to the study's findings, both for Zn and Mn, the extraction equilibrium was reached after 4 min, during which pH modulation regulated the extraction's selectivity [10].

Cu floatation tailings have substantial amounts of Cu, plus other toxic elements such as Pb, Zn, U, and Co. The regularly applied methods in the Cu mining industry often result in great losses of the deficit and valuable metals in the streams of their tailings. The formulation and implementation of hydrometallurgical advancements is a solution that is viable for higher efficiency in recovering various elements [61].

A research investigation on the extraction of Cu and other metals was performed by INCT [62]. An O/A ratio of 1:1 was used in the extraction procedure, and the extractant used comprised a combination of benzoic acid (0.5 M), the aromatic solvent toluene, and the aromatic amine p-toluidine (0.25 M). $Cu^{2+}$ ions were removed using an aqueous sulphate solution. With the help of sodium carbonate, the aqueous phase's pH was kept between 3.6 and 3.8. Other metal ions, such as $Ni^{2+}$ and $Co^{2+}$, could be separated once the Cu ions had been separated. The procedure was carried out by shaking at 25 °C in a mechanical shaker. In the majority of cases, distribution equilibrium was reached with a shaking time below 60 min. Ion separation was continued until metal ions were recovered to a degree greater than 99%. The extraction of V, Mo, and U was studied using D2EHPA in toluene. The concentrations considered in the research were 0.3, 0.2, and 0.1 M. The extraction processes took place for 15 min under O/A ratio of 1:1. With 5% $Na_2CO_3$, the organic phase was back-extracted for 15 min. A 100% extraction from the pregnant solution was confirmed by a total transfer of Cu into the organic phase. The scholars were able to significantly speed up the analytical work using the radiotracer approach, which was a very helpful research tool. The radiotracer was essential in determining the ideal conditions that produced the highest possible recovery efficiency for Cu. The findings could be utilized as a guide when designing procedures for recovering Cu and other supporting components from flotation tailings after Cu ores have been processed, as shown in Figure 4 [62]. The scheme represents two options of hydrometallurgical process application: recovery of metals from flotation tailings or low-grade Cu ore.

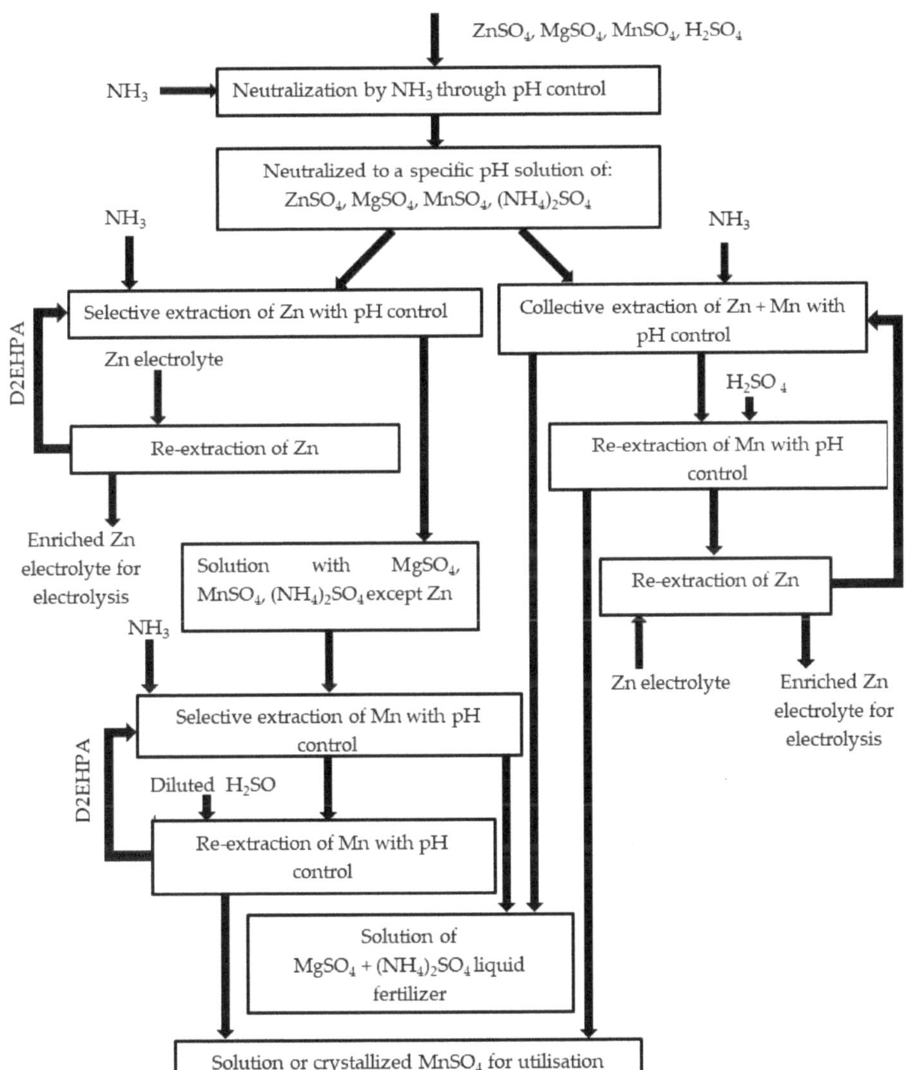


**Figure 3.** Extraction of Zn from electronic wastes.

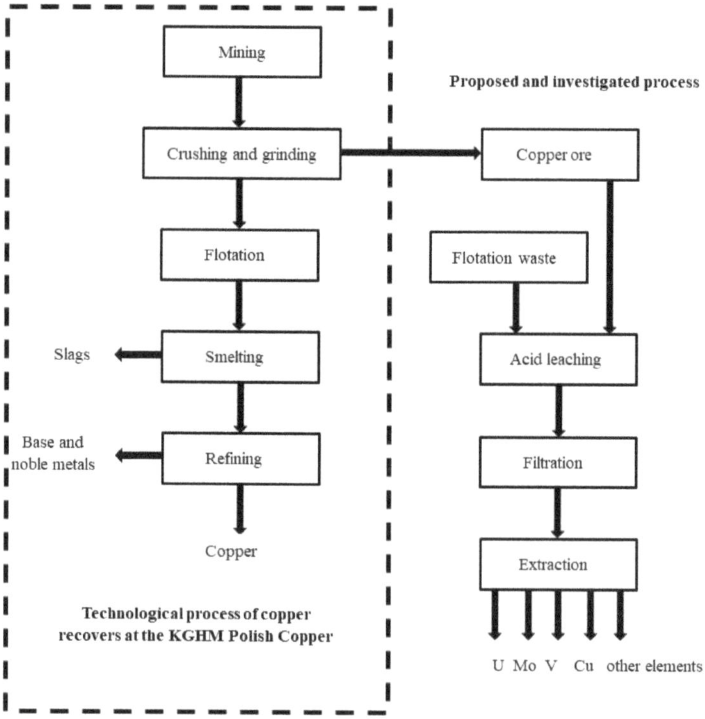


**Figure 4.** Process for pyrometallurgical refining of Cu ore and hydrometallurgical recovery of metals from flotation tailings or low-grade Cu ore [62].

In recent years, the INCT has also been researching the potential for recovering U and other elements created as waste products during the manufacture of Cu concentrate from Cu ore [61]. The utilization of the hydrometallurgical processes unlocks the channels for retrieving all the metals present in the liquor. That can be achieved by using a common chemical engineering unit process that is used to recover U and other metals [38]. In situ leach (ISL) processes, in addition to underground and open pit mining techniques, are frequently used to extract minerals such as U. A small amount of U can be extracted as a byproduct of processing Au and Cu ores or deposits of phosphate rock.

Research was carried out on the Polish Cu mining industry, with the prospect of recovering U and rare metals from industrial wastes [38]. In the recovery of U, the commonly used method of precipitation was assessed. The hydrometallurgical recovery processes of U encompassed the application of a concentrated U mixture for the production of spherical particles of $UO_2$ by the complex sol–gel process (CSGP), Figure 5. The two hydroxyl sets of ascorbic acids were readily present, including the $UO_2^{2+}$ that permits the creation of complexes. The study reported that the combination of both extraction and sol–gel processes had a synergistic impact, which resulted in the design of a reliable scheme for the recovery of U from Cu mining tailings and minerals. For both current and future generations of nuclear reactors, such as pressurized heavy water or fast breeder reactors, a nuclear fuel precursor can be produced using the direct approach of $UO_2$ synthesis (with the potential to eliminate the precipitation phase) [38].

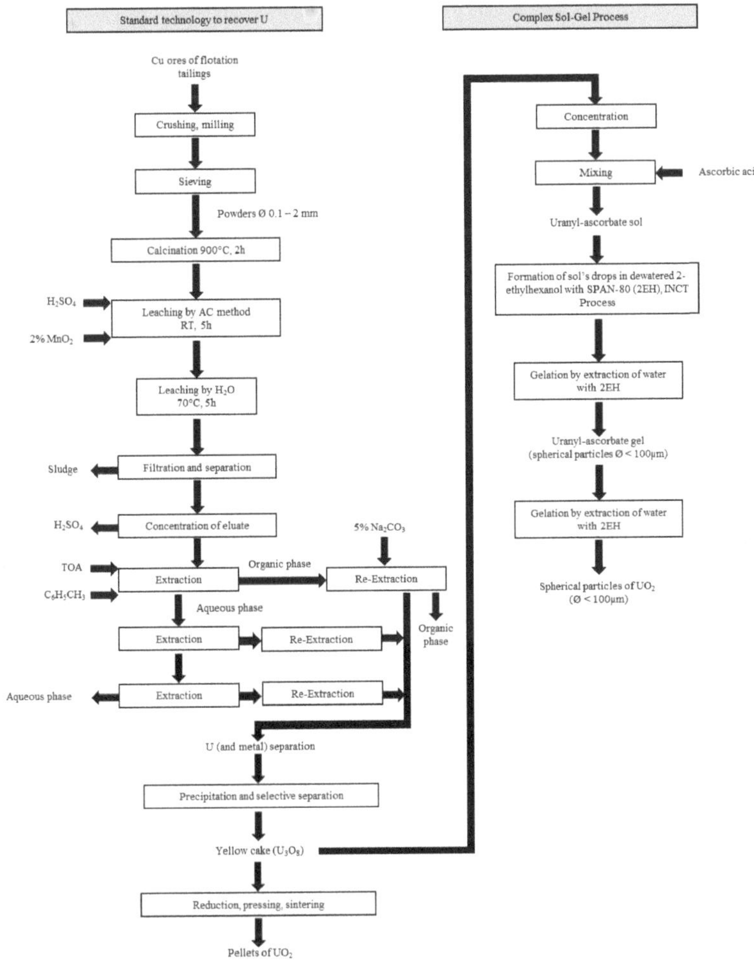


**Figure 5.** Scheme illustrating the hydrometallurgical procedures for extracting ($UO_2$) from various sources as described in the INCT complex sol–gel approach [38].

### 3.3. *Recovery of the Platinum Group Metals*

The use of high-grade platinum group metals ore has increased due to their growing industrial demand. Low-grade ore is now more common due to the increasing shortage of high-grade platinum group metals ore. Unluckily, it is prohibitively expensive to extract them from low-grade ore, and doing so raises significant environmental issues. Pt is used as a catalyst in fuel cells to effectively convert $O_2$ and hydrogen into heat and electrical power. The detection of low Pt levels in biological and environmental samples has been a serious concern with the introduction of Pt-containing catalysts in vehicle exhaust systems. To track the buildup of Pt in the environment, precise and accurate background amounts of Pt in a variety of substances are required. Since Pt is present in sediment samples in such trace amounts, designing techniques to measure it is quite challenging. Developing a process that includes digestion, preconcentration, separation, and Pt detection is an exceptionally effective way to reach incredibly low detection quantities [63]. Utilizing a $^{191}Pt$ radiotracer, the kinetics of the extraction of $Pt^{4+}$ from an HCl solution applying rubeanic acid in TBP, n-butyl alcohol-acetophenone and thenoyltrifluoroacetone (TPA), were studied [63]. Investigating the effects of acidity, mixing duration, Pt quantity, and back-extraction, the most favorable extraction outcomes for TBP and TPA were obtained with 4 and 3 M HCl, respectively. INCT in Warsaw also created a method for recovering Pt from the waste solution. Agua regia was applied as a leaching agent whereas $NH_4Cl$ was used to precipitate Pt for further recovery process, Figure 6.

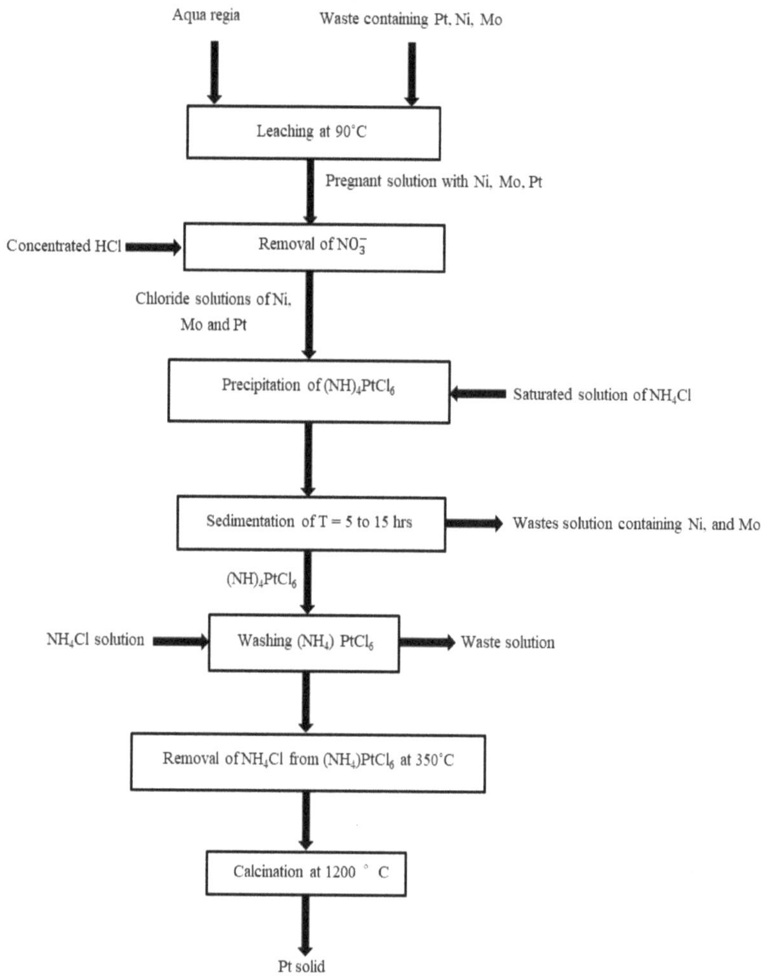


**Figure 6.** The recovery of Pt from the waste solution.

Apart from platinum group metals, precious metals like Au have drawn a lot of interest [10]. Researchers are attempting to use new techniques in both identification and quantitative analysis. A quantity of 0.02 g of Au is often found in a cell phone [10]. The development and use of alternate hydrometallurgical extraction methods has elevated their significance for many research institutions over the past few decades. Societal concern regarding the environmental impacts of cyanidic Au extraction has increased. The Department of Nuclear Methods of Process Engineering, INCT, Warsaw, carried out the development of hydrometallurgical technologies for the separation of metals from waste. The technological line for the solvent extraction of Au from a mixture of Mo-Ni waste was established in Figure 7.

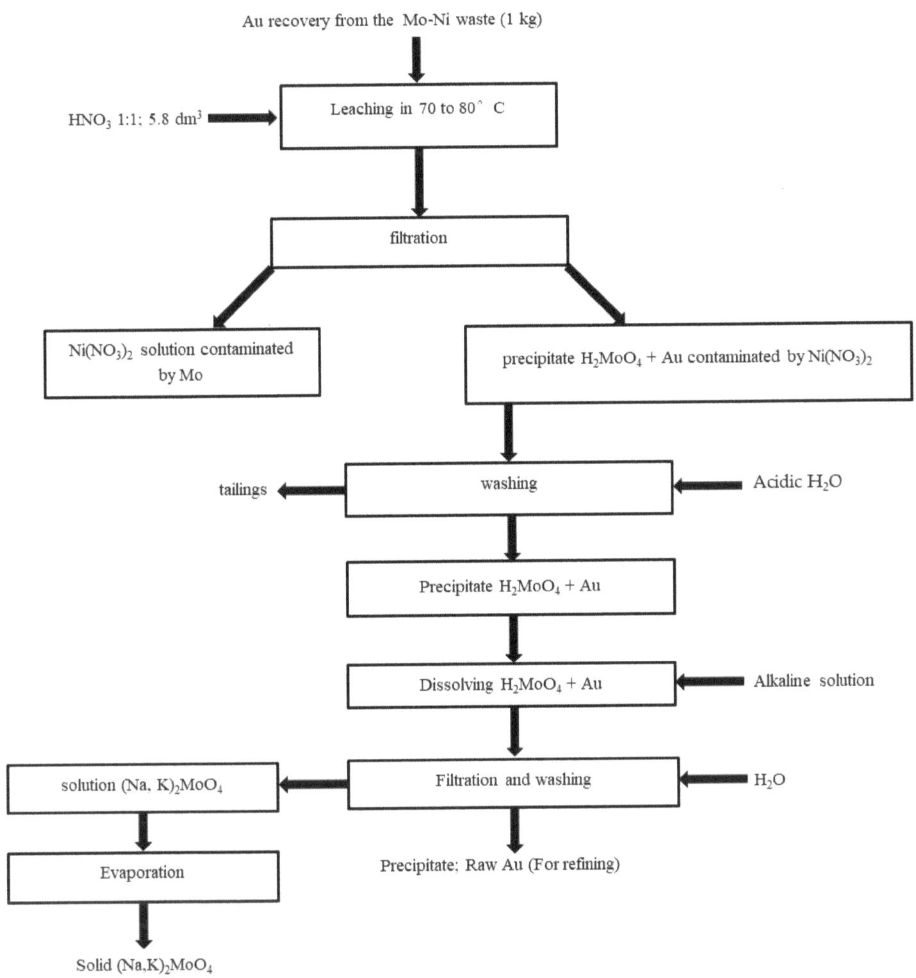


**Figure 7.** Scheme of Au recovery from molybdenum-nickel waste.

### *3.4. Recovery of Rare Earth Elements*

Apatite, xenotime, and monazite are a few of the foremost popular minerals that contain REEs. The extraction and reuse of such minerals from various secondary resources has recently attracted a lot of interest. Considering their secondary origin, various techniques have been carried out to separate them from phosphate or phosphogypsum rocks [64]. Extracting them from the aqueous solution (leachate) is a selective activity. For instance, a mixture of minor actinides (MAs) and REEs was used for the extraction studies. La was removed from MA by employing N,N,N′,N′-tetrakis(4-propenyloxy-2-pyridylmethyl)ethylene-diamine (TPEN) extracting agent. The phase of separation and recovery is intricate and delicate. It is essential to guarantee that each REE separation is highly effective and can be reclaimed at the final stage of the procedure. Knowledge of the extraction characteristics, capacity, phase separation, solubility, selectivity, mass transfer, and economic viability have all been used to develop feasible REE extraction methods [65].

Given their nearly non-volatile nature, low melting point, low flammability, thermal stability, and high ionic conductivity, ionic liquids (ILs) are usually regarded as environmentally friendly solvents and have attracted significant interest. In the extraction of both organic and inorganic materials, ILs made from quaternary ammonium bases and organic acid moieties are frequently utilized [66]. The benefits of ILs, such as high thermal stability and designability, could potentially be introduced through the use of ILs in the solvent extraction of REEs [67]. Numerous factors, such as the kind of acidic medium, diluent, the temperature of the auxiliary agents, metal ions, contact time, organic-to-aqueous phase ratio, pH, and extractant concentration, can have an impact on the process of solvent extraction. Considering various operational conditions that depend on other factors, a particular operational parameter may have various implications on the desired results [68]. It is challenging to selectively extract and purify REEs from a pregnant solution [69].

REEs' distribution can also be explored by utilizing ILs grounded on phosphinic or phosphoric acids and a metal ion. Due to coordination interactions with moieties of organic and inorganic acids, REEs can produce complex anions (like other metals of side groups). The separation of REEs utilizing an IL based on phosphoric acid and an organic base is represented by Equation (5) [66]:

$$3RNH_3^+A^-_{(org)} + Me^{3+}_{(aq)} + 3Cl^-_{(aq)} \leftrightarrow MeA_3{\cdot}3RNH_3^+Cl^-_{(org)} \tag{5}$$

Aliquat 336 can be treated with $KNO_3$ to exchange $Cl^-$ with $NO_3{}^-$ which improves selectivity over REEs, to reduce the viscosity of the extractant and establish a faster mass transfer. Equation (6) can be used to represent reaction equilibria for A336[$NO_3$] extraction [70].

$$Me^{3+}_{(aq)} + y(R_3N^+CH_3)NO^-_{3\,IL} + 3NO^-_{3(aq)} \rightleftharpoons (R_3NCH_3)Me(NO_3)_{y+3\;IL} \tag{6}$$

wherein $Me^{3+}$ stands for the elemental cation, $(R_3NCH_3)M(NO_3)IL$ for the $IL$ and metal cation complex, and $(R_3N^+CH_3)NO_3{}^-\ IL$ for A336[$NO_3$]. $IL$ stands for both the $IL$ phase and the aqueous phase. The equilibrium constant $K_{eq}$ is expressed as follows:

$$K_{eq} = \frac{\left[(R_3NCH_3)Me(NO_3)_{y+3\;IL}\right]}{[Me^{3+}_{(aq)}][(R_3N^+CH_3)NO^-_{3\,IL}]^y[NO^-_{3(aq)}]^3} \tag{7}$$

Equation (8) may be employed to express the reaction equilibrium for D2EHPA separation.

$$Me^{3+} + 3[HA]_2 \rightleftharpoons [MeA_3(HA)_3] + 3H^+ \tag{8}$$

whereby $Me^{3+}$ stands for the metal cation, $[HA]$ for the organic phase's extractant, and $[MeA_3(HA)_3]$ for the complex of the extractant and metal. The equilibrium constant $K_{eq}$ is expressed as follows:

$$K_{eq} = \frac{[MeA_3(HA)_3][H^+]^3}{[Me^{3+}][(HA)_2]^3} \tag{9}$$

The separation of REE such as La can be accomplished using the IL [A336][CA-12] in chloride media. Additionally, it was demonstrated that IL[A336][P507] was suited to separate light REEs in chloride media and would work for the isolation of heavy REEs in nitrate media. By applying IL [A336][P507], the $Ce^{4+}$ and $F^-$ ions have been effectively extracted from $H_2SO_4$ solutions and separated [67]. The recovery of mid-heavy REEs in $H_2SO_4$ solution by [A336][P507] became the subject of a thorough examination. According to the findings, temperature, extractant concentration, and pH all boosted the rate of REE extraction [67]. The results demonstrated that, as both the temperature and extractant concentration increased, so did the distribution ratios of REEs. The influence of pH on the separation behaviors of $Tb^{3+}$ by Cyanex 923, IL [A336][P507], TBP, and P350 (di-(1-methylheptyl)methyl phosphate) is important. Its recovery by Cyanex 923 and TBP in systems of pH = 1.0 to 0.7 improved with rising acidity. At pH < 0, P350 showed a comparable tendency as Cyanex 923 and TBP. Following a reduction in acidity, the $Tb^{3+}$ was extracted more effectively. IL [A336][P507] is more appropriate for the separation of REEs at low acidity, as evidenced by the divergence in recovery performance [67].

Because of the salting-out phenomenon found in the extraction of $Lu^{3+}$, the recovery of REEs decreased as the concentration of $Na_2SO_4$ increased. The process of extraction was hindered by the binding of such cations with $SO_4^{2-}$ in the aqueous medium. However, as the concentration of $NaNO_3$ or NaCl increased, the yield of extraction rose as well. The different behaviors of $NO_3^-$, $SO_4^{2-}$, and $Cl^-$ complexes could be caused by the ion connection between anions and $REE^{3+}$, and the hydration effect. A decline in the recovery of metal ions results from the generation of $REESO_4^+$ ions at high sulphate amounts. It is understood that leaching agents for REEs ores of the ion-adsorbed class once included $NH_4NO_3$, NaCl, $NH_4Cl$, and $(NH_4)_2SO_4$. NaCl was likewise the first leaching solvent to be widely utilized. As the concentration of NaCl grew, so did the effectiveness of REEs' extraction. The above extraction technique is more effective in extracting mid-heavy REEs than the aqueous medium devoid of NaCl. $Al^{3+}$ and $La^{3+}$ experience low extractability as a result, allowing this technique to be employed to retrieve mid-heavy REEs from the pregnant solution. Following a rise in the concentration of $NaNO_3$, the distribution ratios of all REEs rose, particularly for the mid-heavy REEs. The mixture with $NaNO_3$ has a substantially higher extractability than the one employing NaCl [67].

With heating, the precipitated form of Th and REEs such as Ce created after the recovery process can be dissolved in 4 M $HNO_3$. Th can be obtained by extracting it with Aliquat 336 diluted in kerosene, stripping it with HCl solution while agitating, and precipitating it as ammonium hydroxide. $Ce^{3+}$ is, on the contrary, oxidized into its tetravalent form by the incorporation of sodium bromate. It can then be recovered by Aliquat 336 in kerosene that contains 1-octanol, stripped by shaking the organic solution with diluted $HNO_3$ media, and finally precipitated as hydroxide using ammonia. La, Nd, and Y-rich aqueous media can be treated at 25 °C by a synergistic TRPO-TOPO combination in kerosene with a 2:1 O/A phase ratio. $H_2SO_4$ can then be applied to strip the organic media in an equimolar ratio to isolate the majority of the Y, after which, HCl is used as a stripping agent to retrieve Nd and La [71].

In nitrate medium, REEs (Sm and Nd) were extracted from discarded fluorescent lamps using Cyanex 923, which serves as a solvating extractant. Researchers noticed that, although the co-extracted metals took longer than 15 min to reach a state of equilibrium, the REEs did so in just one minute. For greater REE recovery, they recommended low acidity (less than 1 M $HNO_3$), one minute of reaction time in between the two phases, up to 1 M extractant concentration, stripping with 4 M $HNO_3$, and scrubbing by $HNO_3$ or oxalic acid. The REE metal ion was extracted from Nd-Fe-B magnet residue using solvent TBP in IL (Aliquat 336). To improve the distribution coefficient throughout the REE recovery from Ni metal hydride batteries employing chloride medium, $[A336]NO_3^-$ and Cyanex 923 were utilized [52].

To remove REEs and transition metals from Sm/Co and Nd/Fe magnets, Cyphos IL 101 was implemented. Scientists discovered that transition metals are efficiently extracted by the IL, and the maximum distribution rates for Co and Fe, respectively, were attained with 8.5 M and 9 M HCl. The separation coefficients for Sm/Co and Nd/Fe were discovered to be $8 \times 10^5$ and $5 \times 10^6$, respectively. Transition metals including Co, Mn, Fe, and Zn could additionally be extracted using Cyphos 101 IL from an aqueous media that also contains REEs and Ni with 8 M HCl. Following the initial extraction stage, Cyanex 923 and Aliquat 336 nitrate were employed to recover the REEs from the raffinate. Because Aliquat 336 nitrate is more selective for REEs and can handle less Cyanex 923 molecules around it than Cyphos 101, it is preferred over the latter. A336 chloride and A336 thiocyanate were applied to separate Co and Mn. It was possible to recover La from Ni and Sm from Co through the application of trihexyl (tetradecyl) phosphonium nitrate. REEs were successfully recycled using this technique from used permanent magnets and rechargeable batteries. Here, they added Na or ammonium nitrate to the aqueous media, and the consequence of salting out caused the extraction to improve. They discovered that Sm could be extracted as the pentakis complex $[Sm(NO_3)_5]^{3-}$ and La as the hexakis complex $[La(NO_3)_6]^{3-}$. $H_2O$ was the medium used to do the stripping [52].

Synergistic extraction of REEs can offer the benefit of using reduced usage of the main extracting agent, while still being able to achieve a high rate of separation and replenish the extracting agent. TBP and D2EHPA are two of the most popular and commonly employed of these in the commercial process for the extraction of REEs. Although TBP has a higher loading capacity for REEs compared to D2EHPA, D2EHPA possesses a superior separation coefficient for REEs. An assessment was conducted on the impact of the aqueous phase's concentration of $NO_3{}^-$ and $H^+$ ions. Also, an evaluation was conducted on the kind and number of extracting agents present in the organic phase, and on the extraction behavior of La, Nd, Ce, and Y. An organic phase comprising 0.8 M of extracting agents with an equal mole ratio of D2EHPA and TBP in kerosene was employed to test the influence of the $NO_3{}^-$ ion concentration on the recovery of REEs. The concentration of $H^+$ ions in the aqueous media was set at 2 M (pH = −0.3). As the amounts of $NO_3{}^-$ in the aqueous medium increased, the recovery of REE was reported to rise as well. In comparison to the other examined REEs, Y exhibited a higher extraction efficiency. Therefore, when the amount of $NO_3{}^-$ ions rises, so do the distribution ratios of REEs. A sharp increase in the Y distribution ratio over 6 M $NO_3{}^-$ was observed. Because the $REE^{3+}NO_3^-$ species were removed following the recovery mechanism of REEs by $TBP$, which is determined according to the reaction in Equation (10), a common ion effect could potentially be responsible for the rise in extraction caused by an increase in $NO_3{}^-$ ions concentration [69].

$$REE^{3+} + 3NO_3^- + \overline{3TBP} \Rightarrow \overline{REE(NO_3)_3(TBP)_3} \tag{10}$$

With an increase in the concentration of $H^+$ in the aqueous media, the distribution coefficient of REEs dropped. Relative to other REEs, Y is particularly susceptible to changes in the concentration of $H^+$. The cation exchanger extracting agent D2EHPA's recovery mechanism is capable of being used to illustrate how $H^+$ affects REEs extraction. Below is the equation showing the extraction mechanism of REEs from aqueous media using D2EHPA:

$$REE^{3+} + 3\overline{H_2A_2} \Rightarrow \overline{REE(HA_2)_3} + 3H^+ \tag{11}$$

### 3.5. Precipitation

The extracted metals can be recovered in solid form by precipitation. This is realized through the combination of selected ion(s) with an appropriate counter ion in adequate concentrations to surpass the subsequent compound's solubility product and produce a supersaturated solution. Precipitation occurs at lower temperatures, low ion concentration, a high degree of supersaturated solutions, and a minimal stirring of the solutions. The size of the precipitate depends on the type of solvent and pH of the solutions. Metals from sources such as acid mine drainage are normally precipitated as metal oxides, phosphates, sulfides, hydroxides, and carbonates [72].

The precipitation technique is easy to control, simple, inexpensive, and applied in industry. Double-salt precipitation, simple precipitation, and oxidative precipitation are the techniques utilized in hydrometallurgy for secondary resources and primary ores. In simple precipitation, precipitants such as $OH^-$, $C_2O_4{}^{2-}$, $S^{2-}$, and $CO_3{}^{2-}$ are applied [73]. Generally, double-salt precipitation entails the use of salt with more than one anion or cation. Double-salt precipitation is crucial in the recovery of REEs. When some compounds are precipitated or eliminated from a solution by oxidation, the term "oxidative precipitation" is generally used. In other cases, some metal ions can be precipitated by adding complexing agents to form complexes when ligand ions or ligand molecules are available [73]. In most cases, a complexing and precipitating agent can be used together for improved and selective recovery of a metal such as Ni, Mn, Co, Cu, and Fe. For such metals, $OH^-$, $CO_3{}^{2-}$, and $S^{2-}$ precipitants combined with $NH_3$ as a complexing agent are effective in complexation–precipitation.

### 3.6. Ion Exchange

Ion exchange is among the separation techniques that play a big role in the recovery of a wide range of metals with unique properties suitable for the current and future economy. Due to intense mining activities, metals have significantly reduced in ores, and they exist in low concentrations. The ion exchange technique is suitable for the recovery of such metals. That is attributed to its environmental safety, selectivity, powerful resolving ability, high efficiency, moderate cost, and suitability for combination with subsequent techniques in the recovery of noble metals and other applications [74,75]. Ion exchangers with distinct functional groups, due to their good kinetic properties, high mechanical strength, and exchange capacity, increase their suitability in the concentration of microscopic amounts of metals, extracting them in the presence of accompanying components.

Ion exchange resins acting like porous media are essential in separation industries because of the elimination of thermal regeneration, a higher metal ion loading capacity, higher loading rate, and being less prone to fouling by organic materials. The porous property of the resin leads to enhanced adsorption rates because of the improved diffusion of the ions. The metal recovery process's efficiency is firmly reliant on the contact time, pH of the solution, ion concentration, and ion coordination/exchange resin properties such as structure and type of the ligand, crosslinking degree, and swelling [76]. Various types of exchange resins are available commercially in the recovery of metals; for example, Amberlite IR-120 can be used in separating $Cu^{2+}$, $Zn^{2+}$, and $Cd^{2+}$. Chelating ion exchange resins such as Amberlite IRC 748 and Chelex 100 have aminodiacetic acid functional groups to exchange $Zn^{2+}$ and $Cu^{2+}$ from aqueous solutions while macro-porous AMBER JET 1200 Na cation exchange resins are efficient in the separation of metals such as $Ni^{2+}$ and $Pb^{2+}$ [77]. In the recovery of platinum group metals such as $Pa^{2+}$, $Pt^{2+}$, and $Pt^{4+}$ from HCl acid solutions, CYBBER and Purolite grades present high sorption capacity for the anion exchange [74].

## 4. Recovery of Strategic and Critical Metals Using Flotation

Flotation is essentially a surface selectivity-based extraction method that separates the hydrophilic portion of a material from the hydrophobic portion. To create a wide range of separations, this technique selectively modifies the hydrophobicity of the mineral surfaces using several surfactant chemicals [27]. While gangue minerals—unwanted minerals—remain in the water, the target minerals, which are frequently sulphide minerals, are usually hydrophobic and will adhere to the air bubbles [78–82]. The subsequent froth rich in minerals is subsequently scraped off the flotation cell's surface, Figure 8.

If flotation techniques used in other regions are tailored to suit the current minerals, they will be essential for the beneficiation of the minerals in Kenya. Research was conducted to recover Cu and Co by treating the tailings from the flotation of Cu and Co oxidized ores produced at the Kambove Concentrator (DRC). The study focused on determining the dosage of chemicals needed to float the examined tailings and recover the maximum amounts of Cu (44.80%) and Co (88.30%). Centered on the concentrate grade (3.31% Cu and 2.22% Co) and the feasible extraction of the metals of interest, it was determined that flotation was a desirable method for reprocessing the tailings [83]. The tailings from the processing of Cu ores contain rocks that may be recycled as inexpensive raw materials. The overconsumption of reagents due to the particle size distribution in a wide range of slimes is still an obstacle. Such an issue should be taken into consideration for the most effective technical and financial implementation of the aforementioned technique. While the wastewater produced by the flotation of tailings can be reclaimed and fed back into the Kambove Concentrator milling process, the tailings can be added to construction materials as cemented pastes [83].

Slags are also known to have valuable metals, and instead of being an end waste, they can serve as a secondary source of metals [84,85]. They come in a variety of forms, as wastes from combustion processes or as the results of metallurgical operations. They have been used in numerous fields as a resource material [85]. Furthermore, for certain purposes, the characteristics of slags are as good or better than those of competing materials. From nonferrous smelters, a range of nonferrous slags are generated.

Numerous investigations on metal recovery using nonferrous slags have already been conducted. A large number of studies focused on recovering metals from Cu slags [85,86]. By considering variables, including the form of the processed ore, the kind of furnaces employed, and the cooling techniques, the chemical composition of Cu slags from various sources might range significantly. The main phases in crystalline Cu slag are often fayalite and various silicates. Most of the time, Ni and Co are found as oxides where the Co distribution in the slags is extremely homogeneous. On the other hand, many Cu slags have diverse types of Cu minerals. They could take the shape of oxides, sulphides, or a combination of the two. Investigations were conducted on the recovery of Cu and Co from ancient Cu slags collected from the Küre Plant basin of Turkey [86]. The slag was of the fayalitic kind, which contained 1.24% Cu, 53.16% Fe, and 0.53% Co. To retrieve the metal values, two distinct paths were taken. The first method was leaching it after after roasting it with pyrite ($FeS_2$). The second method involved leaching, and roasting the flotation tailings with $FeS_2$, followed by flotation of the slags. It was discovered that the second method worked well for treating the Cu slag. During the flotation process, a Cu concentrate with around 11% Cu was created, and 77% of the Cu was recovered, leaving 93% of the Co in the tailings [86].

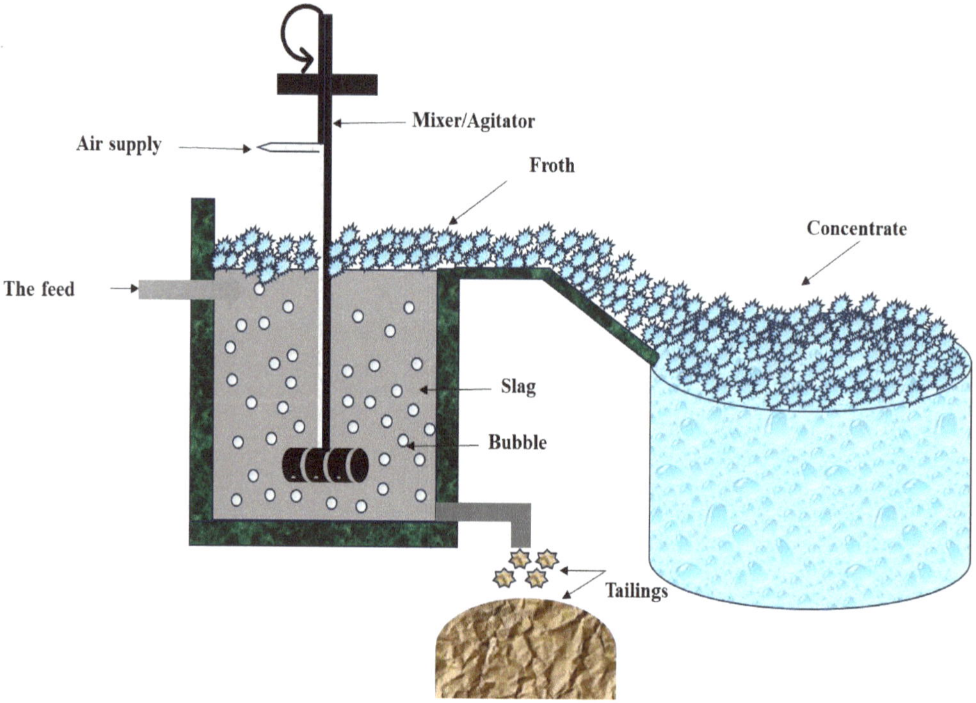


**Figure 8.** A sketch of the froth flotation process.

Due to the small amounts of platinum group metals in primary ores (less than 10 g/ton) and the complex processes involved, their extraction from primary ores is costly. Ores (such as Merensky Reef ore grades, which comprise 3 to 8 g/t platinum group metals tied to Ni, Cu, as sulphides) are mined, ground, processed through gravity separation, and floated. Their concentrate, which is generally 200 to over 2000 g/t platinum group metals alongside 0.4% to 2.8% $Cr_2O_3$, undergoes meltdown at excessive temperatures (>1500 °C). Thereafter, they are purified through hydrometallurgical processes [87]. Traditionally, froth flotation has been used to extract platinum group metals from pure (unweathered) sulphide ores. This method usually yields recoveries of over 85% for Pt. The efficient concentration of such metals through flotation in virgin sulphide ores is dependent upon the presence of highly floating base metal sulphide minerals, such as pyrrhotite, chalcopyrite, and pentlandite. Nevertheless, as an alternate source for keeping up their production, the rapid depletion of sulphide platinum group metal-bearing minerals has sparked a lot of interest from scientists. The focus has been on investigation of their recovery from near-surface oxidized metal ores.

Besides platinum group metals, Au can be extracted using the flotation technique [25,43]. Au floats easily, and it is possible to separate free Au from sulfide-containing ore via selective flotation. A product that is directly smeltable could potentially be produced by removing the floating Au from the sulphides. According to research, selective flotation of Cu sulphides could be observed from $FeS_2$ ores at pH values higher than 11 [88]. The behavior of the free Au in these circumstances was less obvious. Additional collectors (monothiophosphates) could be included for increased Au recovery because they are known to be selective Au accumulators [88,89]. Researchers looked into how several operating factors, such as pH, collector additions, and grind size, affected the efficacy of Au flotation [88]. With a variety of collectors, Au was recovered selectively against $FeS_2$ at elevated pH levels. Nevertheless, in the flotation tests, there was no specificity towards chalcopyrite. Granular Au drifted towards the flotation tail, but fine Au floated more readily. That is because the flotation of Au is highly dependent on some parameters like the type of the collectors, size, structures, and the mineral containing Au, amongst others.

Numerous studies have also been conducted to examine the dynamics of flotation of REEs in terms of pulp chemistry, froth stability, mineral particle size distribution, and collector–mineral surface interactions [90]. The flotation effectiveness of REEs is measured by the flotation rate test, which considers concentrate grade, mass pull, and recovery. The flotation kinetics of the mine is defined as the fluctuation in the recovery of metals over time. The outcomes are utilized to characterize a mineral's flotation behavior in specific flotation conditions. The flotation rate of the REEs depends on the length of flotation, pulp pH, the kind and dose of the collector, the extent of surface–collector contact, and the mineral species. The principal materials that are the primary industrial sources of REE minerals are monazite $(REE)PO_4$ (55 to 60% RE oxides), xenotime $(REE)PO_4$ (55 to 60% RE oxides), and bastnaesite $[REE(CO_3)]F$ (70 to 75% RE oxides). Flotation is considered to be one of the most essential methods for separating REEs from related minerals and producing concentrates. The resulting material often contains 60 to 70% mixed REEs, due to the intricate and fine properties of the REE ore [91].

The assessment was done on the flotation of rare earth oxide (REO) in monazite by combining hematite ($Fe_2O_3$) and quartz ($SiO_2$) with hydroxamic acid [90]. The results of the micro-flotation experiments carried out on individual minerals demonstrated that each model mineral's flotation response depends on pH. The optimal flotation extraction of quartz happened at pH = 3, while that of hematite and monazite was attained at pH = 7. As predicted, adding more hydroxamic acid increased each mineral's flotation yield. When the dose of hydroxamic acid was increased, the specificity coefficients of monazite over quartz remained lower than those of monazite against hematite. In a mixed minerals flotation, it was anticipated that a higher percentage of $Fe_2O_3$ would be associated with REO in flotation concentrates than $SiO_2$. The outcome suggested that depressants are necessary to accomplish selective beneficiation and extraction of REO. Further results on heterogeneous minerals separation experiments demonstrated that, in the absence of depressants, the flotation of REO from $SiO_2$ and $Fe_2O_3$ was unselective. It was reported that 96% $Fe_2O_3$, 99% REO, and 80% $SiO_2$ recoveries of the concentrate were produced by a 2000 g/t hydroxamic acid collector. Whenever 4000 g/t starch was employed, the extraction effectiveness of REO rose from its low average of 6.42% devoid of depressants to 44.78%. Matching recoveries of 93% REO, 38% $SiO_2$, and 81% $Fe_2O_3$ were observed. Contrarily, when $Na_2SiO_3$ was utilized, the resulting concentrates had greater gangue contents ($SiO_2$ > 60% and $Fe_2O_3$

> 90%). That resulted in lower REO separation efficiency. When hydroxamic acid was present, starch provided a better REO upgrade. Nevertheless, $Na_2SiO_3$ was a better choice when REO recovery was the only factor considered. Furthermore, a mixture of starch and $Na_2SiO_3$ demonstrated considerable recoveries [90].

The beneficiation reaction of gangue minerals and REE in the tailings differs noticeably. The separation techniques of froth flotation, wet magnetics, and gravity led to widely differing REE upgrades and recoveries. A flotation study was conducted on the ore by applying sodium oleate acting as a collector. To determine if there was a possibility of extracting and enriching REE minerals in saprolite ore, a study examined three distinct treatment configurations: the relative impacts of pulp pH, depressants, and de-sliming. Most REE minerals (>50%) could be extracted using flotation methods on raw feed. However, the technique was not selective, given that clay minerals and silicate gangue also found their way into the flotation concentrate together with the REE minerals. The specificity of the flotation of REE minerals was enhanced by de-sliming before flotation, using a mixture of depressants ($Na_2SiO_3$ and starch). That resulted in concentrates that had total REO grades of 5.87% and 4.22%. Recovery yields of 45% and 50% at pulp pH = 9 and 10.5 were obtained. Because of their fine-to-ultrafine properties, clay gauge and silicate minerals were collected through the synergistic process of surface activation and entrainment. A random grade–recovery relationship was found when all test results were compared. It indicated that there could be a need for more flotation tests where its process limitations can be looked into. Optimization is also necessary to further maximize both REE recovery and grade. Additionally, the possibility of employing magnetic separation was raised [92].

Apart from the physical methods, ILs have been seen as revolutionary in the flotation of REEs. Tetraethylammonium mono-(2-ethylhexyl)2-ethylhexyl phosphonate ([N2222][EHEHP]) is an IL that has been studied earlier for REE solvent extraction and revealed to be selective and efficient. For the first time, [N2222][EHEHP] was assessed as a collector in bastnäsite flotation, which is the main deposit for producing REEs [93]. The findings were contrasted with two typical gangue minerals found in quartz and hematite. Hematite exhibited an enhanced collectability of [N2222][EHEHP] over bastnäsite, which significantly recovers at pH = 5 with an increased dosage of IL (500 g/t). It implies that these minerals' extraction mechanisms could not be the same. Based on the Fourier transform infrared spectroscopy (FT-IR) data, [EHEHP] moiety adsorption on bastnäsite was verified. It seemed to be most effective in slightly acidic pH conditions. The micro-flotation data, which indicated that bastnäsite recovery was maximum (~50%) at pH = 5, was consistent with this. Furthermore, micro-flotation indicated that a very high dose of IL must be administered to directly float bastnäsite. Protons may interact with $[EHEHP]^-$ and obstruct interactions between REE ions and $[EHEHP]^-$, which explains why bastnäsite flotation recovery is lower at pH = 3. According to FT-IR data, the adsorption of the $[EHEHP]^-$ moiety on hematite is evident in acidic conditions, which matches the micro-flotation recovery pattern. Furthermore, hematite may be more recoverable because of its better yield compared to bastnäsite. The little recovery for quartz at both dosages across the pH range under test signified the absence of reagent adsorption. By using magnetic separation to examine the concentrates and tails, it was discovered that higher collector dosages might result in bastnäsite recovery capacities over 90% and an optimal conversion ratio of 1.7 [93].

## 5. Proposals and Prospects for Future Application of the Hydrometallurgical and Flotation Techniques

In 2016, Kenya implemented the Mining Act meant to modernize and expand the mining industry. Even though the Act was improved from the previous legislation, output in the mining sector remains poor. However, the new regime announced the removal of the moratorium in 2023. The Government emphasized the importance of creating a good business atmosphere for stakeholders and investors. Kenya's future development of the mining sector will be influenced by environmental impact and the adoption of sustainable mining techniques. Given the rising number of international mining firms promising to decarbonize and achieve net zero goals, the mining sector should develop practical approaches to tackle environmental issues in their operations. The approach will compel investors to apply environmentally justifiable techniques for metal recovery.

As Kenya seeks to diversify its mining practices, hydrometallurgical and flotation procedures emerge as a compelling alternative. These innovative approaches will not only address the constraints posed by low-grade ores but also align with global sustainability goals. Through investment in research and regulatory and infrastructure adaptations, Kenya can pave the way for a dynamic and responsible mining sector. That will contribute to both economic development and environmental preservation.

The metal recovery techniques outlined in this article can easily be applied to the needs of the Kenyan mining industry. The development of advanced leaching technologies can enhance the efficiency and selectivity of metal extraction. Continuous research and investment in these technologies can optimize the recovery of strategic and critical metals. One of the challenges in Kenya's mining industry is the presence of low-grade ore deposits. Hydrometallurgy offers a viable solution, enabling the cost-effective extraction of metals from ores that were previously considered economically unviable. Such an option becomes crucial in times of depleted mineral reserves for important metals, when alternative sources must be explored. Beyond primary ores, hydrometallurgy provides an avenue for recovering metals from secondary sources. With the growing demand for recycling and a circular economy, Kenya can leverage hydrometallurgical procedures to extract valuable metals from secondary sources, contributing to resource sustainability. The mining research institutions in Kenya will need to also initiate progressive efforts to enhance the efficacy of the flotation process. That could encompass the implementation of advanced control systems or the formulation of improved surfactants.

There will be a need for the establishment and demonstration of pilot plants to exhibit the effectiveness and feasibility of the described methods. That can be reinforced by introducing capacity-building programs and training dynamism, to equip local experts with the knowledge required to operate flotation and hydrometallurgical processes. Collaboration with research institutions, both domestically and internationally, can fast-track the development of competent and environmentally friendly metal-recovery processes tailored to the country's mineral resources. In addition, regulatory frameworks ought to be adapted to address the drawbacks and opportunities associated with chemical-based techniques for metal recovery. Clear guidelines, environmental impact assessments, and monitoring mechanisms should be established to foster responsible mining practices.

## 6. Conclusions

The vast unexploited mineral resources in Kenya present a major opportunity for economic development. Nonetheless, the environmental impacts of mining in the region cannot be neglected. Mineral extraction processes require a well-adjusted approach that considers both sustainability and economic returns. The adoption of modern flotation and hydrometallurgical methods emerges as an important solution to efficiently extract strategic and critical metals while reducing environmental effects. For full balance in society of a sustainable and new world, we must not forget about people. Skillful management of natural resources and investment in local communities should guarantee stable and peaceful development. The mining industry and government stakeholders must adopt sustainable technologies, while prioritizing responsible mining practices. All-inclusive and collaborative steps are necessary for harnessing Kenya's potential mineral wealth, while protecting the delicate balance of its environment.

**Author Contributions:** Conceptualization, editing, and writing review, N.R.K.; conceptualization and editing, I.H.-K.; conceptualization and editing, M.R.; conceptualization and editing, T.S.; conceptualization and editing, M.S.; editing and conceptualization, A.G.C. All authors have read and agreed to the published version of the manuscript.

**Funding:** TC Project RER1023: Harmonizing Implementation of Radiotracer and Sealed Sources Techniques for Efficient Use of Natural Resources and Environmental Monitoring ME-RER1023-2105581.

**Data Availability Statement:** Not applicable.

**Acknowledgments:** The task was completed with assistance from the Institute of Nuclear Chemistry and Technology's research activities.

**Conflicts of Interest:** The authors declare no conflict of interest.

## References

1. Haneklaus, N.; Sun, Y.; Bol, R.; Lottermoser, B.; Schnug, E. To extract, or not to extract uranium from phosphate rock, that is the question. *Environ. Sci. Technol.* **2017**, *51*, 753–754. [CrossRef] [PubMed]
2. Ascher, W. *Why Governments Waste Natural Resources: Policy Failures in Developing Countries*; JHU Press: Baltimore, MD, USA, 1999.
3. Hilson, G. The Africa Mining Vision: A manifesto for more inclusive extractive industry-led development? *Can. J. Dev. Stud./Rev. Can. d'études Dév.* **2020**, *41*, 417–431. [CrossRef]
4. Kim, J.; Guillaume, B.; Chung, J.; Hwang, Y. Critical and precious materials consumption and requirement in wind energy system in the EU 27. *Appl. Energy* **2015**, *139*, 327–334. [CrossRef]
5. Grandell, L.; Lehtilä, A.; Kivinen, M.; Koljonen, T.; Kihlman, S.; Lauri, L.S. Role of critical metals in the future markets of clean energy technologies. *Renew. Energy* **2016**, *95*, 53–62. [CrossRef]
6. Moss, R.L.; Tzimas, E.; Kara, H.; Willis, P.; Kooroshy, J. Critical Metals in Strategic Energy Technologies: Assessing rare metals as supply-chain bottlenecks in low-carbon energy technologies. In *JRC-Scientific and Strategic Reports, European Commission Joint Research Centre Institute for Energy and Transport*; EU Publications: Luxembourg, 2011. [CrossRef]
7. Zhang, S.; Ding, Y.; Liu, B.; Chang, C.-c. Supply and demand of some critical metals and present status of their recycling in WEEE. *Waste Manag.* **2017**, *65*, 113–127. [CrossRef]
8. Chakhmouradian, A.R.; Smith, M.P.; Kynicky, J. From "strategic" tungsten to "green" neodymium: A century of critical metals at a glance. *Ore Geol. Rev.* **2015**, *64*, 455–458. [CrossRef]
9. Žibret, G.; Lemiere, B.; Mendez, A.-M.; Cormio, C.; Sinnett, D.; Cleall, P.; Szabó, K.; Carvalho, M.T. National Mineral Waste Databases as an Information Source for Assessing Material Recovery Potential from Mine Waste, Tailings and Metallurgical Waste. *Minerals* **2020**, *10*, 446. [CrossRef]
10. Kiprono, N.R.; Smolinski, T.R.; Rogowski, M.; Chmielewski, A.G. The State of Critical and Strategic Metals Recovery and the Role of Nuclear Techniques in the Separation Technologies Development: Review. *Separations* **2023**, *10*, 112. [CrossRef]
11. Ayres, R.U.; Peiró, L.T. Material efficiency: Rare and critical metals. *Philos. Trans. R. Soc. A Math. Phys. Eng. Sci.* **2013**, *371*, 20110563. [CrossRef]
12. Purwadi, I.; van der Werff HM, A.; Lievens, C. Targeting rare earth element bearing mine tailings on Bangka Island, Indonesia, with Sentinel-2 MSI. *Int. J. Appl. Earth Obs. Geoinf.* **2020**, *88*, 102055. [CrossRef]
13. Wang, L.; Liang, T. Geochemical fractions of rare earth elements in soil around a mine tailing in Baotou, China. *Sci. Rep.* **2015**, *5*, 12483. [CrossRef] [PubMed]
14. Campbell, G.A. Rare earth metals: A strategic concern. *Miner. Econ.* **2014**, *27*, 21–31. [CrossRef]
15. Fleming, P.; Orrego, P.; Pinilla, F. Recovery of Rare Earth Elements Present in Mining Tails, by Leaching with Nitric and Hydrochloric Solutions. *World J. Nucl. Sci. Technol.* **2021**, *11*, 1–16. [CrossRef]
16. Government of the Republic of Kenya. Mining and Mineral Policy. 2016. Available online: https://www.idlo.int/sites/default/files/pdfs/highlights/Kenya%20Mining%20Policy%20Popular%20Version-LowRes.pdf (accessed on 17 August 2023).
17. Zhang, W.; Zhu, Z.; Cheng, C.Y. A literature review of titanium metallurgical processes. *Hydrometallurgy* **2011**, *108*, 177–188. [CrossRef]
18. Mwaura, F. An audit of environmental impact assessments for mining projects in Kenya. *J. S. Afr. Inst. Min. Metall.* **2019**, *119*, 485–493. [CrossRef] [PubMed]
19. Opongo, E. Knowledge and Policy Gaps in Extractive Industries in Kenya. *SSRN Electron. J.* **2017**, *12*, 1–138. [CrossRef]
20. Barmao, K.J.; Cherutoi, J.K.; Mitei, C.Y.; Were, M.L.L.; Kiprop, A.; Achieng', O.G. Assessment of Fluoride and selected heavy metals in food chain around Fluorspar mining Plant, Kenya. *Greener J. Environ. Manag. Public Saf.* **2019**, *8*, 15–24. [CrossRef]
21. Kipsang, R.B. Economic and Job Creation Economic and Job Creation Potential of Artisanal and Small-Scale Mining in Taita Taveta County, Kenya. Natural Resources Management for Sustainable Development in Kenya Extractive Industry, UNDP. 2014. Available online: https://www.jkuat.ac.ke/departments/mining/wp-content/uploads/2017/10/Small-Scale-Mining-n-Taita-Taveta-County-Kenya.pdf (accessed on 12 October 2023).
22. Kaniu, I.; Iain, G.D.; Hudson, K.A. Radiological Mapping of the Alkaline Intrusive Complex of Jombo, South Coastal Kenya by In-Situ Gamma-Ray Spectrometry. 2016. Available online: https://meetingorganizer.copernicus.org/EGU2016/EGU2016-17917-1.pdf (accessed on 21 August 2023).
23. African Natural Resources Centre. Rare Earth Elements (REE). Value Chain Analysis for Mineral Based Industrialization in Africa. 2021. Available online: https://www.afdb.org/fr/documents/rare-earth-elements-ree-value-chain-analysis-mineral-based-industrialization-africa (accessed on 21 August 2023).
24. Odumo, B.O.; Nanos, N.; Carbonell, G.; Torrijos, M.; Patel, J.P.; Rodríguez Martín, J.A. Artisanal gold-mining in a rural environment: Land degradation in Kenya. *Land Degrad. Dev.* **2018**, *29*, 3285–3293. [CrossRef]

25. Ogola, J.S.; Mitullah, W.V.; Omulo, M.A. Impact of Gold Mining on the Environment and Human health: A Case Study in the Migori Gold Belt, Kenya. *Environ. Geochem. Health* **2002**, *24*, 141–158. [CrossRef]
26. Government of Kenya. Kenya Mining Investment Handbook. 2016. Available online: https://www.tralac.org/documents/resources/by-country/kenya/1928-kenya-mining-investment-handbook-2016/file.html (accessed on 21 August 2023).
27. Bett, A.K.; Maranga, S.M. Considerations for Beneficiation of Low Grade Iron Ore for Steel Making in Kenya. In Proceedings of the Sustainable Research and Innovation Conference 2014, Nairobi, Kenya, 3–4 May 2012; pp. 263–267. Available online: https://sri.jkuat.ac.ke/jkuatsri/index.php/sri/article/view/481 (accessed on 12 October 2023).
28. Bourdon, E.; Kalt, A.; Meisel, T.; Kaeser, B.; Bourdon, E.; Kalt, A.; Meisel, T.; Kaeser, B. Platinum-group element abundances in xenoliths from Marsabit Volcanic field (Kenya Rift). *AGUFM* **2006**, *2006*, V31D-0609. Available online: https://ui.adsabs.harvard.edu/abs/2006AGUFM.V31D0609B/abstract (accessed on 24 August 2023).
29. Henderson, P.; Pickford, M.; Williams, C.T. A geochemical study of rocks and spring waters at Kanam and Kanjera, Kenya, and the implications concerning element mobility and uptake. *J. Afr. Earth Sci.* **1987**, *6*, 221–227. [CrossRef]
30. Compaore, W.F.; Dumoulin, A.; Rousseau DP, L. Gold Mine Impact on Soil Quality, Youga, Southern Burkina Faso, West Africa. *Water Air Soil Pollut.* **2019**, *230*, 207. [CrossRef]
31. Ceniceros-Gómez, A.E.; Macías-Macías, K.Y.; de la Cruz-Moreno, J.E.; Gutiérrez-Ruiz, M.E.; Martínez-Jardines, L.G. Characterization of mining tailings in México for the possible recovery of strategic elements. *J. S. Am. Earth Sci.* **2018**, *88*, 72–79. [CrossRef]
32. Antwi-Agyei, P.; Hogarh, J.N.; Foli, G. Trace elements contamination of soils around gold mine tailings dams at Obuasi, Ghana. *Afr. J. Environ. Sci. Technol.* **2009**, *3*, 353–359. Available online: https://www.ajol.info/index.php/ajest/article/view/56263 (accessed on 12 April 2023).
33. Matinde, E.; Simate, G.S.; Ndlovu, S. Mining and metallurgical wastes: A review of recycling and re-use practices. *J. S. Afr. Inst. Min. Metall.* **2018**, *118*, 825–844. [CrossRef]
34. Bakatula, E.N.; Cukrowska, E.M.; Straker, C.J.; Weiersbye, I.M.; Tutu, H.; Rüde, T.R.; Freund, A.; Wolkersdorfer, C. Biosorption of heavy metals from gold-mine wastewaters by Penicillium simplicissimum immobilised on zeolite: Kinetic, equilibrium and thermodynamic study. In *Mine Water–Managing the Challenges*; Rüde, T., Freund, A., Wolkersdorfer, C., Eds.; International Mine Water Association: Aachen, Germany, 2011; pp. 271–272.
35. Bridge, G. Contested terrain: Mining and the environment. *Annu. Rev. Environ. Resour.* **2004**, *29*, 205–259. [CrossRef]
36. Oladipo, H.J.; Tajudeen, Y.A.; Taiwo, E.O.; Muili, A.O.; Yusuf, R.O.; Jimoh, S.A.; Oladipo, M.K.; Oladunjoye, I.O.; Egbewande, O.M.; Sodiq, Y.I.; et al. Global Environmental Health Impacts of Rare Earth Metals: Insights for Research and Policy Making in Africa. *Challenges* **2023**, *14*, 20. [CrossRef]
37. Ndlovu, S.; Simate, G.S.; Matinde, E. Waste Production and Utilization in the Metal Extraction Industry. In *Waste Production and Utilization in the Metal Extraction Industry*; CRC Press: Boca Raton, FL, USA, 2017; pp. 1–512. [CrossRef]
38. Chmielewski, A.G.; Wawszczak, D.; Brykała, M. Possibility of uranium and rare metal recovery in the Polish copper mining industry. *Hydrometallurgy* **2016**, *159*, 12–18. [CrossRef]
39. Kiegiel, K.; Gajda, D.; Zakrzewska-Kołtuniewicz, G. Recovery of uranium and other valuable metals from substrates and waste from copper and phosphate industries. *Sep. Sci. Technol.* **2020**, *55*, 2099–2107. [CrossRef]
40. Tunsu, C.; Menard, Y.; Eriksen, D.Ø.; Ekberg, C.; Petranikova, M. Recovery of critical materials from mine tailings: A comparative study of the solvent extraction of rare earths using acidic, solvating and mixed extractant systems. *J. Clean. Prod.* **2019**, *218*, 425–437. [CrossRef]
41. Zhuang, W.Q.; Fitts, J.P.; Ajo-Franklin, C.M.; Maes, S.; Alvarez-Cohen, L.; Hennebel, T. Recovery of critical metals using biometallurgy. *Curr. Opin. Biotechnol.* **2015**, *33*, 327–335. [CrossRef] [PubMed]
42. Blengini, G.A.; Mathieux, F.; Mancini, L.; Nyberg, M.; Cavaco Viegas, H.; Salminen, J.; Garbarino, E.; Orveillion, G.; Saveyn, H. *Recovery of Critical and Other Raw Materials from Mining Waste and Landfills: State of Play on Existing Practices*; EU Publications: Luxembourg, 2019. [CrossRef]
43. Ngure, V.; Kinuthia, G. Health risk implications of lead, cadmium, zinc, and nickel for consumers of food items in Migori Gold mines, Kenya. *J. Geochem. Explor.* **2020**, *209*, 106430. [CrossRef]
44. Belardi, G.; Piga, L.; Quaresima, S.; Shehu, N. Application of physical separation methods for the upgrading of titanium dioxide contained in a fine waste. *Int. J. Miner. Process.* **1998**, *53*, 145–156. [CrossRef]
45. Kyalo, M.N.; Munyerere, I.F.; Rop, B.; Maranga, S.M. Scouring abandoned mines in search for elusive metal (gold) in Kakamega's Rosterman area-A case study in Kenya. In Proceedings of the Sustainable Research and Innovation Conference 2015, Nairobi, Kenya, 6–8 May 2015; pp. 362–366.
46. Richardson, J.F.; Harker, J.H.; Backhurst, J.R. *Coulson and Richardson's Chemical Engineering*, 5th ed.; Butterworth-Heinemann: Oxford, UK, 2002; Volume 2.
47. Bertuol, D.A.; Amado, F.R.; Cruz, E.D.; Tanabe, E.H. Metal recovery using supercritical carbon dioxide. In *Green Sustainable Process for Chemical and Environmental Engineering and Science: Supercritical Carbon Dioxide as Green Solvent*; Elsevier: Amsterdam, The Netherlands, 2020; pp. 85–103. [CrossRef]
48. Kumar, P.A.; Vengatasalam, R. Mineral Beneficiation by Heap Leaching Technique in Mining. *Procedia Earth Planet. Sci.* **2015**, *11*, 140–148. [CrossRef]

49. Thenepalli, T.; Chilakala, R.; Habte, L.; Tuan, L.Q.; Kim, C.S. A brief note on the heap leaching technologies for the recovery of valuable metals. *Sustainability* **2019**, *11*, 3347. [CrossRef]
50. Lommelen, R.; Vander Hoogerstraete, T.; Onghena, B.; Billard, I.; Binnemans, K. Model for Metal Extraction from Chloride Media with Basic Extractants: A Coordination Chemistry Approach. *Inorg. Chem.* **2019**, *58*, 12289–12301. [CrossRef] [PubMed]
51. Alvial-Hein, G.; Mahandra, H.; Ghahreman, A. Separation and recovery of cobalt and nickel from end of life products via solvent extraction technique: A review. *J. Clean. Prod.* **2021**, *297*, 126592. [CrossRef]
52. Swain, N.; Mishra, S. A review on the recovery and separation of rare earths and transition metals from secondary resources. *J. Clean. Prod.* **2019**, *220*, 884–898. [CrossRef]
53. Gajda, B.; Mariusz, B.B. The Effect of Tributyl Phosphate on the Extraction of Nickel(II) And Cobalt(II) ions with Di(2-Ethylhexyl)Phosphoric Acid. *Physicochem. Probl. Miner. Process.* **2007**, *41*, 145–152.
54. Huang, Y.; Guo, H.; Zhang, C.; Liu, B.; Wang, L.; Peng, W.; Cao, Y.; Song, X.; Zhu, X. A novel method for the separation of zinc and cobalt from hazardous zinc–cobalt slag via an alkaline glycine solution. *Sep. Purif. Technol.* **2021**, *273*, 119009. [CrossRef]
55. Choubey, P.K.; Dinkar, O.S.; Panda, R.; Kumari, A.; Jha, M.K.; Pathak, D.D. Selective extraction and separation of Li, Co and Mn from leach liquor of discarded lithium ion batteries (LIBs). *Waste Manag.* **2021**, *121*, 452–457. [CrossRef] [PubMed]
56. Wang, L.Y.; Lee, M.S.; Wang, L.Y.; Lee, M.S. Separation of Co(II) and Ni(II) from chloride leach solution of nickel laterite ore by solvent extraction with Cyanex 301. *IJMP* **2017**, *166*, 45–52. [CrossRef]
57. Niobium, Tantalum, Vanadium and Zirconium Ore in Kenya. The Observatory of Economic Complexity. 2023. Available online: https://oec.world/en/profile/bilateral-product/niobium-tantalum-vanadium-and-zirconium-ore/reporter/ken (accessed on 12 October 2023).
58. Peng, H.; Zhang, C.; Hao, Z.; Jiang, S.; Guo, J.; Huang, H.; Li, B. Vanadium recovery by glycine precipitation. *Environ. Chem. Lett.* **2022**, *20*, 1569–1575. [CrossRef]
59. Cai, Z.; Feng, Y.; Li, H.; Zhou, Y. Selective separation and extraction of vanadium(IV) and manganese(II) from co-leaching solution of roasted stone coal and pyrolusite via solvent extraction. *Ind. Eng. Chem. Res.* **2013**, *52*, 13768–13776. [CrossRef]
60. Pospiech, B. Synergistic Solvent Extraction and Transport of Zn(II) and Cu(II) across Polymer Inclusion Membranes with a Mixture of TOPO and Aliquat 336. *Sep. Sci. Technol.* **2014**, *49*, 1706–1712. [CrossRef]
61. Rogowski, M.; Smolinski, T.; Pyszynska, M.; Brykala, M.; Chmielewski, A.G. Studies on hydrometallurgical processes using nuclear techniques to be applied in copper industry. II. Application of radiotracers in copper leaching from flotation tailings. *Nukleonika* **2018**, *63*, 131–137. [CrossRef]
62. Smolinski, T.; Wawszczak, D.; Deptula, A.; Lada, W.; Olczak, T.; Rogowski, M.; Pyszynska, M.; Chmielewski, A.G. Solvent extraction of Cu, Mo, V, and U from leach solutions of copper ore and flotation tailings. *J. Radioanal. Nucl. Chem.* **2017**, *314*, 69–75. [CrossRef]
63. Parent, M.; Cornelis, R.; Dams, R. Investigation of extraction and back-extraction behaviour of platinum (IV) with rubeanic acid in tributyl phosphate, with tributyl phosphate and with thenoyltrifuoroacetone in n-butyl alcohol-acetophenone by means of platinum-191 radiotracer for platinum-enrichment purposes. *Anal. Chim. Acta* **1993**, *281*, 153–160. [CrossRef]
64. Abbass, M.K.; Jalhoom, M.G.; Kadhim, A.M. Extraction of Rare Earth Elements from Iraqi Phosphate Ore by Using of Tributyl Phosphate. *Eng. Technol. J.* **2020**, *38*, 240–245. [CrossRef]
65. Hidayah, N.N.; Abidin, S.Z. The evolution of mineral processing in extraction of rare earth elements using solid-liquid extraction over liquid-liquid extraction: A review. *Miner. Eng.* **2017**, *112*, 103–113. [CrossRef]
66. Fedorova, M.I.; Levina, A.V. Application of ionic liquid based on Aliquat 336 and D2EHPA in the extraction of transition metals. *IOP Conf. Ser. Mater. Sci. Eng.* **2022**, *1212*, 012021. [CrossRef]
67. Shen, L.; Chen, J.; Chen, L.; Liu, C.; Zhang, D.; Zhang, Y.; Su, W.; Deng, Y. Extraction of mid-heavy rare earth metal ions from sulphuric acid media by ionic liquid [A336][P507]. *Hydrometallurgy* **2016**, *161*, 152–159. [CrossRef]
68. Gorzin, H.; Ghaemi, A.; Hemmati, A.; Maleki, A. Studies on effective interaction parameters in extraction of Pr and Nd using Aliquat 336 from NdFeB magnet-leaching solution: Multiple response optimizations by desirability function. *J. Mol. Liq.* **2021**, *324*, 115123. [CrossRef]
69. Ferdowsi, A.; Yoozbashizadeh, H. Solvent Extraction of Rare Earth Elements from a Nitric Acid Leach Solution of Apatite by Mixtures of Tributyl Phosphate and Di-(2-ethylhexyl) Phosphoric Acid. *Metall. Mater. Trans. B Process Metall. Mater. Process. Sci.* **2017**, *48*, 3380–3387. [CrossRef]
70. Jürjo, S.; Siinor, L.; Siimenson, C.; Paiste, P.; Lust, E. Two-Step Solvent Extraction of Radioactive Elements and Rare Earths from Estonian Phosphorite Ore Using Nitrated Aliquat 336 and Bis(2-ethylhexyl) Phosphate. *Minerals* **2021**, *11*, 388. [CrossRef]
71. El-Nadi, Y.A. Solvent Extraction and Its Applications on Ore Processing and Recovery of Metals: Classical Approach. *Sep. Purif. Rev.* **2016**, *46*, 195–215. [CrossRef]
72. Blais, J.F.; Djedidi, Z.; Cheikh, R.B.; Tyagi, R.D.; Mercier, G. Metals precipitation from effluents: Review. *Pract. Period. Hazard. Toxic Radioact. Waste Manag.* **2008**, *12*, 135–149. [CrossRef]
73. Nie, Z.R.; Ma, L.W.; Xi, X.L. "Complexation-precipitation" metal separation method system and its application in secondary resources. *Rare Met.* **2014**, *33*, 369–378. [CrossRef]
74. Kononova, O.N.; Duba, E.V.; Shnaider, N.I.; Pozdnyakov, I.A. Ion exchange extraction of platinum (IV) and palladium (II) from hydrochloric acid solutions. *Russ. J. Appl. Chem.* **2017**, *90*, 1239–1245. [CrossRef]

75. Cummins, P.M.; Dowling, O.; O'Connor, B.F. Ion-exchange chromatography: Basic principles and application to the partial purification of soluble mammalian prolyl oligopeptidase. *Methods Mol. Biol.* **2011**, *681*, 215–228. [CrossRef]
76. Rodriguez-Freire, L.; Gonzalez-Estrella, J.; Li, G. Technologies for fractionation of wastewater and resource recovery. In *Wastewater Treatment Residues as Resources for Biorefinery Products and Biofuels*; Elsevier: Amsterdam, The Netherlands, 2019; pp. 329–354. [CrossRef]
77. Zewail, T.M.; Yousef, N.S. Kinetic study of heavy metal ions removal by ion exchange in batch conical air spouted bed. *Alex. Eng. J.* **2015**, *54*, 83–90. [CrossRef]
78. Walkowiak, W. Mechanism of Selective Ion Flotation. 1. Selective Flotation of Transition Metal Cations. *Sep. Sci. Technol.* **1991**, *26*, 559–568. [CrossRef]
79. Ejtemaei, M.; Gharabaghi, M.; Irannajad, M. A review of zinc oxide mineral beneficiation using flotation method. *Adv. Colloid Interface Sci.* **2014**, *206*, 68–78. [CrossRef] [PubMed]
80. Önal, G.; Bulut, G.; Gül, A.; Kangal, O.; Perek, K.T.; Arslan, F. Flotation of Aladağ oxide lead–zinc ores. *Miner. Eng.* **2005**, *18*, 279–282. [CrossRef]
81. Crundwell, F.K.; du Preez, N.B.; Knights BD, H. Production of cobalt from copper-cobalt ores on the African Copperbelt—An overview. *Miner. Eng.* **2020**, *156*, 106450. [CrossRef]
82. Kar, B.; Sahoo, H.; Rath, S.S.; Das, B. Investigations on different starches as depressants for iron ore flotation. *Miner. Eng.* **2013**, *49*, 1–6. [CrossRef]
83. Lutandula, M.S.; Maloba, B. Recovery of cobalt and copper through reprocessing of tailings from flotation of oxidised ores. *J. Environ. Chem. Eng.* **2013**, *1*, 1085–1090. [CrossRef]
84. Shen, H.; Forssberg, E. An overview of recovery of metals from slags. *Waste Manag.* **2003**, *23*, 933–949. [CrossRef]
85. Zuo, Z.; Feng, Y.; Dong, X.; Luo, S.; Ren, D.; Wang, W.; Wu, Y.; Yu, Q.; Lin, H.; Lin, X. Advances in recovery of valuable metals and waste heat from copper slag. *Fuel Process. Technol.* **2022**, *235*, 107361. [CrossRef]
86. Bulut, G.; Perek, K.T.; Gui, A.; Arslan, F.; Onal, G. Recovery of metal values from copper slags by flotation and roasting with pyrite. *Min. Metall. Explor.* **2007**, *14*, 13–18. [CrossRef]
87. Yakoumis, I.; Panou, M.; Moschovi, A.M.; Panias, D. Recovery of platinum group metals from spent automotive catalysts: A review. *Clean. Eng. Technol.* **2021**, *3*, 100112. [CrossRef]
88. Forrest, K.; Yan, D.; Dunne, R. Optimisation of gold recovery by selective gold flotation for copper-gold-pyrite ores. *Miner. Eng.* **2001**, *14*, 227–241. [CrossRef]
89. Acarkan, N.; Bulut, G.; Gül, A.; Kangal, O.; Karakaş, F.; Kökkiliç, O.; Önal, G. The effect of collector's type on gold and silver flotation in a complex ore. *Sep. Sci. Technol.* **2011**, *46*, 283–289. [CrossRef]
90. Abaka-Wood, G.B.; Addai-Mensah, J.; Skinner, W. A study of selective flotation recovery of rare earth oxides from hematite and quartz using hydroxamic acid as a collector. *Adv. Powder Technol.* **2018**, *29*, 1886–1899. [CrossRef]
91. Chelgani, S.C.; Rudolph, M.; Leistner, T.; Gutzmer, J.; Peuker, U.A.; Chelgani, S.C.; Rudolph, M.; Leistner, T.; Gutzmer, J.; Peuker, U.A. A review of rare earth minerals flotation: Monazite and xenotime. *IJMST* **2015**, *25*, 877–883. [CrossRef]
92. Abaka-Wood, G.B.; Johnson, B.; Addai-Mensah, J.; Skinner, W. Recovery of Rare Earth Elements Minerals in Complex Low-Grade Saprolite Ore by Froth Flotation. *Minerals* **2022**, *12*, 1138. [CrossRef]
93. Li, R.; Marion, C.; Espiritu ER, L.; Multani, R.; Sun, X.; Waters, K.E. Investigating the use of an ionic liquid for rare earth mineral flotation. *J. Rare Earths* **2021**, *39*, 866–874. [CrossRef]

minerals

MDPI

*Review*

# Control of Non-Ferrous Metal-Sulfide Minerals' Flotation via Pulp Potential

**Marinela Panayotova**

Department of Chemistry, Faculty of Mining Technology, University of Mining and Geology "St. Ivan Rilski", Boyan Kamenov Str. No 1, Sofia 1700, Bulgaria; marichim@mgu.bg

**Abstract:** Studies on the dependence of the technological results of non-ferrous sulfide ore (copper—arsenic-bearing and non-arsenic-bearing—lead–zinc, and polymetallic) flotation on the pulp potential Eh are reviewed. Findings on the relation of Eh and collectorless flotation are presented. Changes in the pulp potential due to different gas applications and various reagent additions are considered. The influence of the grinding medium on the pulp Eh and hence on the flotation results is presented through various examples. The relation between the oxidation–reduction potential and reagent effects is exhibited and explained. pH–Eh ranges of different minerals' flotation, as recorded in various studies, are summarized and visualized jointly for all mentioned ores. It is concluded that the pulp Eh value, considered together with the pH value, is a useful means for flotation selection controlling and deserves further research, especially under industrial conditions. Some problems and difficulties in using pulp Eh for flotation control are discussed.

**Keywords:** pulp oxidation–reduction potential Eh; base non-ferrous metal-sulfide flotation; grinding media effect; flotation potential domains; flotation reagent effects explained

**Citation:** Panayotova, M. Control of Non-Ferrous Metal-Sulfide Minerals' Flotation via Pulp Potential. *Minerals* **2023**, *13*, 1512. https://doi.org/10.3390/min13121512

Academic Editor: Zhiyong Gao

Received: 31 October 2023
Revised: 25 November 2023
Accepted: 27 November 2023
Published: 1 December 2023



## 1. Introduction

The role of oxidation–reduction processes in the flotation separation of sulfide minerals is an accepted fact, with numerous examples, such as: (a) copper–pyrite (Cu–Py) selection, in which pyrite plays the role of the cathode and copper minerals play the role of the anode [1–3]; (b) activation of pyrite by chalcopyrite via Cu ions produced due to galvanic oxidation of chalcopyrite by pyrite [4], and activation of pyrrhotite by $Cu^{2+}$ ions [5]; (c) the interaction of xanthates with sulfide minerals [6–8]; (d) reactions of sulfide minerals and other flotation pulp components (e.g., water, other collectors, modifiers), generally comprising oxidation at the mineral's surface (anodic reaction) coupled with the reduction of oxygen (cathodic reaction) [9]; (e) stability of the froth phase (related to the redox reactions in the pulp) in the flotation of complex ores [10,11]; and (f) redox reactions with the participation of grinding media [9,12,13].

There have been various attempts to estimate the flotation results of minerals, possible processes occurring in the flotation pulp, and mineral separation based on the minerals' mixed potential [14–18]. In short, the mixed potential is the potential measured in an open circuit system, determined by two or more different electrochemical couples available in the slurry and the corresponding anodic and cathodic processes occurring on the same electrode surface, giving equal and opposite currents that result in the zero net charge transfer across the interface. Mixed potential systems are non-equilibrium systems and hence their behavior is widely determined not only by thermodynamics but also by the kinetics of potential-determining reactions proceeding in a given system (such as flotation pulp) [19].

Jørgensen introduced the concept of $p_\varepsilon = -\log$ (electron concentration), similar to the definition of pH, with the idea of using it as a characteristic of the medium, while accounting for the electrical effect of the electrons involved in the reactions [20].

The impossibility of measuring the potential difference over a single interface represents a fundamental electrochemical principle [21]. Therefore, a second electrode is introduced in the system, which is a reference electrode (electrode with a constant and reproducible potential at a given temperature) and, in practice, the electromotive force (e.m.f.) between the two electrodes (indicator and reference) is measured. When the standard hydrogen electrode (SHE), whose potential is accepted as 0.000 V under every temperature, is used as a reference electrode and a Pt electrode as an indicator one, the measured e.m.f. is designated as Eh [20–22].

Later, the parameter $p_e$ was introduced and $p_e = -\log$ (electron activity) [20–23], where $p_e$ is related to the hypothetical activity of electrons at equilibrium. It must be mentioned that $p_e$ is not related to the activity of electrons of the indicator electrode and can be considered as a parameter that characterizes the solution (in our case, the flotation pulp) [20]. That is why it can be applied as a measure of the relative inclination of a solution to transfer or accept electrons. High positive values of Eh denote a strong tendency for oxidation, while very negative Eh values indicate a strong tendency for reduction.

It is widely accepted that the redox potential, Eh relative to SHE, is related to $p_e$ by the following equation [22–24]:

$$Eh = 2.303 \times R \times T \times p_e/F \tag{1}$$

where R = gas constant, T = temperature (K), and F = Faraday's constant.

At T = 298.15 K (25 °C)

$$Eh = p_e \times 0.05916, [V] \tag{2}$$

We must immediately note that the indicator Eh hides two conventions: (a) it is defined by $p_e$ ($p_\varepsilon$)—that is, introduced similarly to pH, it reflects an equilibrium (or quasi-equilibrium) state; (b) the size of the parameter Eh, expressed in volts, depends on the choice of indicator electrode (if an electrode other than platinum is used) [15,24–28] as well as on the temperature.

Some authors have used electrodes prepared by different minerals as indicator electrodes and also refer to the measured potential difference between those electrodes and a standard reference electrode (recalculated with respect to SHE) as Eh [29–31].

In the present paper, the "classical" meaning of the pulp's Eh value will be used, defined as the e.m.f. between the Pt electrode and the standard reference electrode immersed in the pulp, when this difference is expressed with respect to SHE. For this measured value, Woods [19] remarked: "It is presumed that sufficient mineral particles contact the noble metal surface for the measured potential to be that of the mineral of interest", i.e., the potential occurring across the mineral–solution interface. Thus, the defined potential (the pulp potential) represents the conditions in the slurry [32].

Eh can be considered an integral parameter reflecting: (i) the type and concentration of different oxidation–reduction couples that present in the medium (in our case, the flotation pulp) and are capable of participating in electron-exchange reactions; and (ii) the ratio of the various species of the elements in these pairs.

In the 1950s–1960s, Pourbaix [33] created the first pH–Eh diagrams that conveniently show the possible forms of existence of the elements in a water environment under given conditions (the presence of gases in contact with water and their pressure, concentration of dissolved gases, concentration of the various species of the corresponding elements, total concentration of the different elements or their compounds), and, in the diagrams built with modern software (such as HSC Chemistry Software), also temperature. Diagrams have proven to be a handy tool for quickly evaluating the thermodynamic possibility of oxidation–reduction processes involving the corresponding species of the element (including those formed in technological processes, e.g., when grinding the ore) and, from there, for explaining the various interactions with other components of the environment, including with added reagents, leading to a significant change in process indicators in the case of flotation.

At the same time, we should not forget the influence of the pH of the pulp on the floatability of the various minerals and, therefore, on their separation by affecting the equilibrium potentials and the stability of the product(s) available on the mineral surface or formed during the floatation process [29,34–37]. It should also be recalled that, under natural conditions (i.e., without a significant change in the oxidation–reduction conditions of the flotation environment), the dependence between pH and Eh is generally inversely proportional. The introduction into the system of reagents aimed at raising the pH leads to a decrease in Eh [22,29].

When the Eh increase/decrease is not due to a pH change, a sharp "movement" of the Eh value is observed on the pH–Eh diagram without any appreciable change in pH. In this case, the pH–Eh relation does not change in parallel with the oxygen/water and hydrogen/water equilibrium lines. This means that the Eh change is due to alteration in the oxidation–reduction conditions (by gases, chemical additions, or external potential impacts) [38].

Although, from the viewpoint of the basic flotation principles, it would be more useful to monitor the potential on the corresponding mineral interface with the adjacent liquid than the pulp [25,26,29], due to the difficulties in measuring the former and the convenience of measuring the latter under real conditions, in previous years, there have been many studies on the relations between the pulp potential—as influenced by grinding media, mineral nature, and the action of added reagents—and flotation results.

The relatively easy measurement of Eh, and the accumulation of data on known correlations between Eh values (more precisely, ranges of values) measured in the flotation pulp and flotation performance, makes this parameter very attractive for the purpose of predicting and controlling the flotation-selection processes.

The application of potential as a flotation-controlling parameter has been broadly proposed in the literature [30,39–43]. However, the number of publications showing real industrial examples is not very high, although some very useful papers can be referenced [44–48].

This paper aims to review and summarize briefly the studies on relations between pulp Eh values and optimal flotation results, as well as on the application of pulp Eh control in the grinding and selective flotation of non-ferrous sulfide ores of base metals, namely copper and lead–zinc ores. The effect of Eh on collectorless and collector-aided flotation, changes in pulp potential due to the use of different grinding media, the further effect on flotation separation, and the relation between reagent action and effectiveness and pulp Eh are discussed. According to our research, until now, the number of papers that unite the information on flotation Eh ranges, grinding conditions, and reagent action for base non-ferrous metals ores is scarce. Without making any claims to be exhaustive, the present paper aims at filling this gap in the knowledge on mineral flotation to a certain extent and join information given in earlier and more recent studies. In the entire paper, pulp potential Eh is given in mV with respect to the SHE.

## 2. Collectorless Flotation of Cu, Pb, and Zn Sulfides and Eh

Sulfide minerals of Cu, Pb, and (to some extent) Zn exhibit collectorless flotation, due to an acquired surface hydrophobicity, although they do not possess natural floatability, which is characteristic for minerals with a layered structure [49,50]. For example, the natural flotability of molybdenite was used in its collectorless separation from chalcopyrite, galena, and pyrite and from complex sulfide minerals when acetoacetamide (CSD2) is used as a novel depressant [51]. In order to improve the collectorless flotation of Cu, Pb, and Zn sulfides, the hydrophobicity of their surface must be increased by ensuring the exact conditions favorable to the formation of stable hydrophobic species on the minerals surface.

It is now well recognized that collectorless flotation of sulfide minerals requires surface oxidation of the sulfides, resulting in the formation of elemental sulfur ($S^0$), sulfur-excess species (such as $M_{1-x}S$), or polysulfide layers on the minerals' surfaces. The pulp potential must be risen to slightly oxidizing conditions to make collectorless flotation possible [52,53].

Sulfides of copper, lead, and (partially) zinc float without collectors only when are exposed to mild oxidizing conditions [19].

Sulfide is oxidized to elemental sulfur according to Equation (3):

$$MS + 2H_2O \rightarrow M(OH)_2 + S^0 + 2H^+ + 2e^- \tag{3}$$

and the amount of produced sulfur on the sulfide minerals can be correlated with their collectorless flotation behavior. Under alkaline conditions, polysulfides (i.e., $S_x^{2-}$, $2 < x < 8$), rather than elemental sulfur ($S^0$), cause collectorless flotation. Increased floatability at low pH is caused by the greater stability of elemental sulfur, which is more hydrophobic compared to polysulfide [54].

The pulp potential Eh is referenced as a crucial factor in collectorless flotation. Different regions of flotation and non-flotation can be distinguished that are Eh and pH dependent [54–57]. It is established that the collectorless floatability of sulfide minerals is connected to their ease of oxidation and the stability of the hydrophobic surface state generated by oxidation [58]. Collectorless flotation, as well as requiring an oxidizing environment, needs a relatively fresh surface, where hydrophilic oxidation products have not yet been formed [55]. Flotation proceeds as long as the metal oxides/hydroxides formed on the mineral's surface are solubilized [54].

However, excessive oxidation leads to the formation of hydrophilic oxide/hydroxide and sulfoxy species (thiosalts and eventually sulfate). This causes a cessation of collectorless flotation [52].

An increase in the collectorless flotation of sulfides can be also achieved by decreasing the amount of the iron-oxidized species on the mineral surface. On the contrary, adsorption leads to further surface precipitation of some ions, for example, flotation of iron from steel mills results in the formation of a hydroxide layer covering the sulfur-rich chalcopyrite surface and cessation of collectorless flotation [54].

### 2.1. Collectorless Flotation of Chalcopyrite

Heyes and Trahar [59] have found that the collectorless floatability of chalcopyrite depends on the potential—it initiates at ca. −50 mV, reaches 92%–93% at Eh = +200 mV and remains in this range up to +500 mV. In their work, nitrogen was used as the carrier gas. The pulp potential was controlled by the addition of reagents—hypochlorite was used as an oxidizer and dithionite was used for reduction.

The dependence on potential and the mentioned potential range of self-induced flotation of chalcopyrite was supported by the findings of Gardner and Woods [60]. In their experiments, the potential was controlled electrochemically.

Nearly the same potential range (the upper end of the experiment at +400 mV) for chalcopyrite was found by Luttrell and Yoon [55] who applied redox reagents to control the potential. This range of flotation onset correlates with the potential range (0 to +200 mV) in which chalcopyrite is oxidized, iron is removed from its surface, and the remaining iron-deficient copper sulfide increases the surface hydrophobicity of the mineral and ensures flotation in the absence of collectors [61,62].

Hayes and Ralston observed chalcopyrite collectorless flotation in the Eh range from +100 mV to +600 mV (pH 9) in experiments carried out under potentiostatic Eh control [49].

The chalcopyrite lattice was dissolved (Equation (4)), releasing copper and iron atoms and leaving a metal-deficient, sulfur-rich surface that ensures the hydrophobicity needed for collectorless flotation to occur [58]:

$$CuFeS_2 \underset{H_2O}{\overset{H^+}{\rightleftharpoons}} (CuS)_{1-y}(FeS)_{1-z}\left(S^{2-}\right)_{y-z} + yCu^+ + zFe^{+3} \tag{4}$$

where z > y, i.e., the chalcopyrite surface is more deficient in iron than in copper.

The collectorless flotation of chalcopyrite was observed in the Eh range of +260 to +75 mV (corresponding to a pH range from 5 to 9.5). However, excessive oxidation leads to the formation of oxidation products on the sulfide mineral's surface (soluble cations and sulfate in acidic or neutral solutions and metal hydroxides, polythionates, sulfate, and thiosulfate in alkaline solutions), which are hydrophilic and cause a decrease in the collectorless floatability of the chalcopyrite [54,58].

Zheng and Manton found that collectorless flotation of chalcopyrite occurs when the pulp potential is higher than +240 mV at pH 9 [63].

### 2.2. Collectorless Flotation of Chalcocite

Razmjouee and Koleini conducted experiments aimed at studying the collectorless flotation of chalcocite [64]. The addition of varying amounts of potassium permanganate (0.05 M) and dithionite sodium (0.05 M), as oxidizing and reducing reagents, respectively, ensured the control of pulp potential. Under reducing conditions, chalcocite exhibited good floatability with a maximum recovery of 73% at pH = 4 and Eh = −17 mV. The floatability of the mineral decreased to 18% at potential Eh = +807 mV. At pH = 9, the maximum recovery (71%) was obtained at Eh = −296 mV, and, when the potential increased, the floatability of chalcocite was decreased to 10% at Eh = +587 mV.

The increase in floatability of chalcocite is attributed to the production of hydrophobic species ($S^0$) by the reactions (5)–(8) [64,65]:

$$Cu_2S \rightarrow CuS + Cu^{2+} + 2e^- \tag{5}$$

$$CuS \rightarrow Cu^{2+} + S^0 + 2e^- \tag{6}$$

$$Cu_2S + 4H_2O \rightarrow 2Cu^{2+} + SO_4{}^{2-} + 8H^+ + 2e^- \tag{7}$$

$$SO_4{}^{2-} + 8H^+ + 6e^- \rightarrow S^0 + 4H_2O \tag{8}$$

### 2.3. Collectorless Flotation of Galena

Collectorless flotation of galena (PbS) is possible only in moderately oxidizing environments that can be controlled by the pulp potential, but cannot occur under reducing conditions, as shown by Hayes and Ralston [49]. They obtained a maximum galena recovery of 35% at a pH of 9.2 and an Eh of +500 to +550 mV. The authors discussed that the type of oxidation products, and the rate and mechanism of their formation depends on pulp pH and Eh conditions, temperature, and the amount of $O_2$ available to take part in the cathodic process. They found that moderate oxidation facilitates the formation of hydrophobic species, thus enabling collectorless flotation of galena. On the contrary, when PbS is strongly oxidized, the flotation is depressed due to surface covering by hydrophilic oxidation products [49].

Kocabag et al. discussed that collectorless flotation of galena is due to the mineral's natural oleophilicity, which in turn is pH–Eh dependent [66]. A relatively low pH value (but >2) of the flotation pulp favors collectorless flotation of galena (recovery of over 75% with a maximum at a pH of approx. 5.5) due to the formation of elemental sulfur on the mineral's surface, while, in weak alkaline conditions, galena is depressed due to the formation of hydrophilic $Pb(OH)_2$ and $HpbO_2{}^-$ [14,66].

Galena exhibits a collectorless floatability in the pH–Eh region where metastable elemental sulfur exists, indicating that the sulfur species are the hydrophobic entities that render the mineral surface hydrophobic. Clean galena can be rendered hydrophobic only when the samples have been exposed to molecular oxygen dissolved in water. At pH values over 11, galena is depressed, owing to the formation of lead hydroxy species, particularly $Pb(OH)_3{}^-$, which are detained at the mineral surface, rendering it hydrophilic. In addition, when the ore is ground with a mild-steel ball at high pH values, iron hydroxides, possibly coming from the balls corrosion, are formed on the galena surface, inhibiting the collectorless flotation of galena. Due to these reasons, galena exhibits good collectorless flotation at pH 8 in the Eh range +250 to +450 mV [67].

The formation of surface hydrophilic and hydrophobic species that are accountable for depression or flotation of minerals is controlled by the pH–Eh conditions in the pulp [68,69].

Galena oxidation results in products that ensure its collectorless flotation by rendering its surface hydrophobic, such as elemental sulfur ($S^0$), lead-deficient sulfide ($Pb_{1-x}S$, $x < 1$, where some lead ions are removed from the crystal lattice), and lead polysulfide ($PbS_n$, $n > 1$, unbranched sulfur chains attached to the metal-sulfide lattice by ionic or covalent bonds) [8,19,68–70].

Elemental sulfur was found on the galena's surface after treatment in solution of pH 4.9 at potentials above +200 mV [69].

In an acidic environment, $S^0$ may be generated by the reaction [70]:

$$PbS \rightarrow Pb^{2+} + S^0 + 2e^- \quad (9)$$

while in weak alkaline media, the reaction can be presented as

$$PbS + 2H_2O \rightarrow Pb(OH)_2 + S^0 + 2H^+ + 2e^- \quad (10)$$

and in alkaline media

$$PbS + 2H_2O \rightarrow HpbO_2^- + S^0 + 3H^+ + 2e^- \quad (11)$$

The pH–Eh diagram for the PbS–$H_2O$ system at 25 °C shows the presence of elemental sulfur throughout the entire pH range from 0 to 14, provided that the appropriate Eh value of the pulp is achieved. For pH = 9, this Eh range is from approx. +220 to +520 mV. The lower limit of this pH–Eh range is normally achieved by an air set process and explains the good collectorless floatability of galena. When the pH is permanent, the elemental sulfur presents in a corresponding potential range, below which galena is not oxidized, and above which elemental sulfur is further oxidized to hydrophilic species [70,71].

The ratio between the hydrophobic and hydrophilic species that are formed and present on galena surface determines the collectorless flotation recovery of the mineral [70].

The pH–Eh conditions found to be suitable for a collectorless flotation of Cu and Pb sulfides are summarized in Table 1.

**Table 1.** pH–Eh conditions found to be suitable for a collectorless flotation of Cu and Pb sulfides.

| Mineral | Eh Range, mV (SHE) | pH Range | References |
|---|---|---|---|
| Chalcopyrite ($CuFeS_2$) | −50–+500 | 9 | [59] |
| | 0–+400 | 7–10 | [55] |
| | +100–+600 | 9 | [59] |
| | +260–+75 | 5–9.5 | [58] |
| | +201–+488 | 10 | [72] |
| | >+240 | 9 | [63] |
| Chalcocite ($Cu_2S$) | −17 | 4 | [64] |
| | −296 | 9 | [64] |
| Galena (PbS) | +500–+550 | 9.2 | [49] |
| | +250–+450 | 8 | [67] |
| | +220–+520 | 9 | [71] |

### 2.4. Collectorless Flotation of Sphalerite

Sphalerite exhibits poor collectorless floatability on its own. However, when it is in complex sulfide ores where chalcopyrite is present, the floatability of sphalerite can be increased notably due to unintentional activation by copper ions coming from chalcopyrite dissolution (Equations (12) and (13)) [73]:

$$xZnS + Cu^{2+} \rightarrow (ZnS)_{x-1}CuS + Zn^{2+} \quad (12)$$

$$2nCuS + 2e^{-} \rightarrow nCu_2S + S_n{}^{2-} \quad (13)$$

Hydrophobic polysulfide is the main surface species influencing sphalerite flotation. Activated sphalerite possesses similar behavior to chalcopyrite and it is difficult to separate the two minerals. The recovery in the collectorless flotation of sphalerite increases with Cu (II) concentration until an optimum copper concentration for maximum sphalerite flotation is reached; after that, an increase in Cu concentration causes flotation decrease. This decrease is explained by the precipitation of copper hydroxide in neutral to mildly alkaline conditions where copper hydroxide is stable [52].

Sphalerite itself shows a certain collectorless flotation, which can be seen at low pH when mineral bulk cationic impurities, such as copper and iron, diffuse to the surface under acidic conditions, resulting in a "self-activating" mechanism [74]. The impurities are easily oxidized under aerated aqueous conditions at the surface, and elemental sulfur or polysulfides are produced that render the mineral surface hydrophobic and consequently increase sphalerite floatability [75].

The collectorless flotation of chalcopyrite, galena, and sphalerite by applying an Eh control, which was investigated for single minerals and mixtures by Hayes and Ralston, resulted in the conclusion that the "collectorless flotation follows the order chalcopyrite > galena > sphalerite" [49].

It is found that the ease of mineral oxidation was approximately opposite to that of natural floatability and that the order of collectorless flotation dropping is as follows: chalcopyrite > galena > pyrrhotite > pentlandite > covellite > bornite > chalcocite > sphalerite > pyrite > arsenopyrite [76].

In all the cases described above of collectorless flotation of sulfide minerals of non-ferrous base metals, the enhanced floatability is observed in the pH–Eh regions of the corresponding Pourbaix diagrams, where metastable elemental sulfur exists, indicating that metastable elemental sulfur is the main hydrophobic entity [54].

Collectorless flotation decreases the reagent consumption, increases flotation selectivity (grade increases), and impacts the equipment needed positively. However, it can also have negative effects. Woods stressed the fact that self-induced flotation takes place when the potential of the mineral of interest is in the region of its surface oxidation. Therefore, self-induced flotation acts as an obstacle in collector-induced flotation [19].

The natural floatability of chalcopyrite, chalcocite, and galena and the separation efficiency are not sufficient when complex ores are used, and, generally, collectors have to be added. For example, Grano et al. found that both chalcopyrite and galena show only moderate collectorless flotation properties, and the selectivity and recovery are increased when collectors are used [77]. That is why the next and major part of this paper is devoted to the effect of Eh on collector-aided flotation of sulfide minerals of non-ferrous base metals.

## 3. Eh as an Index in Copper Ore Flotation

### 3.1. Studies on Optimum Flotation Ranges

#### 3.1.1. Non-Arsenic-Bearing Ores

##### Monominerals or Monomineral Mixtures

Initially, studies were carried out mainly with monominerals or monomineral–quartz mixtures.

Nearly 45 years ago, Heyes and Trahar, based on batch flotation studies with chalcocite–quartz mixtures, raised the idea that if the Eh is in the correct domain, chalcocite could be floated, as long as the process is not carried out under measurable concentrations of dissolved oxygen [78]. According to their findings, the lower-limiting pulp potential for chalcocite flotation with potassium ethyl xanthate (KEX) is the range from −200 to −100 mV, and the pH had no influence on the potential of the lower flotation boundary, while in a progressively oxidizing system, the range is between −100 and 0 mV. The upper flotation range is +200 to +300 mV at pH 11, while at pH 8, over 85% recovery is observed up to +600 mV.

Nearly at the same time, Chander and Fuerstenau proposed electrochemical flotation in order to separate chalcocite from molybdenite by choosing properly the potential of flotation [79].

Changing the Eh value in a real process, i.e., achieving oxidizing or reducing conditions in a flotation system, optimizing the flotation response is usually achieved by the control of pH, dissolved oxygen levels in the system, the addition of modifiers such as hydrogen peroxide ($H_2O_2$), sodium hypochlorite (NaClO), and sodium hydrosulfide (NaHS), as well as by imposing an electrochemical impact.

Richardson and Walker showed that the flotation of monomineral chalcocite, chalcopyrite, bornite, and pyrite, with KEX as the collector, is controlled by the conditioning of the redox potential Eh [80]. The identified Eh ranges, at a pH of 9.2, for the flotation of chalcocite, bornite ($Cu_5FeS_4$), chalcopyrite, and pyrite ($FeS_2$) are correspondingly −250 to 0 mV, −50 to +150 mV, +50 to +200 mV, and +200 to +400 mV. At potentials spanning the upper 50 mV of the listed ranges, a recovery of 98%–100% was reached for $Cu_2S$, $Cu_5FeS_4$, and $CuFeS_2$, while for pyrite, a maximum recovery of 85%–87% was achieved at a potential of approx. 330 mV. The determined ranges for floatation occurring for the mentioned minerals are similar to the ranges found by other researchers, as can be seen below in Table 2.

Grano et al. discussed that the flotation results, when suitable reagents (collectors and depressants) were used, strongly depend on Eh for copper and lead–zinc ores [77].

Göktepe [29] studied the relation between the flotation of chalcopyrite, pyrite, sphalerite, and galena minerals and pulp potential (measured with Pt as the indicator electrode), using sodium isopropyl xanthate (SIPX) as the collector and Aerofroth 65 as the frother (Figure 1). In chalcopyrite flotation with xanthate, the pulp potential varied from +642 mV to +220 mV within the pH range of 2 to 13 and the maximum recovery was observed in the range from +500 to +300 mV, at pH = 5–11. For pyrite, the flotation range was determined as −150 to +300 mV and the maximum recovery was obtained in the pH range from 5 to 9, in which the pulp potential was approx. +280 mV.

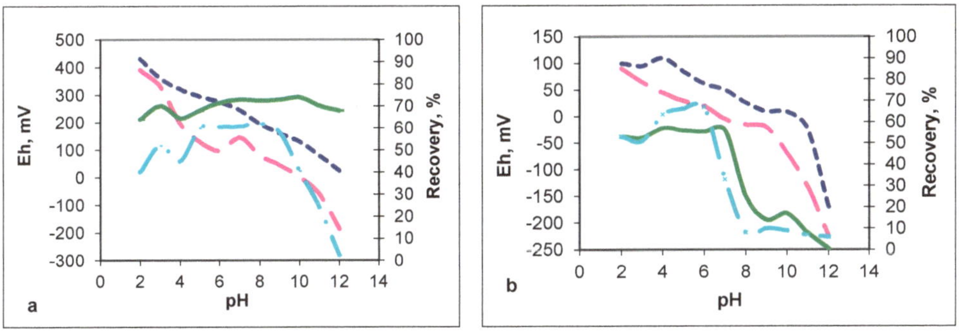


**Figure 1.** Relation between the flotation recovery, pulp potential Eh, and pulp pH values for (**a**) chalcopyrite and pyrite (Eh chalcopyrite - - - -, Eh pyrite — — — —, recovery chalcopyrite —— recovery pyrite — · — · —), and (**b**) galena and sphalerite (Eh galena - - - -, Eh sphalerite — — —, recovery galena ——— recovery pyrite — · — · —), with SIPX as the collector (compiled based on the data from [29]).

Although the author has mentioned that differences are observed when comparing single-mineral flotation and complex-ore flotation, the author's final conclusion is that the pulp potential represents an important parameter that can be associated with flotation results and can be used to regulate the condition of the sulfide surface and prognosticate the regions of optimal flotation.

Guo and Yen also studied the relation between the chalcopyrite floatability and pulp potential [72]. Using contact-angle measurements, the authors determined the floatable potential range as +41 to +441 mV at pH 10 and −9 to +541 mV at pH 7, with $7 \times 10^{-4}$ M potassium amyl xanthate (PAX). The micro- flotation tests conducted showed that the natural chalcopyrite floated without a collector in the range from +154 to +481 mV (pH 10), and with $7 \times 10^{-4}$ M PAX, the most suitable range was +41 to +446 mV.

From studies with mixtures of individual minerals and quartz, it has been found that it is possible to achieve flotation-separation of $CuFeS_2$ from PbS by controlling the redox potential [41]. High recoveries of chalcopyrite were found at a potential of +450 mV, while galena recovery was low at that potential. Galena was found to float effectively at a potential of −150 mV. Nevertheless, the minerals interact with each other when they are present together in the flotation pulp, which leads to a decrease in their inherent differences. At +450 mV, chalcopyrite recovery was decreased by a factor of four when galena was present, while galena recovery was nearly doubled in the presence of chalcopyrite. At a potential of −150 mV, chalcopyrite recovery was enhanced by approximately five times, while galena flotation recovery was decreased by 40% [41].

On the other hand, studies of He et al. [81] in mixed-mineral separation experiments of chalcopyrite from pyrite, at pH 9, using SIPX as the collector, showed that the chalcopyrite flotation behavior with Eh is similar for single-mineral and mixed-mineral systems. The flotation recovery of these minerals was Eh dependent. The maximum recovery of pyrite and chalcopyrite was obtained at Eh values of +135 mV and +275 mV, respectively. The maximum separation was achieved at +275 mV [81].

Kalegowda et al. studied the effect of pulp potential on sodium ethyl xanthate (NaEX) induced chalcopyrite flotation [82]. They used ultraviolet visible spectroscopy (UV-VIS), X-ray photo-electron spectroscopy (XPS), synchrotron-based X-ray photoemission electron microscopy (X-PEEM), and time-of-flight secondary ion mass spectrometry (ToF-SIMS) to study the flotation process. At Eh = +385 mV and pH 4, the chalcopyrite recovery was 97%, while under the reducing conditions, Eh = −100 mV and pH 9, it was only 41%.

Liu et al. studied chalcopyrite flotation with butyl xanthate as the collector. The optimum floatability was achieved in the pulp pH range of 4–10 and a pulp potential between +380 and +450 mV [83].

Woods summarized the findings of three research groups with respect to Eh ranges suitable for flotation recovery of chalcocite with ethyl xanthate; the flotation does not occur at Eh below approximately −200 mV and efficient flotation occurs at approx. 0 V [19].

Complex-Ore Systems

Investigations showed that the flotation behavior of mixtures could not be predicted precisely from single-mineral studies, owing to mineral–mineral interactions [84]. However, the separation of chalcocite and pyrite from their mixture was obtained by controlling the potential, although the flotation of the mixture did not exactly follow the performance predicted from measurements on the individual mineral beds [84]. This shows that the flotation separation and recovery of minerals in pulps containing more than one sulfide mineral is still dependent on potential. However, obtaining the dependence of mineral flotation on potential in mixtures is not as simple as assuming they are the sum of the potentials of the individual components determined separately [19].

Although there have been studies claiming that a generalized connection between pulp potential and flotation behavior could not be found for the complex-ore system [85], the numbers of studies showing the opposite is rising.

Sepalveda-Suarez and Castro studied the effect of Eh on the selective flotation of molybdenite by using Anamol-D (a Cu depressant), and $N_2$ as a flotation gas [86]. In order to prevent recirculation of co-floating minerals, it is recommended that in Cu–Mo byproduct plants, the rougher flotation is carried out as selectively as possible. It is found that Cu mineral depression started effectively at potentials below −180 to −200 mV. The high effect of Eh on Mo recovery was observed in the potential range from +20 to −280 mV, with a fast recovery of Mo in concentrate with a high grade in the potential range from −180 to −280 mV. It worth noting that the experiments were carried out with fresh bulk Cu–Fe–Mo concentrate taken directly from the "El Salvador" industrial plant.

Studies were carried out on the selective flotation of copper and zinc from three pyrite-bearing complex ores from Grong (Norway), Pyhiisalmi (Finland), and Aljustrel (Portugal) deposits. Alkyl xanthate was used as the collector and rougher flotation was conducted at pH 11.5. It was found that copper flotation with the highest selectivity for the investigated ores was achieved under aerated conditions, in the potential range from +170 to +240 mV (measured using Pt as an indicator electrode) [87].

Chimonyo et al. studied the effect of the addition of oxidizing agents ($H_2O_2$ and NaClO—0.1 M solutions) on the recoveries and grades of chalcopyrite and pentlandite from a PGM-bearing Merensky ore. Batch- flotation experiments were carried out at the natural pH (of 9.2) and the natural dissolved-oxygen (DO) levels of the system, in the presence and absence of Eh modifiers and a xanthate collector [11]. At pH 9, the natural Eh of the pulp was between +100 and +200 mV. When NaClO was added, the Eh value increased to achieve values between +500 and +600 mV, and any further additions of NaClO did not increase the potential further. When the pulp pH was initially adjusted to pH 11 by the addition of NaOH, the natural Eh decreased to between 0 and −100 mV. When NaClO was added at pH 11, the measured Eh value increased to +250 mV. In both cases of NaClO addition, the pulp pH value was practically unchanged, thus showing that the Eh increase was independent of pH under the conditions tested. When $H_2O_2$ was added (at pH 9), the pulp potential increased from +300 mV up to +400 mV. In all studied cases, the pulp potential after the addition of the oxidizing reagent exceeded +150 mV (the critical potential for dixanthogen formation), and it was expected to improve the tion results. Indeed, the potential increase led to an increase in the recovery of solids and water, compared to the tests without the addition of oxidants. The higher recoveries of solids and water at pH 9 were found with $H_2O_2$. However, this was not the case with copper recovery and grade. The addition of $H_2O_2$ led to a significant decrease in the final recoveries and grades of copper in comparison to the base case, both in the absence and presence of xanthate. These effects were attributed to the possible significant oxidation of the chalcopyrite and the considerably greater recovery of gangue solids in the concentrate, respectively. Chalcopyrite recovery of approximately 80% was observed in the entire Eh range (+100 to +600 mV) in the presence of xanthate, which, according to the authors, demonstrated that the low recoveries obtained in the absence of xanthate did not result from an Eh effect, but most likely resulted from chemical reactions that render the mineral particles more hydrophilic. Low recoveries were also obtained for nickel. The results were ascribed to the formation of different hydrophilic hydroxides on the pentlandite surface. The addition of NaClO in the absence of xanthate improved the copper recovery slightly and the recovery and grade of nickel significantly. According to the authors, NaClO probably selectively increased the hydrophobicity and, consequently, the recovery of pentlandite. The addition of NaClO in the presence of xanthate did not significantly change the copper recovery and grade, but the recovery and grade of nickel were decreased. When the oxidants were added at considerably lower concentrations (0.001 M), copper recovery increased notably, and nickel recovery was slightly increased. The results indicated that an attentive control of the $H_2O_2$ and NaClO concentrations at suitably low levels (and corresponding Eh values) can increase sulfide mineral recoveries.

Further studies by the same research group reached the conclusion that using higher amounts of NaClO to change the pulp potential in the flotation of a Merensky ore did not have any positive effect on the recoveries of copper or nickel minerals. The result of NaClO addition was a significant increase in the recoveries of solids and water [88]. In studies of the above-mentioned ore at natural values of pH 9, the Eh was made more negative by the addition of a depressant, carboxymethyl cellulose (CMC). This ensured that a wide Eh range was covered—from −120 to +660 mV (the maximum obtained by NaClO addition). It was found that an increase in Eh caused increased gangue recovery and was deleterious to valuable mineral recoveries and grades [89].

Mehrabani et al. studied the effect of pulp potential on the separation of chalcopyrite and molybdenite [90]. The potential was changed by the addition of sodium sulfide ($Na_2S$) and/or by using nitrogen. It was discovered that the addition of 16 kg/t $Na_2S$ into the flotation pulp under aeration conditions decreased the pulp potential from +450 to −375 mV in 30 s. Then, by using air as the flotation gas, the pulp potential gradually increased to −178 mV. However, a further increase in the $Na_2S$ concentration did not decrease the pulp potential further. A molybdenum recovery of 91.74% at only 0.44% floatability of copper minerals was achieved at a $Na_2S$ consumption of 16 kg/t. The use of $N_2$ in the flotation frothing stage helped to decrease the consumption of $Na_2S$ to 6 kg/t (pH = 10.5, pulp potential remained stable in the range −30 to −80 mV), where molybdenite recovery of 92.88% was achieved at a separation efficiency of 90.03%. Studies confirmed that the Eh value is a useful parameter in Mo flotation separation from Cu and that the pulp Eh value can be changed and maintained by simultaneous use of $Na_2S$ and $N_2$.

Bahrami et al. studied the relation between flotation results and pulp potential measured in real flotation pulp samples taken from the flotation cells of different processing circuits (Cu and Mo) and the different phases of copper and molybdenum flotation in Sungun complex [48]. According to the Sungun copper–molybdenum processing-complex flow sheet, the copper flotation is carried out in two phases, each one of them consisting of rougher, cleaner, and re-cleaner, as well as scavenger stages. The molybdenum flotation is carried out in one rougher and seven cleaner stages. For copper concentration plants, the initial Eh value (measured with a Pt electrode as an indicator electrode) was nearly −40 mV before the addition of collectors and frothers. Further, the measured Eh values in rougher flotation cells were in the range of −77 to −86 mV in phase 1 and from −59 to −86 mV in phase 2. Studies showed that, at the measured potential range, valuable minerals were completely floated and separated from gangue. In copper concentration plants, the measured Eh values for cleaner and re-cleaner flotation in phases 1 and 2 were nearly the same as in rougher flotation: −65 to −66 mV and −77 to −102 mV for cleaner cells in phases 1 and 2, respectively, and −64 and −71 to −85 for re-cleaner in phases 1 and 2, respectively. At that Eh range, all free and locked copper sulfides were recovered. In copper concentration plants, the measured Eh values in the scavenger circuits were in the ranges of −67 to −73 mV and −76 to −88 mV, respectively, for phases 1 and 2. The feed of the scavenger circuit had a copper grade of 1.69%. After the scavenger flotation, a concentrate containing >6% Cu and tailings containing approx. 0.2% Cu were obtained. The measured Eh values in the rougher flotation circuit of the molybdenum plant were in the range of −585 to −600 mV. The reducing conditions were achieved by using $Na_2S$ as a depressor for copper sulfide minerals, and gasoil was utilized as a collector. Molybdenite floated well at that potential, while the Cu minerals were depressed. Thus, good separation of liberated copper and molybdenite minerals was achieved in the rougher stage of flotation. When the flotation of molybdenite was conducted at Eh values more positive than −580 mV, some chalcopyrite floated and was recovered in the molybdenum rougher concentrate. In the molybdenum plant, good results in the cleaner circuit were obtained when the Eh in the cleaner cells was on average less than −640 mV. All copper sulfide minerals were depressed and molybdenite floated in the concentrate. The flotation behavior of the fine particles (−25 μm) deviated from the above results. It was found that copper and molybdenum concentrates with high quality can be produced by fine particle flotation—achieved by tuning the pulp potential to over +100 mV in copper plants and to less than −600 mV in molybdenum plants. Finally, the authors summarized that a pulp potential higher than −100 mV in the Cu concentration plants ensured the proper conditions for separation of copper sulfide and molybdenite minerals from pyrite and other gangue minerals. The pulp potential in the range −500 to −700 mV in the molybdenum processing plant ensured molybdenite flotation and separation from depressed copper minerals.

#### 3.1.2. Arsenic-Bearing Ores

The worldwide growing need for copper forces the mineral-processing industries to process low-grade copper ores which, besides chalcopyrite and bornite, contain arsenic-bearing copper minerals (tennantite—$Cu_{12}As_4S_{13}$, enargite—$Cu_3AsS_4$) and arsenopyrite (FeAsS) [91]. Arsenic (As) is a penalty element in many base-metal concentrates prepared for smelting. This raises the need to find optimal conditions for selective separation of Cu and As minerals, and the pulp Eh value plays an important role.

Monominerals or Monomineral Mixtures

Kantar determined that the most suitable potential range of the solution is +140 to +240 mV (over 98% recovery) for enargite (pure mineral sample) flotation and separation from copper minerals, at pH 10.5 and NaEX used as the collector [92]. In the micro-floatation experiments, the Eh value of the solution was controlled using hydrogen peroxide ($H_2O_2$) as an oxidizing agent and $Na_2S$ as a reducing agent.

Guo and Yen discussed the possibility of separating enargite from chalcopyrite by controlling the pulp potential (via NaOCl or $Na_2S$ addition), conducting studies with single minerals and a synthetic composite ore sample containing both minerals [93]. They found that the chalcopyrite and enargite floatabilities with PAX were almost the same in the pulp potential range from −160 to +440 mV. The chalcopyrite floatability decreased significantly whilst the floatability of enargite was unchanged when the pulp potential was increased from +440 to +540 mV. Flotation of chalcopyrite was not observed at a pulp potential higher than +540 mV. Enargite showed good floatability even when the Eh value was increased up to +740 mV. Enargite flotation was decreased at potentials higher than +740 mV. Further, the results from experiments with synthetic composite ore showed that chalcopyrite can be successfully separated from enargite at a pulp potential of +740 mV. Enargite concentrate with a grade of 92.8% was obtained at a recovery of 93.4%. Chalcopyrite with a grade of 93.2% was concentrated in the tailings with 92.6% recovery. The work of Guo and Yen showed that, though single-mineral studies may not give exact data for mixtures, they can be employed to identify the conditions for selective flotation in polymetallic ores.

The floatabilities (with KEX as the collector) of chalcopyrite, enargite, pyrite, and chalcocite at pH 8 and pH 11 were studied and the pulp potential was monitored continuously during the experiments [94]. Figure 2 shows that the pulp-potential dependence of enargite flotation is dissimilar from that of chalcopyrite, chalcocite, cuprite, and pyrite. A threshold (verge) potential was defined by Senior et al. as the potential at which recovery after 1 min is 50%. The experimental results have shown that $CuFeS_2$ had a verge potential of approx. +160 mV at pH 8, while at the same pH value, as well as at pH 11, the verge potential of $Cu_2S$ was approx. −155 mV. The threshold potential, below which enargite does not float and above which it floats, was established as approximately −70 mV at pH 8 and approximately −25 mV at pH 11. The enargite flotation at this pulp potential was practically independent of pH over the range tested. At pH 11, cuprite ($Cu_2O$) started to float at Eh values more negative than −350 mV. Pyrite flotation was observed in the range from −170 to +380 mV at pH 8, and from −40 to +350 mV at pH 11. However, its recovery was approx. two times lower when compared to other studied minerals at both pH values. Chalcocite and cuprite showed an upper limiting potential at pH 11 (>+270 and >+290 mV, respectively), above which their floatability decreased. This was not the case with enargite and chalcopyrite (up to +400 mV).

Comparison of the flotation results for chalcopyrite and enargite at pH 8 (Figure 2a) points at the possibility of separating these minerals in the potential range −25 mV to +50 mV, where enargite floats ahead of chalcopyrite. The comparison of the flotation response of enargite and chalcocite at pH 11 shows the availability of two possible windows for separation of chalcocite and enargite. The first is between −150 mV and −75 mV, at which chalcocite floats ahead of enargite. The second is above +270 mV, at which chalcocite does not float but enargite floats. Finally, a conclusion was drawn that enargite could be floated from chalcopyrite and other copper–iron sulfides at pH 8 to 11 at 0 V.

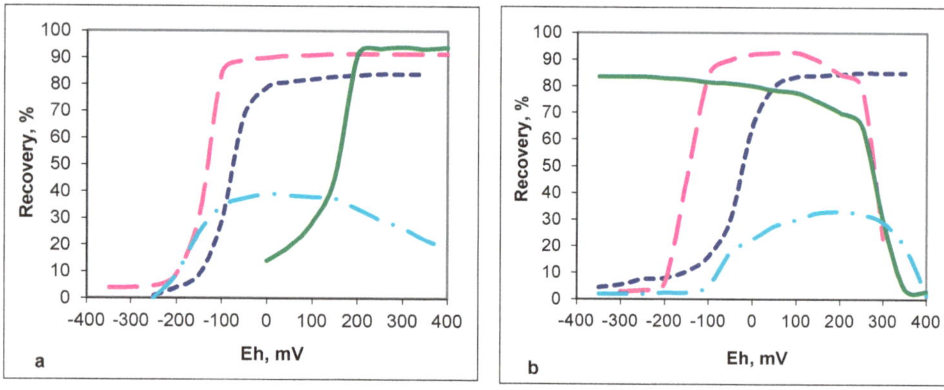


**Figure 2.** Relation between the flotation recovery and pulp potential Eh: (**a**) pH 8 (enargite - - - -, chalcocite — — — —, chalcopyrite ——— and pyrite — · — · —); and (**b**) pH 11 (enargite - - - -, chalcocite — — — —, cuprite ——— and pyrite — · — · —), with KEX as the collector (compiled based on the data from [94]).

Most of the studies aimed to relate the flotation recovery of minerals and the pulp potential have been conducted with fresh water. However, a shortage of freshwater leads to the use not only of recycled water but also of sea water in mineral processing. The flotation of chalcopyrite and enargite in seawater, as a function of Eh, has been studied and compared to the flotation in buffer solution with pH 10 (prepared with freshwater) [95]. Monomineral samples were used in the micro-flotation tests. Potassium amyl xanthate was used as the collector, the Eh value was adjusted with $H_2O_2$ and sodium hydrosulfide solutions (NaHS), and the flotation gas was nitrogen (Figure 3).

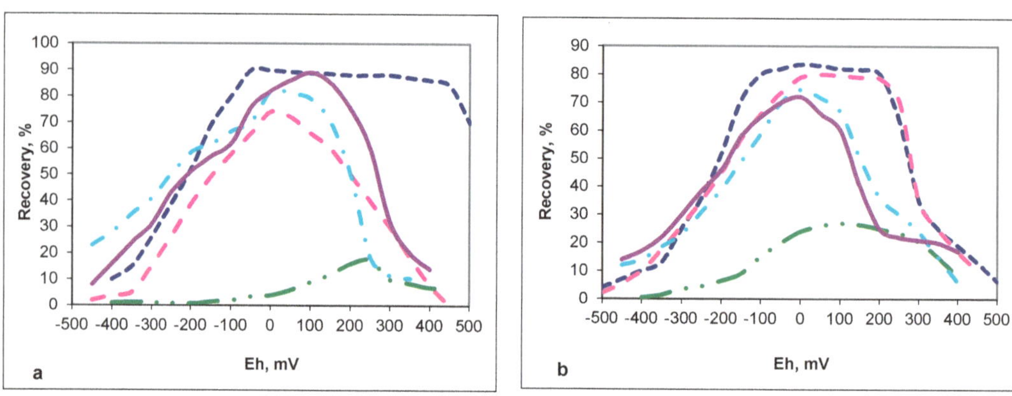


**Figure 3.** Relation between the flotation recovery and pulp potential Eh: (**a**) enargite (buffer at pH 10 - - - -, seawater at pH 9 — — — —, seawater at pH 10 — ·· — ·· —, seawater at pH 9 + Ca — · — · —, seawater at pH 9 + Mg ———); and (**b**) chalcopyrite (buffer at pH 10 - - - -, seawater at pH 9 — — — —, seawater at pH 10 — ·· — ·· —, seawater at pH 9 + Ca — · — · —, and seawater at pH 9 + Mg ———), with PAX as the collector (compiled based on the data from [95]).

In buffer solution, enargite recovery stared at −200 mV and approx. 90% recovery was achieved at Eh values from −50 to +450 mV. The recovery sharply decreased at Eh values below −150 mV and above +500 mV. The chalcopyrite recovery also started at −200 mV and reached 84%–85% at Eh values from −100 to +200 mV. A steep depression of chalcopyrite started at potentials higher than +200 mV.

When seawater is used as the flotation medium, at pH 9, enargite recovery is observed in the Eh range −200 to +250 mV. However, the recovery was approx. 15% lower compared to that achieved in the buffer solution. The enargite recovery steeply decreased at Eh > +100 mV. In the reducing range of Eh values, enargite recovery was lower than that achieved in the buffer solution. In short, in contrast to that observed in the buffer solution at pH 10, the enargite flotation in seawater at pH 9 was depressed for both oxidizing and reducing conditions. The flotation data obtained in seawater at pH 10.5 exhibits a strong depression of enargite throughout the whole studied Eh range, which is attributed to the adhesion of magnesium hydrolysis products (from the water) on the enargite surface, rendering it hydrophilic.

The flotation behavior of chalcopyrite in seawater at pH 9 is similar to that in the buffer solution, while at pH 10.5, a strong depression of chalcopyrite is observed in the whole range of studied Eh values. The results for enargite recoveries, as a function of Eh at pH 9, in presence of 1300 mg/L Mg and 300 mg/L Ca, show that enargite flotation is depressed when both solutions are used. This is in agreement with the enargite flotation results obtained with seawater. It was found that, under oxidizing conditions, the depressing effect of the calcium ions was stronger compared to the effect of the magnesium ions, although the concentration of calcium was much lower than the concentration of magnesium. This fact may be related to the presence of highly stable calcium compounds, such as hydrophilic arsenates, on the enargite surface. Chalcopyrite recovery as a function of Eh at pH 9, in solutions of magnesium and calcium, is strongly depressed in the oxidizing range of Eh. Magnesium ions exhibit a slightly stronger depressing effect than calcium. The results obtained under oxidizing conditions in seawater at pH = 9 may be explained by the availability of hydrophilic arsenates of calcium/magnesium on the enargite surface.

A non-exhaustive summary of the empirical information available on the probable Eh and pH ranges that are most suitable for the flotation of major copper ores, as well as on pyrite and arsenopyrite as the major accompanying minerals, with xanthate as the collector (except for tennantite, for which thionocarbamates were used), is presented in Table 2 (compiled mainly based on the works [48,76] and with additional information from the papers referred to in this manuscript).

**Table 2.** Eh and pH values and ranges suitable for collector-aided flotation of major copper ores and pyrite.

| Mineral | Eh Range, mV (SHE) | pH (Range) | References |
|---|---|---|---|
| Bornite ($Cu_5FeS_4$) | −100–+100 | 5–10 | [76] |
| | −50–+150 | 9.2 | [80] |
| Chalcopyrite ($CuFeS_2$) | 0–+200 | 6–10 | [76] |
| | +50–+200 | 9.2 | [80] |
| | +300–+500 | 5–11 | [29] |
| | +88–+493 | 10 | [72] |
| | +450 | 11 | [41] |
| | −160–+440 | 10 | [93] |
| | +275 | 9 | [81] |
| | +160–+360 | 8 | [94] |
| | +385 | 4 | [82] |
| | −100 | 9 | [82] |
| | −100–+200 | 10 | [95] |
| | −100–+200 | 9, seawater | [95] |
| | +380–+450 | 4–10 | [83] |
| | 0–+400 | - | [48] |
| Chalcocite ($Cu_2S$) | −200–0 | 8–11 | [76] |
| | −200–0 | - | [78] |
| | −250–0 | 9.2 | [80] |
| | −296 | 9 | [64] |
| | >−155–+400 | 8 | [94] |
| | >−155–+290 | 11 | [94] |
| | −100–+300 | - | [48] |

**Table 2.** *Cont.*

| Mineral | Eh Range, mV (SHE) | pH (Range) | References |
|---|---|---|---|
| Covellite (CuS) | +50–+150 | 5–10 | [76] |
| Arsenopyrite (FeAsS) | +300–+600 | 2–7 | [76] |
| Enargite ($Cu_3AsS_4$) | −50–+520 | 8–11 | [76] |
| | +140–+240 | 10.5 | [92] |
| | +440–+740 | 10 | [93] |
| | −75–+400 | 8 | [94] |
| | −25–+400 | 11 | [94] |
| | −100–+500 | 10 | [95] |
| | −200–+250 | 9, seawater | [95] |
| | −100–+400 | - | [48] |
| Tennantite $Cu_{12}As_4S_{13}$ | +150–+250 | 8–12 | [76] |
| Cuprite ($Cu_2O$) | −400–+290 | 11 | [94] |
| Pyrite ($FeS_2$) | −150–+300 | 4–8 | [76] |
| | +200–+400 | 9.2 | [80] |
| | −150–+300 | 5–9 | [29] |
| | +135 | 9 | [81] |
| | −170–+380 | 8 | [94] |
| | −40–+350 | 11 | [94] |
| | 0–+300 | - | [48] |

As can be seen from Table 2, studies devoted to chalcopyrite, chalcocite, enargite, and pyrite are prevailing, corresponding with the main copper minerals available and exploited for industrial purposes. There are some inevitable differences in the Eh and pH ranges for good floatation found by the different authors that may be related to the studied mineral samples' (micro)chemistry, preparation, and exact flotation conditions. However, generally, pH–Eh areas may be delineated where a given mineral is expected to float.

Complex-Ore Systems

Many years ago, Huch patented a method for flotation selective separation of chalcocite from enargite by concentrating enargite from a slurry containing both minerals. According to the inventor, the separation was achieved by adjusting the pulp potential to the range of +50 to +200 mV, by the addition of $H_2O_2$ [96]. However, the invention does not mention the reference electrode used in the potential measurements that were carried out.

Based on the results obtained by studies using monominerals, Senior et al. concluded that enargite could be rejected from ore from Tampakan deposits in the Philippines by controlling the flotation pulp potential [94]. A conceptual flow sheet was proposed, in which the bulk Cu–As flotation concentrate was reground and subjected to further treatment. The first stage in the flow sheet utilizes the difference in the beginnings of flotation for chalcocite and cuprite from those of enargite and chalcopyrite. The pulp pH was adjusted to 11 (under an $N_2$ atmosphere) and the Eh to −125 mV. This way, chalcocite and cuprite were separated by flotation from enargite. The second stage (Eh of approx. 0 V, pH of 11) uses the difference in the potentials of the beginning of flotation of enargite and that of copper–iron sulfides. Chalcocite and cuprite that are not floated in the first stage float with enargite because the potential in the second stage is in the Eh range for the efficient flotation of these minerals. Chalcopyrite does not float. In the third stage, enargite is separated from chalcocite and cuprite at pH 11 (under an air atmosphere) after establishing a potential of approx. +290 mV. Further, Smith and Bruckard [97] reported on the successful use of one of the separation methods proposed in the work [94] to treat Tampakan ore and obtained copper concentrate with sufficiently low arsenic content that allows sending it to a conventional smelter for copper metal production.

In order to selectively separate chalcopyrite and bornite from tennantite, Smith and Bruckard applied Eh control of the flotation pulp, prepared from a composite sample of high-arsenic drill cores from Northparkes copper–gold ore, Australia [97]. They used bulk copper–arsenic concentrate in the experiments for the separation of As-bearing and As-non-bearing minerals. The separation was based on findings that the lower-limiting pulp-potential threshold for tennantite is lower than that for chalcopyrite. So, the potential window used in the reducing region is that in which tennantite is floatable but chalcopyrite is not. In the oxidizing pulp, potential-region selectivity between tennantite and chalcopyrite has not been established. A pulp-potential region exists (between −200 and −130 mV) in which tennantite can be floated from the other non-tennantite copper minerals (NTCu)—chalcopyrite and bornite. Arsenic recovery in this Eh region is between 80 and 90%, while recovery of NTCu is only 30%. Tennantite floatability decreased below −200 mV and the floatability of the other copper minerals increased above −130 mV, and thus, selectivity between copper and arsenic was decreased. Decreasing the pulp potential of a bulk copper–arsenic concentrate to −150 mV at pH 12 (collectors—NaEX and AP 208, a dithiophosphate) achieved tennantite flotation, and thus, its separation from the chalcopyrite and bornite that remained unfloated. After the tennantite was removed by flotation under reducing conditions, the depressed non-tennantite copper minerals were floated by changing the flotation gas from nitrogen to air at pH 12. When the pulp potential reached the air-set potential (approx. +120 mV), the non-tennantite copper minerals were floated, and the froth product represented a low-arsenic copper concentrate. As a final result, a low-arsenic high-copper concentrate containing 52% of the NTCu and assaying 2600 ppm As was obtained from the drill core tested, with a head grade of 0.11% As and 1.2% Cu.

Further, the same working group presented results of several selected case-studies in which the separation of copper and arsenic from sulfide high-arsenic copper ores was achieved based on flotation with a controlled pulp potential [46]. Low-arsenic copper concentrates that can be smelted without penalty, and high-arsenic copper concentrates, that require further processing, were obtained. Four types of raw materials were used (Table 3). Initially, a rougher-scavenger concentrate was obtained, by applying a standard industrial procedure used for the corresponding ore. Further pulp-potential control was applied during a subsequent flotation stage of the corresponding concentrate to achieve a copper–arsenic separation. The results are summarized in Table 3, along with data on flotation process selectivity. A value greater than one designates the selectivity of copper over arsenic.

As can be seen from Table 3, for all ores studied, no selectivity between copper and arsenic minerals was achieved by conventional flotation. A separation between the arsenic-bearing copper minerals and the non-arsenic-bearing copper minerals can be achieved by using a controlled-potential flotation at a suitable potential, specific for the given case. In addition, the authors noticed that the efficiency of the pulp potential's impact on the separation and selection depends on the ore's mineralogy, the degree of liberation of the arsenic-bearing minerals, as well as on the reagent's scheme, especially the collector type, used in producing the bulk copper–arsenic rougher concentrate.

Long et al. carried out a survey on the flotation behavior of arsenic and non-arsenic copper minerals at the Rosebery plant copper circuit [98]. Samples from the pulp and concentrates (where relevant) were taken from the grinding, rougher, and cleaner circuits, from each flotation cell. Different parameters characterizing the pulp, such as pH, Eh, and DO were measured in the pulp samples. Simultaneously, the concentrate samples were analyzed to determine the concentrations of Cu and As. It was found that the pulp potential changed from +321 mV to +155 mV in the conditioning tank. This is related to the addition of sodium meta-bi-sulfite (SMBS) to control lead and zinc concentrations in copper concentrate. Further, pulp potential was in the range +225 to +265 mV in the flotation cells. The copper-to-arsenic recovery ratio was 9:1, i.e., much higher than found by Smith et al. [46] for a similar type of ore at +250 mV (Table 3). The difference is explained with the use of a different potential controlling agent and with the predominance of tetrahedrite in Rosebery ore compared to tennantite in the ore used in the work [46].

**Table 3.** Results of the selected case-studies of separation of copper and arsenic from sulfide arsenic copper ores [46].

| Ore Sample; Parameter | Conventional Low-As Cu Ore | Conventional Low-As Cu-Pb-Zn Ore | Conventional low-As Cu Concentrate | Conventional Low-As Cu-Mo Concentrate |
|---|---|---|---|---|
| Major Cu minerals | Chalcopyrite, Bornite | Chalcopyrite | Chalcopyrite | Covellite |
| Major As minerals | Tennantite | Tennantite | Tennantite | Enargite |
| Cu, % | 1.21 | 0.95 | 22.5 | 0.48 |
| As, % | 0.11 | 0.061 | 0.34 | 0.073 |
| Fe, % | n.d. | 8.49 | 21.2 | 5.4 |
| S, % | 0.68 | 13.0 | 27.3 | 8.7 |
| Pb, % | n.d. | n.d. | 0.58 | n.d. |
| Zn, % | n.d. | 7.24 | n.d. | 0.53 |
| Mo, % | n.d. | n.d. | 0.10 | 0.025 |
| Collector type used in rougher | Xanthate/ dithiophosphate | Thionocarbamate | Thionocarbamate | Thionocarbamate/ xanthate ester |
| Pulp potential, mV in conventional | +250 | +230 | +150 | +180 |
| Selectivity, in conventional | 1.0 | 1.0 | 1.1 | 1.0 |
| Pulp potential, mV in controlled | −200 | +400 | +350 | +500 |
| Selectivity, in Eh controlled | 4.7 | 2.8 | 2.4 | 2.8 |

n.d.—not determined, Cu/As selectivity = Cu recovery/As recovery.

Tayebi-Khorami et al. studied the selective separation of other copper sulfide minerals from enargite in a rougher flotation cycle under a controlled pulp potential, using porphyry copper ores with a different As content, from a deposit situated in the SE part of Mindanao, the Philippines [99]. The ore samples represented mixtures of several drill-core intersections with a high arsenic content (HAS) or low arsenic content (LAS). Flotation experiments were carried out at different pulp-potential values (Table 4). Nitrogen gas was used to maintain the Eh values when they had to be lower than the air-set value. Dilute solutions of sodium dithionite ($Na_2S_2O_4$) and NaClO were used as a reducing and an oxidizing reagent, respectively. The collector was KEX, the frother was Aerofroth 65, and the pulp pH was adjusted with NaOH solution to 11. The overall recovery results for 10 min flotation of non-enargite Cu minerals (NECu), enargite, pyrite, and non-sulfide gangue minerals (NSGM) at different Eh values (−200, 0, +200, and +400 mV) for the HAS and the LAS samples are presented in Table 4 (compiled based on the data from [99]).

For both HAS and LAS samples, the NECu recovery achieves a maximum of approx. 82%–83% in the Eh range between +100 and +200 mV. Similar recoveries of NECu and enargite are found at reducing potentials; there is no separation between NECu and enargite at the pulp potential between −200 mV and −130 mV. The recoveries of NECu and enargite at Eh +400 mV decreased in both samples. The enargite did not float significantly in the HAS sample and its recovery changed little with the increase in Eh. In addition, it is found that, at pH 11 and Eh +400 mV, pyrite is depressed without any depressant addition. The study in a real ore system proved that flotation at a reducing pulp potential ($\leq -200$ mV) is suitable to separate enargite from other copper minerals. Based on the results obtained, a flow sheet is proposed for a processing method aimed at As removal from Cu concentrate. It comprises: (i) production of a bulk copper–arsenic flotation concentrate at air-set potential and pH 11; (ii) regrinding the concentrate in a closed circuit with a cyclone and using the overflow as a feed for the arsenic rejection stage; and (iii) conducting a separation flotation in two steps—(a) enargite flotation at −200 mV, under nitrogen gas and the non-enargite copper mineral's depression, and (b) copper flotation from the tail obtained at the first step at pH 11 at the air-set potential using air to float the copper minerals and obtain a low-arsenic high-copper concentrate suitable for direct smelting.

**Table 4.** Overall recovery results for 10 min flotation of non-enargite Cu minerals (NECu), enargite, pyrite, and non-sulfide gangue minerals (NSGM) at different Eh values for the HAS and the LAS samples, compiled based on [99].

| Recovery, %, Eh | −200 mV | 0 mV | +200 mV | +400 mV |
|---|---|---|---|---|
| HAS ore | | | | |
| NECu | 45 | 76 | 83 | 77 |
| Enargite | 47 | 60 | 63 | 51 |
| Pyrite | 74 | 81 | 77 | 40 |
| NSGM | 5 | 6 | 8 | 9 |
| LAS ore | | | | |
| NECu | 46 | 77 | 78 | 55 |
| Enargite | 47 | 82 | 82 | 60 |
| Pyrite | 76 | 81 | 78 | 22 |
| NSGM | 5 | 6 | 7 | 8 |

Summarizing the above-described data, it can be stated that pulp potential Eh is an important parameter governing the flotation process of Cu-sulfide minerals. Its adjustment and monitoring can be a useful tool for improving the flotation separation results. However, it has not been forgotten that the Eh ranges of the best recovery and separation and the efficiency of the pulp potential's impact on minerals depends also on the ore's mineralogy, on the reagent scheme used, as well as on grinding conditions.

### *3.2. Research on Grinding Environment and Its Effect on Flotation*

Studies on the effect of grinding conditions (conventional or autogenous circuits) or grinding medium (mild-, stainless-, or forged-steel) on the Eh values and Eh changes, and thus, on the Cu minerals' flotation results started in the 1980s–1990s [12,59,77,100].

Heyes and Trahar [59] found that strongly reducing conditions were achieved by grinding in an iron mill, which led to the suppressed flotation of chalcopyrite.

Yelloji and Natarajan [100] found that oxyhydroxide species of iron were present on the surface of the sulfide minerals that were in contact with the grinding medium. Those iron species' formation was found to be dependent on the minerals' relative electrochemical activity.

Grano and Ralston [77] measured Eh values of −220 mV and +325 mV when Mount Isa copper ore was ground in a cast iron and a ceramic mill, respectively. The non-reducing environment and low iron concentration of the autogenous mill resulted in enhanced chalcopyrite flotability when the Mount Isa copper concentrator changed normal-steel-rod/ball milling to autogenous-ball milling.

Subrahmanyam and Forssberg [12] studied different pulp parameters, including Eh measured directly in the pulp, in conventional and autogenous circuits of the plant processing of a low-grade copper ore. They measured higher Eh values in the autogenous circuit compared to the conventional circuit.

Yuan et al. investigated the effect of grinding media (mild-steel and stainless-steel), pre-conditioning, and flotation gas type on the flotation results [15]. They found the grinding environment to be the controlling variable. They observed that the Eh of the pulp at the end of grinding and the beginning of the conditioning stage was approx. +250 mV and approx. −150 mV with a stainless-steel and a mild-steel medium, respectively. The grinding medium influences the pulp chemistry after the pre-conditioning stage and the influence is more pronounced in the case of mild-steel grinding. A strong dependence is found of Cu/Zn selectivity on the pre-conditioning type and applied flotation gases when mild-steel grinding was applied, but a dependence was not found for stainless-steel grinding. According to the authors, the galvanic reactions during or after mild-steel grinding make the surface oxidation of sulfides and sulfur mobilizing into the pulp easier. At the same time, iron is leached in the pulp due to corrosion, thus affecting the pulp potential [15].

Leppinen et al. also reported the effect of grinding media (mild-steel and stainless-steel) on the copper floatation results of three different pyrite-bearing complex ores from Finland, Norway, and Portugal [87]. More reducing conditions were generated by grinding in a non-alloyed steel mill than by grinding in a stainless-steel mill. As a result, copper recovery was low at low potentials after grinding with a mild-steel medium and increased at higher oxidizing potentials, achieved by contact with the air during the conditioning and flotation. Due to an excessive recovery of pyrite, the copper grade was low at low potentials. The copper grades and recoveries were raised to that obtained by grinding carried out with stainless steel when lime was added into the grinding stage. When a stainless-steel medium was used as the grinding medium, copper recovery was high from the beginning. On the other hand, the flotation selectivity was low due to highly oxidizing conditions.

The effects of the grinding media (mild-steel and stainless-steel rods) and the addition of NaHS in the conditioning stage immediately after the grinding on the flotation of copper minerals from the Northparkes Mines copper–gold ore was studied by Freeman et al. [9]. The Eh values at the beginning of conditioning were +40 mV and +415 mV for mild steel and stainless steel, respectively. When the ore was ground using mild-steel grinding media, the copper flotation recovery without the addition of NaHS was 37% in the rougher stage and 72% when rougher and scavenger were combined. The process was fast. The addition of NaHS (100 g/t) increased the recovery to 83% (combined rougher and scavenger) and the grade achieved was 22.6%. The addition of NaHS to the pulp, using mild-steel rods, increased froth mineralization and stability, thus improving the copper minerals' flotation kinetics. When the ore was ground using stainless-steel media, the copper flotation recovery was over 90% in the rougher stage and over 94% (at a grade of 15.8%) when rougher and scavenger were combined. The use of NaHS increased the recovery and grades by less than 1% in this case and can be omitted. The findings are attributed to the reaching of the optimum potential for bornite and chalcopyrite flotation in the pulp obtained after grinding with a stainless-steel grinding medium.

Furthermore, Peng et al. reported the role of the iron oxidation species formed during grinding with mild steel in the flotation separation of chalcopyrite and pyrite [101]. Grinding changed the oxidation–reduction conditions and facilitated the reduction of copper(II) to copper(I) and the formation of $Cu_2S$ on the pyrite surface, which resulted in pyrite activation. On the contrary, these effects were not observed when chromium stainless steel was used as the grinding medium. For that reason, a stainless-steel grinding medium led to a higher chalcopyrite recovery and separation from pyrite than a mild-steel grinding medium.

Grano reported that when a mild-steel ball mill is used, iron is oxidized, and corresponding iron species are formed, which is reflected by a low Eh = −200 mV at pH 7. However, when the gro((und material slurry is transferred to the flotation cells by pumping, air is introduced into the pulp (while there is no direct contact with the iron balls), and another oxidation–reduction couple, namely $O_2/H_2O$, is introduced to the system. Aeration causes an increase in the Eh value above the critical value for collector adsorption [38].

Since it is more economically viable to control the electrochemical conditions during grinding via the addition of reducing or oxidizing reagents, than to change the mill's materials, a study was carried out in which the potential during grinding was adjusted (by the addition of dithionite and peroxide) to be −185 mV, −10 mV, or +260 mV at the end of the grinding period [102]. The Eh was measured directly in the pulp. It was found that the more reducing grinding conditions resulted in a higher pyrite recovery (that would result in a worse Cu/Py selection). The values obtained for Py recovery were 16, 32, and 72 wt.% at +260 mV, −10 mV, and −185 mV, respectively. By using Visual MINTEQ (ver. 3.0) software, it was found that at −185 mV, over 99 wt.% of aqueous copper was present as $Cu^+$ ions, at −10 mV, these species were 28 wt.% of all dissolved Cu, while at +260 mV, practically no $Cu^+$ was found. This confirmed that the reducing grinding conditions can lead to increased concentrations of $Cu^+$ species that facilitate formation of $Cu_2S$ on the pyrite surface and activate it.

Corin et al. [103] studied the effect of the grinding conditions on the floatation behavior of a Nkomati ore. They used a Magotteaux Mill® with a 21% Cr grinding medium, applied mill purging with either nitrogen or air, corresponding to the required concentration of dissolved oxygen (in the pulp), and measured the corresponding Eh value. Contrary to other similar studies, this study found that chalcopyrite recoveries and grades were barely affected by the change in chemical characteristics, including Eh, of the grinding environment. Nevertheless, the conclusion was that, despite many literature sources reporting that the results of base-metal-sulfide flotation plants can be optimized by controlling the Eh, the Eh value is not a regulating means but an indicator of more fundamental chemical reactions proceeding on the mineral's surface, involving the dissolved oxygen concentration and/or the hydroxyl ion concentration (via the pH). However, the grinding medium used and the method of changing the Eh actually does not account for reactions leading to the possible formation of iron hydroxide/oxide species that would result in the Eh changing, as well as affecting further flotation processes. Actually, in the same paper, the authors mention that if other means are used to control the Eh, different surface reactions may occur and the effect of Eh on the further flotation process would be different.

Nooshabadi et al. also found that a mild-steel grinding medium produced more reducing conditions compared to stainless steel [104]. When grinding pyrite, the Eh value changed in 60 min from the initial value corresponding to air-saturated water to +290 and +140 mV for stainless-steel and mild-steel grinding media, respectively. They proposed another explanation for the effect of oxidation–reduction conditions, resulting from grinding, on the subsequent flotation results, besides the explanations of the impact of galvanic interactions taking place in the system "grinding media—sulfide mineral" that lead to unselective surface coatings of mineral surfaces by iron-oxidation products. The scientists proved the formation of $H_2O_2$ in pulp liquid from water molecules during wet grinding and when solids obtained by dry grinding were placed in water immediately after the grinding. They found that the pyrite surface acted as catalyst for this process. A mild-steel grinding medium created more $H_2O_2$ compared to stainless steel. The authors raised the idea that $H_2O_2$, instead of oxygen, may be the main oxidizing agent in the pulp, leading to the unintended activation and lack of selectivity in the sulfide ore floatation. Further studies by the same authors found that chalcopyrite also facilitated the generation of $H_2O_2$ in pulp liquid during wet grinding as well as when placing the freshly ground minerals in water immediately after dry grinding [105]. The results of Nooshabadi were in line with the findings of other authors [106].

Jacques et al. studied the influence of grinding medium (forged and stainless (chromium 21%) steel) on the flotation behavior of partially oxidized sulfide minerals [107]. Experiments were carried out with a hypogene copper-sulfide ore that was artificially oxidized for different lengths of time up to and including 16 months by percolating a small quantity of acidic ferric-sulfate solution inoculated with a mixed bacterial culture through columns packed with the ore. It is found that, generally, stainless steel ensured a more stable pulp chemistry and produced pulp with a higher dissolved oxygen content and higher pulp potential Eh than forged steel, both at the ball-mill discharge and at the flotation-feed measuring points. The Eh values (for samples oxidized for different time periods) at mill discharge were in the range +16 to +238 mV for a forged-steel mill and +157 to +263 mV for a stainless-steel mill. The Eh values at the flotation-feed point were in the range +93 to +183 mV for a forged-steel mill and +159 to +221 mV for a stainless-steel mill. The comparison of the two grinding environments showed that stainless-steel grinding led to better copper recovery and flotation rate constants (considering the flotation as a first order reaction) at each oxidation stage. The recoveries of Cu of 73.4 to 88.8% for a forged-steel mill and 82.4 to 92.4% for a stainless-steel mill were obtained. Flotation rate constants of 0.6 to 1.9 $min^{-1}$ for a forged-steel mill and 3.2 to 4.3 $min^{-1}$ for a stainless-steel mill were found.

In addition, stainless steel led to better selectivity in the flotation—the copper/iron-sulfide selectivity index was in the range 1.3 to 5.5 for forged steel and 4.5 to 9.4 for stainless steel.

Peng et al. studied the effect of the addition of a collector (sodium butyl xanthate (SBX)) to the mill before grinding on the pulp Eh value and flotation of chalcopyrite at pH 10 [108]. They found that collector addition before grinding led to a higher pulp Eh value. Immediately after grinding, it was higher compared to the case when the xanthate was added after grinding (+236 vs. +226 mV). In both cases, the Eh value increased with increased collector dosage and at a SBX concentration of 0.002 mol/L, it reached +290 mV—a value suitable for chalcopyrite flotation. A recovery of 93.6% was achieved when the collector was added before grinding compared to 90.03% in the case when the collector was added after grinding. XPS analysis revealed that collector addition before grinding ensured a higher collector adsorption and a lower amount of detrimental iron species on the chalcopyrite surface.

Liu et al. investigated the effect of grinding conditions (wet or dry, and with or without lime addition) on the pulp Eh values and reported flotation results for two copper-sulfide ores from the Lujiu (LJ) mine situated in north-eastern China and the Dexing (DX) mine situated in south-eastern China [109]. Chalcopyrite was the main copper-bearing mineral in both ores and pyrite was the other main sulfide mineral. Chalcopyrite in the ore was mainly associated with pyrite and gangue minerals. Quartz, mica, chlorite, calcite, and illite were the main gangue minerals in the LJ ore. Quartz, mica, calcite, chlorite, and feldspar were the main gangue minerals in the DX ore. Lime was used as the pH regulator and pyrite was depressed by lime addition. Three different approaches were followed for each ore: (a) wet milling and subsequent lime addition in the flotation cell; (b) wet milling with lime added in the mill before grinding; and (c) dry milling and subsequent lime addition in the flotation cell. A Magotteaux mill with mild-steel balls was used. It was found that for the LJ ore, the Eh value after 45 min of grinding changed from approximately +80 mV to approximately +30 mV when no lime was added, and to +45 mV when lime was added. The flotation tests (with sodium isobutyl xanthate (SIBX) as the collector and DOW 200 as the frother) showed a higher Cu recovery for wet- or dry-ground ore with the addition of lime compared to the case of wet grinding without lime (approx. 93% vs. approx. 90%). In addition, the Cu grade was considerably higher in the case of lime addition (for example, for 90% recovery, approximately 15% vs. approximately 7%). The Eh value after 45 min decreased from approximately +75 mV to approximately +25 mV when lime was not added before the grinding of DX ore but increased from an initial value of +45 mV to approximately +65 mV after lime addition. This was attributed to the differences in the ore mineralogy. The floatation results showed that wet grinding with lime led to a slightly lower Cu recovery when compared to the other two cases (with approximately 87% recovery). However, the grade of the produced rougher Cu concentrate was considerably lower (4%–6%) than that of the concentrate obtained from LJ ore. The results are attributed mainly to better depression of pyrite and to less entrainment of gangue minerals in the case when LJ ore is used.

While most of the studies have found better flotation results when a stainless-steel grinding medium was used, there are results that support the opposite case as well, which may be related to the influence of the mineralogical composition of the ores.

Corin et al. used flotation results as a judgment tool in studying the effects of grinding media [110]. Cu–Ni base-metal ore was used, containing nearly 5% total sulfide (Ni 0.67 wt%, Cu 0.36 wt%, S 5.67 wt%) and 7% talc. It was ground without reagent addition, using as the grinding media mild-steel rods (RMMS), forged-steel balls (BMFS), stainless-steel rods (RMSS), and stainless-steel (21% Cr) balls (BMCS). When mild steel was used, the measured Eh values were low after grinding, i.e., +32 mV. Stainless steel gave higher values of the pulp potential of +86 to +89 mV. However, the use of stainless-steel grinding media led to the lower flotation recovery of base metals. Rougher Cu concentrates (batch experiments, with SIBX used as a collector at a dose of 50 g/t and DOW 200 as a frother at a dose of 20 g/t) with grades of 1.75% and 1.79% were obtained at recoveries of 80% and 77%, respectively, when RMMS and BMFS were applied. Under the same flotation conditions, Cu recovery was 75 and 74% when using RMSS and BMCS, respectively. These findings were attributed to the formation of higher iron amounts in the pulp obtained by grinding with RMMS, causing a decrease in talc hydrophobicity and the concentrate dilution.

Very recent studies have compared Eh values and dissolved oxygen (DO) concentrations, obtained when forged steel and stainless steel (containing 30% Cr) were used as the grinding media [111]. A copper–molybdenum ore sample (with chalcopyrite and pyrite as the main Cu- and Fe-bearing minerals) was ground with fresh and saline water. Grinding with forged steel in saline water led to a significantly lower pulp Eh value and DO concentration compared to stainless steel. Immediately after grinding in fresh water, the pulp Eh values of +241.7 mV and +303.9 mV were measured for forged steel and stainless steel, respectively. The respective DO values were 1.28 ppm and 5.00 ppm. When the grinding discharge was transferred to the flotation cell with additional fresh water, the pulp Eh values were +298.5 mV and +307.0 mV (DO 7.19 ppm and 7.42 ppm), respectively, for forged and stainless steel. The more oxidizing conditions ensured by stainless-steel grinding inhibited copper activation on pyrite and allowed effective pyrite depression [112]. Similarly, in high-salinity water, immediately after grinding, the forged-steel media produced a lower Eh value of +85.4 mV compared to +303.3 mV for stainless steel. The respective DO concentrations were 0.48 ppm and 4.47 ppm. When the grinding discharge was transferred to the flotation cell with additional fresh water, the pulp Eh values were +233.4 mV and +305.7 mV (DO 3.63 ppm and 6.68 ppm, respectively). Nevertheless, the high pulp Eh ensured by stainless steel in high-salinity water was not able to satisfactorily inhibit copper-activation action on pyrite due to the fact that more copper ions were released from chalcopyrite oxidation in high-salinity water. That is why the authors proposed the addition of sodium metabisulfite ($Na_2S_2O_5$) to selectively depress pyrite in chalcopyrite flotation while keeping the chalcopyrite recovery in high-salinity water [111].

Another very recent study reported that not only the grinding media (mild-steel (MS) and stainless-steel (SS)) but also wet- or dry-grinding conditions affect the pulp potential after the grinding and the subsequent results of Cu-sulfide ore flotation [113]. The redox potential of the pulp, prepared with dry-ground ore, was considerably higher (in the range of +100 to +200 mV) than that of pulp prepared with ore ground under wet conditions, irrespective of the type of milling media. Dry grinding prevents the galvanic interaction between sulfide minerals and the grinding media. The lowest redox potential (−400 mV) was measured as a result of the wet-MS-grinding process. The pulp potential was significantly decreased due to the release of ferrous ions from the grinding media. These ions consume all of the oxygen dissolved in the pulp. The redox potential measured in pulp prepared with wet-SS-ground material was −200 mV. The rate of Cu flotation was high for dry-ground ore with SS media, resulting in a higher selective concentrate in terms of grade. Nonetheless, Cu recovery was lower due to the instability of the froth structure.

Generally, it can be concluded that when carbon steel is used as the grinding medium, galvanic interactions causing oxidation of steel-grinding media (that is less noble than the sulfide minerals) and different oxidation–reduction reactions are proceeding in the pulp that are leading to the precipitation of iron oxyhydroxy species at the sulfide surfaces, decreasing the amount of oxygen dissolved in pulp, and thus, lowering the pulp Eh values, which generally has as a consequence poor copper-sulfide floatability.

## 4. Eh as an Index in Galena Flotation

### 4.1. Studies on Optimum Flotation Ranges

Research into the potential control in galena flotation began practically at the same time as that of the potential control in the flotation of copper-sulfide ores, although the latter proved to be more numerous over time.

Woods pointed out that for galena ground in a normal mill (iron balls), the beginning of flotation with xanthate starts at 0 V (pH 9) [114]. Gardner and Woods noted that significant flotation of galena initiated at +100 mV in the particle-bed electrode cell [115].

Johnson et al. showed that galena from lead–zinc ore from Mt Isa Mines floated with ethyl xanthate in the Eh range 0 mV to +150 mV (at pH 8.3) [116].

Guy and Trahar found that the flotation of galena with an ethyl xanthate system proceeds in the Eh range from 0 to +350 mV (at pH 8) [117].

Grano et al. [77] also conducted studies on the potential dependence of the flotation of galena from Mt Isa lead-zinc ore—using ethyl xanthate at pH 8.1. They found significant dependency of the recovery on Eh—the flotation initiated at 0 mV and a recovery over 60% was observer in the Eh range from 0 to +400 mV with a maximum (over 80%) at Eh = +60 to +140 mV. Flotation was not observed in the Eh range from −200 mV to −400 mV.

Investigations carried out in the research group of Fuerstenau—a pioneer in the studies on sulfide flotation under potential control [41]—showed that galena flotation separation from mixed oxide–sulfide lead ores can be improved by controlling the pulp potential during the ore sulfidization by addition of $Na_2S$ in suitable concentrations [118]. High galena recovery and good separation from cerussite can be achieved at pH 9.5 using amyl xanthate as collector in the Eh range of 0 to +290 mV. Galena flotation falls drastically above a certain sulfide dosage that decreases the pulp Eh to reducing values therefore inhibiting the xanthate chemisorption onto the mineral surface.

Göktepe [29] pointed the Eh range +100 to +50 mV (pH 2 to 7) as optimal for galena flotation—Figure 1b.

Woods [41] pointed that galena can float at recovery in the range 50%–88% with ethyl xanthate in the Eh range +150 to +400 mV (pH 12.8 to 8.6), with the highest recovery at Eh of approx. +150 mV.

In small scale single-mineral flotation tests, Liu et al. [83] achieved, with butyl xanthate as the collector, flotation recovery of galena over 95% with pulp Eh in the range +440 to +360 mV (while the pulp pH was in the range 2 to 9). On the other hand, under alkaline conditions (pH > 9), the recovery of galena decreased sharply and the minimum recovery of 15.51% was achieved at pH 12. At Eh in the range +75 to +100 mV (and pH 8 to 9), galena recovery in the range 85%–93% was achieved. At Eh lower than +75 mV (obtained by $Na_2S$ addition), a sharp decrease in the galena recovery was found, reaching only 18% at Eh = −242 mV. The authors explained the sharp decrease in the floatation recovery with the removal of BX from the mineral surface by the $Na_2S$ via a competitive adsorption between the anions $BX^-$ and $HS^-$, under which the reagents exist at the studied pH–Eh conditions.

Hintikka and Leppinen [119] applied nitrogen and air in different amounts in order to control the potential during the flotation of a complex sulfide ore with the aim to improve the selectivity in the flotation of copper minerals. Galena was the main Pb mineral sphalerite was the main Zn mineral, and chalcopyrite was the main Cu mineral. A relatively high recovery of Pb minerals was observed, together with chalcopyrite (i.e., low selectivity for copper minerals), at Eh (measured with a Pt indicator electrode) = −20 to −70 mV and at Eh = +120 to +190 mV.

Pälson et al. studied the effect of redox control in a pilot flotation column for lead–zinc ore containing 3.6% Pb and 0.8% Zn as galena and sphalerite in a sandstone hoist rock [24]. The potential was adjusted by varying the oxygen activity of the flotation gas fed to the spargers and nitrogen was used as an oxygen-free gas. The highest recovery of Pb minerals (approx. 95%) was observed in the Eh range of +348 to +284 mV for particle sizes < 75 μm when air was used as the flotation gas (the frother was SIBX—25 kg/t, pH 7.7 to 7.9). For the studied gases, better selectivity between lead and zinc was observed when pure nitrogen was used, with Eh values in the range from +6 to −34 mV, and particles < 75 μm; the average Zn recovery was approx. 25%, while the Pb recovery was in the range 80 to 95%.

Uribe-Salas et al. studied the flotation recovery of galena from fine-grain, complex pyritic ore. They reported good results at potentials of approx. +300 mV, when the collector used was a mixture based on sec-butyl dithiophosphate (SBDTPH), mercaptobenzothiazole (Aeroflot 404), diisobutyl dithiophosphinate (Aerophine 3418A), and amyl xanthate (AX). When PAX was used as collector, the Eh range found to be suitable was +150 to +200 mV (pH 11–12) [120,121].

Gu et al. studied the flotation of Pb–Zn–Py ores with the collector sodium diethyldithiocarbamate (DDTC). For different ores, PbS recovery was in the range 76%–92% at an Eh range from +350 to +150 mV (corresponding to pH values from 9.5 to 12.9). Under these conditions, the content of pyrite and sphalerite in the lead rougher concentrate was minimal [122]. The application of the optimum pulp potential found in the conditions in the plants that process ores from different mines led to an increase in Pb concentrate recoveries (compared to traditional technology) in the range of 2.2 to 5.1% with the grade increase in the range of 0.7 to 4.6%.

Further studies of the same research group carried out at pH values over 12.5 and Eh values below +200 mV, with DDTC as the collector, found, as the optimum conditions for galena flotation from Pb–Zn–Fe sulfide ores, Eh = +150 to +180 mV and pH = 12.5–12.8 [44,71]. A stage potential control flotation was developed and proposed for Mengzi lead–silver–zinc ore, where N,N′diphenylamino-dithiolphosphoric acid(NNDDC) is used as a prime collector and DDTC is used as secondary one. The pulp-potential values significantly affected the floatability from real mixed-mineral pulp. Under a controlled pulp potential (+330 mV, pH 8.8) flotation recovery of galena reached 85%. The grades of Pb and Ag of galena concentrate were 55% and 1800 g/t, respectively. When pulp-potential control was ensured during grinding and flotation, the recoveries of Pb and Ag were 86.5% and 65%, respectively. The recovery of Zn in marmatite concentrate was 91.25% and the grade of Zn was 42.5%. The authors reported that the potential-controlled flotation of galena has been applied successfully in many complex lead–zinc–iron sulfide mines in China [123].

Hu et al. [57] studied the flotation of ore from the Nanjing lead–zinc mine. In the generally used flow sheet, $ZnSO_4$ and $Na_2SO_3$ are applied as the depressants of ZnS and $FeS_2$ under neutral pH conditions, and the flotation separation of galena from sphalerite and pyrite is poor. They proposed flotation separation carried out in lime-modified alkaline pulp (pH = 12.4–12.5), at a potential kept under +170 mV using DDTC as a collector in galena flotation. In comparison with the common flow sheet, the recovery and grade of lead concentrate increased from 85.9% to 88.9% and from 52.1% to 60.0%, respectively. The same research group achieved good flotation separation with ore from the Fankou lead–zinc mine in Guangdong Province, China, where the pH was modified to 12 by lime addition, and pulp potential was again kept less than +170 mV, using a mixture of xanthate and DDTC as collectors in galena flotation [57].

A non-exhaustive summary of the information available on the most-probably suitable Eh and pH ranges for flotation of galena is summarized in Table 5.

**Table 5.** Probable Eh and pH values and ranges that are most suitable for collector-aided flotation of galena.

| Eh Range, mV (SHE) | pH Range | Collector | References |
|---|---|---|---|
| 0—flotation starts | 9 | Xanthate | [114] |
| +100—flotation starts | 8.3 | Ethylxanthate | [115] |
| 0–+150 | | Ethyl xanthate | [116] |
| 0–+350 | 8 | Ethyl xanthate system | [117] |
| +100–+50 | 2–7 | Sodium isopropyl xanthate | [29] |
| +0–+400 | 8.1 | Ethyl xanthate | [77] |
| −20–−70 and +120–+190 | 11.5 | Potassium amyl xanthate | [119] |
| +348–+284 | 7.7–7.9 | Na-isopropyl xanthate | [24] |

**Table 5.** *Cont.*

| Eh Range, mV (SHE) | pH Range | Collector | References |
|---|---|---|---|
| 0–+290 | 9.5 | Amyl xanthate | [118] |
| +150–+400 | 12.8–8.6 | Ethyl xanthate | [41] |
| +150–+200 | 11–12 | Potassium amyl xanthate | [120] |
| approx. +300 | 11–12 | SBDTPH + Aeroflot 404 + Aerophine 3418A + AX | [120] |
| +350–+150 | 9.5–12.9 | Diethyldithioncarbamate | [122] |
| +150–+180 | 12.5–12.8 | Diethyldithioncarbamate | [71] |
| <+170 | 12.4–12.50 | Diethyldithiocarbamate, lime addition | [57] |
| <+170 | 12 | Xanthate + DDTC | [57] |
| +440–+360 | 2–9 | Butyl xanthate | [83] |
| +75–+100 | 8–9 | Butyl xanthate + $Na_2S$ | [83] |

### *4.2. Research on Grinding Environment and Its Effect on Flotation*

Similar to the studies on the effect of grinding environment on Cu mineral flotation, the studies on the effect of grinding conditions on galena flotation results started in the 1980s–1990s of the previous century.

Hayes and Ralston [49] reported that galena and chalcopyrite can be separated by Eh control if they are ground together in a reducing environment. However, if they are ground together in an oxidizing environment, the selectivity and floatability is decreased due to the formation of hydrophilic oxidation products on the mineral's surface.

The effect of different grinding media (forged-steel and high-Cr-steel) on galena flotation performance in the lead circuit of the Mount Isa Mines lead–zinc concentrator was studied in laboratory experiments [124]. Galena recovery of the 3-micron fraction was 63% when forged-steel media was used and when stainless-steel media was applied the recovery was 79%. Also, better selectivity for galena against sphalerite and iron sulfides was found with a high-Cr steel medium. The corresponding Eh and pH values after grinding were Eh = +72 mV and pH = 7.8 for a forged-steel grinding medium and Eh = +275 mV and pH = 8.2 for stainless steel.

The lower Eh value for the forged-steel grinding medium was related to the corrosion of forged steel and thus formation of higher levels of oxidized iron hydrophilic species that covered the galena's surface and impeded its flotation [70,125]. During grinding with forged steel, the main anodic reaction is:

$$Fe \rightarrow Fe^{2+} + 2e^-,\ E^0 = 0.447\ V \quad (14)$$

In contact with air, $Fe^{2+}$ may be further oxidized:

$$Fe^{2+} \rightarrow Fe^{3+} + e^-,\ E^0 = 0.771\ V \quad (15)$$

The proceeding main cathodic reaction is:

$$O_2 + 2H_2O + 4e^- \rightarrow 4OH^-,\ E^0 = 0.401\ V \quad (16)$$

With the water in the flotation pulp, hydrolysis reactions proceed:

$$Fe^{2+} + 2H_2O \rightarrow Fe(OH)_2 + 2H^+ \quad (17)$$

$$Fe^{3+} + 3H_2O \rightarrow Fe(OH)_3 + 3H^+ \quad (18)$$

The produced iron hydroxides are hydrophilic and may partially or completely cover the sulfide mineral surface, impeding the interaction between collectors and minerals and thus reducing flotability.

Peng and Grano found a linear relation of galena or chalcopyrite flotation recovery on the quantity of iron-oxidation species on the mineral, and the galena flotation was more susceptible to that species than chalcopyrite flotation [125]. In addition, the concentration of dissolved $O_2$ was decreased further when grinding with forged- or mild-steel media, thus hindering the adsorption of xanthate on the sulfide mineral surfaces, preventing the flotability of these minerals [70].

Furthermore, besides iron-oxidation species, $H_2O_2$ is formed in the grinding system during wet grinding and from mixing freshly ground galena with water immediately after dry grinding. Dissolved $O_2$ can form superoxide anions $(O_2^\bullet)^-$ that react with $H^+$ to form $H_2O_2$ (Equations (19) and (20)). In the presence of dissolved molecular oxygen, at pH < 4, galena participates in the formation of hydrophobic sulfur and $H_2O_2$ (Equations (21) and (22)) [104,126,127]:

$$O_2 + e^- \rightarrow (O_2^\bullet)^- \tag{19}$$

$$2(O_2^\bullet)^- + 2H^+ \rightarrow H_2O_2 + O_2 \tag{20}$$

$$PbS + H_2O_2 \rightarrow Pb^{2+} + S^0 + 2e^- + 2^\bullet OH \tag{21}$$

$$2^\bullet OH \rightarrow H_2O_2 \tag{22}$$

When $O_2$ is not present to participate in reactions, $H_2O_2$ is formed by combining two hydroxyl radicals obtained by the reaction of galena with water (Equations (23) and (24)) [123]:

$$PbS + H_2O \rightarrow {}^\bullet OH + H^\bullet + Pb^{2+} + S^0 + 2e^- \tag{23}$$

$$2^\bullet OH \rightarrow H_2O_2 \tag{24}$$

$H_2O_2$ is a stronger oxidizing agent than $O_2$; it oxidizes galena and thus depresses its flotation [126]. This depressing action of $H_2O_2$ on galena is attributed to its strong oxidizing action on lead xanthate that presents on the galena's surface, resulting in a decomposition of lead xanthate (Equation (25)):

$$Pb(EX)_2 + H_2O_2 \rightarrow Pb(OH)_2 + (EX)_2 \tag{25}$$

Nooshabadi and Rao [126] found that the amount of produced $H_2O_2$ increased with decreasing pH (from 7 to 2), thus leading to an increase in Eh (from +275 to +380 mV). Increased grinding time and galena loading also increased the amount of produced $H_2O_2$ because of the increase in surface area and its interaction with water. Increased $H_2O_2$ production caused decreased galena flotation—76%, 70%, and 65% recovery at pH 6.3, 4, and 3, respectively.

In addition, $H_2O_2$ facilitates the corrosion of galena, thereby promoting the formation of lead oxide, thiosulfate, and sulfate [128].

Generally, when carbon/forge steel are used as grinding media, more reducing conditions are formed in the pulp, leading to an increase in the amount of hydrophilic species on the galena's surface and decreased flotation. In addition, PbS facilitates the formation of $H_2O_2$, which depresses galena flotation.

## 5. Eh as an Index in Zinc Mineral Flotation

Compared to studies related to copper-bearing minerals and galena, the investigations into zinc minerals are scarce and are mainly pointed at complex sulfide ores with the aim to improve flotation selectivity.

Hintikka and Leppinen [119] used nitrogen and air in different orders and proportions to control the potential during the flotation of a complex sulfide ore (containing chalcopyrite, galena, and sphalerite) with the aim to improve the selectivity in the flotation of copper minerals. A relatively high recovery of Zn minerals was found at Eh = +120 to +190 mV (pH 11.5, PAX 150 g/t, $SO_2$ 400 g/t, flotation time 10 min). However, this was accompanied by a high chalcopyrite recovery (i.e., low selectivity for copper minerals). The EDTA analysis of the ore slurry showed that the high recoveries of lead and zinc corresponded to high amounts of EDTA-soluble lead and zinc. Therefore, the conclusion was that oxidation products of zinc and lead are formed that improve the interaction with amyl xanthate.

The same authors continued their studies with three different complex sulfide ore samples [87]. All the studied ores contained over 50% pyrite, with the highest content in the Adjuster ore, 84.7%. Besides pyrite, the Adjuster ore contained chalcopyrite, bournonite, tetrahedrite, galena, and arsenopyrite. The Grong ore contained chalcopyrite, sphalerite, pyrrhotite, and pyrite. The Pyhäsalmi ore contained chalcopyrite, sphalerite and pyrite. The experiments showed that the grinding medium had a strong effect on the pulp potential and the flotation results of the three ore types studied. Immediately after the grinding, the potential difference between the two grinding media (mild-steel and stainless-steel) was between 100 and 250 mV depending on the ore type; the biggest difference was found for the Aljustrel ore. The mild-steel media grinding resulted in more reducing potential. The Eh values for good zinc flotation of the three ore types were obtained at full aeration; these were in the range +130 to +230 mV (pH 11.5 to 12, with xanthate as the collector). The highest recovery was found at Eh = +230 mV. Conditions in the activation and conditioning stages affect the zinc floatation of the complex ores. The highest Zn grades for the Aljustrel and the Grong ores were obtained at the activation stage at a potential of Eh = −70 mV, whereas the Pyhäsalmi ore yielded the best recovery and grade at "air-set" potentials (Eh ca. +220 mV).

Pälson et al., using a pilot flotation column, investigated the impact of redox control on the flotation of lead–zinc ore containing 3.6% Pb and 0.8% Zn as galena and sphalerite in a sandstone hoist rock [24]. They adjusted the potential by varying the activity of oxygen in the flotation gas fed to system; nitrogen was used as an oxygen-free gas. In the Eh range of +348 to +284 mV, when air was used as the flotation gas (the frother was SIPX at 25 kg/t, pH 7.7 to 7.9), Zn mineral recovery was approx. 82% for particles < 53 μm and approximately 60% for particles < 75 μm. The lowest Zn recovery, of approx. 25%, was observed in the Eh range from +6 to −34 mV, in which pure nitrogen was used, and the solid phase consisted of particles < 75 μm. Under these conditions, the Pb recovery was in the range 80 to 95%, and thus better selectivity between lead and zinc was achieved.

Göktepe [29] reports the Eh range +45 to +20 mV (pH 4 to 6) as optimal for sphalerite floatation (Figure 1b).

Gu et al. studied the flotation of Pb–Zn–Py ores, with DDTC used as the collector. For different ores, ZnS recovery was in the range 45%–55% at Eh from +250 to +350 mV (corresponding to pH values from 11.2 to 9.5) [122]. The authors explained their findings with the activating role of the $Cu^{2+}$ ion that, under these conditions, produces copper sulfides able to prevent the sphalerite surface from over-oxidation [129]. The application of the optimum pulp-potential conditions found in the plants that process ores from different mines, led to an increase (compared to traditional technology) in Zn-concentrate recoveries in the range of 2.3 to 4.6% with a grade increase in the range 0.3 to 2.1%.

Hu et al. [56] investigated the flotation of ore from the Nanjing lead–zinc mine. They carried out flotation separation in alkaline pulp (pH = 12.4–12.5, achieved by lime addition), at a potential kept less than +170 mV. After galena separation, $CuSO_4$ was used as a sphalerite activator. The grade and recovery of zinc concentrate increased from 52.6% to 53.0% and from 87.0% to 91.9%, respectively, compared to the classically used flow sheet in the factory.

The flotation of Zn minerals from polymetallic-sulfide complex ore from the Eureka Mine was carried out in the Eh region of +50 to +100 mV (pH 11.2, using xanthate as the collector) with a recovery of approximately 92% at a grade of 59.4% [130].

The zinc-reverse process is used to upgrade Zn concentrate by pyrite removal at Mount Isa Mines, Mt Isa. The process was not successful at a pulp Eh of approximately +200 mV (pH of approx. 5). Experiments showed that the pulp Eh should be less than +100 mV to achieve a successful sphalerite depression [38].

Nooshabadi and Rao found that $H_2O_2$ was formed in the sphalerite grinding system at pH < 6 or when freshly dry-ground sphalerite was mixed with water [131]. When dissolved molecular oxygen is available in the pulp, $Zn^{2+}$ ions produced via Equation (26) can catalyze the generation of the superoxide anion $(O_2^{\bullet})^-$ to form $H_2O_2$ (according to Equations (19) and (20)):

$$ZnS \rightarrow Zn^{2+} + e^- + S^0 \quad (26)$$

The same authors studied the influence of the addition of a depressant and a collector during grinding on the generation of $H_2O_2$ by metal sulfides and its effect on the flotation of a complex (Cu-, Pb-, and Zn-sulfide) ore containing pyrite and silicates as the main gang materials [132]. The collectors used were PAX and Danaflot 871 (dialkyl dithiophoshate mercaptobenzothiazole), and the frother was Dowfroth 250 (polypropylene oxide methanol). Copper sulfate was the source of copper ions for the activation of sphalerite flotation. In the reference experiment, the reagents were added to the ground sulfide pulp. In other experiments, either only a collector or a collector, an activator, and a depressant ($NaHSO_3$ and dextrin) were added to the mill before grinding. The researchers found that the addition of only a depressant during grinding decreased the formation of $H_2O_2$, thus leading to a lower depression of Py. The addition of a collector and a depressant together during ore grinding increased $H_2O_2$ generation in the flotation pulp, which led to the oxidation of the pyrite surface, causing better Py depression and higher metal recovery of Cu, Pb, and Zn in concentrates with higher grades. Generally, the addition of reagents during grinding decreased the Eh and increased the pH values of the pulp. The areas in which Zn minerals floated well were in the Eh range from +220 to +300 mV and the corresponding pH range from 10.5 to 8.8, with optimum values of Eh = +260 to +265 mV and pH 9.5 (in which a recovery of 43% was achieved at a grade of 42.2% in the first concentrate from Zn rougher flotation).

Long et al. studied the effects of different grinding media (balls of common steel and stainless steel (SS)) on the flotation behavior of copper-activated marmatite [133]. After grinding, the Eh potential of the pulp prepared with material ground with the SS medium was 54 mV higher than that of the pulp prepared with material ground with normal-steel balls. Although during the conditioning, collector (butyl xanthate) addition, and flotation stages, the Eh values of both types of pulp become close, the flotation recovery of marmatite in the presence of copper ions acting as an activator was higher from the pulp prepared with material ground with the SS medium—approx. 87% vs. approx. 80% for pulp prepared with material ground with normal-steel balls. Based on data from FTIR, XPS, and ion-concentration measurements, the authors explain the flotation difference observed with the different products formed on the marmatite surface—$Cu_2S$ and $Fe(OH)_3$ in the case of normal-steel grinding and $Cu(OH)_2$ and hydrophobic $S^0$ in the case of the SS grinding medium.

A non-exhaustive summary of the information available on the most probably suitable Eh and pH ranges for flotation of sphalerite is shown in Table 6.

Finally, we have to point out that the use of pulp potential as a parameter to control flotation of base non-ferrous metal-sulfide ores has been widely proposed in the literature, including by a few of the examples already discussed in the above text [30,41,46,57,119,122].

However, we should not forget the other point of view, expressed by some authors [41,134], that it is almost impossible to practically control the flotation process only by controlling the potential in sulfide flotation plants, due to the complexity of maintaining the state of electrode probes so that they respond correctly to plant conditions, or due to the many relationships between Eh and added reagents, galvanic interactions between minerals, grinding media effects, and conditioning procedures.

**Table 6.** Probable Eh and pH values and ranges that are most suitable for collector-aided flotation of sphalerite.

| Eh Range, mV (SHE) | pH Range | Collector | References |
|---|---|---|---|
| +120–+190 | 11.5 | Potassium amyl xanthate | [119] |
| +130–+230 | 11.5–12 | Xanthate | [87] |
| +250–+350 | 11.2–9.5 | Diethyldithioncarbamate | [122] |
| +50–+100 | 11.2 | Xanthate | [130] |
| +220–+300 | 10.5–8.8 | PAX and Danaflot 871 | [132] |
| +45–+20 | 4–6 | Sodium isobutyl xanthate | [29] |

## 6. Floatation Reagent Action and Pourbaix (pH–Eh) Diagrams

### 6.1. pH–Eh Diagrams in Copper Mineral Flotation Separation

As described above (Sections 2 and 3.1), different copper minerals float best in different ranges of Eh values. Floating sulfide minerals are semiconductors, i.e., they can take part in electrochemical reactions and either donate or accept electrons in an electrochemically active system [88,135–137]. This means that the potential across the mineral/solution interface is an essential factor in determining flotation recovery [19,138].

When the mineral's near-surface area (space charge area) is rich in holes (positively charged), a negatively charged collector radical can be adsorbed onto the mineral's surface to render it hydrophobic. When the mineral's near-surface area is rich in electrons (negatively charged), it will not permit a negatively charged radical to be adsorbed. Rather, the collector would form a hydrogen bond on this surface to hydrophobize the mineral [88,138–140].

Pioneering studies on the mechanism of the interaction of flotation collectors with sulfide minerals were made by Woods (galena/ethyl xanthate) [141], Kowal and Pomianowski (chalcocite/ethyl xanthate) [142], and Chander and Fuerstenau (chalcocite/diethyldithiophosphate) [143].

The adsorption of collector(s) onto most of the sulfide's surface is controlled by redox potential, as reported many years ago [138,144] and reconfirmed during years [76,82,140], including by recent publications [37,134].

It is established that the adsorption of xanthate species can be realized in three forms [134]: (a) specific adsorption of a xanthate ion, which does not make the mineral's surface hydrophobic (Equation (27)); (b) chemisorption of a radical (further developing to formation of a complex with metal ions), which renders the surface slightly hydrophobic (Equations (28)–(30)); and (c) physical adsorption of dixanthogen, formed via reaction (31), which makes the mineral surface strongly hydrophobic. Path (a) occurs only in the absence of oxygen. Paths (b) and (c) occur in the presence of oxygen and the surface becomes hydrophobic. Generally, chemisorption (in which a monolayer on the mineral surface is formed) is the thermodynamically favorable process, and because of this, it is considered the most optimal use of the collector [18,19,145–147]:

$$X^- \rightarrow X_{ads} + e^- \tag{27}$$

$$MeS + 2X^- \rightarrow MeX_2 + S^0 + 2e^- \tag{28}$$

$$MeS + 2X^- + 4H_2O \rightarrow MeX_2 + 8H^+ + SO_4^{2-} + 8e^- \tag{29}$$

$$2MeS + 4X^- + 3H_2O \rightarrow 2MeX_2 + 6H^+ + S_2O_3^{2-} + 8e^- \tag{30}$$

$$2X^- \rightarrow X_2 + 2e^- \quad (31)$$

where $X^-$ represents the xanthate ion and Me denotes the corresponding metal in the sulfide ore.

In the flotation pulp, the described anodic reactions (Equations (27)–(31)) are commonly joint with the cathodic reduction of oxygen (Equation (16)), and in an acidic environment (Equation (32)):

$$\frac{1}{2}O_2 + 2H^+ + 2e^- \rightarrow H_2O \quad (32)$$

Generally, the cathodic reduction of oxygen, proceeding simultaneously with specific anodic reaction(s) occurring on the mineral's surface and depending on the mineral's properties, facilitates the establishment of the "mixed potential". At the mixed potential, the collector transfers electrons to the mineral surface via an anodic oxidation reaction, and a cathodic reaction of oxygen reduction returns the charge back to the solution [19].

It has to be mentioned that some small (and actually, apparent) discrepancies appear in the literature regarding the potential at which the equilibrium between xanthate and dixanthogen is established. Lotter et al. [76] and Zanin et al. [148] give a potential of +163 mV as the value for xanthate oxidation on pyrite's surface. Wonder Chimonyo et al. [11] give a value of approx. +150 mV as the critical potential for dixanthogen formation, while Lui et al. [83] calculate this value to be +136 mV. Some authors claimed that the xanthate/dixanthogen equilibrium potential depends on the type of xanthate (the number of carbon atoms in a hydrocarbon chain and the corresponding cation); the lower equilibrium potential of the xanthate/dixanthogen couple is observed with a higher number of carbon atoms in the hydrocarbon chain [149,150].

Actually, the equilibrium potential for reaction (31) is given by the Nernst equation—Equation (33) [92]:

$$Eh = E^0 - 0.0591 \times \log[X^-],\ E^0 = -0.07\ V \quad (33)$$

and depends on the concentration of the used xanthate. Kantar [92] reported that the equilibrium potential is +66 mV for $5 \times 10^{-3}$ M xanthate, and this corresponded with the observed lower-limiting flotation potential of enargite in alkaline solutions.

For commonly used xanthate concentrations in the range $10^{-5}$ to $10^{-3}$, the equilibrium potential of xanthate oxidation is expected to be in the range from +107 to +226 mV.

In addition, different reference sources give different values for $E^0$, for example, Xu et al. [151], following Winter and Woods [152], use $E^0 = -0.158$ V, which would result in a different potential for xanthate oxidation at the same concentrations of the reagent. This confirms that Eh areas of eventual activation or depression of a given mineral under given conditions have to be considered, not exact Eh values. On the other hand, it should be remembered that the $E^0$ value reflects standard equilibrium conditions for a given oxidation–reduction pair, i.e., usually in clear solution.

Many authors explain the experimentally determined most-susceptible Eh regions for flotation of different minerals with the xanthate/dixanthogen adsorption potential being reached [8,103,153,154].

Due to the mixed-potential formation, there is a particular potential at which the anodic reaction transforms the xanthate to produce the metal xanthate or adsorbed dixanthogen, coupled with the oxygen cathodic reduction. Dixanthogen is formed if the mineral surface reaches a mixed-potential value higher than the xanthate/dixanthogen equilibrium potential. The transformation is mineral-specific—dixanthogen formation is preferential on pyrite, while on chalcopyrite, both the formation of a metal complex and adsorption of dixanthogen are proceeding, and $PbX_2$ formation is favored on galena [134]. According to the mixed-potential mechanism, the flotation will not start if the potential is below the region in which a collector oxidation reaction can take place, and at potentials inside this region, flotation is possible.

Chemisorption of ethyl xanthate on chalcocite's surface and the formation of a monolayer on the mineral is reported at Eh = −150 mV, where the onset of flotation occurs. At a potential over 0 V, xanthate interacts with the mineral to form copper(I) xanthate and efficient flotation still occurs [19].

Potassium amyl xanthate adsorption on pyrite has an optimum Eh range of +100 to +300 mV at pH 9 [155], while sodium isopropyl xanthate adsorbs best on pyrite at pH 9 in the range from +35 to +260 mV [156].

For chalcopyrite, the maximum adsorption of sodium ethyl xanthate was observed at +385 mV at pH 4, at which the mineral surface was mildly oxidized and covered with adsorbed CuEX. Under reductive conditions (−100 mV, pH 9), a certain desorption of xanthate from chalcopyrite was initiated; it was higher at −200 mV, and pH 9 and an increase in concentration of Fe (III) and Cu (I) oxides and hydroxides was noticed [82].

Tolley et al. [4] found decreased adsorption of ethyl xanthate on chalcocite's surface at reducing potentials (approximately 95% of the xanthate that was present in the system at +245 mV was adsorbed, vs. approximately 65% at −155 mV). The latter yielded good copper grades but poor recoveries.

In addition to desorption, oxidation-decomposition of metal xanthates formed on the mineral's surface can take place, for example, of copper xanthate [93,94]:

$$CuX + H_2O \rightarrow CuO + X^- + 2H^+ + e^- \quad (34)$$

The experimentally observed upper flotation range with xanthates may be explained as the potential being reached at which the metal xanthate is decomposed (Equation (34)), the potential being described with the expression (35) [92]:

$$Eh = E^0 + 0.0591 \times \log[X^-] - 0.1182\,pH,\ E^0 = 1.787\ V \quad (35)$$

At pH 10.5, for a xanthate concentration of $5 \times 10^{-3}$ M, the Eh is calculated at 410 mV, using Equation (35), which is in good agreement with the findings of Kantar for the upper flotation limit of enargite [92].

Pulp pH–Eh conditions determine the possibility of the presences of different metal and non-metal (for example, sulfur) species on the mineral's surface, which influence the mineral's floatability [4,76,148], as well as the adsorption of metal species (for example, copper species) from the solution on the mineral's surface in order to activate the surface for the attachment of hydrophobic collectors [156,157].

Different Cu-bearing minerals have been found to be hydrophobized (and consequently rendered floating) by different species of the used reagents. For example, for chalcocite, bornite, and covellite, metal xanthates are found to be active surface entities; dixanthogen is the hydrophobizing agent for pyrite, while metal xanthate is needed for initial floatation of chalcopyrite and dixanthogen for complete flotation of the same mineral [76,145,148].

Arsenic-bearing minerals float with dixanthogen adsorption at a potential above approximately +160 mV. At these Eh values, for arsenopyrite, the suitable pH range is below 7, while for enargite and tennantite, the range of pH is 8–11/12 [76].

As described above (Sections 3.2 and 4.2), the impact of grinding medium on further flotation results is explained by different authors as resulting from the development of oxidizing or reducing conditions (i.e., high or low Eh value) in the pulp [9,15,59,104,111].

The above-mentioned established phenomena, and the relationships between them, are explained by the pH–Eh diagrams, as proposed by Fuerstenau [158], confirmed in subsequent years by: (i) comparing the predictions made from the pH–Eh diagrams and the experimentally observed flotation recovery of chalcocite as a function of potential [159]; (ii) explaining the convincing dependence of enargite flotation on Eh and pH values of the flotation solution [92]; and (iii) elucidating the collectorless floatability of some minerals [57,160]—and reconfirmed recently [70,83,148] by showing the forms of existence of the reactant(s) and the corresponding metal (copper, iron, lead, etc.) species in the given flotation medium, i.e., in a given pH–Eh region.

For example, the potential at which xanthogenate is oxidized to dixanthogen (Eh = +163 mV) gives the theoretical boundary between the two forms on the pH–Eh diagrams. The line describing the equilibrium between xanthate radical and xanthic acid is at pH = 1.6 on these diagrams. The boundary between di-ethyl dithiophosphate and dithiolate is at +255 mV [76]. The boundary between sulfur ions originating from minerals and the formation of $S^0$ is not only Eh but also pH dependent, however, at pH 7, the transformation occurs at approximately −180 mV, while at pH 5 it occurs at approx. −30 mV [83].

Woods et al. showed that metastable copper sulfides and sulfur counted for electrochemical reactions at chalcocite surfaces and for results obtained in an aqueous system considering chalcocite, covellite, anilite, and djurleite [161]. Latter Aghazadeh et al. confirmed that the region where metastable elemental sulfur exists is exactly the pH–Eh area where chalcopyrite exhibited a self-induced floatability [54].

The usefulness of using Porbaix diagrams is demonstrated by Grano [38] by explaining the finding in a flotation plant which normally treats porphyry copper ore (chalcocite and chalcopyrite as the main Cu minerals) but with high sulfide skarn content. The pH range in which the plant works is 11 to 11.5, achieved by lime addition. The pH–Eh diagram for the CuS–$H_2O$ system shows that at pH 11, chalcocite is stable only to an Eh approx. 0 mV and above this Eh value it will begin to oxidize. The pulp Eh measured in the beginning of the flotation (cyclone overflow) was over +150 mV, signifying that chalcocite oxidation may occur. When NaHS was added to the cyclone overflow to achieve an Eh of −50 mV, the system entered the stability region of $Cu_2S$ and the copper recovery increased.

Lotter et al. [76] actually visualized most clearly the relationship between the different species of xanthate (xanthic acid (HBX), free xanthate radical ($BX^-$), and dixanthogen ($X_2$)) at the corresponding pH–Eh values and the most suitable Eh ranges for the flotation of the various copper minerals (Figure 4). Empirically observed flotation domains of galena and sphalerite, as described above, are added in the figure.

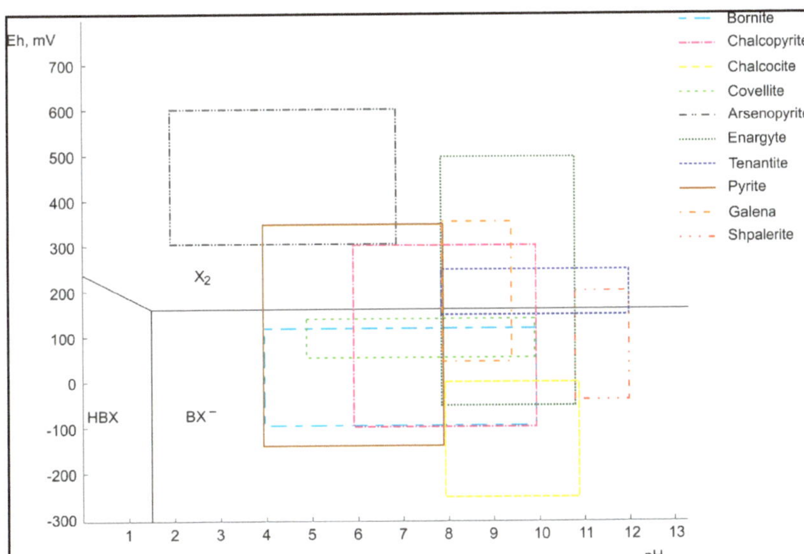


**Figure 4.** Flotation domains of bornite, chalcopyrite, chalcocite, covellite, arsenopyrite, enargite, tennantite, pyrite, galena, and sphalerite, plotted based on empirical data on Eh and pH values, with butyl xanthate as the collector (adapted from [76] with added data).

Figure 4 shows that the flotation domains of chalcocite, bornite, and covellite fall in the butyl-xanthate radical domain, while chalcopyrite crosses the Eh equilibrium line between free-xanthate radical and dixanthogen domains. This is consistent with the above-mentioned literature and established by spectral analysis findings about the xanthate species that are responsible for the hydrophobicity of those Cu sulfides.

Pyrite floats in the domain with pH from 4 to 8 and Eh from −150 to +350 mV, with higher recoveries at an Eh over +150 mV, which is in line with findings that dixanthogen is the main species causing hydrophobicity in pyrite flotation. Figure 4 exhibits a clear boundary for pyrite flotation at pH 8. This is consistent with Py depression at higher pH values of the pulp and is the reason to commonly use lime as a pH modifier in a copper circuit to reject pyrite.

As can be seen in Figure 4, the domain for arsenopyrite flotation is at Eh values higher than that of the free-xanthate radical oxidation to dixanthogen. This is consistent with the findings that dixanthogen is the species that makes arsenopyrite's surface hydrophobic [162].

On the other hand, Lui et al. used pH–Eh diagrams to elucidate the mechanism (competitive adsorption) of removal of butyl-xanthate radical from the chalcopyrite and galena surfaces using sodium sulfide (existing in the form of $HS^-$) in the bulk flotation of copper–lead mixed concentrate [83]. The authors explained the influence of the grinding medium as reaching a corresponding pH–Eh region of the formation of iron oxyhydroxides, which impart hydrophilic properties to minerals.

Pyrite activation by copper can be also clarified as reaching the pH–Eh region in which reduction of $Cu^{2+}$ to $Cu^+$ ions, capable of forming CuS with hydrophobizing properties on the pyrite surface, takes place. This explains the findings of He et al. that the pulp potential Eh is a significant factor in determining recoveries and Cu speciation on the pyrite surface in copper-activated pyrite flotation [156]. They observed that a conditioning potential of +35 mV at pH 9 increased pyrite recovery by facilitating the formation of Cu(I)–S and Cu(I)–xanthate hydrophobic species on the Py surface. At high Eh values, the formation of hydrophilic-oxide/hydroxide species on the pyrite surface decreased the recovery.

Diagrams, such as those proposed by Lotter et al. [76] and Lui et al. [83] can serve both to explain the phenomena observed in experimental practice and to justify the necessary change in the used reagents and floatation gas and/or the need to add an oxidizer or reducer to the floatation medium.

The different thermodynamically stable pH–Eh ranges of the collector(s) and metal species (in minerals) permits the selective flotation of metal sulfides by manipulating both pH and redox potential. Some examples of such manipulations can be mentioned:

1. Usage of mixed and new collectors:

Lotter et al. overlapped the Pourbaix stability diagrams of xanthate and dithiophosphate species in water solution with the outcomes of preceding studies on the different sulfide minerals [76]. Thus, pH–Eh regions are found where the corresponding species of a given collector are able or unable to adsorb on the minerals, therefore leading to the possibility of achieving selective floatation. Based on the drawn diagrams, and due to the fact that the different minerals float best with different forms of different collectors, the authors proposed the use of a mixed-collector system of xanthate and dithiophosphate to recover a bulk concentrate as a mixture of sulfides.

New types of collectors are proposed that might replace xanthates in the flotation, such as N-propyl-N0-ethoxycarbonyl thiourea (PECTU) [163], 1-hydroxyethylidene-1, 1-diphosphonic acid (HEDP) [164], and even nanocollectors [165,166].

2. Change in floatation gas:

It is well known that the pulp redox potential correlates with pulp pH and dissolved oxygen by the relation [145]:

$$Eh = 1.23 + 0.015 \times \log P_{O2} \times 0.059pH, V \tag{36}$$

The presence of oxidizing agents (such as $O_2$, $H_2O_2$, NaOCl) dissolved in the pulp changes the Eh of the flotation pulp and creates conditions for the existence and/or appearance of the different species of the reactants and/or mineral components. Because the dissolved $O_2$ will always shift the potential in the direction of higher values, in some cases it is probably better for $N_2$ to be used as a carrier flotation gas, instead of air.

3. Control of pulp pH value:

   In flotation plants, usually lime is used to control the pulp pH.

4. Addition of reducing reagents to the floatation pulp:

   A decrease in the amount of oxygen and, especially, availability in the pulp of reducing agents (such as sodium sulfide, hydrazine, sodium dithionite, NaHS) also changes the Eh of the flotation pulp and creates conditions for the existence and/or appearance of the different species of the reactants and/or mineral components.

Due to the fact that dixanthogen is the main species accountable for pyrite flotation, reagents that form a more-reducing environment than the $X^-/X_2$ couple would act as pyrite depressants by avoiding the oxidation of xanthate. This explains the depressing effect of sulfide/sulfite ions, which are easily oxidized (Equations (37) and (38)), with standard half-cell potentials which are considerably lower than that of the $X^-/X_2$ couple [167]:

$$S^{2-} + 2e^- \rightarrow S^0,\ E^0 = -0.48\ V \quad (37)$$

$$SO_3{}^{2-} + 2OH^- \rightarrow SO_4{}^{2-} + H_2O + 4e^-,\ E^0 = -0.93\ V \quad (38)$$

Another example is the use of cyanide, a well-known depressor in the flotation of sulfide minerals. Cyanide can act in different ways:

(1) It is preferentially adsorbed on the sulfide-mineral surfaces compared to xanthate, and thus inhibits xanthate adsorption and the oxidation of xanthate on the mineral surface practically in the entire Eh region where xanthate would be adsorbed and oxidized [23,150].

(2) As a strong reducing agent, cyanide reacts with oxygen dissolved in the pulp to form cyanogen and/or cyanate:

$$2CN^- + H_2O + 1/2\ O_2 \rightarrow (CN)_{2(g)} + 2OH^- \quad (39)$$

$$2CN^- + O_2 \rightarrow 2CNO^- \quad (40)$$

$$(CN)_2 + 1/2O_2 + 2OH^- \rightarrow 2CNO^- + H_2O \quad (41)$$

(3) The xanthate oxidation to dixanthogen is suppressed due to the large consumption of oxygen by cyanide. In this way, the redox potential of the flotation pulp is decreased to values unfavorable for xanthate adsorption on the mineral's surface.

(4) Additionally, a direct reduction of dixanthogen by cyanide may also occur:

$$(EX)_2 + 2CN^- \rightarrow 2EX^- + (CN)_2 \quad (42)$$

5. The addition of novel depressants [51,168].
6. Changing Eh and pH, similarly to the reagents' addition, achieved by using an external voltage source with electrodes inserted in the pulp [4,99,119,169,170] with the anticipation that mineral particles in the pulp will attain the potential of the control electrode and the measuring electrodes will receive the potential of the particles [28], or by using water that was previously subjected to electrolysis as the pulp preparation water [171].

Hintikka [119] reported that the disadvantages of potential controlling by using chemicals include chemical side-reactions which modify mineral surfaces in the pulp or bring about decomposition of flotation collectors and/or high consumption of reagents.

However, it should be mentioned that some authors consider that the chemical control of Eh is more easily implemented in plants and leads to a more unchanging electrochemical environment around the mineral particles in the flotation pulp compared to potentiostatic control that depends on the efficiency of electrode/particle contact [19,49,54].

### 6.2. pH–Eh Diagrams in Galena Flotation Separation

Xanthate is a commonly used collector in galena flotation. pH–Eh diagrams were created to explain equilibriums in a galena–ethyl xanthate system and the anodic dissolution of galena in acidic conditions nearly 30 years ago [172], and have been used in subsequent years to explain the phenomena and results obtained in galena floatation [24,67,123], as well as recently to explain the removal of the adsorbed xanthate species from the mineral's surface in copper–lead mixed bulk concentrate by sodium sulfide addition [70,83].

By using in situ FTIR-ATR measurements of ethyl xanthate adsorption on galena at pH 9.2, Leppinen et al. found that lead ethyl xanthate-like species are formed on the galena surface [145]. Nowadays, it is commonly accepted that lead dixanthate is the hydrophobic entity on the galena surface, ensuring xanthate-facilitated flotation. As a semiconductor, galena allows electron transfer in the solid phase [138,140,169]. Thus, formation of lead dixanthate is possible via reaction (29) [19,70,114,117,173,174] or via oxidation of the mineral, followed by ion exchange with the collector [19] (Equation (43)):

$$PbS + 2O_2 \rightarrow PbSO_4 \text{ and further } PbSO_4 + 2X^- \rightarrow PbX_2 + SO_4{}^{2-} \tag{43}$$

Other reactions, namely the ones described by Equations (28) and (30), are also possible [70].

The reactions described in Equations (27)–(30) and (43) are anodic and the coupled cathodic reaction under air conditions is the oxygen reduction (Equations (16) and (32)) depending on a medium pH value.

The interaction of xanthates with a galena surface starts at approx. 0 V, with thermodynamically favored chemisorption (Equation (27)) forming a monolayer of adsorbed xanthate. In the chemisorption, the sulfur entities of xanthate are equally bonded to lead atoms in the mineral surface, similarly to the bonding in lead dixanthate molecules, but in chemisorption, the metal atoms remain part of the mineral lattice [19].

Chemisorption is considered the most efficient use of the collector, due to the fact that a monolayer is formed before the formation of multilayers of bulk lead dixanthate [41]. In addition, it is reported that the solubility of chemisorbed xanthate is smaller than that of bulk lead dixanthate [146].

Some authors consider chemisorbed xanthate as a direct precursor of lead dixanthate and dixanthogen, which are produced via the following reactions [175]:

$$PbS + 2X_{ads} \rightarrow PbX_2 + S^0 \tag{44}$$

$$PbS + 4X_{ads} \rightarrow PbX_2 + X_2 + S^0 \tag{45}$$

According to Fredriksson and Holmgren, at low xanthate concentrations, formation of the adsorbed species of PbX is favored. Infrared spectra showed that, at high xanthate concentrations, lead dixanthate precipitate was formed on the surface of PbS [176].

O'Dea et al. showed that the balance between monolayer adsorption and $PbX_2$ precipitation depends on Eh and the concentration of $X^-$ ions in the solution. At low Eh (at pH = 8.7) and low xanthate concentrations ($10^{-5}$ M), the overall concentration of adsorbed xanthates is higher than at high Eh and higher xanthate concentrations ($10^{-2}$ M) [177]. The investigators found that the most beneficial conditions to produce chemically-adsorbed, continuous xanthate layers, on an unoxidized PbS surface are Eh = +130 mV, pH 8.7, and $10^{-5}$ M xanthate. They confirmed that the xanthate adsorption onto oxidized galena surfaces is unfavorable even at high concentrations of KEX.

Some authors mention that although lead dixanthate is the principal hydrophobic species produced on the galena surface, dixanthogen and chemisorbed xanthate may also be found and they might contribute to the xanthate-induced flotation of galena. Increased addition of potassium ethyl xanthate and Eh is helpful to the generation of $X_2$ [70].

Woods reports that the formation of $PbX_2$ takes place in the range from 0 to +200 mV, where the galena flotation with xanthate (pH = 8.1) is observed [19]. Generally, the stability of lead xanthates increases with an increase in the chain length. Multilayers of lead xanthate can form on a mineral surface.

The pH–Eh domain of galena flotation with xanthate, constructed on the basis of the empirical data mentioned in this paper, is shown in Figure 4. As can be seen, the domain crosses the Eh equilibrium line between free xanthate radical and dixanthogen domains.

The upper Eh edge for galena flotation is observed. The upper boundary is near to the potential at which the metal xanthate oxidizes to the metal oxide and diethyl dixanthogen and these processes are pH dependent. Dixanthogen is more hydrophobic than metal xanthates, so the oxidation to it and to a metal oxide would not inhibit flotation significantly. However, when the mineral itself is not sufficiently covered and passivated by a metal xanthate, mineral oxidation takes place, leading to the appearance of considerable amounts of hydrophilic metal oxide on the mineral surface and consequently to mineral depression [19]. At lower potentials (<−50 mV), lead xanthate is converted back to galena [8,146] and the process is the most intensive at −300 mV [19].

For example, when low Eh is achieved by $Na_2S$ addition, due to the lower insolubility of lead sulfide, compared to that of $PbX_2$, the latter is transformed back to galena:

$$PbX_{2(s)} + S^{2-} \rightarrow PbS_{(s)} + 2X^{-} \quad (46)$$

The depression effect of $S^{2-}$ is pH dependent (higher at higher pH values). Galena exhibits the highest sensitivity to sulfide iron, compared to the other sulfide minerals [19].

In addition to xanthate, other collectors are used for the floatation of sulfide minerals. Leppinen and Mielczarski [178] were probably the first who studied the interaction of galena with potassium diphenyldithiophosphate and sodium diethyldithiophosphate. They concluded that the main type of adsorption product formed by collectors on the galena surface is analogous to lead dixanthate.

Buckley and Woods were probably the first who studied the interaction of galena with a diethyldithiophosphate system using voltammetry. They found that the collector was chemisorbed at potentials below that at which lead dithiophosphate was formed, thus ensuring the surface hydrophobicity. At high potentials, lead dithiophosphate dissolved [179].

In the case when diethyldithioncarbamate (DDTC) is used as a collector, hydrophobic $PbD_2$ species are formed in the Eh range from 0 to 200 mV, according to the reaction:

$$PbS + 2D^{-} \rightarrow PbD_2 + S^0 + 2e^{-} \quad (47)$$

where D stands for DDTC.

In addition, DDTC reacts with hydrophilic $HPbO_2^-$ species formed by PbS oxidation in this pH–Eh (reaction 11), thus increasing the mineral's hydrophobicity, as described by reaction (48) [71]:

$$HPbO_2^- + 2D^- + 3H^+ \rightarrow PbD_2 + 2H_2O \quad (48)$$

Studies have also shown that $PbD_2$ is firmly adsorbed on the galena surface in the range from −900 to +600 mV.

It can be summarized that although Pourbaix pH–Eh diagrams are based on equilibrium (thermodynamically defined) Eh and pH values that differ from the non-equilibrium values in the real pulp, they still can be very useful. The joint measurement of Eh and pH and the use of the corresponding pH–Eh diagram(s) can help to define which species are present in a given flotation system and their thermodynamic stability. This is possible because most equilibriums are predominantly controlled by Eh (e.g., oxidation/reduction) and pH (e.g., hydrolysis). Knowledge on the species' stability would give information on which minerals from a given complex sulfide ore would float and which would be depressed. Even more, merging two or more diagrams of species of metals that are present in the pulp discloses more clearly the possibility of the reactions involved [180]. Further, the pH–Eh diagrams can reveal directions for the improvement of the process by changing the Eh and/or pH values of the pulp using different means (reagent selection, their amounts, optimization and selection of points of additions, gas use, an external impact changing the potential). Thus, the Eh regions (at suitable pH values) can be reached in which enough hydrophobic species of a given mineral are present on its surface to improve its recovery, while at the same time, enough hydrophilic species are present on the surface of other mineral(s), thus enhancing their depression, leading to an increase in the flotation selectivity, yield, and quality of concentrates.

A generalized scheme of the proposed complex methodology for using Eh and pH data to guide the industrial process control is given in Figure 5. For complex ores, the last three steps may be applied for each metal flotation circuit.

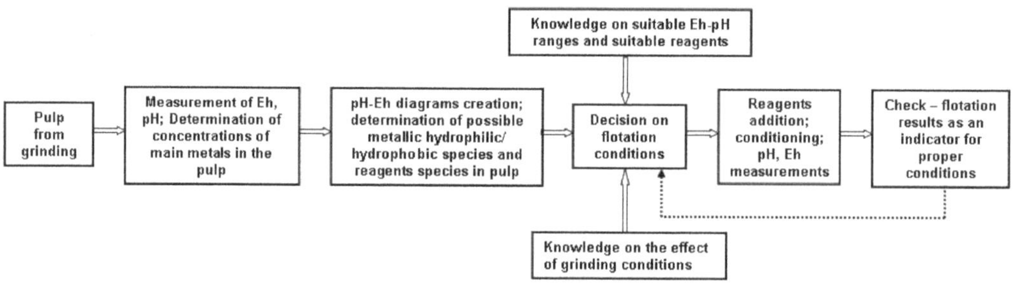


**Figure 5.** A generalized scheme of the proposed complex methodology for using Eh and pH measurements in the control of industrial processes.

## 7. Conclusions

Research on the electrochemistry of non-ferrous metal-sulfide minerals flotation has been carried out for half a century. Based on the studies described in the worldwide literature, it can be stated that a consensus is reached that: (i) sulfide minerals float only under suitable oxidation–reduction conditions; and (ii) the Eh and pH values of the flotation pulp govern the reactions occurring between the various minerals in the pulp and collectors, and thus the flotation results.

Interest in measuring the Eh value of flotation pulp and its use in flotation has not abated in the last 30–40 years, with the focus generally being in the following directions: (i) searching for suitable Eh ranges for selective separation of minerals; (ii) changing the conditions and, accordingly, the Eh by adding reagents or imposing an electrochemical impact with the aim to improve the flotation results; and (iii) explanation of the interaction of the minerals with the reagents and/or the effect of the grinding medium on the efficiency of the flotation process.

Eh is a complex indicator—it reflects the specificity of minerals in the pulp, the interactions between them, the addition of reagents, grinding media, and conditioning procedure effects, and the changing nature and concentration of the flotation gas, reagents, and minerals. Its complexity is both the advantage and the disadvantage of using it to control flotation processes. The disadvantage is expressed in the fact that it cannot be calculated for the real case when there are many different oxidation–reduction pairs in the pulp and at changing concentrations of the participating species, and that it is relatively difficult to adjust. The advantage can be seen in the opportunity that, based on measurements in a real environment and established correlations between the complex indicator Eh and the flotation yield, selectivity, and quality of concentrate, the Eh value (jointly with pH) can be used as a reliable unified indicator for flotation control.

The creation of pH–Eh diagrams that include the main species of the elements forming the minerals in a given flotation pulp and the main reagent species, requires an exact knowledge of their concentrations in the pulp. Finding reliable methods for the real-time determination of concentrations in industrial flotation systems is one of the challenges and a mandatory direction for future research.

Solving the problems with long-term stability and fouling of electrodes under the harsh conditions in the flotation pulp is another compulsory direction for further research.

The establishment of reliable "predictive" correlations between Eh and pH values measured in the pulp under real industrial conditions and flotation process parameters represents the most important issue requiring the focus of future research.

In conclusion, it can be noted that despite some known mistrust and difficulties in using the pulp Eh value to control the flotation processes, this indicator, especially considered together with the pH of the pulp, appears to be quite useful and promising, deserving further research, particularly under industrial conditions.

**Funding:** This research received no external funding.

**Conflicts of Interest:** The author declares no conflict of interest.

## References

1. Chander, S. Oxidation-reduction effects in depression of sulfide minerals—A review. *Min. Metall. Explor.* **1985**, *2*, 26–35. [CrossRef]
2. Ekmekçi, Z.; Demirel, H. Effects of galvanic interaction on collectorless flotation behaviour of chalcopyrite and pyrite. *Int. J. Miner. Process.* **1997**, *52*, 31–48. [CrossRef]
3. Wang, X.; Qin, W.; Jiao, F.; Wu, J. The influence of galvanic interaction on the dissolution and surface composition of galena and pyrite in flotation system. *Miner. Eng.* **2020**, *156*, 106525. [CrossRef]
4. Tolley, W.; Kotlyar, D.; Van Wagoner, R. Fundamental electrochemical studies of sulfide mineral flotation. *Miner. Eng.* **1996**, *9*, 603–637. [CrossRef]
5. Buswell, A.M. An Electrochemical Investigation into the Floatability of Pyrrhotite. Ph.D. Thesis, University of the Witwatersrand, Johannesburg, South Africa, 1998.
6. Janetski, N.D.; Woodburn, S.I.; Woods, R. An electrochemical investigation of pyrite flotation and depression. *Int. J. Miner. Process.* **1977**, *4*, 227–239. [CrossRef]
7. Gebhardt, J.E.; Dewsnap, N.F.; Richardson, P.E. *Electrochemical Conditioning of a Mineral Particle Bed Electrode for Flotation*; Report of Investigations 8951; Bureau of Mines Report of Investigations; United States Department of Inferior: Washington, DC, USA, 1985; pp. 6–10.
8. Buckley, A.N.; Woods, R. Chemisorption—The thermodynamically favoured process in the interaction of thiol collectors with sulphide minerals. *Int. J. Miner. Process.* **1997**, *51*, 15–26. [CrossRef]
9. Freeman, W.A.; Newell, R.K.; Quast, B. Effect of grinding media and NaHS on copper recovery at Northparkes Mines. *Miner. Eng.* **2000**, *13*, P1395–P1403. [CrossRef]
10. Van Deventer, J.S.J.; Ross, V.E.; Dunne, R.C. The effect of galvanic interaction on the behaviour of the froth phase during the flotation of a complex sulphide ore. *Miner. Eng.* **1993**, *6*, 1217–1229. [CrossRef]
11. Chimonyo, W.; Wiese, J.; Corin, K.; O'Connor, C. The use of oxidising agents for control of electrochemical potential in flotation. *Miner. Eng.* **2017**, *109*, 135–143. [CrossRef]
12. Subrahmanyam, T.V.; Forssberg, K.S.E. Grinding and flotation pulp chemistry of a low grade copper ore. *Miner. Eng.* **1995**, *8*, 913–921. [CrossRef]
13. Zhai, Q.; Liu, R.; Li, J.; Sun, W.; Hu, Y. Changing the pulp properties and surface hydrophilicity of galena and pyrite by selecting the appropriate grinding media towards their selective separation. *Minerals* **2023**, *13*, 1213. [CrossRef]
14. Cheng, X.; Iwasaki, I. Pulp potential and its implications to sulfide flotation. *Miner.Process. Extr. Metall. Rev.* **1992**, *11*, 187–210. [CrossRef]
15. Yuan, X.-M.; Pälsson, B.I.; Forssberg, K.S.E. Flotation of a complex sulphide ore Cu/Zn selectivity control by adjusting potential with different gases. *Int. I. Miner. Process.* **1996**, *46*, 155–179. [CrossRef]
16. Moslemi, H.; Shamsi, P.; Habashi, F. Pyrite and pyrrhotite open circuit potentials study: Effects on flotation. *Miner. Eng.* **2011**, *24*, 1038–1045. [CrossRef]
17. Wang, J.; Zeng, H. Recent advances in electrochemical techniques for characterizing surface properties of minerals. *Adv. Colloid. Interfac.* **2021**, *288*, 102346. [CrossRef]
18. Tafirenyika, T.P.; O'Connor, C.T.; Corin, K.C. Investigating the Influence of the Electrochemical Environment on the Flotation of Bornite and Chalcocite. *Minerals* **2022**, *12*, 1527. [CrossRef]

19. Woods, R. Electrochemical aspects of sulfide mineral flotation. In *Flotation Plant Optimisation—A Metallurgical Guide to Identifying and Solving Problems in Flotation Plants*; Greet, C.J., Ed.; The Australasian Institute of Mining and Metallurgy: Melbourne, Australia, 2010; pp. 123–136.
20. Thorstenson, D.C. *The Concept of Electron Activity and Its Relation to Redox Potentials in Aqueous Geochemical Systems*; U.S. Geological Survey, Open-File Report 84 072; US Government Printing Office: Washington, DC, USA, 1984; pp. 1–38.
21. Bockris, J.O'.M.; Reddy, A.K.N. *Modern Electrochemistry*; Plenum Press: New York, NY, USA, 1970; p. 659.
22. Stumm, W.; Morgan, J.J. *Aquatic Chemistry*; John Willey & Sons Inc.: New York, NY, USA, 1970; p. 318.
23. Wang, H.; Forssberg, K.S.E. The solution electrochemistry of sulfide-xanthate-cyanide systems in sulfide mineral flotation. *Miner. Eng.* **1996**, *9*, 527–546. [CrossRef]
24. Pålsson, B.I.; Persson, H. Redox control in a pilot flotation column. *Miner. Eng.* **1996**, *9*, 73–83. [CrossRef]
25. Rand, D.A.J.; Woods, R. Eh measurements in sulphide mineral slurries. *Int. J. Miner. Process.* **1984**, *13*, 29–42. [CrossRef]
26. Finch, A.; Labonté, G. Verification of electrodes for pulp potential measurement. *Miner. Eng.* **1989**, *2*, 557–564. [CrossRef]
27. Zhou, R.; Chander, S. Comparison of gold, platinum and sulfide ion selective electrodes as sensors for Eh measurement in sulfide solutions. *Miner. Metall. Process.* **1991**, *8*, 91–96. [CrossRef]
28. Kocabağ, D.; Güler, T. A comparative evaluation of the response of platinum and mineral electrodes in sulfide mineral pulps. *Int. J. Miner. Process.* **2008**, *87*, 51–59. [CrossRef]
29. Göktepe, F. Effect of pH on pulp potential and sulphide mineral flotation. *Turkish J. Eng. Env. Sci.* **2002**, *26*, 309–318.
30. Chander, S. A brief review of pulp potentials in sulfide flotation. *Int. J. Miner. Process.* **2003**, *72*, 141–150. [CrossRef]
31. Heimala, S.; Jounela, S.; Saari, M. Flotation control with mineral electrodes. In Proceedings of the XVIth International Mineral Processing Congress, Stockholm, Sweden, 5–10 June 1988; pp. 1713–1718.
32. Yi, G.; Macha, E.; Van Dyke, J.; Macha, R.; McKay, T.; Free, M.L. Recent progress on research of molybdenite flotation: A review. *Adv. Colloid. Interfac.* **2021**, *295*, 102466. [CrossRef]
33. Pourbaix, M. *Atlas of Electrochemical Equilibrium in Aqueous Solutions*, 2nd ed.; National Association of Corrosion Engineers: Houston, TX, USA; Cebelcor, Brussels, 1974; pp. 29–55.
34. Foroutan, A.; Abadi, M.A.Z.H.; Kianinia, Y.; Ghadiri, M. Critical importance of pH and collector type on the flotation of sphalerite and galena from a low-grade lead–zinc ore. *Sci. Rep.* **2021**, *11*, 3103. [CrossRef]
35. Joseph-Soly, S.; Quast, K.; Connor, J.N. Effects of Eh and pH on the oleate flotation of iron oxides. *Miner. Eng.* **2015**, *83*, 97–104. [CrossRef]
36. Tao, D.; Wang, Y.; Li, L. An electrochemical study of surface oxidation and cCollectorless flotation of pyrite. *Int. J. Electrochem. Sci.* **2018**, *13*, 5971–5982. [CrossRef]
37. Park, I.; Hong, S.; Jeon, S.; Ito, M.; Hiroyoshi, N.A. Review of recent advances in depression techniques for flotation separation of Cu–Mo sulfides in porphyry copper deposits. *Metals* **2020**, *10*, 1269. [CrossRef]
38. Grano, S. Chemical measurements during plant surveys and their interpretation. In *Flotation Plant Optimisation—A Metallurgical Guide to Identifying and Solving Problems in Flotation Plants*; Greet, C.J., Ed.; The Australasian Institute of Mining and Metallurgy: Melbourne, Australia, 2010; pp. 107–121.
39. Plaksin, I.N.; Shafeev, R.S.; Chanturiya, V.A. The relationship between the energy structure of mineral crystals and their flotation properties. In *Proceedings of the VIII International Mineral Processing Congress*; Mekhanobr: St. Petersburg, Russia, 1969; Volume 2, pp. 235–245. (In Russian)
40. Ralston, J. Eh and its consequences in sulphide mineral flotation. *Miner. Eng.* **1991**, *4*, 859–878. [CrossRef]
41. Woods, R. Electrochemical potential controlling flotation. *Int. J. Miner. Process.* **2003**, *72*, 151–162. [CrossRef]
42. Chanturiya, V.A.; Krasavtseva, E.A.; Makarov, D.V. Electrochemistry of sulfides: Process and environmental aspects. *Sustainability* **2022**, *14*, 11285. [CrossRef]
43. Rao, S.R.; Labonté, G.; Finch, J.A. Electrochemistry in the plant. In *Innovations in Flotation Technology. NATO ASI Series*; Mavros, P., Matis, K.A., Eds.; Springer: Dordrecht, The Netherlands, 1992; Volume 208, pp. 57–100. [CrossRef]
44. Gu, G.; Hu, Y.; Wang, H.; Qui, G.; Wang, D. Original potential flotation of galena and its industrial application. *J. Cent. South Univ. Technol.* **2002**, *9*, 91–94. [CrossRef]
45. Orwe, D.; Grano, S.R.; Lauder, D.W. Increasing fine copper recovery at the Ok Tedi concentrator, Papua New Guinea. *Miner. Eng.* **1998**, *11*, 171–187. [CrossRef]
46. Smith, L.K.; Davey, K.J.; Bruckard, W.J. The use of pulp potential control to separate copper and arsenic—An overview based on selected case studies. In Proceedings of the XXVI International Mineral Processing Congress (IMPC), New Delhi, India, 24–28 September 2012; Paper No. 321. pp. 05057–05067.
47. Cruz, C.; Botero, Y.L.; Jeldres, R.I.; Uribe, L.; Cisternas, L.A. Current status of the effect of seawater ions on copper flotation: Difficulties, opportunities, and industrial experience. *Miner. Process. Extr. Metall. Rev.* **2022**, *43*, 545–563. [CrossRef]
48. Bahrami, A.; Kashani, R.H.; Kazemi, F.; Ghorbani, Y. Oxidation-reduction effects in the flotation of copper sulfide minerals and molybdenite—A proof of concept at industrial scale. *Miner. Eng.* **2022**, *180*, 107505. [CrossRef]
49. Hayes, R.A.; Ralston, J. The collectorless flotation and separation of sulphide minerals by Eh control. *Int. J. Miner. Process.* **1988**, *23*, 55–84. [CrossRef]
50. Lekki, J.; Drzymala, J. Flotometric analysis of the collectorless flotation of sulphidematerials. *Colloids Surf.* **1990**, *44*, 179–190. [CrossRef]

51. Zhang, W.; Jin, X.; Feng, Z.; Zheng, R.; Cao, J.; Chen, J.; Sun, W.; Xu, S.; Gao, Z. Collectorless flotation separation of molybdenite from complex sulfide minerals employing a bi-carbonyl depressant. *Sep. Purif. Technol.* **2023**, *322*, 124207. [CrossRef]
52. Fornasiero, D.; Ralston, J. Effect of surface oxide/hydroxide products on the collectorless flotation of copper-activated sphalerite. *Int. J. Miner. Process.* **2006**, *78*, 231–237. [CrossRef]
53. Abramov, A.A.; Avdohin, V.M. *Oxidation of Sulfide Minerals in Benefication Processes*; Gordon and Breach Science Publishers: Amsterdam, The Netherlands, 1997; pp. 191–272.
54. Aghazadeh, S.; Mousavinezhad, S.K.; Gharabaghi, M. Chemical and colloidal aspects of collectorless flotation behavior of sulfide and non-sulfide minerals. *Adv. Colloid. Interfac.* **2015**, *225*, 203–217. [CrossRef] [PubMed]
55. Luttrell, G.; Yoon, R. Surface studies of the collectorless flotation of chalcopyrite. *Colloids Surf.* **1984**, *12*, 239–254. [CrossRef]
56. Ceylan, H.; Hicyilmaz, C.; Guler, T. Collectorless flotation of lead and zinc sulphide from Dereköy oredeposits. *Physicochem. Probl. Miner. Process.* **2002**, *36*, 197–208.
57. Hu, Y.; Sun, W.; Wang, D. *Electrochemistry of Flotation of Sulphide Minerals*; Tsinghua University Press: Beijing, China; Springer: Berlin/Heidelberg, Germany, 2009. [CrossRef]
58. Fairthorne, G.; Fornasiero, D.; Ralston, J. Effect of oxidation on the collectorless flotation of chalcopyrite. *Int. J. Miner. Process.* **1997**, *49*, 31–48. [CrossRef]
59. Heyes, G.W.; Trahar, W.J. The natural flotatability of chalcopyrite. *Int. J. Miner. Process.* **1977**, *4*, 317–344. [CrossRef]
60. Gardner, J.R.; Woods, R. An electrochemical investigation of the natural floatability of chalcopyrite. *Int. J. Miner. Process.* **1979**, *6*, 1–16. [CrossRef]
61. Buckley, A.N.; Woods, R. An X-ray photoelectron spectroscopic study of the oxidation of chalcopyrite. *Aust. J. Chem.* **1984**, *37*, 2403–2413. [CrossRef]
62. Buckley, A.N.; Hamilton, I.C.; Woods, R. Investigation of the surface oxidation of sulphide minerals by linear potential sweep voltammetry and X-ray photoelectron spectroscopy. In *Flotation of Sulphide Minerals*; Forsberg, K.S.E., Ed.; Elsevier Science: Amsterdam, The Netherlands, 1985; pp. 41–60.
63. Zheng, X.; Manton, P. A potential application of collectorless flotation in a copper/gold operation. *Miner. Eng.* **2010**, *23*, 895–902. [CrossRef]
64. Razmjouee, S.M.A.; Koleini, S.M.J. Collectorless flotation of chalcocite by controlling redox potential. *J. Min. Environ.* **2012**, *3*, 99–102. [CrossRef]
65. Walker, G.; Stout, J.; Richardson, P. Electrochemical flotation of sulfides: Reactions of chalcocite in aqueous solution. *Int. J. Miner. Process.* **1984**, *12*, 55–72. [CrossRef]
66. Kocabag, D.; Kelsall, G.H.; Shergold, H.L. Natural oleophilicity/hydrophobicity of sulphide minerals, I. Galena. *Int. J. Miner. Process.* **1990**, *29*, 195–210. [CrossRef]
67. Ralston, J. The chemistry of galena flotation: Principles & practice. *Miner. Eng.* **1994**, *7*, 715–735. [CrossRef]
68. Chernyshova, I.V.; Andreev, S.I. Spectroscopic study of galena surface oxidation in aqueous solutions I. Identification of surface species by XPS and ATR/FTIR spectroscopy. *Appl. Surf. Sci.* **1997**, *108*, 225–236. [CrossRef]
69. Kartio, I.; Wittstock, G.; Laajalehto, K.; Hirsch, D.; Simola, J.; Laiho, T.; Szargan, R.; Suoninen, E. Detection of elemental sulphur on galena oxidized in acidic solution. *Miner. Process.* **1997**, *51*, 293–301. [CrossRef]
70. Hu, Y.; Wu, M.; Liu, R.; Sun, W. A review on the electrochemistry of galena flotation. *Miner. Eng.* **2020**, *150*, 106272. [CrossRef]
71. Gu, G.; Hu, Y.; Qui, G.; Wang, H.; Wang, D. Potential control flotation of galena in strong alkaline media. *J. Cent. South Univ. T.* **2002**, *9*, 16–20. [CrossRef]
72. Guo, H.; Yen, W.-T. Pulp potential and floatability of chalcopyrite. *Miner. Eng.* **2003**, *16*, 247–256. [CrossRef]
73. Khmeleva, T.; Skinner, W.; Beattie, D.A. Depressing mechanisms of sodium bisulphite in the collectorless flotation of copper-activated sphalerite. *Int. J. Miner. Process.* **2005**, *76*, 43–53. [CrossRef]
74. Chandra, A.P.; Gerson, A.R. A review of the fundamental studies of the copper activation mechanisms for selective flotation of the sulfide minerals, sphalerite and pyrite. *Adv. Colloid. Interfac.* **2009**, *145*, 97–110. [CrossRef]
75. Tayebi-Khorami, M.; Manlapig, E.; Forbes, E.; Edraki, M.; Bradshaw, D. Effect of surface oxidation on the flotation response of enargite in a complex ore system. *Miner. Eng.* **2018**, *119*, 149–155. [CrossRef]
76. Lotter, N.O.; Bradshaw, D.J.; Barnes, A.R. Classification of the major copper sulphides into semiconductor types, and associated flotation characteristics. *Miner. Eng.* **2016**, *96–97*, 177–184. [CrossRef]
77. Grano, S.; Ralston, J.; Smart, R.S.C. Influence of electrochemical environment on the flotation behavior of Mt. Isa copper and lead-zinc ore. *Int. J. Miner. Process.* **1990**, *30*, 69–97. [CrossRef]
78. Heyes, G.W.; Trahar, W.J. Oxidation-reduction effects in the flotation of chalcocite and cuprite. *Int. J. Miner. Process.* **1979**, *6*, 229–252. [CrossRef]
79. Chander, D.W.; Fuerstenau, D.W. Electrochemical flotation separation of chalcocite from molybdenite. *Int. J. Miner. Process.* **1983**, *10*, 89–94. [CrossRef]
80. Richardson, P.E.; Walker, G.W. The flotation of chalcocite, bornite, chalcopyrite and pyrite in electrochemical flotation cell. In *Proceedings of the 15th International Mineral Processing Congress, Cannes, France, 2–9 June 1985*; Edition GEDIM: St Etienne, France, 1985; Volume 2, pp. 198–210.
81. He, S.; Skinner, W.; Fornasiero, D. Effect of oxidation potential and zinc sulphate on the separation of chalcopyrite from pyrite. *Int. J. Miner. Process.* **2006**, *80*, 169–176. [CrossRef]

82. Kalegowda, Y.; Chan, Y.-L.; Wei, D.-H.; Harmer, S.L. X-PEEM, XPS and ToF-SIMS characterisation of xanthate induced chalcopyrite flotation: Effect of pulp potential. *Surf. Sci.* **2015**, *635*, 70–77. [CrossRef]
83. Liu, R.; Lu, H.; Xu, Z.; Wang, C.; Sun, W.; Wu, M.; Dong, Y.; Bai, L. New insights into the reagent-removal mechanism of sodium sulfide in chalcopyrite and galena bulk flotation: A combined experimental and computational study. *J. Mater. Res. Technol.* **2020**, *9*, 5352–5363. [CrossRef]
84. Gebhardt, J.E.; Richardson, P.E. Differential flotation of a chalcocite-pyrite particle bed by electrochemical control. *Min. Metall. Explor.* **1987**, *4*, 140–145. [CrossRef]
85. Göktepe, F.; Williams, K.P. Electrochemical effects in flotation of a Turkish complex sulphide ore. *Miner. Eng.* **1995**, *8*, 1035–1048. [CrossRef]
86. Sepalveda-Suarez, C.; Castro, S.H. The effect of redox potential on the selective flotation of molybdenite by using Anamol-D as copper depressant. In *Proceedings International Symposium on Electrochemistry in Mineral and Metal Processing*; Woods, R., Doyle, F.M., Richardson, P., Eds.; The Electrochemical Society: Pennington, NJ, USA, 1996; Volume 96-6, pp. 98–107.
87. Leppinen, J.O.; Hintikka, V.V.; Kalapudas, R.P. Effect of electrochemical control on selective flotation of copper and zinc from complex ores. *Miner. Eng.* **1998**, *11*, 39–51. [CrossRef]
88. Chimonyo, W.; Corin, K.C.; Wiese, J.G.; O'Connor, C.T. Redox potential control during flotation of a sulphide mineral ore. *Miner. Eng.* **2017**, *110*, 57–64. [CrossRef]
89. Sheni, N.; Corin, K.; Wiese, J. Considering the effect of pulp chemistry during flotation on froth stability. *Miner. Eng.* **2018**, *116*, 15–23. [CrossRef]
90. Mehrabani, J.V.; Pourghahramani, P.; Asqarian, H.; Bagherian, A. Effects of pH and pulp potential on selective separation of molybdenite from the Sungun mine Cu-Mo concentrate. *Int. J. Min. Geo-Eng.* **2017**, *51*, 147–150.
91. Filippou, D.; St-Germain, P.; Grammatikopoulos, T. Recovery of metal values from copper–arsenic minerals and other related resources. *Miner. Process. Extr. Metall. Rev.* **2007**, *28*, 247–298. [CrossRef]
92. Kantar, C. Solution and flotation chemistry of enargite. *Colloid. Surf. A* **2002**, *210*, 23–31. [CrossRef]
93. Guo, H.; Yen, W.-T. Selective flotation of enargite from chalcopyrite by electrochemical control. *Miner. Eng.* **2005**, *18*, 605–612. [CrossRef]
94. Senior, G.D.; Guy, P.J.; Bruckard, W.J. The selective flotation of enargite from other copper minerals—A single mineral study in relation to beneficiation of the Tampakan deposit in the Philippines. *Int. J. Miner. Process.* **2006**, *81*, 15–26. [CrossRef]
95. Yepsen, R.; Gutierrez, L. Effect of Eh and pH on the flotation of enargite using seawater. *Miner. Eng.* **2020**, *159*, 106612. [CrossRef]
96. Huch, R.O. Method for Achieving Enhanced Copper-Containing Mineral Concentrate Grade by Oxidation and Flotation. US Patent 5295585, 22 March 1994.
97. Smith, L.K.; Bruckard, W.J. The separation of arsenic from copper in the Northparkes copper-gold ore using controlled-potential flotation. *Int. J. Miner. Process* **2007**, *84*, 15–24. [CrossRef]
98. Long, G.; Peng, Y.; Bradshaw, D. Flotation separation of copper sulphides from arsenic minerals at Rosebery copper concentrator. *Miner. Eng.* **2014**, *66–68*, 207–214. [CrossRef]
99. Tayebi-Khorami, M.; Manlapig, E.; Forbes, E.; Bradshaw, D.; Edraki, M. Selective flotation of enargite from copper sulphides in Tampakan deposit. *Miner. Eng.* **2017**, *112*, 1–10. [CrossRef]
100. Yelloji Rao, M.K.; Natarajan, K.A. Effect of electrochemical interactions among sulfide minerals and grinding medium on chalcopyrite flotation. *Min. Metall. Explor.* **1989**, *6*, 146–151. [CrossRef]
101. Peng, Y.; Grano, S.; Fornasiero, D.; Ralston, J. Control of grinding conditions in the flotation of chalcopyrite and its separation from pyrite. *Int. J. Miner. Process.* **2003**, *69*, 87–100. [CrossRef]
102. Peng, Y.; Wang, B.; Gerson, A. The effect of electrochemical potential on the activation of pyrite by copper and lead ions during grinding. *Int. J. Miner. Process.* **2012**, *102–103*, 141–149. [CrossRef]
103. Corin, K.C.; Mishra, J.; O'Connor, C.T. Investigating the role of pulp chemistry on the floatability of a Cu–Ni sulfide ore. *Int. J. Miner. Process.* **2013**, *120*, 8–14. [CrossRef]
104. Nooshabadi, A.J.; Larsson, A.-C.; Kota, H.R. Formation of hydrogen peroxide by pyrite and its influence on flotation. *Miner. Eng.* **2013**, *49*, 128–134. [CrossRef]
105. Nooshabadi, A.J.; Kota, H.R. Formation of hydrogen peroxide by chalcopyrite and its influence on flotation. *Miner. Metall. Process.* **2013**, *30*, 212–219. [CrossRef]
106. Ikumapayi, F.; Sis, H.; Johansson, B.; Hanumantha, R.K. Recycling process water in sulfide flotation, Part B: Effect of $H_2O_2$ and process water components on sphalerite flotation from complex sulfide. *J. Miner. Metall. Process.* **2012**, *29*, 192–198. [CrossRef]
107. Jacques, S.; Greet, C.J.; Bastin, D. Oxidative weathering of a copper sulphide ore and its influence on pulp chemistry and flotation. *Miner. Eng.* **2016**, *99*, 52–59. [CrossRef]
108. Peng, H.; Wu, D.; Abdelmonem, M. Flotation performances and surface properties of chalcopyrite with xanthate collector added before and after grinding. *Results Phys.* **2017**, *7*, 3567–3573. [CrossRef]
109. Liu, J.; Han, L.; Corin, K.C.; O'Connor, C.T. A study of the effect of grinding environment on the flotation of two copper sulphide ores. *Miner. Eng.* **2018**, *122*, 339–345. [CrossRef]
110. Corin, K.C.; Song, Z.G.; Wiese, J.G.; O'Connor, C.T. Effect of using different grinding media on the flotation of a base metal sulphide ore. *Miner. Eng.* **2018**, *126*, 24–27. [CrossRef]

111. Mu, Y.; Peng, Y. Maximise pyrite depression in copper ore flotation using high salinity water. *Miner. Eng.* **2023**, *196*, 108060. [CrossRef]
112. Mu, Y.; Peng, Y. The effect of saline water on copper activation of pyrite in chalcopyrite flotation. *Miner. Eng.* **2019**, *131*, 336–341. [CrossRef]
113. Can, N.M.; Başaran, Ç. Effects of different grinding media and milling conditions on the flotation performance of a copper-pyrite ore. *Minerals* **2023**, *13*, 85. [CrossRef]
114. Woods, R. The anodic oxidation of ethylxanthate on galena electrodes. *Aust. J. Chem.* **1972**, *25*, 2329–2335. [CrossRef]
115. Gardner, J.R.; Woods, R. The use of a particulate bed electrode for the electrochemical investigation of metal and sulphide flotation. *Aust. J. Chem.* **1973**, *26*, 1635–1644. [CrossRef]
116. Johnson, N.W.; Jowett, A.; Heyes, G.W. Oxidation-reduction effects in galena flotation: Observations on Pb-Zn-Fe sulphides separation. *T. I. Min. Metall.* **1982**, *91*, C32–C37.
117. Guy, P.J.; Trahar, W.J. The influence of grinding and flotation environments on the laboratory batch flotation of galena. *Int. J. Miner. Process.* **1984**, *12*, 15–38. [CrossRef]
118. Herrera-Urbina, R.; Sotillo, F.J.; Fuerstenau, D.W. Effect of sodium sulfide additions on the pulp potential and amyl xanthate flotation of cerussite and galena. *Int. J. Miner. Process.* **1999**, *55*, 157–170. [CrossRef]
119. Hintikka, V.V.; Leppinen, J.O. Potential control in the flotation of sulphide minerals and precious metals. *Miner. Eng.* **1995**, *8*, 1151–1158. [CrossRef]
120. Uribe-Salas, A.; Martínez-Cavazos, T.E.; Nava-Alonso, F.C.; Méndez-Nonell, J.; Lara-Valenzuela, C. Metallurgical improvement of a lead/copper flotation stage by pulp potential control. *Int. J. Miner. Process.* **2000**, *59*, 69–83. [CrossRef]
121. Uribe-Salas, A.; Nava-Alonso, F.; Martínez-Cavazos, T.E.; Pérez-Garibay, R. The effect of pulp potential on the metallurgical performance of a lead/copper bulk flotation stage. In *Developments in Mineral Processing, 13. Proc XXI International Mineral Processing Congress*; Massacci, P., Ed.; Elsevier: Amsterdam, The Netherlands, 2000; Volume B, pp. B8b96–B8b102.
122. Gu, G.-h.; Hu, Y.H.; Qui, G.Z.; Wang, D. Potential control flotation of galena and its industrial application. In *Electrochemistry in Mineral and Metal Processing V: Proceedings of the International Symposium*; Woods, R., Doyle, F.M., Eds.; The Electrochemical Society: Pennington, NJ, USA, 2000; pp. 131–149.
123. Qing, W.-q.; He, M.-f.; Chen, Y.-p. Improvement of flotation behavior of Mengzi lead-silver-zinc ore by pulp potential control flotation. *T. Nonferr. Metal. Soc.* **2008**, *18*, 949–954. [CrossRef]
124. Cullinan, V.J.; Grano, S.R.; Greet, C.J.; Johnson, N.W.; Ralston, J. Investigating fine galena recovery problems in the lead circuit of Mount Isa Mines lead/zinc concentrator, Part 1: Grinding Media Effects. *Miner. Eng.* **1999**, *12*, 147–163. [CrossRef]
125. Peng, Y.; Grano, S. Inferring the distribution of iron oxidation species on mineral surfaces during grinding of base metal sulphides. *Electrochim. Acta* **2010**, *55*, 5470–5477. [CrossRef]
126. Nooshabadi, A.J.; Rao, K.H. Formation of hydrogen peroxide by galena and its influence on flotation. *Adv. Powder Technol.* **2014**, *25*, 832–839. [CrossRef]
127. Nooshabadi, A.J.; Rao, K.H. Formation of hydrogen peroxide by sulphide minerals. *Hydrometallurgy* **2014**, *141*, 82–88. [CrossRef]
128. Wang, D.; Jiao, F.; Qin, W.; Wang, X. Effect of surface oxidation on the flotation separation of chalcopyrite and galena using sodium humate as depressant. *Sep. Sci. Technol.* **2017**, *53*, 961–972. [CrossRef]
129. Gu, G.-h.; Hu, Y.; Qui, G.; Liu, R. Electrochemistry of sphalerite activated by $Cu^{2+}$ ion. *Trans. Non-Ferr. Met. Soc. China* **2000**, *10*, 64–67.
130. Greet, C. The Eureka Mine—An example of how to identify and solve problems in a flotation plant. In *Flotation Plant Optimisation—A Metallurgical Guide to Identifying and Solving Problems in Flotation Plants*; Greet, C.J., Ed.; The Australasian Institute of Mining and Metallurgy: Melbourne, Australia, 2010; pp. 1–34.
131. Nooshabadi, A.J.; Rao, K.H. Formation of hydrogen peroxide by sphalerite. *Int. J. Miner. Process.* **2013**, *125*, 78–85. [CrossRef]
132. Nooshabadi, A.J.; Rao, K.H. Complex sulphide ore flotation: Effect of depressants addition during grinding on $H_2O_2$ formation and its influence on flotation. *Int. J. Miner. Process.* **2016**, *157*, 89–97. [CrossRef]
133. Long, T.; Chen, Y.; Shi, J.; Chen, W.; Zhu, Y.; Zhang, C.; Bu, X. Effect of grinding media on the flotation of copper-activated marmatite. *Physicochem. Probl. Miner. Process.* **2020**, *56*, 229–237. [CrossRef]
134. O'Connor, C. Review of important developments since the 1st IMPC in 1952 in the understanding of the effects of chemical factors on flotation. *Miner. Eng.* **2021**, *170*, 106960. [CrossRef]
135. Plaksin, I.N.; Shafeev, R.S. On the effects of the electrochemical potential on the distribution of xanthate across the surface of sulfides. *Dokl. Akad. Nauk. USSR* **1958**, *118*, 546–548. (In Russian)
136. Plaksin, I.N. Interaction of minerals with gases and reagents in flotation. *Min. Eng.* **1959**, *214*, 319–324.
137. Lotter, N.O.; Bradshaw, D.J. The formulation and use of mixed collector systems in sulphide flotation. *Miner. Eng.* **2010**, *23*, 945–951. [CrossRef]
138. Plaksin, I.N.; Shafeev, R.S. Effects of some semiconductor properties of the surface on the interaction of xanthate with galena. *Dokl. Akad. Nauk. USSR* **1960**, *132*, 399–401. (In Russian)
139. Bard, A.J.; Faulkner, L.R. *Electrochemical Methods: Fundamentals and Applications*, 2nd ed.; John Wiley & Sons: New York, NY, USA, 2001; pp. 534–575.
140. Panayotov, V.; Habashi, F.; Panayotova, M. *Chemical Technology of Ore Processing*; Publishing House"St. Ivan Rilski"of the University of Mining and Geology "St. Ivan Rilski": Sofia, Bulgaria, 2014; pp. 14–62.

141. Woods, R. The oxidation of ethyl xanthate on platinum, gold, copper, and galena electrodes: Relation to the mechanism of mineral flotation. *J. Phys. Chem.* **1971**, *75*, 354–362. [CrossRef]
142. Kowal, A.; Pomianowski, A. Cyclic voltammetry of ethyl xanthate on a natural copper sulphide electrode. *J. Electroanal. Chem.* **1973**, *46*, 411–420. [CrossRef]
143. Chander, S.; Fuerstenau, D.W. The effect of potassium diethyldithiophosphate on the electrochemical properties of platinum, copper and copper sulfide in aqueous solutions. *J. Electroanal. Chem.* **1974**, *56*, 217–247. [CrossRef]
144. Woods, R. Electrochemistry of sulfide flotation. In *Principles of Mineral Flotation: The Wark Symposium*; Jones, M.H., Woodcock, J.T., Eds.; Australasian Institute of Mining and Metallurgy: Parkville-Melbourne, VIC, Australia, 1984; Volume 40, pp. 91–115.
145. Leppinen, J.O.; Basilio, C.I.; Yoon, R.H. In-situ FTIR study of ethyl xanthate adsorption on sulfide minerals under conditions of controlled potential. *Int. J. Miner. Process.* **1989**, *26*, 259–274. [CrossRef]
146. Buckley, A.N.; Woods, R. Xanthate chemisorption on lead sulfide. *Colloid. Surface.* **1994**, *89*, 71–76. [CrossRef]
147. Rao, S.R. *Surface Chemistry of Froth Flotation. Fundamentals, Reagents and Mechanisms*; Springer: Berlin/Heidelberg, Germany, 2003; pp. 314–325.
148. Zanin, M.; Lambert, H.; du Plessis, C.A. Lime use and functionality in sulphide mineral flotation: A review. *Miner. Eng.* **2019**, *143*, 105922. [CrossRef]
149. Mielczarski, J.A.; Mielczarski, E.; Cases, J.M. Influence of chain length on adsorption of xanthates on chalcopyrite. *Int. J. Miner. Process.* **1998**, *52*, 215–231. [CrossRef]
150. Guo, B.; Peng, Y.; Espinosa-Gomez, R. Cyanide chemistry and its effect on mineral flotation. *Miner. Eng.* **2014**, *66–68*, 25–32. [CrossRef]
151. Xu, S.; Zanin, M.; Skinner, W.; Brito-Abreu, S. Influence of grinding conditions on the pulp chemistry and flotation of oxidised pyrite. *Miner. Eng.* **2022**, *177*, 107385. [CrossRef]
152. Winter, G.; Woods, R. The relation of collector redox potential to flotation efficiency: Monothiocarbonates. *Separat. Sci.* **1973**, *8*, 261–267. [CrossRef]
153. Wang, X.H.; Forssberg, K.S.E. Mechanisms of pyrite flotation with xanthates. *Int. J. Miner. Process.* **1991**, *33*, 275–290. [CrossRef]
154. Dzinza, L. An Investigation into the Effect of Potential Modifiers on the Flotation of a Copper Sulphide Ore. Master Thesis, University of Cape Town, Cape Town, South Africa, 2018.
155. Huai, Y.; Peng, Y. The formation of iron sulphide on oxidised pyrite during sulphidisation and subsequent interactions with xanthate. *Miner. Eng.* **2020**, *157*, 106564. [CrossRef]
156. He, S.; Fornasiero, D.; Skinner, W. Correlation between copper-activated pyrite flotation and surface species: Effect of pulp oxidation potential. *Miner. Eng.* **2005**, *18*, 1208–1213. [CrossRef]
157. Laajalehto, K.; Leppinen, J.; Kartioa, I.; Laiho, T. XPS and FTIR study of the influence of electrode potential on activation of pyrite by copper or lead. *Colloid. Surface. A* **1999**, *154*, 193–199. [CrossRef]
158. Fuerstenau, D.W. Thermodynamic of sulphide mineral flotation system. In *The Physical Chemistry of Mineral-reagent Interactions in Sulfide Flotation, Proceedings of Symposium, College Park, MD, USA, 6–7 April 1978*; Department of the Interior, Bureau of Mines: Washington, DC, USA, 1979; pp. 35–62.
159. Woods, R.; Young, C.A.; Yoon, R.H. Ethyl xanthate chemisorption isotherms and Eh-pH diagrams for the copper/water/xanthate and chalcocite/water/xanthate systems. *Int. J. Miner. Process.* **1990**, *30*, 17–33. [CrossRef]
160. Fuerstenau, M.C.; Chander, S.; Woods, R. Sulfide Mineral Flotation. In *Froth Flotation: A Century of Innovation*; Fuerstenau, M.C., Jameson, G.J., Yoon, R.H., Eds.; Society for Mining, Metallurgy, and Exploration: Littleton, CO, USA, 2007; pp. 425–464.
161. Woods, R.; Yoon, R.H.; Young, C.A. Eh -pH diagrams for stable and metastable phases in the copper-sulfur-water system. *Int. J. Miner. Process.* **1987**, *20*, 109–120. [CrossRef]
162. López Valdivieso, A.; Sánchez López, A.A.; Ojeda Escamilla, C.; Fuerstenau, M.C. Flotation and depression control, of arsenopyrite through pH and pulp redox potential using xanthate as collector. *Int. J. Miner. Proc.* **2006**, *81*, 27–34. [CrossRef]
163. Qiu, X.; Huang, Z.; Cao, F.; Sun, D.; Wang, P.; Chen, C. Flotation separation of chalcopyrite from pyrite using a novel O-n-butyl-N-isobutyl thionocarbamate as the selective collector. *Colloid Surf. A* **2023**, *661*, 130890. [CrossRef]
164. Huang, Z.; Wang, J.; Sun, W.; Hu, Y.; Cao, J.; Gao, Z. Selective flotation of ohalcopyrite from pyrite using diphosphonic acid as oollector. *Miner. Eng.* **2019**, *140*, 105890. [CrossRef]
165. Hoseinian, F.S.; Rezai, B.; Kowsari, E.; Chinnappan, A.; Ramakrishna, S. Synthesis and characterization of a novel nanocollector for the removal of nickel ions from synthetic wastewater using ion flotation. *Sep. Purif. Technol.* **2020**, *240*, 116639. [CrossRef]
166. Castellón, C.I.; Toro, N.; Gálvez, E.; Robles, P.; Leiva, W.H.; Jeldres, R.I. Froth flotation of chalcopyrite/pyrite ore: A critical review. *Materials* **2022**, *15*, 6536. [CrossRef]
167. Mu, Y.; Peng, Y.; Lauten, R.A. The depression of pyrite in selective flotation by different reagent systems—A Literature review. *Miner. Eng.* **2016**, *96–97*, 143–156. [CrossRef]
168. Khoso, S.A.; Lyu, F.; Meng, X.; Hu, Y.; Sun, W. Selective separation of chalcopyrite and pyrite with a novel and non-hazardous depressant reagent scheme. *Chem. Eng. Sci.* **2019**, *209*, 115204. [CrossRef]
169. Chanturiya, V.A.; Lunin, V.D. *Elektrokhimicheskie Metody Intensifikatsii Protsessa Flotatsii (Electrochemical Methods of Flotation Process Intensification)*; Nauka: Moscow, Russia, 1983; pp. 21–136.

170. Panayotov, V.; Kovachev, K.; Önal, G.; Dovan, J.; Diner, H.; Bulut, G.; Panayotova, M.; Ninova, V. Technology for copper–zinc flotation by electrochemical treatment. In *Developments in Mineral Processing, 13. Proc XXI International Mineral Processing Congress*; Massacci, P., Ed.; Elsevier: Amsterdam, The Netherlands, 2000; Volume B, pp. B8b103–B8b108.
171. Chanturiya, E.L.; Chanturiya, V.A.; Zhuravleva, E.S. Prospects of application of water preparation of electrochemical technology in copper-zinc ores flotation. *Tsvetnye Met.* **2016**, *1*, 13–19. [CrossRef]
172. Pritzker, M.D.; Yoon, R.H. Thermodynamic calculations on sulfide flotation systems: I. Galena-ethyl xanthate system in the absence of metastable species reagents explanation. *Int. J. Miner. Process.* **1984**, *12*, 95–125. [CrossRef]
173. Leppinen, J.O.; Rastas, J.K. The interaction between ethyl xanthate ion and lead sulfide surface. *Colloids Surf.* **1986**, *20*, 221–237. [CrossRef]
174. Mhonde, N.; Johansson, L.-S.; Corin, K.; Schreithofer, N. The effect of sodium isobutyl xanthate on galena and chalcopyrite flotation in the presence of dithionite ions. *Miner. Eng.* **2021**, *169*, 106985. [CrossRef]
175. Chernyshova, I.V. In Situ FTIR-spectroelectrochemical study of the anodic processes on a galena (PbS) electrode under open-air conditions in the absence and presence of n-butyl xanthate. *Langmuir ACS J. Surf. Colloids* **2002**, *18*, 6962–6968. [CrossRef]
176. Fredriksson, A.; Holmgren, A. An in situ ATR-FTIR investigation of adsorption and orientation of heptyl xanthate at the lead sulphide/aqueous solution interface. *Miner. Eng.* **2008**, *21*, 1000–1004. [CrossRef]
177. O'Dea, A.R.; Prince, K.E.; Smart, R.S.C.; Gerson, A.R. Secondary ion mass spectrometry investigation of the interaction of xanthate with galena. *Int. J. Miner. Process.* **2001**, *61*, 121–143. [CrossRef]
178. Leppinen, J.; Mielczarski, J. Spectroscopic study of the adsorption of thiol collectors on lead sulphide in the presence of sodium sulphide. *Int. J. Miner. Process.* **1986**, *18*, 3–20. [CrossRef]
179. Buckley, A.N.; Woods, R. An electrochemical and electron spectroscopic investigation of the galena/diethyl dithiophosphate system. *Colloids Surf.* **1991**, *59*, 307–319. [CrossRef]
180. Huang, H.-H. The Eh-pH diagram and its advances. *Metals* **2016**, *6*, 23. [CrossRef]

minerals

MDPI

*Article*

# Bioleaching of Chalcopyrite by a New Strain *Leptospirillum ferrodiazotrophum* Ksh-L Isolated from a Dump-Bioleaching System of Kashen Copper-Molybdenum Mine

**Anna Khachatryan [1], Narine Vardanyan [1], Sabine Willscher [2], Garegin Sevoyan [3], Ruiyong Zhang [4,*] and Arevik Vardanyan [1,*]**

1 Laboratory of Metals Bioleaching, Institute of Microbiology, SPC "Armbiotechnology" NAS of Armenia, 14 Gyurjyan str., Yerevan 0056, Armenia; anna.khachatryan@asnet.am (A.K.); nvard@sci.am (N.V.)
2 Faculty of Natural Sciences I, Martin-Luther-Universität Halle-Wittenberg, Hoher Weg 8, 06120 Halle, Germany; sabine.willscher@gmail.com
3 Department of Ecology, Faculty of Biology, Yerevan State University, 1 Alex Manoogian str., Yerevan 0025, Armenia; armbiosevoyan@yahoo.com
4 Key Laboratory of Marine Environmental Corrosion and Biofouling, Institute of Oceanology, Chinese Academy of Sciences, No. 7 Nanhai Road, Qingdao 266071, China
* Correspondence: ruiyong.zhang@qdio.ac.cn (R.Z.); arevik.vardanyan@asnet.am (A.V.); Tel.: +86-13210842959 (R.Z.); +374-94900931 (A.V.)

**Abstract:** A new strain of *Leptospirillum* sp. Ksh-L was isolated from a dump-bioleaching system of the Kashen copper-molybdenum mine (South Caucasus). Ksh-L is an obligate chemolithoautotroph, capable of oxidizing ferrous iron ($Fe^{2+}$). Cells are Gram-negative and vibrio- or spirillum-shaped of a 0.5–3 μm size. The optimal conditions for the growth are 35 °C and pH 1.6–1.8. $Cu^{2+}$ and $Zn^{2+}$ have different effects on the oxidizing ability of the *Leptospirillum* sp. Ksh-L culture depending on the phase of growth and concentration of $Fe^{2+}$. Under the conditions of gradually increasing the concentration of copper in the medium, during 4–5 successive subculturing experiments, it was possible to obtain an adapted culture of *Leptospirillum* sp. Ksh-L, capable of growing in the medium in the presence of up to 400 mM $Cu^{2+}$. A bioleaching experiment indicates that Ksh-L can efficiently oxidize chalcopyrite. However, the bioleaching of copper from chalcopyrite by *Leptospirillum ferrodiazotropum* Ksh-L increased about 1.8 times in association with *At. thiooxidans* ATCC 19377. Phylogenetic analysis based on 16S rRNA gene sequences (GenBank ID ON226845) shows that strain Ksh-L forms a single cluster into Group III. The strain possesses 99.59%, 99.52%, and 96.60% sequence similarity with the strains YTW-96-06, YTW-66-06, and *Leptospirillum ferrodiazotrophum* 5C in Group III, respectively.

**Keywords:** *Leptospirillum ferrodiazotrophum*; isolation; characterization; phylogenetic analysis; bioleaching of sulfide minerals

**Citation:** Khachatryan, A.; Vardanyan, N.; Willscher, S.; Sevoyan, G.; Zhang, R.; Vardanyan, A. Bioleaching of Chalcopyrite by a New Strain *Leptospirillum ferrodiazotrophum* Ksh-L Isolated from a Dump-Bioleaching System of Kashen Copper-Molybdenum Mine. *Minerals* **2024**, *14*, 26. https://doi.org/10.3390/min14010026

Academic Editors: Naoko Okibe and Elizabeth Watkin

Received: 18 October 2023
Revised: 14 December 2023
Accepted: 22 December 2023
Published: 25 December 2023



## 1. Introduction

Bioleaching is a technique that uses microorganisms to remove metals from ore where traditional extraction methods are not economically viable. This method is frequently applied to sulfide mineral ores, which are the source of many valuable and precious metals, including copper, gold, and silver. Traditional methods of metal extraction from sulfide minerals, such as pyrometallurgy, are costly, energy-intensive, and environmentally damaging [1,2]. Bioleaching, using microbial metabolisms to break down metal ores, provides a low-cost solution to this issue. Microbes produce energy during the bioleaching process by oxidizing iron and sulfur from sulfide minerals.

Sulfide ores can be dissolved by ferrous-oxidizing acidophiles to create ferric iron, which then attacks minerals to form sulfur or polysulfide on the ore surface [3,4]. The target metals are released as a result of the oxidants' attack on the sulfide minerals [2]. Due to the intimate connections between functional oxidizers, environmental conditions, and bioleaching performances, many researchers have recently focused on microbial communities [5,6].

Approximately 70% of the copper reserves in the world come from chalcopyrite [7,8]. It is one of the most resistant ores in hydrometallurgical processing and the most common copper ore. Modern research on the possibility of increasing total recoveries of metal values from such mineral resources plays a significant role in the bio-hydrometallurgical processing of complex low-grade non-ferrous metal concentrates [9–13].

Mesoacidophiles are the most frequently used microorganisms in industrial low-grade chalcopyrite heap-bioleaching operations because of their ambient temperature optimum and ranges, which have been shown to have a significant impact on the rate of copper extraction from chalcopyrite [14–16].

*Leptospirilla* was discovered to be the predominant iron-oxidizing bacteria in gold-arsenopyrite and pyrite bio-oxidation reactors working at 40 °C, although *Acidithiobacillus ferrooxidans* had long been thought to be the most significant microorganism in the bioleaching of metals [17–22].

*Leptospirillum* spp. bacteria are vibrio- and spiral-shaped chemolithotrophic organisms. They have been formally recognized as coherent bacteria since they fix carbon utilizing the Benson–Calvin cycle, employing oxygen as their only electron acceptor and ferrous iron as their sole electron donor [20,23–26]. The genus *Leptospirillum* has been classified into three groups, I, II, and III based on the 16S rRNA gene phylogeny [27]. *Leptospirillum ferrooxidans* is a representative of group I [28], *Leptospirillum ferriphilum* and *Leptospirillum rubrum* are representatives of group II, and *Leptospirillum ferrodiazotrophum* is a representative of group III [29–31]. In addition, microbial community genomics has identified further species, "*Leptospirillum* sp. group IV UBA BS" [30,31].

Notably, proteomics research has revealed that *L. ferrodiazotrophum* (group III) carbohydrate metabolism is significantly better than that of *Leptospirillum* group II and *Ferroplasma* type II [25]. Also, because the nif genes were only found in this bacterium's genome sequence, it is thought to be a minor member of the genus and the only one that is capable of fixing nitrogen [30–33]. Based on this, one representative was isolated in a nitrogen-free liquid medium from AMD biofilm in which *Leptospirillum* group III was abundant [31]. This Fe(II)-oxidizing, free-living diazotroph was tentatively named "*L. ferrodiazotrophum*". Hence, despite its low quantity in the microbial community, *L. ferrodiazotrophum* played a central role in biofilm formation and nitrogen fixation.

Gene complement variations highlight significant physiological variances that may have played a crucial role in the *Leptospirillum* groups' likely sympatric separation. *Leptospirillum* group II is more capable of producing potentially important polymers for the establishment of floating biofilms, such as cellulose, cellobiose, and starch/amylose, than *Leptospirillum* group III in dealing with the osmotic challenges brought on by the near-molar $FeSO_4$ solutions. With the potential completion of the glycolysis and TCA pathways, *Leptospirillum* group III appears to be more suited for energy production and nitrogen fixation. These results are in line with the descriptions of *Leptospirillum* group II as an early colonist and *Leptospirillum* group III as a component of late-stage biofilms [34,35]. Interestingly, the complements of signal transduction and chemotaxis genes in *Leptospirillum* groups II and III are quite different, as are many regulatory genes, pointing to adaptation to various microenvironments (such as those with particular levels of oxygen, redox potential, and availability of fixed nitrogen). Signal transduction, motility, and chemotaxis appear to be more crucial in *Leptospirillum* group III than in group II, according to genomic and proteomic evidence [33,35]. *Leptospirillum* group III is identified by biofilm characterization investigations as distributed cells and microcolonies in interior biofilm zones [34], where geochemical gradients are predicted to be prominent. Combining this distribution with the presumed metabolic traits may suggest that *Leptospirillum* group III is a microaerophile that prefers to grow in nutrient-poor areas of biofilms, where its capacity to fix nitrogen may be essential. Where oxygen availability is certainly low and *Leptospirillum* group II is present, a few studies on anaerobic metabolism (for example, making use of a within-biofilm nitrogen cycle) are available [33].

Most metals are soluble in acidic pH, and extremely acidophilic microorganisms should be tolerant to high concentrations of metals. Acidophiles are metal-tolerant by both active and passive mechanisms. The passive metal tolerance mechanism is based on the internal positive membrane potential by creating a chemiosmotic gradient that the cations should travel against to enter the cytoplasm [36,37]. Active systems include the efflux of metals from the cytoplasm to the periplasmic space, carried out by ATPases located in the internal membrane of the bacteria [38]. Some microorganisms may pump metal from the cytoplasm directly to the extracellular space by systems of the RND (resistance nodulation cell division) family of carriers, the Cus system of *Escherichia coli* being the best known of this kind of detoxification organization [39]. The capacity of some species to bind the metal in the periplasmic space using metal chaperones has also been reported for copper [40].

It should be noted that bacterial growth in the form of biofilms significantly increases the resistance of bacteria to metals [41–43]. Acidophiles attached to surfaces, such as sulfide minerals, form biofilms that usually include extracellular polymeric substances (EPSs) that can sorb metals to provide a further degree of metal tolerance [44,45].

Several mechanisms of resistance to Mo (VI) have been identified in acidophiles, including the putative Mo (V) resistance mechanism in *At. ferrooxidans* strain Funis 2–1 [46]. Mo (VI) is chemically reduced by Fe(II), and Mo(V) forms binds to the plasma membrane, probably to the cytochrome-c oxidase (lowering its activity), inhibiting Fe(II) oxidation and consequently growth. Resistance is based on a combination of a cytochrome-c oxidase that is tolerant to higher concentrations of Mo (V) and on Mo (V)-oxidizing activity six-fold greater than that detected in the sensitive *At. ferrooxidans* strain AP19-3 [46].

With the development of sequencing technology, many genomic-analysis-based studies have investigated the mechanisms underlying heavy metal resistance under extremely acidic conditions [47–50]. Several genes associated with EPS formation have been identified thus far [51]. Genome/transcriptome analyses showed the presence of genes involved in biofilm formation in *Leptospirillum* spp. [51,52]. In *Acidithiobacillus* and *Leptospirillum* spp. a membrane efflux pump encoded by the czcCBA cluster is responsible for resistance to cadmium, zinc, and cobalt [48,52–54]. The resistance of bacteria to copper is very important from the point of view of their application in biotechnological processes, where the concentration of copper ions can vary in the range from 15 to 100 mM $CuSO_4$ [55]. Copper resistance systems in some acidophiles include a copper P-Type transporter identified in *L. ferriphilum* ML-04 [48]. In addition, earlier descriptions of acidophile copper resistance and the potential role of inorganic polyphosphates in metal resistance are also available [56,57].

This study addressed the characterization and identification of *Leptospirillum* Ksh-L, isolated from an acid mining drainage of copper ore in Armenia, its ability to degrade sulfide minerals, and its tolerance to metal ions.

## 2. Materials and Methods

### 2.1. Culture Conditions and Isolation

To obtain an enrichment culture of iron-oxidizing bacteria, 9K medium [58] with ferrous iron as a source of energy was inoculated with a sample of ore material from a dump-bioleaching system of Kashen copper-molybdenum mine (South Caucasus) and incubated at 35 °C and 150 rpm. The serial dilution method was used to obtain a pure culture of *Leptospirillum* spp.

### 2.2. Morphology Studies

Gram staining was performed using the Huker method [59]. The morphology of cells was studied with Motic BA310 trinocular (×1000) microscope supplied by Moticam A16 Camera (Barcelona, Spain, MoticIncorporation Ltd.).

*2.3. Optimal pH and Temperature for Growth*

Studies of the influence of temperature and pH on the growth of strain Ksh-L were carried out in 100 mL flasks containing 50 mL of 9K medium with ferrous iron ($Fe^{2+}$) and 10% inoculum. Cultivation was performed on the orbital shaker-incubator ES-20/60 (Biosan, Riga, Latvia) at 150 rpm. The growth ranges for temperature and pH were set as 25–50 °C and 1.6 to 3.0, respectively.

*2.4. Influence of Copper and Zinc*

The influence of copper and zinc ions on oxidation of $Fe^{2+}$ by *Leptospirillum* sp. Ksh-L was studied in Mackintosh (MAC) medium [60] in the concentration range from 10 to 300 mM and at different concentrations of the source of energy ($Fe^{2+}$). The data presented in the text are formed on the average from repeated experiments with ±2% variation of $Fe^{2+}$.

*2.5. EPS Analysis, Extraction, and Determination*

2.5.1. EPS Extraction

For EPS extraction experiments, culture *Leptospirillum* Ksh-L was grown in 10 L MAC medium supplemented with $FeSO_4 \times 7H_2O$ as a source of energy at 35 °C with shaking at 160 rpm. In the stationary growth phase, cells were collected by centrifugation (10,000 rpm, 10 min) at 4 °C. To get rid of any remaining bacteria, the supernatant was collected and filtered through 0.2 m pore size filters under sterile conditions. Colloidal EPS was present in the fraction obtained. The pellet was re-suspended in 10 mL of 20 mM EDTA at pH 7 and centrifuged for 10 min at 7500 rpm (9900× *g*) and 4 °C to release the bound EPS. The washed fraction was applied to the supernatant. The fresh pellet was re-dissolved in 10 mL of 20 mM EDTA at pH 7, and the suspension was incubated at 4 °C for 1 h while being shaken. The remaining cells were then removed by centrifuging the bacterial suspension and extracting agent mixtures for 10 min at 7500 rpm (9900× *g*) and 4 °C [61].

2.5.2. Determination of EPS Composition

Total carbohydrate values were determined by spectrophotometry using the phenol-sulfuric acid method with D-glucose as the standard, as described by [62]. Total protein quantification was conducted by spectrophotometry using the Bradford procedure [63]. A calibration curve was developed using a series of bovine serum albumin (BSA) standards. Uronic acids were quantified using the protocol of Blumenkrantz and Asboe-Hansen [64]. Meta-hydroxydiphenyl solution in 0.5% NaOH, ortho-hydroxydiphenyl, $H_2SO_4$/sodium tetraborate solutions, and cetyltrimethylammonium bromide solution were used as a reaction mixture. Absorbance measurements were performed at 520 nm. The content of proteins, carbohydrates, and uronic acids in EPS was expressed in μg/mL culture medium.

*2.6. DNA Extraction, PCR of 16s rRNA, Sequencing, and Phylogenetic Analysis*

2.6.1. DNA Extraction

The identification of isolated strain Ksh-L was performed based on 16S rRNA gene nucleotide sequence analysis. DNA extraction was carried out according to the Macherey-Nagel™ (NucleoSpin™) protocol. Bacterial cultures in the logarithmic phase were centrifuged to obtain the bacterial biomass and frozen for DNA purification and taxonomic classification of the strain based on 16S rRNA gene sequences. For DNA extraction, the following reagents were used: BE buffer (Tris-HCl pH-8.0), proteinase K, MG-lysozyme solution, BS salt buffer, 96% ethanol, and distilled water.

2.6.2. Extraction and Purification of PCR Product

A 1.5% agarose gel Tris-HCl buffer solution was prepared. The agarose aqueous solution was heated in a microwave oven for 1–2 min before boiling and then cooled to 45–60 °C. A 1% 10 mM Tris-HCl, 1 mM EDTA aqueous solution (1 μL per 10 mL agarose solution) was added to the cooled agarose solution. A total of 5 μL of the test DNA solution was mixed with a 3 μL 6 μL 6X Thermo Scientific TriTrack DNA Loading Dye and 5 μL Thermo Scientific GeneRuler buffer containing 1 kb DNA Ladder markers. Electrophoresis was performed on GE Healthcare/Amersham Pharmacia EPS-601 Electrophoresis Power Supply for 25 min at a voltage of 104 V at 100 mA. The electropherograms and images were obtained using the GeneMarkerHID® Spectrum GeneMarkerHID® and Spectrum ProView™ Sequencing Software TM625 package (Promega Corporation 2800, Fitchburg, MA, USA).

The PCR amplification was carried out according to the Macherey-Nagel™ (NucleoSpin™) Extract II protocol. PCR amplification was carried out in 50 μL reaction mixture, which contained 10 mg of DNA sample, 1 μL universal bacterial PCR primers a-fD1 (27F) (AGAGTTTGATCCTGGCTCAG) and rP2 (ACGGCTACCTTGTTACGAG), b- 908 fwd and 796 rev, 5 μL Taq-DNA polymerase, 1 μL dNTPs, and 40 μL double distilled water. PCR temperature–time process is as follows: 1st cycle 95 °C × 5 min, further 30 cycles of denaturation −94 °C × 40 s, connection 50/53 °C × 55–60 s, synthesis 72 °C × 1.5 min, final cycle −72 °C × 3 min, and then standby at 8–10 °C. PCR products were tested by 1.5% agarose gel electrophoresis and sequenced with primers 908fwd (16Sfwd) (GTGCCAGCAGCCGCG) and 796rev (16Srev) (GGGTTGCGCTCGTTG) by Microsynth AG (Balgach, Switzerland). PCR product purification was performed using the QIAquick PCR purification kit following the manufacturer's instructions. PCR fragments were sequenced by a 454 GS-FLX Titanium sequencer, using the Sanger method [65].

#### 2.6.3. Construction of Phylogenetic Tree

Close relative and phylogenetic affiliation of the obtained 16S rRNA sequences were determined by submitting to the NCBI 16S ribosomal RNA GenBank database using NCBI blastn search analyses (www.ncbi.nlm.nih.gov) performed with Geneious prime 2022.0.2. (https://www.geneious.com) and the 16S Biodiversity tool (RDP tool version 2.12) [66,67]. The construction of phylogenetic trees was performed with MEGA 11 software using the neighbor-joining method [68,69].

### 2.7. Leaching Experiments

Chalcopyrite ($CuFeS_2$) from Shamlugh ore deposit (Armenia) was tested in the bioleaching experiments. The chemical composition of minerals is presented in Table 1. Feed minerals were ground to a particle size $\leq$ 63 μm.

**Table 1.** Chemical composition of the analyzed chalcopyrite (wt%).

| Sample | Fe | Cu | S |
|---|---|---|---|
| Chalcopyrite | 29.7 | 30.2 | 33.8 |

Bioleaching of chalcopyrite was performed using a pure culture of *L. ferrodiazotropum* Ksh-L as well as its associations with *At. thiooxidans* ATCC 19377 obtained from DSMZ. Bioleaching experiments were carried out in 250 mL Erlenmeyer flasks containing 100 mL of MAC medium without iron at 30 °C. Pulp density (PD) was 4% and pH 1.8. The inoculum of used cultures was 10%, and all experiments were carried out in triplicate. Chemical control with the same conditions and without inoculum was included. Copper, total iron, ferric (Fe(III)), and ferrous (Fe(II)) ions in leachate were analyzed for 30 days. pH was measured using Hi2211-01 Benchtop pH/mV Meter (Hanna Instruments, Vöhringen, Germany). The value of oxidation/reduction potential (ORP) was measured using a standard hydrogen electrode (SHE) in relation to an Ag/AgCl reference electrode (mV vs. Ag/AgCl). Copper and total iron were determined using atomic absorption spectrophotometer AAS SP-IAA1800H (Bioevopeak, Qingdao, China). Concentrations of ferric (Fe(III)) and ferrous (Fe(II)) ions were determined using the complexometric method with EDTA [70].

Consumption of Fe(II) was calculated as a difference between initial Fe(II) and concentration of Fe(II) determined at certain times of experiment (Equation (1)). Consumption is expressed by g/L and %.

$$[Fe(II)]_c = [Fe(II)_i - [Fe(II)]_t / [Fe(II)]_i \times 100\% \quad (1)$$

where $[Fe(II)]_c$ is the consumption of Fe(II); $[Fe(II)_i$– is initial concentration of Fe(II) in the medium; and $[Fe(II)]_t$—concentration of Fe(II) determined at certain point of time.

Inhibition of iron oxidation in the presence of metal ions was determined according to Equation (2).

$$\text{Inhibition (\%)} = Fe(II)_{cm} / Fe(II)_c \times 100 \quad (2)$$

where $Fe(II)_{cm}$ is consumption of Fe in the presence of metal ions (g/L), and $Fe(II)_c$ is consumption of Fe in the absence of metal ions (g/L).

## 3. Results

### 3.1. Isolation of Strain Ksh-L

The selected sample from a dump-bioleaching system of the Kashen copper-molybdenum mine was transferred in a 9K liquid medium and incubated at 35 °C at 150 rpm for 5–7 days. As a result of the active growth of bacterial cells, the medium color changes from pale green to orange-red due to iron oxidation.

### 3.2. Morphology

Cells of *Leptospirillum* sp. Ksh-L are Gram-negative, motile, and vibrio- or spiral-shaped. Cells have a diameter of 0.5 μm and a length of 1.0–3.0 μm (Supplementary Figure S1). This is typical for the genus of *Leptospirillum*. The morphology results show that strain Ksh-L seems to be consistent with previously described *Leptospirillum* species [71].

### 3.3. Optimal pH and Temperature for Growth

The growth of the newly isolated Khs-L strain was evaluated according to the increase in the number of cells and the results of their biological activity, increasing the amounts of $Fe^{3+}$ in the medium. As shown in Figure 1, the growth curves of the *Leptospirillum* sp. Ksh-L strain in the initial pH range of 1.6–2.0 are S-shaped. The duration of the lag phase was about 48 h, but the growth of bacteria was more active at pH values of 1.6 and 1.8 (Figure 1).

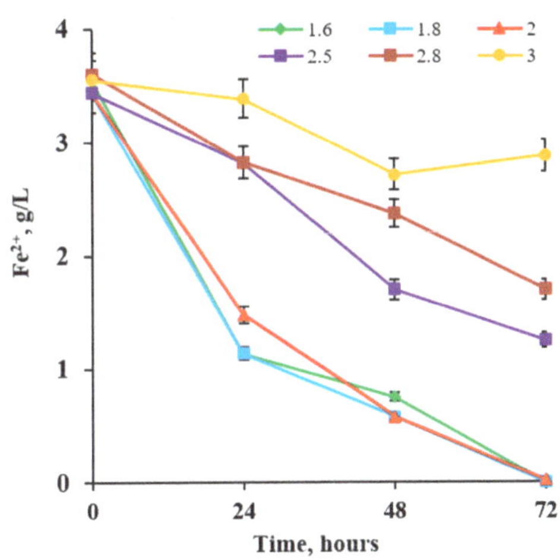


**Figure 1.** Dynamics of ferrous iron oxidation by *Leptospirillum* sp. Ksh-L at different pH values ($Fe^{2+}$—4 g/L, temperature 35 °C, 160 rpm).

The most intensive oxidation of $Fe^{2+}$ was observed at pH values of 1.6 and 1.8. As can be seen from the presented data, at the initial pH values of 2.8 and 3.0, $Fe^{2+}$ oxidation was sharply suppressed, and the duration of the lag phase was increased to 20 and 48 h, respectively (Figure 1). It can be explained that the pH variation affects enzyme activity because changes in ionization affect the system components. This indicates that the Ksh-L strain is sensitive to pH, and a much higher pH value will inhibit the activity of bacteria.

As can be seen from Figure 2, the optimum growth temperature of the Ksh-L strain was observed at 35 °C. Thus, optimal conditions for the growth of isolated strain *Leptospirillum* sp. were observed at 35 °C and a pH range from pH 1.6 to pH 2.0.

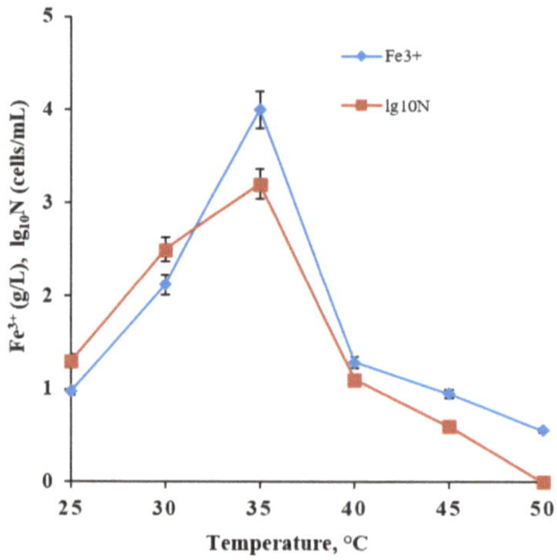


**Figure 2.** Effect of temperature on growth and iron oxidation by *Leptospirillum* sp. Ksh-L (pH 2.0, duration 48 h). N represents cell number (cells/mL).

### 3.4. *Influence of Copper and Zinc*

Like most bioleaching microorganisms, strain *Leptospirillum* sp. Ksh-L has been isolated from an environment (dump-bioleaching system) that has unusually high concentrations of potentially toxic metals (e.g., copper and iron) as well as high concentrations of heavy metals (e.g., arsenic and silver). These metals can exert harmful effects on microorganisms. Their toxic effects include the blocking of biologically important functional groups and the denaturation of enzymes [72].

Cu (II) inhibition of growth and Fe(II) oxidation have also been demonstrated in *Sulfobacillus thermosulfidooxidans* subsp. *asporogenes* via competitive inhibition of Fe(II) oxidation [73]. The influence of $Cu^{2+}$ and $Zn^{2+}$ ions on the oxidation of $Fe^{2+}$ by strain Ksh-L was studied in a concentration range from 25 to 200–300 mM. As can be seen from Figure 3a, copper in all tested concentrations inhibits the oxidation of $Fe^{2+}$.

Furthermore, the higher the copper concentration, the higher the extent of inhibition of iron oxidation. Thus, iron oxidation by *Leptospirillum* sp. Ksh-L for 24 h was suppressed by 36.7, 65.3, and 89.8% at copper concentrations of 50, 100, and 150 mM, respectively, with a content of 3.2 g/L $Fe^{2+}$ in the medium. $Cu^{2+}$ in the concentration of 200–300 mM almost completely (90%–96%) inhibits iron oxidation by *Leptospirillum* sp. Ksh-L (Figure 3b). It can be noted that the extent of inhibition of iron oxidation by $Cu^{2+}$ ion reduces for 48 h along with bacterial growth and is 17.4, 44.9, and 71.5%, respectively (Figure 3b).

The effect of copper on oxidation of $Fe^{2+}$ *Leptospirillum* sp. Ksh-L was studied depending on substrate concentrations in the medium (Figure 4a,b). As shown in Figure 4, the degree of oxidation of $Fe^{2+}$ by *Leptospirillum* sp. Ksh-L in the presence of tested copper concentrations was slightly higher when the content of $Fe^{2+}$ in the medium increased from 4.0 g/L (71.4 mM) to 7.0 g/L (125 mM). Thus, the increase in the concentration of substrate from 4.0 to 7.0 g/L leads to the enhancement of the amount of oxidized iron by bacteria in the presence of the tested concentration of copper.

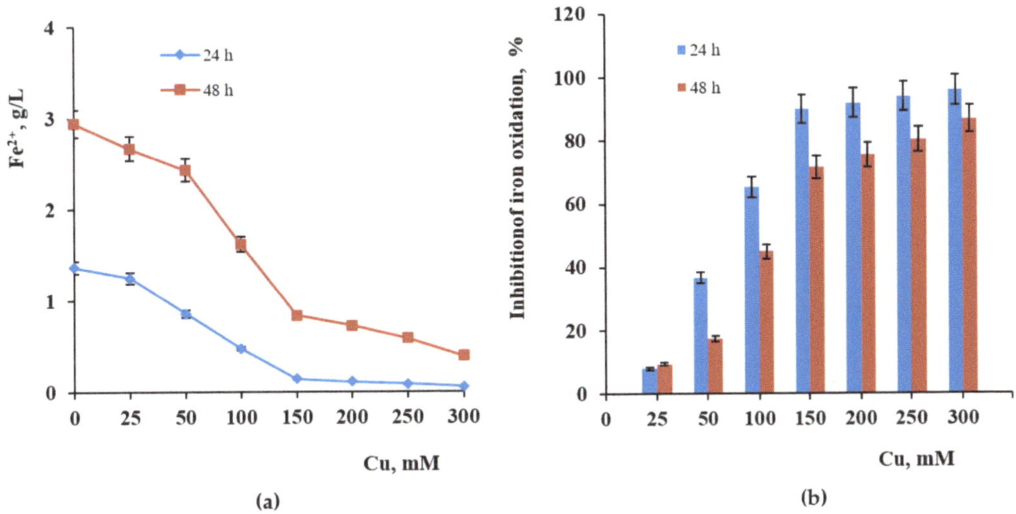


**Figure 3.** Oxidation of $Fe^{2+}$ g/L (**a**) and inhibition of iron oxidation % (**b**) by *Leptospirillum* sp. Ksh-L at different concentrations of copper during 24 and 48 h of growth.

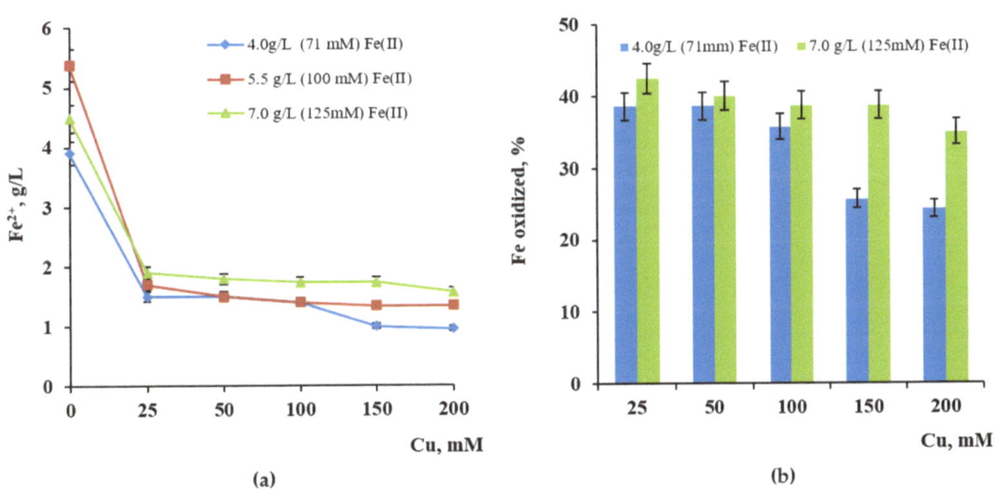


**Figure 4.** Effect of Cu on oxidation of $Fe^{2+}$ g/L (**a**) and % (**b**) by *Leptospirillum* sp. Ksh-L at different initial concentrations of $Fe^{2+}$ in the medium (pH 1.95, temperature 35 °C, 180 rpm, duration—43 h).

From the data presented, it can be seen that the oxidation of $Fe^{2+}$ by *Leptospirillum* sp. Ksh-L is suppressed by about 25% at 25 mM of $Zn^{2+}$ and is sharply inhibited (up to 65–77%) at zinc concentrations in the range from 50 to 200 mM (Figure 5a,b).

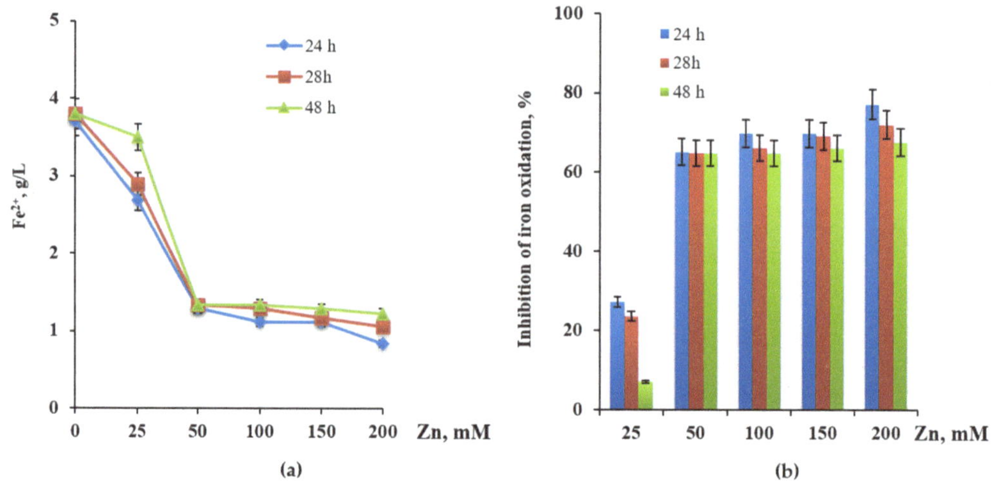


**Figure 5.** Influence of different concentrations of $Zn^{2+}$ ions on oxidation of $Fe^{2+}$ (**a**) and % (**b**) by *Leptospirillum* sp. Ksh-L depending on cultivation period ($Fe^{2+}$—3.8 g/L, pH 1.95, temperature 35 °C, 180 rpm).

As shown in Figure 6, an increase in the substrate concentration led to the enhancement of the amount of oxidized iron by Ksh-L in the presence of 25 mM $Zn^{2+}$, while it did not facilitate iron oxidation by bacteria in concentrations of Zn above 100 mM. Moreover, no decrease in the inhibitory effect of zinc ions was observed along with the growth of bacteria, as shown in the case of copper (Figure 5).

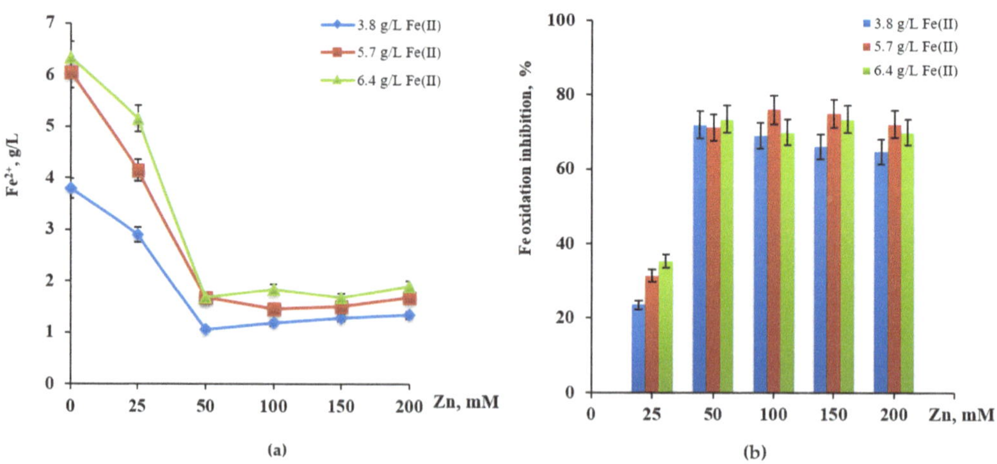


**Figure 6.** Influence of $Zn^{2+}$ on oxidation of $Fe^{2+}$ g/L (**a**) and Fe oxidation inhibition % (**b**) by *Leptospirillum* sp. Ksh-L at different concentrations of substrate ($Fe^{2+}$) (pH 1.95, temperature 35 °C, 180 rpm, 24 h).

Along with the growth of *Leptospirillum* sp. Ksh-L, the amount of oxidized iron in the presence of 50 mM $Cu^{2+}$ increased. It is assumed that with the growth of bacteria, the cells form EPSs and create accordingly a less toxic and more favorable environment for the growth of cells in the presence of copper.

### 3.5. Adaptation of L. ferrodiazotrophum Ksh-L

Although bacteria do not react well to sudden and significant changes in heavy metal ion concentrations, they can be adapted to gradually increased concentrations over a while, to increase their tolerance to such metals. In the next series of the experiment under conditions of gradually increasing the concentration of copper in the medium, during 4–5 successive subculturing experiments, it was possible to obtain an adapted culture of *Leptospirillum* sp. Ksh-L capable of growing in the medium in the presence of up to 400 mM $Cu^{2+}$ (Figure 7).

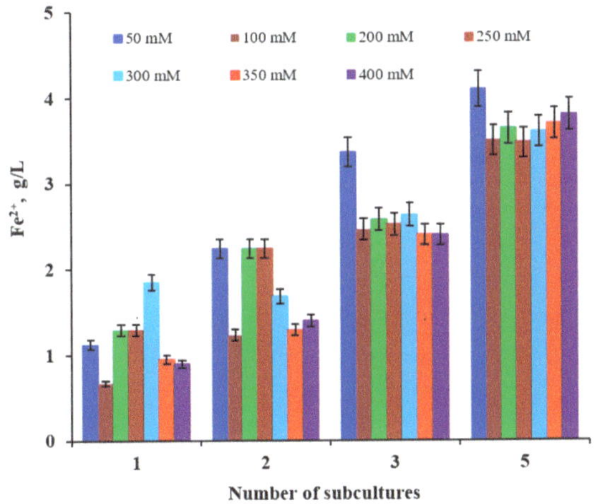


**Figure 7.** Oxidation of $Fe^{2+}$ during an adaptation of *Leptospirillum* sp. Ksh-L under conditions of gradually increased concentrations of Cu in the medium (50–400 mM $Cu^{2+}$).

### 3.6. EPS Analysis

EPSs play an essential role in the formation of a biofilm, which mediates the adhesion of cells to the mineral surface and forms a cohesive three-dimensional polymer, interconnecting and immobilizing cells in the process of bioleaching by iron- and sulfur-oxidizing bacteria [74–76]. An important role of capsular polysaccharides as a fundamental structural element of the EPS, determining the mechanical stability of biofilm was disclosed. One of the objectives of the present study was to investigate the chemical composition of a colloidal polysaccharide of the newly isolated iron-oxidizing chemolithotrophic bacteria *L. ferrodiazotrophum* Ksh-L. The studies carried out showed that the total amounts of colloidal and capsular EPSs are 474.4 and 208.58 μg/mL, respectively.

As shown in Table 2, the amount of carbohydrates in the capsular EPS is considerably higher than that in the colloidal EPS. The amounts of protein in colloidal and capsular EPSs are approximately the same, 24.41 μg/mL and 27.33 μg/mL, respectively. Uronic acids were not detected in both EPSs. The obtained data are comparable and agree with the corresponding data of other species of the genus *Leptospirillum*, *L. ferriphilum* CC and *L. ferrooxidans* ZC studied by us previously [77,78].

**Table 2.** EPS composition of *L.ferrodiazotrophum* Ksh-L grown on ferrous iron (Fe (II)) as a source of energy (pH 1.95, temperature 35 °C, 180 rpm, cultivation 72 h).

| EPS Composition | Protein (μg/L) | Carbohydrates (μg/L) | Uronic Acids |
|---|---|---|---|
| Capsular | 16.3 | 6.0 | BDL * |
| Colloidal | 8.3 | 3.9 | BDL * |

* BDL: below the detection limit.

### 3.7. Phylogenetic Analysis of 16S rRNA

A PCR-amplified 16S rRNA product was detected by 1% agarose gel electrophoresis (Supplementary Figure S2). The length of the 16S rRNA of *Leptospirillum* sp. Ksh-L is about 1.5 kb and is sequenced sequentially. The sequence of the 16S rRNA of *Leptospirillum* sp. Ksh-L was submitted to GenBank, and the accession number ON226845 was obtained. The length of sequenced fragments of the gene encoding 16S rRNA is 1469 bp. The nucleotide sequence of the strain *Leptospirillum* sp. Ksh-L was phylogenetically compared with the *Leptospirillum* species (Figure 8).

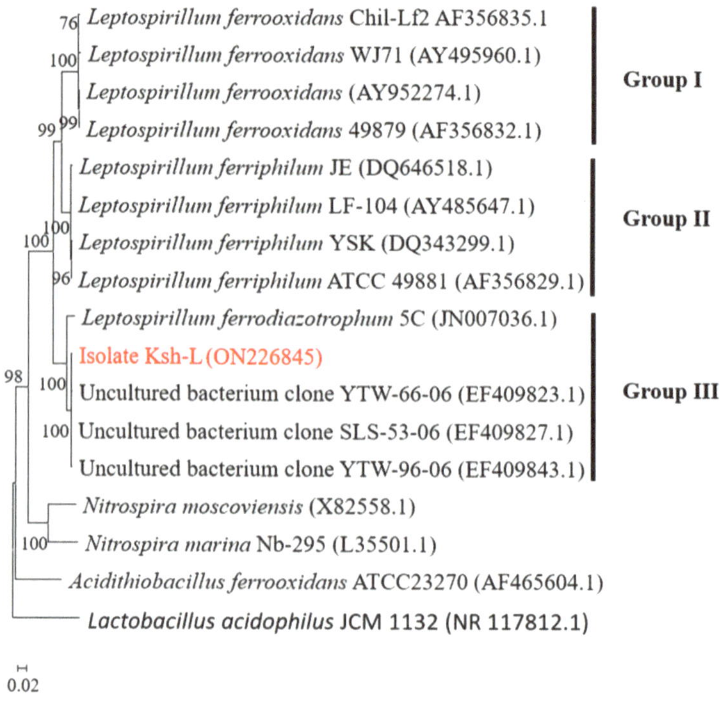


**Figure 8.** Phylogenetic position of strain *Leptospirillum* sp. Ksh-L. The evolutionary history was inferred using the neighbor-joining method [79]. The percentage of replicate trees in which the associated taxa clustered together in the bootstrap test (10,000 replicates) are shown next to the branches [80]. New isolate highlighted in red.

Preliminary screening of the GenBank database was performed using BLAST (http:www.ncbi.nlm.nih.gov/blast). Based on the homology of 16S rRNA, the phylogenetic development tree was built as shown in Figure 8. The sequence was divided into three groups: Group I—*L. ferrooxidans*; Group II—*L. ferriphilum*; and Group III—*Leptospirillum ferrodiazotrophum* and uncultured clones. The isolated *Leptospirillum* sp. Ksh-L strain formed a single cluster into Group III and possessed 99.66% sequence similarity with uncultured bacterium clone SLS-53-06 (Table 3). As shown in Figure 8, the isolate Ksh-L compared to other strains in Group III with strains YTW-96-06, YTW-66-06, and *Leptospirillum ferrodiazotrophum* 5C and possessed 99.59%, 99.52%, and 96.60% sequence similarity, respectively (Table 3).

In Figure 8, the *Lactobacillus acidophilus* JCM 1132 strain is used as an out-group to root the tree, and the database accession numbers of the gene sequences used are given in parentheses.

Thus, the newly iron-oxidizing strain Ksh-L obtained from a dump-bioleaching system of the Kashen copper-molybdenum mine was identified as *Leptospirillum ferroodiazotrophum*. Type strain: *Leptospirillum ferroodiazotrophum* Ksh-L.

**Table 3.** Identity of 16S rRNA gene of isolated *Leptospirillum* sp. Ksh-L with other strains.

| Isolated Strain | Strain Name, GenBank (Accession Number) | Identity, % | Reference |
|---|---|---|---|
| *L. ferrodiazotrophum* Ksh-L | Uncultured bacterium clone YTW-66-06 (EF409823.1) | 99.52 | [81] |
| | Uncultured bacterium clone SLS-53-06 (EF409827.1) | 99.66 | [81] |
| | Uncultured bacterium clone YTW-96-06 (EF409843.1) | 99.59 | [81] |
| | *Leptospirillum ferrodiazotrophum* 5C (JN007036.1) | 96.60 | [82] |
| | *Leptospirillum ferriphilum* ATCC 49,881 (AF356829.1) | 91.02 | [20] |
| | *Leptospirillum ferrooxidans* WJ71 (AY495960.1) | 89.60 | [83] |
| | *Nitrospira moscoviensis* (X82558.1) | 81.22 | [84] |

### 3.8. Bioleaching of Chalcopyrite

A comparative study was carried out on the bioleaching of chalcopyrite by a pure culture *L. ferrodiazotrophum* Ksh-L and its association with sulfur-oxidizing bacteria *At. thiooxidans* ATCC 19377 at 30 °C.

The data presented in Figure 9a,b show that compared to uninoculated control, isolated bacterium *L. ferrodiazotropum* Ksh-L stimulated the extraction of copper and iron by 1.7 and 2.3 times, respectively. However, in association with sulfur-oxidizing *At. thiooxidans*, *L. ferrodiazotropum* Ksh-L oxidizes chalcopyrite much more actively than in pure culture. This finding corresponds with the literature data from similar studies conducted using *Leptospirillum ferooxidans* with the association of *At. thiooxidnas* or *At. caldus*. According to some researchers, mixed cultures of mesophiles have been reported to oxidize sulfide minerals more efficiently than pure cultures [85–87].

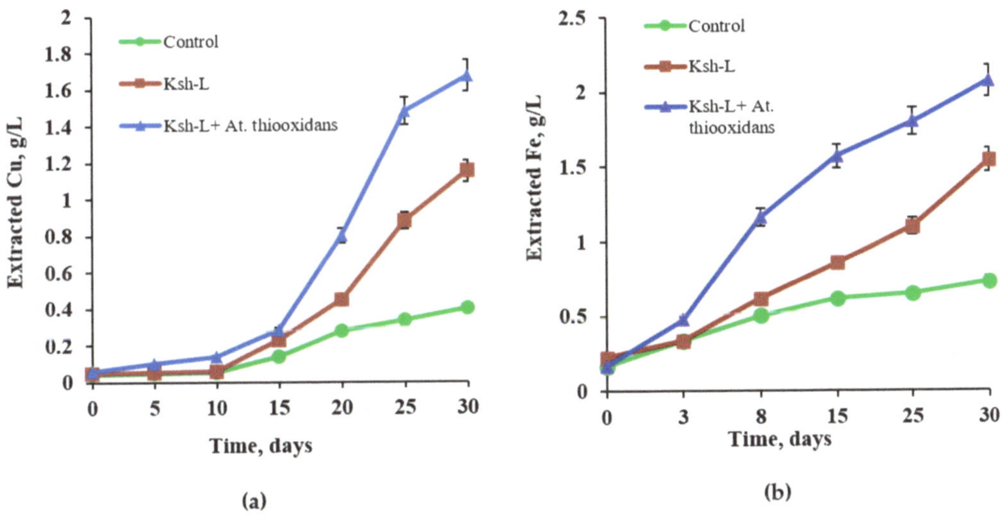


**Figure 9.** Bioleaching of copper (**a**) and iron (**b**) from chalcopyrite by pure-culture *L. ferrodiazotrophum* Ksh-L and association with *At. thiooxidans* ATCC 19377 ($CuFeS_2$—4%, pH 1.8, temperature 30 °C, 180 rpm).

Thus, in the presence of *At. thiooxidans*, the extraction of copper and iron from chalcopyrite by *L. ferrodiazotropum* Ksh-L for 30 days increases approximately 1.8 and 1.9 times, respectively (Figure 9a,b, Table 4).

**Table 4.** Leaching of iron and copper from chalcopyrite by *L. ferrodiazotropum* Ksh-L and association with sulfur-oxidizing bacteria *At. thiooxidans* ATCC 19377.

| Bacteria | Extraction of Fe 30 Days | | | | Extraction of Cu | | Final | |
|---|---|---|---|---|---|---|---|---|
| | g/L $Fe^{3+}$ | $Fe^{2+}$ | Fe Total | % Fe Total | g/L | % | pH | ORP, mV |
| Control (uninoculated) | 0 | 0.672 | 0.672 | 5.6 | 0.52 | 4.1 | 1.8 | 520 |
| *L. ferrodiazotropum* Ksh-L | 1.096 | 0.448 | 1.544 | 12.8 | 0.884 | 6.9 | 1.7 | 600 |
| *L. ferrodiazotropum* Ksh-L + *At. thiooxidans* | 2.016 | 0.616 | 2.744 | 23.0 | 1.67 | 13.4 | 1.5 | 720 |

The data presented in Table 1 show that chalcopyrite leaching is correlated with the pH and ORP of the solution. When *L. ferrodiazotropum* Ksh-L was used in monoculture, the final pH was 1.7, and the ORP was 600 mV, while in the *L. ferrodiazotropum* Ksh-L variant with *At. thiooxidans* ATCC 19377, the pH value was comparatively lower (1.5), and the ORP was significantly higher (720 mV) (Table 4).

Chalcopyrite is an acid-soluble sulfide mineral and is therefore subject to attack by both ferric iron ($Fe^{3+}$) and protons ($H^+$) (Equations (3) and (4)) [55,88].

$$CuFeS_2+2Fe_2(SO_4)_3 \rightarrow CuSO_4+5FeSO_4+2S^0 \quad (3)$$

$$CuFeS_2+4H^+ \rightarrow Fe^{2+}+Cu^{2+}+2H_2S \quad (4)$$

$$0.125S_8+1.5\ O_2+H_2O \xrightarrow{At.thiooxidans\ ATCC\ 19377} SO_4{}^{2-}+2H^+ \quad (5)$$

*At. thiooxidans* ATCC 19377 in a mixed culture oxidizes sulfide and sulfur to sulfuric acid and contributes to the decrease in pH (pH 1.5), thereby preventing the formation of jarosite and a hydrophobic layer of sulfur on the surface of chalcopyrite (Equation (5)), removes the effect of passivation of the mineral, and promotes intense oxidation of chalcopyrite. According to Christel and Dopson, 2016, sulfur-oxidizing *A. caldus* seemed to have a supporting role in the early stages of mineral dissolution [89]. The attached sulfur-oxidizing bacteria on the mineral surface utilized the reduced inorganic sulfur compounds (RISCs) released from the mineral by a direct mechanism [90]. In a study conducted by Tao et al. (2021), six artificial communities with varying functions or biodiversity were recreated using six common bioleaching species for chalcopyrite leaching. Communities with low diversity also performed somewhat poorly in bioleaching, and the absence of sulfur oxidizers greatly decreased copper extraction rates in those communities [91].

## 4. Conclusions

A new strain of iron-oxidizing strain *Leptospirillum* sp. Ksh-L was isolated from a dump-bioleaching system of the Kashen copper-molybdenum mine. The cells of the strain Ksh-L are Gram-negative, motile, and vibrio- or spiral-shaped, with a 0.5 μm width and a 1.0–3.0 μm length.

The optimal temperature for the growth of *Leptospirillum* sp. Ksh-L is 35 °C, and the optimal pH is 1.6–1.8. The current study showed how different concentrations of commercially important metals ($Cu^{2+}$, $Zn^{2+}$) have various effects on the oxidizing ability of the strain *Leptospirillum* sp. Ksh-L, depending on the phase of growth and concentration of ferrous iron as a source of energy.

Based on the homology of 16S rRNA, the isolated strain *Leptospirillum* sp. Ksh-L formed a single cluster into Group III and, compared to other strains in Group III, possessed 99.59%, 99.52%, and 96.60% sequence similarity with strains YTW-96-06, YTW-66-06, and *Leptospirillum ferrodiazotrophum* 5C, respectively. Thus, the newly iron-oxidizing strain Ksh-L obtained from a dump-bioleaching system of the Kashen copper-molybdenum mine was identified as *Leptospirillum ferroodiazotrophum*. Type strain: *Leptospirillum ferroodiazotrophum* Ksh-L.

It was also shown that the bioleaching of copper and iron from chalcopyrite by the association of *L. ferrodiazotropum* Ksh-L and *At. thiooxidans* ATCC 19377 in comparison with pure-culture *L. ferrodiazotropum* Ksh-L for 30 days increases about 1.8 and 1.9 times, respectively. Thus, it is supposed that the association of isolated *L. ferrodiazotropum* Ksh-L with sulfur-oxidizing *At. thiooxidans* ATCC 19377 can be successfully used to enhance the efficiency of copper extraction from chalcopyrite.

**Supplementary Materials:** The following supporting information can be downloaded at https://www.mdpi.com/article/10.3390/min14010026/s1, Figure S1: Microphotography of iron-oxidized bacteria (Motic BA 310 trinocular (×1000) microscope with Digital Camera Moticam A16 (×0.5)). Figure S2: Gel electrophoresis of PCR amplification of 16S rRNA amplified product of 16S rDNA for bacterial isolate Leptospirillum sp. Ksh-L.

**Author Contributions:** Conceptualization, A.K., N.V., S.W., G.S., R.Z., and A.V.; methodology, N.V., S.W., G.S., and R.Z.; software, A.K. and A.V.; validation, N.V., S.W., G.S., and R.Z.; formal analysis, A.K. and A.V.; investigation, A.K. and A.V.; resources, N.V., S.W., G.S., R.Z., and A.V.; data curation, A.V.; writing—original draft preparation, A.K. and A.V.; writing—review and editing, N.V., S.W., G.S., and R.Z.; visualization, A.V. and R.Z.; supervision, N.V., S.W., and R.Z.; project administration, N.V. and A.V.; funding acquisition, S.W., R.Z., and A.V. All authors have read and agreed to the published version of the manuscript.

**Funding:** This research was funded by the Higher Education Science Committee of the Ministry of Education, Science, Culture and Sports of Republic of the Armenia, grant number 22rl-031.

**Data Availability Statement:** Data are contained within the article.

**Acknowledgments:** All authors express gratitude to Martin Herzberg and Diana Galea from the Institute for Biology/Microbiology Martin-Luther University Halle-Wittenberg for molecular biology analysis.

**Conflicts of Interest:** The authors declare no conflict of interest. The funders had no role in the design of the study; in the collection, analyses, or interpretation of data; in the writing of the manuscript; or in the decision to publish the results.

## References

1. Zhao, H.; Zhang, Y.; Zhang, X.; Qian, L.; Sun, M.; Yang, Y.; Zhang, Y.; Wang, J.; Kim, H.; Qiu, G. The dissolution and passivation mechanism of chalcopyrite in bioleaching: An overview. *Miner. Eng.* **2019**, *136*, 140–154. [CrossRef]
2. Jones, S.; Santini, J.M. Mechanisms of bioleaching: Iron and sulfur oxidation by acidophilic microorganisms. *Essays Biochem.* **2023**, *67*, 685–699. [CrossRef]
3. Rohwerder, T.; Gehrke, T.; Kinzler, K.; Sand, W. Bioleaching review part A: Progress in bioleaching: Fundamentals and mechanisms of bacterial metal sulfide oxidation. *Appl. Microbiol. Biotechnol.* **2003**, *63*, 239–248. [CrossRef]
4. Huang, C.; Qin, C.; Feng, X.; Liu, X.; Yin, H.; Jiang, L.; Liang, Y.; Liu, H.; Tao, J. Chalcopyrite bioleaching of an in situ leaching system by introducing different functional oxidizers. *RSC Adv.* **2018**, *8*, 37040. [CrossRef]
5. Watling, H.R.; Watkin, E.L.; Ralphe, D.E. The resilience and versatility of acidophiles that contribute to the bio-assisted extraction of metals from mineral sulphides. *Environ. Technol.* **2010**, *31*, 915–933. [CrossRef]
6. Spolaore, P.; Joulian, C.; Gouin, J.; Morin, D.; D'Hugues, P. Relationship between bioleaching performance, bacterial community structure and mineralogy in the bioleaching of a copper concentrate in stirred-tank reactors. *Appl. Microb. Biotechnol.* **2011**, *89*, 441–448. [CrossRef]
7. Ghorbani, Y.; Franzidis, J.P.; Petersen, J. Heap leaching technology—Current State, innovations, and future directions: A review. Miner. Process. *Extr. Metall. Rev.* **2016**, *37*, 73–119. [CrossRef]
8. Veloso, T.C.; Peixoto, J.J.M.; Pereira, M.S.; Leao, V.A. Kinetics of chalcopyrite leaching in either ferric sulphate or cupric sulphate media in the presence of NaCl. *Int. J. Miner. Process* **2016**, *148*, 147–154. [CrossRef]
9. Rawlings, D.E.; Johnson, D.B. The microbiology of biomining: Development and optimization of mineral-oxidizing microbial consortia. *Microbiology* **2007**, *153*, 315–324. [CrossRef] [PubMed]
10. Qin, W.Q.; Li, W.Z.; Lan, Z.Y.; Qiu, G.Z. Simulated small-scale pilot heap leaching of low-grade oxide zinc ore with integrated selective extraction of zinc. *Miner. Eng.* **2007**, *20*, 694–700. [CrossRef]
11. Giaveno, A.; Lavalle, L.; Chiacchiarini, P.; Donati, E. Bioleaching of zinc from low-grade complex sulphide ores in an airlift by isolated *Leptospirillum ferrooxidans*. *Hydrometallurgy* **2007**, *89*, 117–126. [CrossRef]
12. Gericke, M.; Muller, H.H.; van Staden, P.J.; Pinches, A. Development of a tank bioleaching process for the treatment of complex Cu-polymetallic concentrates. *Hydrometallurgy* **2008**, *94*, 23–28. [CrossRef]

13. Patel, B.C.; Tippe, D.R.; Dave, S.R. Optimization of copper and bioregeneration under metallic stress. *Hydrometallurgy* **2012**, *117–118*, 18–23. [CrossRef]
14. Pradhan, N.; Nathsarma, K.C.; Srinivasa Rao, K.; Sukla, L.B.; Mishra, B.K. Heap bioleaching of chalcopyrite: A review. *Miner. Eng.* **2008**, *21*, 355–365. [CrossRef]
15. Abhilash; Mehta, K.D.; Pandey, B.D. Bacterial Leaching Kinetics for Copper Dissolution from a Low- Grade Indian Chalcopyrite Ore. *Rev. Esc. Minas* **2013**, *66*, 245–250. [CrossRef]
16. Panda, S.; Akcil, A.; Pradhan, N.; Deveci, H. Current scenario of chalcopyrite bioleaching: A review on the recent advances to its heap-leach technology. *Bioresour. Technol.* **2015**, *196*, 694–706. [CrossRef]
17. Lawson, E.N. The composition of mixed populations of leaching bacteria active in gold and nickel recovery from sulphide ores. In *ternational Biohydrometallurgy Symposium. In Biomine 97*; Australian Mineral Foundation: Glenside, Australia, 1997; pp. QP4.1–QP4.10.
18. Rawlings, D.E.; Tributsch, H.; Hansford, G.S. Reasons why *'Leptospirillum'*-like species rather than *Thiobacillus ferrooxidans* are the dominant iron-oxidizing bacteria in many commercial processes for the biooxidation of pyrite and related ores. *Microbiol. Read.* **1999**, *145*, 5–13. [CrossRef]
19. Rawlings, D.E.; Coram, N.J.; Gardner, M.N.; Deane, S.M. Thiobacillus caldus and *Leptospirillum ferrooxians* are widely distributed in continuous-flow biooxidation tanks used to treat a variety of metalcontaining ores and concentrates. In *Biohydrometrallurgy and the Environment toward the Mining of the 21st Century, Part A*; Amils, R., Balleste, A., Eds.; Elsevier Press: Amsterdam, The Netherlands, 1999; pp. 773–778.
20. Coram, N.J.; Rawlings, D.E. Molecular relationship between two groups of the genus *Leptospirillum* and the finding that *Leptospirillum ferriphilum* sp. nov. dominates in South African commercial bio-oxidation tanks that operate at 40 °C. *App Environ. Microbiol.* **2002**, *68*, 838–845. [CrossRef]
21. Schippers, A. Microorganisms involved in bioleaching and nucleic acid-based molecular methods for their identification and quantification. In *Microbial Processing of Metal Sulfides*; Donati, R.E., Sand, W., Eds.; Springer: New York, NY, USA, 2007; pp. 3–33.
22. Dave, S.R. Selection of *Leptospirillum ferrooxidans* SRPCBL and development for enhanced ferric regeneration in stirred tank and airlift column reactor. *Bioresour. Technol.* **2008**, *99*, 7803–7806. [CrossRef]
23. Markosyan, G.E. New iron oxidizing bacteria *Leptospirillum ferrooxidans* nov. gen. nov. sp. *Biol. J. Armen.* **1972**, *25*, 26–29. (In Russian)
24. Vardanyan, N.; Akopyan, V. *Leptospirillum*-like bacteria and evaluation of their role in pyrite oxidation. *Microbiology* **2003**, *72*, 438–442. [CrossRef]
25. Ram, R.J.; Verberkmoes, N.C.; Thelen, M.P.; Tyson, G.W.; Baker, B.J.; Blake, R.C., II; Shah, M.; Hettich, R.L.; Banfield, J.F. Community proteomics of a natural microbial biofilm. *Science* **2005**, *308*, 1915–1920. [CrossRef] [PubMed]
26. Liu, J.S.; Xie, X.H.; Xiao, S.M.; Wang, X.M.; Zhao, W.J.; Tian, Z.L. Isolation of *Leptospirillum ferriphilum* by single-layered solid medium. *J. Centr South. Univ. Technol.* **2007**, *14*, 467–473. [CrossRef]
27. Bond, P.L.; Banfield, J.F. Design and performance of rRNA targeted oligonucleotide probes for in situ detection and phylogenetic identification of microorganisms inhabiting acid mine drainage environments. *Microb. Ecol.* **2001**, *41*, 149–161. [CrossRef] [PubMed]
28. Hippe, H. *Leptospirillium* gen. nov (ex Markoysan 1972), nom. rev., including *Leptospirillium ferrooxidans* sp. nov. (ex Markoysan 1972), nom. rev. and *Leptospirillium thermoferrooxidans* sp. nov. (Golovacheva et al. 1992). *Int. J. Syst. Evol. Microbiol.* **2000**, *50*, 501–503. [CrossRef] [PubMed]
29. Bond, P.L.; Smriga, S.P.; Banfield, J.F. Phylogeny of microorganisms populating a thick, subaerial predominantly lithitrophic biofilm at an extreme acid mine drainage site. *Appl. Environ. Microbiol.* **2000**, *66*, 3842–3849. [CrossRef]
30. Tyson, G.W.; Chapman, J.; Hugenholtz, P.; Allen, E.E.; Ram, R.J.; Richardson, P.M.; Solovyev, V.V.; Rubin, E.M.; Rokhsar, D.S.; Banfield, J.F. Community structure and metabolism through reconstruction of microbial genomes from the environment. *Nature* **2004**, *428*, 37–43. [CrossRef]
31. Tyson, G.W.; Lo, I.; Baker, B.J.; Allen, E.E.; Hugenholtz, P.; Banfield, J.F. Genome-directed isolation of the key nitrogen fixer *Leptospirillum ferrodiazotrophum* sp. nov. from an acidophilic microbial community. *Appl. Environ. Microbiol.* **2005**, *71*, 6319–6324. [CrossRef]
32. Allen, E.E.; Tyson, G.W.; Whitaker, R.J.; Detter, J.C.; Richardson, P.M.; Banfield, J.F. Genome dynamics in a natural archaeal population. *Proc. Natl. Acad. Sci. USA* **2007**, *104*, 1883–1888. [CrossRef]
33. Goltsman, D.S.; Denef, V.J.; Singer, S.W.; VerBerkmoes, N.C.; Lefsrud, M.; Mueller, R.S.; Dick, G.J.; Sun, C.L.; Wheeler, K.E.; Zemla, A.; et al. Community genomic and proteomic analyses of chemoautotrophic iron-oxidizing *"Leptospirillum rubarum"* (group II) and *"Leptospirillum ferrodiazotrophum"* (group III) bacteria in acid mine drainage biofilms. *Appl. Environ. Microbiol.* **2009**, *75*, 4599–4615. [CrossRef]
34. Wilmes, P.; Remis, J.P.; Hwang, M.; Auer, M.; Thelen, M.P.; Banfield, J.F. Natural acidophilic biofilm communities reflect distinct organismal and functional organization. *ISME J.* **2009**, *3*, 266–270. [CrossRef] [PubMed]
35. Goltsman, A.D.S.; Dasari, M.; Thomas, B.C.; Shah, M.B.; VerBerkmoes, N.C.; Hettich, R.L.; Banfield, J.F. New group in the *Leptospirillum* clade: Cultivation-independent community genomics, proteomics, and transcriptomics of the new species *"Leptospirillum* group IV UBA BS". *Appl. Environ. Microbiol.* **2013**, *79*, 5384–5393. [CrossRef] [PubMed]

36. Baker-Austin, C.; Dopson, M. Life in acid: pH homeostasis in acidophiles. *Trends Microbiol.* **2007**, *15*, 165–171. [CrossRef] [PubMed]
37. Slonczewski, J.L.; Fujisawa, M.; Dopson, M.; Krulwich, T.A. Cytoplasmic pH measurement and homeostasis in bacteria and archaea. *Adv. Microb. Physiol.* **2009**, *55*, 1–79. [PubMed]
38. Rensing, C.; Grass, G. Escherichia coli mechanism of copper homeostasis in a changing environment. *FEMS Microbiol. Rev.* **2003**, *27*, 197–213. [CrossRef] [PubMed]
39. Outten, F.W.; Huffman, D.L.; Hale, J.A.; Ó Halloran, T.V. The independent cue and cus system confer copper tolerance during aerobic and anaerobic growth in Escherichia coli. *J. Biol. Chem.* **2001**, *276*, 30670–30677. [CrossRef]
40. Puig, S.; Thiele, D.J. Molecular mechanisms of copper uptake and distribution. *Curr. Opin. Chem. Biol.* **2002**, *6*, 171–180. [CrossRef]
41. Costerton, J.W.; Lewandowski, Z.; Caldwell, D.E.; Korber, D.R.; Lappin-Scott, H.M. Microbial biofilms. *Ann. Rev. Microbiol.* **1995**, *49*, 711–745. [CrossRef]
42. Sutherland, I.W. The biofilm matrix--an immobilized but dynamic microbial environment. *Trends Microbiol.* **2001**, *9*, 222–227. [CrossRef]
43. Harneit, K.; Goksel, A.; Kock, D.; Klock, J.H.; Gehrke, T.; Sand, W. Adhesion to metal sulfide surfaces by cells of *Acidithiobacillus ferrooxidans, Acidithiobacillus thiooxidans* and *Leptospirillum ferrooxidans*. *Hydrometallurgy* **2006**, *83*, 245–254. [CrossRef]
44. Teitzel, G.M.; Parsek, M.R. Heavy metal resistance of biofilm and planktonic *Pseudomonas aeruginosa*. *Appl. Environ. Microbiol.* **2003**, *69*, 2313–2320. [CrossRef] [PubMed]
45. Harrison, J.J.; Ceri, H.; Turner, R.J. Multimetal resistance and tolerance in microbial biofilms. *Nat. Rev. Microbiol.* **2007**, *5*, 928–938. [CrossRef] [PubMed]
46. Yong, N.K.; Oshima, M.; Blake, R.C.; Sugio, T. Isolation and some properties of an iron-oxidizing bacterium *Thiobacillus ferrooxidans* resistant to molybdenum ion. *Biosci. Biotechnol. Biochem.* **1997**, *61*, 1523–1526. [CrossRef]
47. Valdés, J.; Pedroso, I.; Quatrini, R.; Dodson, R.J.; Tettelin, H.; Blake, R., II; Eisen, J.A.; Holmes, D.S. *Acidithiobacillus ferrooxidans* metabolism: From genome sequence to industrial applications. *BMC Genom.* **2008**, *9*, 597.
48. Mi, S.; Song, J.; Lin, J.; Che, Y.; Zheng, H.; Lin, J. Complete genome of *Leptospirillum ferriphilum* ML-04 provides insight into its physiology and environmental adaptation. *J. Microbiol.* **2011**, *49*, 890–901. [CrossRef] [PubMed]
49. Cárdenas, J.P.; Lazcano, M.; Ossandon, F.J.; Corbett, M.; Holmes, D.S.; Watkin, E. Draft genome sequence of the iron-oxidizing acidophile *Leptospirillum ferriphilum* type strain DSM 14647. *Genome Announc* **2014**, *2*, e01153-14. [CrossRef] [PubMed]
50. Jiang, H.; Liang, Y.; Yin, H.; Xiao, Y.; Guo, X.; Xu, Y.; Hu, Q.; Liu, H.; Liu, X. Effects of arsenite resistance on the growth and functional gene expression of *Leptospirillum ferriphilum* and *Acidithiobacillus thiooxidans* in pure culture and coculture. *Biomed. Res. Int.* **2015**, *2015*, 203197. [CrossRef] [PubMed]
51. Valenzuela, L.; Chi, A.; Beard, S.; Orell, A.; Guiliani, N.; Shabanowitz, J.; Hunt, D.F.; Jerez, C.A. Genomics, metagenomics and proteomics in biomining microorganisms. *Biotechnol. Adv.* **2006**, *24*, 197–211. [CrossRef]
52. Moreno-Paz, M.; Gómez, M.J.; Arcas, A.; Parro, V. Environmental transcriptome analysis reveals physiological differences between biofilm and planktonic modes of life of the iron oxidizing bacteria *Leptospirillum* spp. in their natural microbial community. *BMC Genom.* **2010**, *11*, 404. [CrossRef]
53. Mangold, S.; Potrykus, J.; Björn, E.; Lövgren, L.; Dopson, M. Extreme zinc tolerance in acidophilic microorganisms from the bacterial and archaeal domains. *Extremophiles* **2013**, *17*, 75–85. [CrossRef]
54. Zhang, Y.; Wu, X.; Liu, D.; Duan, H.; Fan, H. Sequencing and bioinformatics analysis of the metal-related genes in *Acidithiobacillus ferrooxidans* strain DC. *Folia Microbiol.* **2013**, *58*, 551–560. [CrossRef] [PubMed]
55. Watling, H.R. The bioleaching of sulphide minerals with emphasis on copper sulphides—A review. *Hydrometallurgy* **2006**, *84*, 81–108. [CrossRef]
56. Orell, A.; Navarro, C.A.; Arancibia, R.; Mobarec, J.C.; Jerez, C.A. Life in blue: Copper resistance mechanisms of bacteria and Archaea used in industrial biomining of minerals. *Biotechnol. Adv.* **2010**, *28*, 839–848. [CrossRef] [PubMed]
57. Orell, A.; Navarro, C.A.; Rivero, M.; Aguilar, J.S.; Jerez, C.A. Inorganic polyphosphates in extremophiles and their possible functions. *Extremophiles* **2012**, *16*, 573–583. [CrossRef] [PubMed]
58. Silverman, M.P.; Lundgren, D.G. Studies on the chemoautotrophic iron bacterium *Ferrobacillus ferrooxidans*. I. An improved medium and a harvesting procedure for securing high cell yields. *J. Bacteriol.* **1959**, *77*, 642–647. [CrossRef] [PubMed]
59. Gerhardt, P.; Murrayn, R.G.E.; Costilow, R.N.; Nestar, E.W.; Wood, W.A.; Krieg, N.R.; Phillips, G.B. Manual of Methods for General Bacteriology. American Society of Microbiology: Washington, DC, USA, 1981.
60. Mackintosh, M.E. Nitrogen fixation by *Thiobacillus ferrooxidans*. *J. Gen. Microbiol.* **1987**, *105*, 215–218. [CrossRef]
61. Castro, L.; Zhang, R.; Munoz, J.A.; Gonzalez, F.; Blazquez, M.L.; Sand, W. Characterization of exopolymeric substances (EPS) produced by Aeromonas hydrophila under reducing conditions. *Biofouling* **2014**, *30*, 501–511. [CrossRef]
62. Dubois, M.; Gilles, K.A.; Hamilton, J.K.; Rebers, P.; Smith, F. Colorimetric method for determination of sugars and related substances. *Anal. Chem.* **1996**, *28*, 350–356. [CrossRef]
63. Bradford, M.M. A rapid and sensitive method for the quantitation of microgram quantities of protein utilizing the principle of protein-dye binding. *Anal. Biochem.* **1976**, *72*, 248–254. [CrossRef]
64. Blumenkrantz, N.; Asboe-Hansen, G. New method for quantitative determination of uronic acids. *Anal. Biochem.* **1973**, *54*, 484–489. [CrossRef]

65. Sanger, F.; Nicklen, S.; Coulson, A.R. DNA sequencing with chain-terminating inhibitors. *Proc. Natl. Acad. Sci. USA* **1977**, *84*, 5463–5467. [CrossRef] [PubMed]
66. Altschul, S.F.; Madden, T.L.; Schäffer, A.A.; Zhang, J.; Zhang, Z.; Miller, W.; Lipman, D.J. Gapped BLAST and PSI-BLAST: A new generation of protein database search programs. *Nucleic Acids Res.* **1997**, *25*, 3389–3402. [CrossRef] [PubMed]
67. Kearse, M.; Moir, R.; Wilson, A.; Stones-Havas, S.; Cheung, M.; Sturrock, S.; Buxton, S.; Cooper, A.; Markowitz, S.; Duran, C.; et al. Geneious Basic: An integrated and extendable desktop software platform for the organization and analysis of sequence data. *Bioinformatics* **2012**, *28*, 1647–1649. [CrossRef]
68. Tamura, K.; Nei, M.; Kumar, S. Prospects for inferring very large phylogenies by using the neighbor-joining method. *Proc. Natl. Acad. Sci. USA* **2004**, *101*, 11030–11035. [CrossRef] [PubMed]
69. Kumar, S.; Stecher, G.; Li, M.; Knyaz, C.; Tamura, K. MEGA X: Molecular evolutionary genetics analysis across computing platforms. *Mol. Biol. Evol.* **2018**, *35*, 1547–1549. [CrossRef] [PubMed]
70. Lucchesi, C.A.; Hirn, C.F. EDTA Titration of total Iron in Iron(II) and Iron(III) mixtures. Application to Iron driers. *Anal. Chem.* **1960**, *32*, 1191–1193. [CrossRef]
71. Zhang, R.Y.; Xia, J.L.; Peng, J.H.; Zhang, Q.; Zhang, C.G.; Nie, Z.Y.; Qui, G.Z. A new strain *Leptospirillum ferriphilum* YTW315 for bioleaching of metal sulfides ores. *Trans. Nonferrous Met. Soc. China* **2010**, *20*, 135–141. [CrossRef]
72. Valix, M.; Loon, L. Adaptive tolerance behaviour of fungi in heavy metals. *Miner. Eng.* **2003**, *16*, 193–198. [CrossRef]
73. Vartanyan, N.S.; Karavaiko, G.I.; Pivovarova, T.A.; Dorofeev, A.G. Resistance of *Sulfobacillus thermosulfidooxidans* subspecies *asporogenes* to $Cu^{2+}$, $Zn^{2+}$ and $Ni^{2+}$ ions. *Microbiology* **1990**, *59*, 399–404, (English translation of Mikrobiologiya).
74. Sand, W.; Gehrke, T. Analysis and function of the EPS from the strong acidophile *Thiobacillus ferrooxidans*. In *Microbial Extracellular Polymeric Substances*; Wingender, J., Neu, T.R., Flemming, H.C., Eds.; Springer: Berlin/Heidelberg, Germany, 1999; pp. 127–140.
75. Sheng, G.P.; Yu, H.Q.; Li, X.Y. Extracellular polymeric substances (EPS) of microbial aggregates in biological wastewater treatment systems: A review. *Biotechnol. Adv.* **2010**, *28*, 882–894. [CrossRef]
76. Vardanyan, A.; Vardanyan, N.; Markosyan, L.; Sand, W.; Vera, M.; Zhang, R. Biofilm formation and extracellular polymeric substances (EPS) analysis by new isolates of *Leptospirillum*, *Acidithiobacillus* and *Sulfobacillus* from Armenia. *Adv. Mater. Res.* **2015**, *1130*, 153–156. [CrossRef]
77. Vardanyan, A.; Vardanyan, N.; Khachatryan, A.; Zhang, R.; Sand, W. Adhesion to mineral surfaces by cells of *Leptospirillum*, *Acidithiobacillus* and *Sulfobacillus* from Armenian sulfide ores. *Minerals* **2019**, *9*, 69. [CrossRef]
78. Vardanyan, N.; Badalyan, H.; Markosyan, L.; Vardanyan, A.; Zhang, R.; Sand, W. Newly isolated *Acidithiobacillus* sp. Ksh from Kashen copper ore: Peculiarities of EPS and colloidal exopolysaccharide. *Front. Microbiol.* **2020**, *11*, 1–11. [CrossRef] [PubMed]
79. Saitou, N.; Nei, M. The neighbor-joining method: A new method for reconstructing phylogenetic trees. *Mol. Biol. Evol.* **1987**, *4*, 406–425. [PubMed]
80. Felsenstein, J. Confidence limits on phylogenies: An approach using the bootstrap. *Evolution* **1985**, *39*, 783–791. [CrossRef] [PubMed]
81. Yin, H.; Cao, L.; Qiu, G.; Wang, D.; Kellogg, L.; Zhou, J.; Liu, X.; Dai, Z.; Ding, J.; Liu, X. Molecular diversity of 16S rRNA and gyrB genes in copper mines. *Arch. Microbiol.* **2008**, *189*, 101–110. [CrossRef] [PubMed]
82. Corbett, M.K. *Leptospirillum*: A Study of the Nitrogen Fixing Capabilities. Ph.D. Thesis, Philosophy of Curtin University, Perth, Australia, 2011; pp. 1–202. Available online: https://espace.curtin.edu.au/handle/20.500.11937/81 (accessed on 31 December 2011).
83. Hallberg, K.B.; Johnson, D.B. Microbiology of a wetland ecosystem constructed to remediate mine drainage from a heavy metal mine. *Sci. Total Environ.* **2005**, *338*, 53–66. [CrossRef] [PubMed]
84. Ehrich, S.; Behrens, D.; Lebedeva, E.; Ludwig, W.; Bock, E. A new obligately chemolithoautotrophic, nitrite-oxidizing bacterium, *Nitrospira moscoviensis* sp. nov. and its phylogenetic relationship. *Arch. Microbiol.* **1995**, *164*, 16–23. [CrossRef]
85. Akcil, A.; Ciftci, H.; Deveci, H. Role and contribution of pure and mixed cultures of mesophiles in bioleaching of a pyritic chalcopyrite concentrate. *Miner. Eng.* **2007**, *20*, 310–318. [CrossRef]
86. Fu, B.; Zhou, H.; Zhang, R.; Qiu, G. Bioleaching of chalcopyrite by pure and mixed cultures of *Acidithiobacillus* spp. and *Leptospirillum ferriphilum*. *Int. Biodeterior. Biodegrad.* **2008**, *62*, 109–115. [CrossRef]
87. Wang, Y.; Zeng, W.; Qiu, G.; Chen, X.; Zhou, H. A moderately thermophilic mixed microbial culture for bioleaching of chalcopyrite concentrate at high pulp density. *Appl. Environ. Microbiol.* **2014**, *80*, 741–750. [CrossRef] [PubMed]
88. Schippers, A.; Sand, W. Bacterial Leaching of Metal Sulfides Proceeds by Two Indirect Mechanisms via Thiosulfate or via Polysulfides and Sulfur. *Appl. Env. Microbiol.* **1999**, *65*, 319–321. [CrossRef] [PubMed]
89. Christel, S.; Dopson, M. Less may be more: Improving chalcopyrite bioleaching kinetics via sequential inoculation of acidophilic model species. Book of abstracts of the 10th International Conference on Establishment of Cooperation between Companies and Institutions in the Nordic Countries, the Baltic Sea Region and the World. In Proceedings of the Linnaeus Eco-Tech, Kalmar, Sweden, 21–23 November 2016; p. 169.

90. Watling, H.R.; Collinson, D.M.; Li, J.; Mutch, L.A.; Perrot, F.A.; Rea, S.M.; Reith, F.; Watkin, E.L.J. Bioleaching of a low-grade copper ore, linking leach chemistry and microbiology. *Miner. Eng.* **2014**, *56*, 35–44. [CrossRef]
91. Tao, J.; Liu, X.; Luo, X.; Teng, T.; Jiang, C.; Drewniak, L.; Yang, Z.; Yin, H. An integrated insight into bioleaching performance of chalcopyrite mediated by microbial factors: Functional types and biodiversity. *Bioresour. Technol.* **2021**, *319*, 124219. [CrossRef]

minerals

MDPI

Article

# Automated Mineralogy and Diagnostic Leaching Studies on Bulk Sulfide Flotation Concentrate of a Refractory Gold Ore

**Mustafa K. Guner [1,2,*], Gülay Bulut [2], Ahmad Hassanzadeh [1,3], Stefanie Lode [1] and Kurt Aasly [1]**

1 Department of Geoscience and Petroleum, Faculty of Engineering, Norwegian University of Science and Technology, 7031 Trondheim, Norway; ahmad.hassanzadeh@ntnu.no (A.H.); stefanie.lode@ntnu.no (S.L.); kurt.aasly@ntnu.no (K.A.)
2 Department of Mineral Processing Engineering, Istanbul Technical University, 34469 Istanbul, Turkey; gbulut@itu.edu.tr
3 Maelgwyn Mineral Services Ltd., Ty Maelgwyn, 1A Gower Road, Cathays, Cardiff CF24 4PA, UK
* Correspondence: mustafa.k.guner@ntnu.no; Tel.: +0047-46744966

**Abstract:** Obtaining detailed and precise information from a classified refractory gold ore has been a long-standing challenge in mineral processing and process mineralogy. Although the concept of diagnostic leaching has been extensively addressed in the literature, very little information is available linking this approach with current advanced characterization techniques such as automated mineralogy. The present research study aims to characterize the flotation concentrate of refractory gold ore by combining diagnostic leaching and automated mineralogy to examine its processability. The diagnostic leaching process was applied stepwise at five stages, and the automated mineralogy was performed on different size fractions of the flotation concentrate. The chemical (X-ray fluorescence (XRF), atomic absorption spectroscopy (AAS), and inductively coupled plasma-optical emission spectroscopy (ICP-OES)) and mineralogical (X-ray diffraction (XRD)) analyses of the feed sample confirmed that the ore is a low-grade gold ore with 0.7 g/t Au. Initially, bottle roll tests were carried out to investigate leaching behavior, and the ore's refractory nature, and gold recoveries of bottle roll tests in different sizes were below 40 wt%; it is classified as a high refractory gold ore as a result of direct cyanide leaching. Bulk sulfide flotation was applied to increase the gold content of the material. Automated mineralogy results demonstrated that most of the gold in the concentrate is present as an invisible gold form, and 63.7 wt% of gold was distributed in pyrite. Diagnostic leaching results showed 39.7 wt% of total gold was leachable using direct cyanide leaching, and around 33 wt% of undissolved gold was located in pyrite and arsenopyrite.

**Keywords:** automated mineralogy; refractory gold ore; gold flotation; gold mineralogy; cyanide leaching; diagnostic leaching

**Citation:** Guner, M.K.; Bulut, G.; Hassanzadeh, A.; Lode, S.; Aasly, K. Automated Mineralogy and Diagnostic Leaching Studies on Bulk Sulfide Flotation Concentrate of a Refractory Gold Ore. *Minerals* **2023**, *13*, 1243. https://doi.org/10.3390/min13101243

Academic Editors: Elizabeth Watkin, Vladko Panayotov and Marinela Ivanova Panayotova

Received: 4 July 2023
Revised: 12 September 2023
Accepted: 20 September 2023
Published: 23 September 2023



## 1. Introduction

A sharp reduction in the cut-off grade of gold mines and increasing demand for this element in modern technologies have led metallurgists to deal with refractory gold-type ores more than before. As many different newly opened mineral deposits are typically more polymineralic and complex, the easy-to-treat mineral deposits become rarer. Similarly, processing polymineralic and complex gold ore bodies has become relatively challenging and requires sophisticated solutions [1,2]. The easily accessible free-milling gold deposits gradually decline, and therefore, the interest in refractory gold ores, and the processing of their tailings, increases globally. For such ore bodies, a very good understanding of the ore characteristics is crucial with respect to the selection of an appropriate methodology for downstream processes.

In mineral processing, gold ores are commonly classified as (i) free-milling, (ii) refractory gold, and (iii) complex gold ore types, where the free-milling ore type has high gold recovery (>90%), while the refractory one has a recovery of lower than 90% [3–7] and the gold recovery of the third class is only economically feasible under excessive reagent consumption [8]. There are several properties that lead to the classification of gold ore as a refractory gold ore as given below:

- Gold is physically locked in the host mineral [3,8];
- Gold is chemically locked in the form of an alloy with other metals [3,8];
- Gold is present as sub-microscopic-size invisible gold or a solid solution in sulfide structures [7–11];
- The ore contains gangue minerals whose properties lead to high consumption of oxygen or cyanide [9];
- Carbonates and clays present in the ore act as activated carbon (preg-robbing effect) [3,12];
- The presence of passivation layers on the gold grains [8].

If one or more of these conditions leads to gold recovery below 90%, the ore is classified as refractory gold ore and general conclusion in the literature is that the most important properties that make the ore refractory are the first three bullet points mentioned above [3–11]. The ore type specified in the first bullet point can achieve high recovery with ultrafine grinding, but oxidation processes are required for the ore structures identified in the second and third subjects [3,8,9]. Gold defined as invisible gold mentioned in the third bullet point is invisible using conventional optical and electron microscopes, and typically occurs as inclusions in pyrite found together with arsenopyrite [10]. However, it is believed that such statements should be revised and updated based on improvements in technologies utilized for the characterization of such ores. For this reason, the current study applies a supplementary method to the diagnostic leaching test as a commonly used approach for the characterization of gold ores.

Automated mineralogy has been proven as a key technology for the characterization of primary and secondary raw materials, plant optimization, feasibility study, and process design [13–21]. Scanning electron microscopy (SEM) combined with energy dispersive spectrometry (EDS) and specific software [17] has been applied to provide detailed information about ore properties, such as bulk mineralogy, liberation degree, grain and particle size distribution, mineral associations, mineral compositions, and metal deportment. This technology is capable of providing invaluable information on types of gold-bearing minerals, gold exposure, mineral associations of locked gold grains, and assays of invisible gold [22,23]. Although the automated mineralogy method provides detailed information about the ore, it is generally used with other chemical and mineralogical analysis methods. Chryssoulis and Cabri [24] used an elemental analysis for trace elements, which indicate minerals potentially containing gold, as a part of the study, which they defined as "gold mineralogical balance" in addition to diagnostic cyanidation and an electron probe micro-analysis (EPMA). Also, heavy liquid separation (HLS) or gravity separation is generally used to increase the chance of detecting gold before sample preparation for a SEM analysis [19,22,25].

Diagnostic leaching is not an extraction method but has been widely accepted and applied as an alternative analytical approach. Through this, one can investigate gold deportment and its distribution in different mineral matrices. The technique applies sequential leaching based on acidic and oxidative reagents to measure the gold recovery. Diagnostic leaching is a simple and relatively rapid method that enables metallurgists to determine which mineral prevents gold dissolution with the cyanidation process [26]. This information is very essential for process design in gold concentration plants.

The standard procedure for diagnostic leaching was initially developed by Anglo American Research Laboratories in the 1980s [27]. The main purpose of each leaching stage during the diagnostic leaching test is to apply specific reagents to dissolve selected target minerals, liberate the gold, and then recover it with the cyanide leaching process. The first stage starts with conventional or intensive direct cyanide leaching. All stages, except the first one, occur in two leaching conditions, initially, acid treatment with a selected reagent, followed by cyanide leaching. NaCN leaching is commonly applied to dissolve free gold from destroyed minerals using oxidative/acidic leaching. The initial stage aims to dissolve as much as possible free-milling gold (leachable gold) under conventional or intensive leaching conditions such as high cyanide (CN) concentration, long leaching time, high pH, and high concentration of dissolved oxygen. The undissolved gold, which does not respond to cyanide dissolution in the first stage, is mostly in a refractory structure. This indeed shows that some mineralogical structures prevent interaction between cyanide and gold, and some oxidative reagents are required to provide this interaction. Acidic reagents are used for destroying target minerals in each stage. Lorenzen and Tumilty [27] addressed which reagent should be employed under specific conditions and published a guideline. According to their stepwise description, applied reagents and removed minerals are given in Table 1 [26,28] and chemical dissolution reactions of sphalerite and pyrite with specified reagents have been provided as examples in Equations (1) and (2) [28,29]. Test conditions, including solid rate, leaching time, reagent type, and concentration, can be varied depending on the material and operator as long as the target mineral(s) are entirely destroyed and the gold recovery is maximum for each stage. Also, it is not mandatory to apply all oxidative leaching reagents in Table 1. According to the guidelines, some oxidative leaching can be eliminated [27,28,30,31], and it is suggested to use a minimum number of acidic reagents for more accurate assessment [26–33]. In addition to Table 1, carbon in leach (CIL) can be applied to determine the preg-robbing effect [22]. Although acidic reagents can be selectable, almost all the studies in the literature have used the following three reagents: hydrochloric acid (HCl), sulfuric acid ($H_2SO_4$), and nitric acid ($HNO_3$) [26,28–35]. In addition to these common three reagents, hydrofluoric acid (HF) has been used in the last stage to dissolve silicates [25,36–38], and in some procedures, the remaining residue was dissolved with the aqua-regia solution instead of HF after $HNO_3$ leaching [34,39] or applied roasting at a high temperature to remove carbonaceous compounds in the residue [40].

$$ZnS + H_2SO_4 \rightarrow ZnSO_4 + H_2S \quad (1)$$

$$2FeS_2 + 10HNO_3 \rightarrow 2Fe^{3+} + 2H^+ + 4SO_4^{2-} + 10NO + 4H_2O \quad (2)$$

**Table 1.** Diagnostic leaching reagents and target materials of these [27,28].

| Reagent | Removed Material |
|---|---|
| Sodium cyanide | Gold |
| Sodium carbonate | Gypsum and arsenates |
| Hydrochloric acid | Pyrrhotite, calcite, dolomite, galena, and goethite |
| Hydrochloric acid/Tin(II) chloride | Hematite, calcine, and ferrites |
| Sulfuric acid | Uraninite, sphalerite, unstable copper sulfides, unstable base metal sulfides, and unstable pyrite |
| Iron(III) chloride | Sphalerite, galena, unstable sulfides, and tetrahedrite |
| Nitric acid | Pyrite, arsenopyrite, and marcasite |
| Oxalic acid | Oxide coatings |
| Hydrofluoric acid<br>Acetonitrile elution | Silicates<br>Loaded carbons |

The most critical challenge of diagnostic leaching is that the reagent dissolves not only one target mineral but also several associated minerals together with a different portion in one stage [35]. Therefore, combined diagnostic leaching with the mineralogical results allows mineral processors to interpret the ore characteristics more accurately and choose the appropriate processes for ore treatments.

In the literature, as shown in Table 2, studies are conducted using different types of materials and various methods. As seen, most of these studies were mainly focused on the chemical treatments disregarding the mineralogical information that can play a complementary role in an ore characterization. In addition, most of these studies used primary resources such as refractory gold ore or complex gold ore. This is a matter of presentation for the current study. This research initially explains how an ore can be classified as refractory gold ore through experimental activities. It also includes practical studies on increasing the gold content. Furthermore, the research aims to characterize the bulk sulfide flotation concentrate of refractory gold ore using automated mineralogy and diagnostic leaching. To enhance the accuracy of characterization, the results obtained from diagnostic leaching are correlated and compared with automated mineralogy results. Thus, for the first time in the literature, diagnostic leaching and automated mineralogy are applied together to understand the processability of the flotation concentrate of refractory gold.

**Table 2.** Literature review about studies using various SEM-based analyses and diagnostic leaching.

| Material | Method | Reference |
|---|---|---|
| Refractory gold ore | QEMSCAN and diagnostic leaching | [25] |
| Complex gold ore | PIXE and diagnostic leaching | [26] |
| Flotation tailing | Diagnostic leaching | [27] |
| Roasted calcine sample | Diagnostic leaching | [27] |
| Refractory arsenic–silver ore | MLA and diagnostic leaching | [32] |
| Refractory gold–silver ore | Diagnostic leaching | [29] |
| Refractory gold ore | Diagnostic leaching | [30] |
| Complex gold ore | QEMSCAN and diagnostic leaching | [33] |
| Flotation concentrate | Diagnostic leaching | [34] |
| Refractory gold–silver ore | Diagnostic leaching | [36] |
| Refractory arsenic–gold ore | X-ray spectrometer (EDS), electron probe microanalysis (EPMA), and diagnostic leaching | [37] |
| Refractory gold ore | Diagnostic leaching | [38] |
| Carbonaceous gold ore | Diagnostic leaching | [39] |

## 2. Materials and Methods

### *2.1. Materials and Sample Preparation*

In this study, sample materials were collected from drill cores representing different depths of the sulfur zone of a low-sulfidation epithermal-type Au-Ag mineralization. The deposit studied in this research is located in north-western Turkey. However, the name of the mining company, the procedure of drilling, and details on the location are preserved due to confidentiality.

Each drill core sample was crushed with a lab-scale jaw crusher separately to a $d_{80}$ of approximately 1.7 mm. The samples were later mixed with a cement mixer to create a final composite. Representative splits of the initial sample were collected with the grid sampling method for wet chemical and mineralogical studies. Gold and silver contents for each sample were analyzed using atomic absorption spectroscopy (AAS, PinAAcle 900f); inductively coupled plasma optical emission spectroscopy (Agilent 511 ICP-OES) and XRF were employed for other elemental compounds in the final composite. The amount of total S and C was determined with an induction furnace/infrared (IR) carbon sulfur analyzer (LECO CS230SH). The XRD analysis showed that detected phases were quartz, kaolinite, illite, pyrite, dolomite, and siderite. Elemental content of the initial material that was the feed of flotation before automated mineralogy and diagnostic leaching is detailed in Section 3.1.

The ore was divided into two parts using a riffle and split into 1 kg bags with Retsch Sample Divider PT100; 1 kg bags were grouped for grinding tests, sieve analyses, flotation, and leaching tests. The grinding optimization test was performed at the different grinding times with Lab Scale ESSA Rod Mill. The material was prepared with dry grinding in three different sizes, as $d_{80}$ = 45 μm, 73 μm, and 106 μm, for direct cyanide leaching. The rest of the ore was stored to produce sufficient flotation concentrate as a requirement for starting diagnostic leaching tests. All prepared materials were sealed and stored in a fridge below −16 °C.

### 2.2. Methods

#### 2.2.1. Cyanide Leaching Tests

In the cyanide leaching experiments, conventional bottle roll tests were performed to investigate leaching behaviors and understand the ore's refractory nature. Experimental tests were performed with 450 g of initial amount material per test in three different particle sizes (i.e., $d_{80}$ = 45 μm, $d_{80}$ = 73 μm, and $d_{80}$ = 106 μm), five different leaching time intervals (i.e., 2, 8, 24, 48, and 72 h), constant solid ratios (45% ($w/w$)), and an initial cyanide (Merck GmbH, ≥95.0 purity and 49.01 g/mol) concentration of 1000 kg/t at an alkaline medium (pH = 10.5–11.0). The pH, dissolved oxygen (DO) level, and free cyanide consumption into the solution were monitored at certain time intervals (2, 8, 24, and 48 h). If necessary, NaCN and lime were used for cyanidation and pH adjustments, respectively. A pH meter (WTW 315i) was used for its measurements, and a DO-meter (WTW Oxi 3205) was utilized for dissolved oxygen measurements. A silver nitrate ($AgNO_3$) titration mechanism was performed using the Bottle-Top Burette Titrette procedure to determine the free cyanide concentration in the saturated solution. In addition to understanding the refractory nature of the ore, the optimum leaching condition was determined based on the cyanide leaching tests for further applications, such as flotation and diagnostic leaching.

#### 2.2.2. Flotation Tests

Flotation is a well-known method for upgrading sulfide ores with low gold content [41–46]. As the gold grade of the ore was very low, it was planned to upgrade the gold content with bulk sulfide flotation. For this purpose, several flotation experiments under different operating conditions were performed. The experimental condition to achieve the highest gold content in concentrate was selected and repeated to produce enough bulk concentrate as feed material for the diagnostic leaching tests; optimizing the flotation process for current material is not the scope of this study. The flotation experiments were performed in a 2.5 L mechanically agitated Denver® flotation machine with 850 g of material (30% $w/w$ solid ratio). The pulp was agitated at a rate of 1400 rpm, pH was adjusted to 8–9 using $Ca(OH)_2$, tap water was used for all experiments, and 25 g/t of thionocarbamate (Solvay AERO® MX980) and 25 g/t of a potassium amyl xanthate (PAX) mixture were added at each stage as collectors. Aeration was regulated at a rate of 2 L/min for 5 min during the flotation with a conditioning time of 3 min in two rougher stages. The concentrate was floated in two stages for 5 min.

#### 2.2.3. Automated Mineralogy Analysis

In this study, the main focus of automated mineralogical analyses (AM) was to locate, identify, and define textural properties of the gold and the different particles carrying gold. Minor focus was given to the overall mineralogical properties of the flotation concentrate. Gold deportment was investigated in terms of liberation/locking and grain sizes of host gold minerals, the mineral association of gold-containing particles, gold distribution using mineral grains, and indicator elements for gold. It is also important to emphasize that the material used for both AM and diagnostic leaching is flotation concentrate produced with the method explained in Section 2.2.2., not the initial material created with mixed drill core samples. It is worth noting that, because of the low concentration of gold, a Knelson centrifugal concentrator was used for preconcentration in order to increase the possibility to detect gold during the analysis.

The material was sieved into three different size fractions, −106 + 74 μm, −74 + 38 μm, and −38 μm, and each fraction was split to an appropriate mass and blended with graphite at a particular portion following the procedure given by Røisi and Aasly [47]. Further, the sample/graphite mix was mixed with epoxy resin (EPO-TEK® 301, Billerica, MA, USA) to produce a regular 25 mm polished section. After curing, each 25 mm section was cut into four cross sections that were re-molded into one 30 mm block per fraction using epoxy resin; 30 mm polished sections were created and carbon-coated before the SEM analyses. Automated mineralogy analyses were conducted at the Norwegian Laboratory for Minerals and Materials Characterisation (MiMaC, Piovene Rocchette, Italy) facility at the Department of Geoscience and Petroleum, NTNU, Norway. A ZEISS Sigma 300 FE SEM was used for BSE imaging, a Bruker XFlash 6–60 EDS detector for EDS analyses, and ZEISS Mineralogic Mining© software for quantitative automated mineralogical analyses. Mapping a full polished section at a small step size would require an unfeasibly long acquisition time. To locate and identify gold and host minerals, a bright phase search (BPS) routine was first applied to the sample by utilizing grey-level thresholding on the backscattered electron (BSE) images to discriminate gold, which has a high atomic mass and accordingly, light grey levels, from other mineral phases with lower atomic masses (low grey levels). Setting the threshold limits to include only these targeted brighter sulfide phases, analysis time is significantly reduced. However, knowing the minerals hosting the bright phases is of great importance, and therefore, as an additional step of enlarging the analyzed area around the grains of interest, the dilation routine was added to the image processing set up. This dilation routine adds a defined number of pixels to the periphery of the selected area. Hence, it enables measurement to be expanded to a rim of the minerals that host the bright phases. To avoid unnecessary measuring of epoxy, thresholding and other arithmetic image processing scripts are applied to exclude epoxy resin and fractures in the analyzed area.

Microscope settings were set to a 20 kV acceleration voltage, 0.009 s dwell time, minimum of 2000 spectrum counts, and 2.5 μm step size/pixel size for the full grain mapping and 0.6 μm for BPS analyses. the degree of liberation of grain was classified as follows: if it is greater than 90%, it is considered "liberated"; if it falls between 30% and 90%, it is categorized as "middling"; and if it is less than 30%, it is classified as "locked". This classification is based on the mineral liberation determined with the partial perimeter.

### 2.2.4. Diagnostic Leaching Tests

The XRD analysis of feed material provides important preliminary information to determine the minimum number of stages, select reagents, and design the whole experiment for the diagnostic leaching test. The XRD analysis shows that the studied sample, which was feed of flotation, comprises quartz, kaolinite, illite, pyrite, dolomite, and siderite. Table 1 shows minerals dissolved using which reagents [27,28]. According to Table 1, and the known mineralogical and chemical properties of the run-of-mine (ROM) ore, HCl was selected due to its effectiveness against a wide range of carbonatite minerals, and $H_2SO_4$ was beneficiated to dissolve unstable metallic sulfides and specifically labile pyrite. The diagnostic leaching test conditions are given in Table 3. Ferric leaching is not commonly used, but it is suggested when the sample has sulfide concentrates. According to Lorenzen's guideline, when the sample contains over 10% of any of these minerals—tetrahedrite, galena, sphalerite, or concentrated pyrite—it is recommended to apply a ferric solution (100 g/L $FeCl_3$ + 2 M HCl) before cyanide leaching [28]. Since the material applied in this study was a sulfide flotation concentrate, ferric leaching was added to the leaching sequence. $HNO_3$ was selected as the most critical reagent because the ROM ore contained pyrite and a significant amount of pyrite was floated during the flotation. Therefore, it was estimated that the highest increase in gold extraction should be in the last two stages.

**Table 3.** Experimental conditions considered for the diagnostic leaching tests.

| Stage | Feed Product | Solid Ratio (%, *w*/*w*) | Leaching Time (h) | Reagent | Temperature (°C) | pH | DO |
|---|---|---|---|---|---|---|---|
| 1st stage | Flotationconcentrate | 45 | 48 | NaCN 5 kg/t | 25 | 11.5–12.0 | 20–30 |
| 2nd stage | Residue of the previous leaching | 33 | 24 | HCl %12 *v*/*v* | 70 | - | |
| | Residue of the previous leaching | 45 | 48 | NaCN 1 kg/t | 25 | 11.5–12.0 | 20–30 |
| 3rd stage | Residue of the previous leaching | 33 | 24 | $H_2SO_4$ %48 *v*/*v* | 80 | - | |
| | Residue of the previous leaching | 45 | 48 | NaCN 1 kg/t | 25 | 11.5–12.0 | 20–30 |
| 4th stage | Residue of the previous leaching | 33 | 24 | $FeCl_3$ 100 mg/L, 2 M HCl | 95 | - | |
| | Residue of the previous leaching | 45 | 48 | NaCN 1 kg/t | 25 | 11.5–12.0 | 20–30 |
| 5th stage | Residue of the previous leaching | 33 | 24 | $HNO_3$ %55 *v*/*v* | 90 | - | |
| | Residue of the previous leaching | 45 | 48 | NaCN 1 kg/t | 25 | 11.5–12.0 | 20–30 |

In the first stage of diagnostic leaching, a 24 h cyanide leaching process was carried out with a 5 kg/t NaCN concentration at a pulp with a solid ratio of 45% (*w*/*w*) and pH of about 11.5–12.0. The concentration of dissolved oxygen was in the range of 25–30 ppm. The reason for the aggressive leaching condition was to achieve maximum dissolution at room temperature (25 ± 3 °C) within 24 h of leaching time. High cyanide and oxygen concentrations shortened the leaching time and increased the leaching kinetics. The main purpose of diagnostic leaching was not to indicate an optimum condition for the gold extraction, but rather to transfer as much gold as possible to the liquid phase. The dissolved gold of the first stage represents leachable gold (free-milling gold).

All acidic leaching tests were conducted in an IKA LR-2.ST laboratory reactor with a 2 L glass reactor vessel surrounded by a heating bath. A mechanical stirrer was placed in the middle of the vessel at the reactor's top using a remote controller. All acidic treatments were conducted at 100 rpm, and the temperature was controlled with an IKA HBR-4 heating bath. The reactor setup was located under a laboratory fume hood. After completing the leaching process, the residue was dewatered with a lab-scale press filter, and the sample solution was taken from the pregnant solution for the chemical analysis. The residue cake was dried at 50 °C for 24 h and washed with water two times and dried again at the same condition; it is important to remove all the remained acidic reagents from the cake to avoid a potential reaction between acidic reagents during subsequent leaching. The final dried solid residue was powdered, and the representative samples were taken for chemical analyses. The material that was applied to acidic leaching was a sulfide concentrate. Therefore, it was necessary to be careful of possible gas emissions and overflowing of the pulp during the contact between the acidic leach solution and the material, especially during $HNO_3$ leaching. This attention was important for both work safety and the consistency of test results.

## 3. Results and Discussions

### 3.1. Feed Properties

The feed sample analyses using the ICP-OES and LECO showed that the sample contained 0.7 g/t of Au, 1.25 g/t of Ag, and 2.5% of total sulfides, where the detailed chemical composition is given in Table 4. The XRF analysis of the sample indicated the presence of $SiO_2$ as the most abundant phase; kaolinite, illite, pyrite, dolomite, siderite, and other minerals were encountered using a Rigaku X-ray diffractometer (XRD, D/Max-IIIC).

**Table 4.** Chemical composition of the final mixed sample using ICP-OES and XRF analyses.

| Method | Element | Unit | Content | Element | Unit | Content |
|---|---|---|---|---|---|---|
| ICP-OES | Au | g/t | 0.7 | Fe | wt% | 3.4 |
| | Ag | g/t | 1.2 | S | wt% | 2.5 |
| | Cu | g/t | 73.1 | C | wt% | 0.4 |
| | As | g/t | 449.8 | Ni | g/t | 93.1 |
| | Pb | g/t | 94.5 | Zn | g/t | 95.1 |
| | Mo | g/t | 13.0 | Cr | g/t | 28.3 |
| | Sn | g/t | <10 | Sb | g/t | 21.6 |
| XRF | $Al_2O_3$ | wt% | 10.8 | $SiO_2$ | wt% | 73.7 |
| | $K_2O$ | wt% | 2.1 | $SO_3$ | wt% | 2.5 |
| | $CO_2$ | wt% | 4.8 | MgO | wt% | 0.8 |
| | $TiO_2$ | wt% | 0.4 | $Na_2O$ | wt% | 0.2 |

### 3.2. *Cyanide Leaching and Flotation*

Chemical (XRF, AAS, and ICP-OES) and mineralogical (XRD) analyses of the feed sample confirmed that the ore is a low-grade gold ore with 0.7 g/t of Au (Table 4). Cyanide leaching tests in different conditions were performed to understand ROM ore's leaching properties. As Table 5 shows, it was observed that the maximum gold recovery was below 40% as a result of cyanide leaching tests performed in three different particle sizes ($d_{80}$ = 45, 73, and 106 μm) and up to 72 h. Therefore, the ore was classified as high-refractory gold ore according to La Brooy's classification [8]. It was observed that both the gold extraction recovery and the cyanide consumption of the ores with $d_{80}$ sizes of 45 μm and 73 μm were very close to each other, and the gold recovery in tests with a sample $d_{80}$ of 106 μm was the lowest compared to two other sizes. Table 5 shows that the gold recovery and reagent consumption increased as the particle size became finer. The rise in the recovery in gold extraction after 24 h is relatively low for all three sizes. Therefore, the optimum leaching time was decided as 24 h, and the leaching size 73 μm was chosen as a consequence of the cyanide leaching tests.

**Table 5.** Cyanide leaching results in different leaching size and time.

| Leaching Size, $d_{80}$ (μm) | Leaching Time (h) | Au Recovery (%) | NaCN Consumption (kg/t) |
|---|---|---|---|
| 45 | 2 | 32.2 | 0.28 |
| | 8 | 35.5 | 0.69 |
| | 24 | 37.0 | 0.82 |
| | 48 | 38.6 | 0.90 |
| | 72 | 39.0 | 0.98 |
| 73 | 2 | 35.8 | 0.37 |
| | 8 | 35.8 | 0.56 |
| | 24 | 39.2 | 0.75 |
| | 48 | 39.2 | 0.86 |
| | 72 | 39.7 | 0.97 |
| 106 | 2 | 29.8 | 0.38 |
| | 8 | 33.4 | 0.49 |
| | 24 | 33.4 | 0.54 |
| | 48 | 36.8 | 0.62 |
| | 72 | 36.8 | 0.69 |

After ensuring that the ore is a high-refractory gold ore and contains relatively high S content, bulk sulfide flotation was employed as a pretreatment method in order to increase

Au content. It was recognized that 92% of the total sulfide was recovered with the flotation tests with a mass pull (also known as yield defined as concentrate mass divided by the feed mass) of 15.9%. The initial content of S was 2.5% and increased to 18.4%. Gold content, which was 0.7 ppm, increased to 2.5 ppm with a 59% recovery. The application of the flotation is not to optimize the flotation parameters but to increase the Au content of the feed. According to these conditions, the final flotation concentrate, which is the feed of the diagnostic leaching, included the 2.5 ppm Au content. The flotation test results are given in Table 6.

**Table 6.** Results of flotation experiments.

| Product | Mass (wt%) | Au Content (ppm) | Au Recovery (wt%) |
|---|---|---|---|
| Concentrate | 15.9 | 2.5 | 59.7 |
| Tailing | 84.1 | 0.3 | 40.3 |
| Total | 100.0 | 0.7 | 100.0 |

### 3.3. *Automated Mineralogy (AM)*

As a result of the automated mineralogy study, a bulk mineralogy, BSE image, colored mineral map, liberation degree for every mineral, trace element analysis for gold, and elemental gold deportment in minerals were obtained. The bulk mineralogical composition of the flotation concentrate (feed material of diagnostic leaching) gathered as a result of the based-on AM analysis is given in Figure 1. According to the results, the most dominant mineral in the sample is pyrite for each size fraction. In addition to pyrite, galena, arsenopyrite, skinnerite, barite, and quartz were encountered in different quantities. Pyrite was found at 51.4 wt% in the −106 + 74 µm, 48 wt% in the −74 + 38 µm, and 28 wt% in the −38 µm size fractions. Overall, pyrite constitutes 34.9 wt% of the total, while the portion of other minerals in the material is as follows: galena, 14.4 wt%; skinnerite, 14.9 wt%; barite, 8.5 wt%; arsenopyrite, 5.6 wt%; and quartz, 1.1 wt%. After comparing the XRD result of ROM ore with flotation concentrate, it was observed that the mineralogy of the materials changed drastically after flotation; sulfides are the most dominant fraction in the concentrate while most of the silicates remained in tailing. Quartz content was around 59 wt% before flotation, and the highest quartz content after flotation was 3.5 wt% in the range of −106 + 74 µm. It was presented that 2.9 wt% of gold is dispersed in quartz; on the other hand, the amount of gold remaining in silicates after diagnostic leaching is below 1 wt%. These two results show that gold is not dispersed within the quartz in a significant portion.

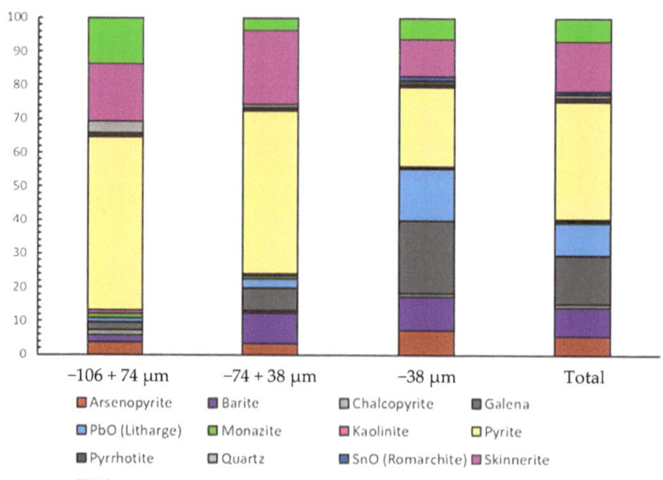


**Figure 1.** Bulk mineralogy (wt%) of the flotation concentrate (feed of diagnostic leaching).

In automated mineralogy using ZEISS Mineralogic, minerals are classified based on energy-dispersive X-ray spectroscopy (EDS) and are quantified based on their chemical composition. Additionally, mineral maps were generated for each sample. In Figure 2, examples of mapped particles are given for pyrite, feldspar, quartz, muscovite, ilmenite, and chalcopyrite grains.

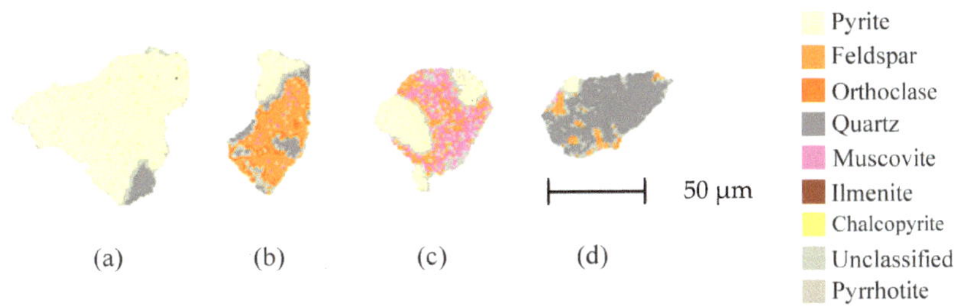


**Figure 2.** Automated mineralogy mineral maps for some particles as examples, (**a**) disseminated pyrrhotite and chalcopyrite in pyrite, (**b**) pyrite, quartz, and feldspar association, (**c**) sericitic alteration of feldspar associated with pyrite, and (**d**) pyrite and sericitic feldspar associated with quartz.

In terms of the mineral liberation analysis (Figure 3), pyrite was chosen as the target mineral for three reasons; the material contains a very limited amount of visible gold, the dominant mineral after flotation is pyrite, and a large portion of the invisible gold that is sub-microscopic-size gold as explained in the Introduction section disseminated as inclusion in pyrite. It was observed that a significant amount of pyrite in the concentrate was liberated based on the liberation definition, which is a partial perimeter explained in Section 2.2.3. Pyrite is free with 47.7 wt%, 46.8 wt%, and 69.8 wt% liberation degrees between −106 + 74 μm, −74 + 38 μm, and −38 μm size fractions, respectively.

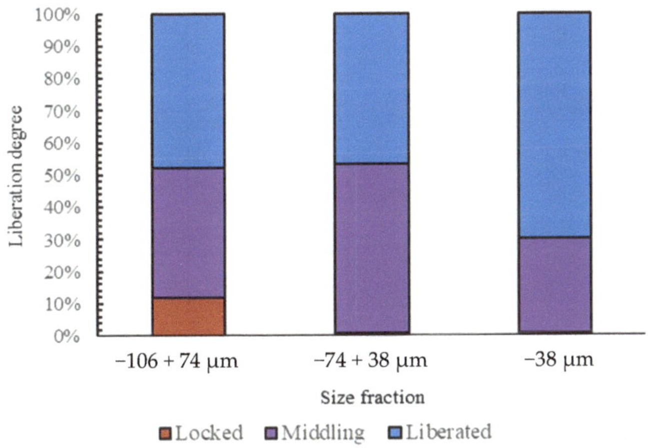


**Figure 3.** The liberation degree of pyrite in the size fraction based on the partial perimeter.

As galena has a relatively similar average atomic number to gold, its grains were frequently picked up using the BPS. On the other hand, BPS as an analyzing method worked very well to detect micron-sized grains of bright phases; however, a limited amount of the visible electrum and Au grains was detected (Figure 4). Invisible gold occurring as traces in different mineral grains is given in Figure 5.

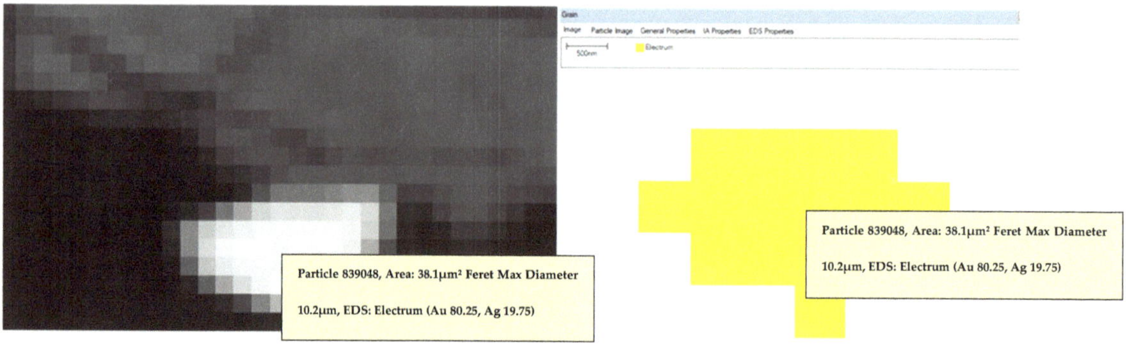


**Figure 4.** Detected electrum using BPS.

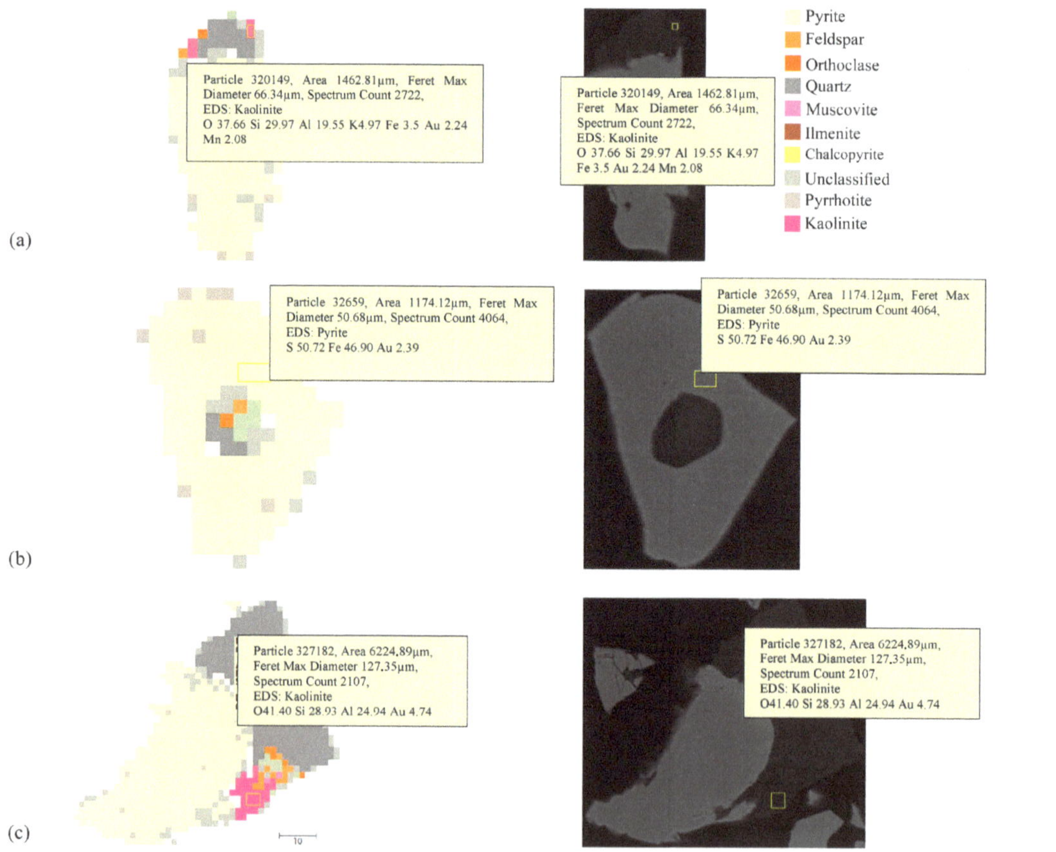


**Figure 5.** Occurrence of invisible gold inside mineral grains; (**a**) gold in kaolinite hosted using pyrite grain, 10 µm scale; (**b**) gold-inclusion pyrite, 5 µm scale; (**c**) invisible gold hosted using kaolinite, 10 µm scale.

Automated mineralogy provides detailed particle and grain maps and tables based on elemental content. It can be determined for a particular mineral in what quantities Au is detected with other elements. This detailed analysis provides an opportunity to make a trace element analysis for gold. After examining the correlation of Au with other elements in the samples, the most detected trace elements along with Au are copper (Cu), aluminum (Al), cobalt (Co), antimony (Sb), tin (Sn), and arsenic (As).

According to the AM results, 63.7 wt%, 3.5 wt%, 2.0 wt%, 5.2 wt%, 10.8 wt%, and 2.9 wt% of the gold are deported in pyrite, galena, arsenopyrite, skinnerite, barite, and quartz (Figure 6); also, an important amount of the gold exists in the form of invisible gold. This result clarifies the cause of the ore's refractoriness and low gold solubility. Also, a low portion of the gold (1.8 wt%) was locked inside the grouped phases with mixed spectra dominated with Fe and Si that was grouped as an Fe-Sil mix (only a few gold and electrum grains could be detected, and cyanide could still dissolve around 40 wt% of gold in the material after flotation). Although the current investigation did not cover it, Mössbauer spectroscopy can provide additional information about invisible gold using electronic interactions of gold atoms with various materials.

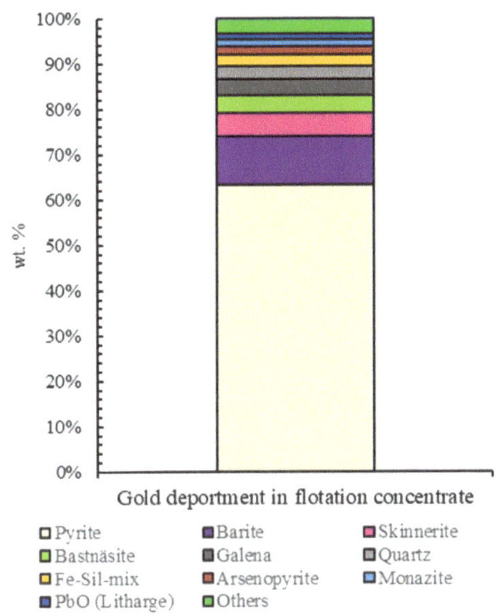


**Figure 6.** Gold deportment with minerals in the concentrate.

### 3.4. Diagnostic Leaching

In the first stage of diagnostic leaching, NaCN leaching was completed with 39.7% Au extraction, and the residue of NaCN leaching contained 2.0 ppm of Au. HCl (12 vol%) was applied in the second stage where its residue was leached using the NaCN. Through this stage, 14.9% of total gold was liberated from its host mineral. Since carbonates like dolomite were detected with the XRD analysis of the ROM sample, the presence of such minerals could potentially cause gold extraction of 14.9% at this stage, where the gold extraction increased up to 54.6% at the end of the second stage.

$H_2SO_4$ leaching was performed as the third stage, and it was aimed to dissolve labile (unstable) metal-bearing minerals, labile copper sulfides, and pyrite to release gold locked inside these matrices. Cumulative gold extraction increased to 67.1% after this stage, while the gold content in the residue decreased to 1.5 ppm with the NaCN leaching after $H_2SO_4$ treatment. This result showed that 12.5% of the gold in this concentrate could not interact with the cyanide due to the presence of minerals like uraninite, sphalerite, and labile metal sulfides.

At the fourth stage of diagnostic leaching, $FeCl_3$ treatment was applied to dissolve pyrite. The critical point of ferric leaching was that the Eh had to be above 500 mV. Hydrogen peroxide ($H_2O_2$) was used to increase the Eh of the system, but ferric chloride (iron (III) chloride) treatment was not enough to dissolve pyrite in the concentrate. Only

8.2% of total gold was liberated with $FeCl_3$ and lower than estimates made before the experimental study. As mentioned in Section 2.2.4, $HNO_3$ is the key reagent to dissolve various metal sulfides that tend to be associated with gold, such as pyrite and arsenopyrite, as it was mentioned in Section 2.2.4., because material was a sulfide flotation concentrate. According to the results, the highest mass loss and gold liberation happened during stage five ($HNO_3$ + NaCN leaching), 24.4% of the material left the system, and 24.6% of gold was dissolved and extracted with the last cyanide leaching step. Figure 7 shows the gold extraction of each stage and the Au content of all products. After stage five, the final residue included 0.02 ppm of Au, meaning that silicates and other undissolved entities embedded only 0.02 ppm of total Au. Thus, HF treatment was not required.

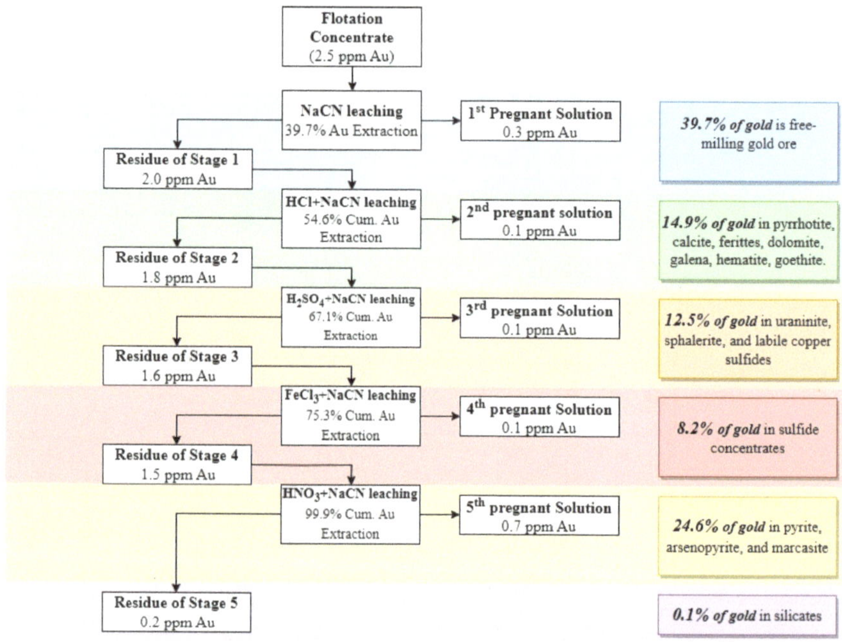


**Figure 7.** Results of diagnostic leaching tests in each stage (wt%).

The results shown in Figure 7 indicate that 39.7% of the total gold was leachable with direct cyanide leaching. In addition, approximately 14.9% of the gold could not be dissolved with direct cyanide leaching in stage two because it was enclosed inside minerals such as pyrrhotite, calcite, dolomite, hematite, and goethite. Stage three showed that sphalerite and unstable copper sulfides hosted approx. 12% of the gold to prevent dissolution with cyanide. The portion of the gold located in the unstable sulfide concentrates, which was expected to be predominantly pyrite, was approximately 8% based on stage four. The most dominant host mineral was indicated as pyrite, as approximately 25% of the gold was released after $HNO_3$ dissolution of pyrite and arsenopyrite; when considering the AM results based on stage four and stage five, it can be summarized that approximately 33% of the gold was enclosed in the pyrite. One thing that is worth emphasizing is that diagnostic leaching alone does not give accurate results, but shows groups with more than one mineral, based on Table 1. That is why mineralogical characterization is necessary for more accurate results. Otherwise, a metallurgist cannot indicate whether the gold is in pyrite, arsenopyrite, or marcasite, as in stage five.

## 4. Conclusions

The main purpose of the study was to evaluate flotation concentrate in terms of refractoriness by performing diagnostic leaching and correlating results with automated mineralogy. This study successfully demonstrated that diagnostic leaching could be used as a fast and practical alternative analysis method not only on the ROM ore but also for flotation concentrate. The diagnostic leaching revealed the refractory property of the material with the first stage, which was direct cyanide leaching, and the automated mineralogy explained the reasons for this refractoriness. In the case of Au in S-bearing minerals, flotation was introduced as a very simple and effective method to concentrate Au, but the increase in the Au content in the material did not indicate that the material is easy to process.

Diagnostic leaching should be performed in a particular order, but it was a very flexible method; the mineralogical analysis was the key point for reagent selection in designing diagnostic leaching sequences and conditions. Acidic reagents could dissolve more than one mineral simultaneously; therefore, using only this method can be misleading. Applying advanced and detailed mineralogical studies such as automated mineralogy with diagnostic leaching was found to be useful for gold deportment study. Although diagnostic leaching provides practical and direct indicators of the material's processability, automated mineralogy provides very detailed information not only about gold deportment but also minerals present in the gangue other than the gold-bearing minerals, and it is able to affect further processing of concentrate.

Since refractory structures of this concentrate were defined with both diagnostic leaching and automated mineralogy, further studies, including pretreatment processes such as pressure oxidation, Nitrox/Redox, or bioleaching, should be examined, and integration between flotation and those pretreatment methods should be investigated in terms of technical and economic aspects as well. Moreover, an alternative reagent to cyanide, meaning a solvent with a lower environmental impact, should be integrated into the diagnostic leaching test procedure.

**Author Contributions:** Conceptualization, M.K.G. and G.B.; methodology, M.K.G., G.B., S.L. and K.A.; software, M.K.G. and S.L.; validation, G.B., S.L. and K.A.; formal analysis, M.K.G. and A.H.; investigation, M.K.G.; resources, M.K.G.; data curation, M.K.G. and S.L.; writing—original draft preparation, M.K.G. and A.H.; writing—review and editing, M.K.G., G.B., S.L., A.H. and K.A.; visualization, M.K.G.; supervision, G.B. and K.A.; project administration, K.A.; Funding acquisition, K.A. All authors have read and agreed to the published version of the manuscript.

**Funding:** This project was partially funded by the Norwegian Laboratory for Mineral and Materials Characterisation, MiMaC, with a project number of 269842/F50.

**Data Availability Statement:** The presented data will be available per request.

**Acknowledgments:** The Research Council of Norway is acknowledged for the support to the Norwegian Laboratory for Mineral and Materials Characterisation, MiMaC, project number: 269842/F50. The first author thanks Sercan Sevgul, who is a metallurgical superintendent from Koza Gold Co-operation, for his theoretical and technical help. In addition, the first author would like to thank Przemyslaw B. Kowalczuk (Pshem), his PhD supervisor at NTNU, for his constant support of this manuscript. We are truly thankful to Istanbul Technical University and the staff of the Mineral Processing Engineering Department for their continuous assistance.

**Conflicts of Interest:** The authors declare no conflict of interest.

## References

1. Lakshmanan, V.I.; Gorain, B. *Innovations and Breakthroughs in the Gold and Silver Industries*; Springer: Cham, Switzerland, 2019; p. 288, ISBN 978-3-030-32551-0. [CrossRef]
2. Hassanzadeh, A.; Safari, M.; Hoang, D.H.; Khoshdast, H.; Albijanic, B.; Kowalczuk, P. Technological assessments on recent developments in fine and coarse particle flotation systems. *Miner. Eng.* **2022**, *180*, 107509. [CrossRef]
3. Afenya, P. Treatment of carbonaceous refractory gold ores. *Miner. Eng.* **1991**, *4*, 1043–1055. [CrossRef]
4. Akçıl, A.; Çiftçi, H. Pretreatments Applied to Refractory Gold Ores. *Sci. Min. J.* **2009**, *48*, 17–30.

5. Asamoah, R.; Amankwah, R.; Addai-Mensah, J. Cyanidation of refractory gold ores: A review. In Proceedings of the 3rd UMaT Biennial International Mining and Mineral Conference, Tarkwa, Ghana, 30 July 2014. [CrossRef]
6. Yüce, A.E. *Cevher Hazırlama el Kitabı*, 1st ed.; Önal, G., Ateşok, G., Perek, K., Eds.; Yurt Madenciliğini Geliştirme Vakfı: Istanbul, Turkey, 2014; pp. 425–450.
7. Marsden, J.; House, I. *The Chemistry of Gold Extraction*, 2nd ed.; SME: Denver, CO, USA, 2006; pp. 19–68.
8. La Brooy, S.R.; Linge, H.G.; Walker, G.S. Review of gold extraction from ores. *Miner. Eng.* **1994**, *7*, 1213–1241. [CrossRef]
9. Komnitsas, C.; Pooley, F. Mineralogical characteristics and treatment of refractory gold ores. *Miner. Eng.* **1989**, *2*, 449–457. [CrossRef]
10. Chen, T.; Cabri, L.; Dutrizac, J. Characterizing gold in refractory sulfide gold ores and residues. *JOM* **2002**, *54*, 20–22. [CrossRef]
11. Reich, M.; Kesler, S.E.; Utsunomiya, S.; Palenik, C.S.; Chryssoulis, S.L.; Ewing, R.C. Solubility of gold in arsenian pyrite. *Geochim. Cosmochim. Acta* **2005**, *69*, 2781–2796. [CrossRef]
12. Yannopoulos, J.C. (Ed.) Treatment of Refractory Gold Ores. In *The Extractive Metallurgy of Gold*, 1st ed.; Springer: New York, NY, USA, 1991; pp. 79–110. [CrossRef]
13. Butcher, A.R.; Helms, T.A.; Gottlieb, P.; Bateman, R.; Ellis, S.; Johnson, N.W. Advances in the quantification of gold deportment by QEMSCAN. In Proceedings of the Seventh Mill Operators Conference, Kalgoorlie, Australia, 12–14 October 2000.
14. Baum, W.; Lotter, N.O.; Whittaker, P.J. Process Mineralogy–A New Generation for Ore Characterization and Plant Optimization. In Proceedings of the SME Annual Meeting, Denver, CO, USA, 23–25 February 2004; pp. 23–25.
15. Fandrich, R.; Gladys, O. Modern SEM based quantitative mineralogy. *Acta Microsc.* **2007**, *16*, 44–45.
16. Malvik, T. History and growth of modern process mineralogy. *Mineralproduksjon* **2014**, *5*, A1–A19.
17. Zhou, J.; Gu, Y. Geometallurgical characterization and automated mineralogy of gold ores. In *Gold Ore Processing*; Elsevier: Amsterdam, The Netherlands, 2016; pp. 95–111.
18. Fernandes, I.B.; Rudolph, M.; Hassanzadeh, A.; Bachmann, K.; Meskers, C.; Peuker, U.; Reuter, M.A. The Quantification of entropy for multicomponent systems: Application to microwave-assisted comminution. *Miner. Eng.* **2021**, *170*, 107016. [CrossRef]
19. Gu, Y. Automated scanning electron microscope based mineral liberation analysis: An introduction to JKMRC/FEI mineral liberation analyser. *J. Miner. Mater. Charact. Eng.* **2003**, *2*, 33–41. [CrossRef]
20. Graham, S.D.; Brough, C.; Cropp, A. An introduction to ZEISS mineralogic mining and the correlation of light microscopy with automated mineralogy: A case study using BMS and PGM analysis of samples from a PGE-bearing chromite prospect. In Proceedings of the Precious Metals '15, Falmouth, UK, 11–12 May 2015. volume 11.
21. Becker, M. The contribution of applied mineralogy to sustainability in the mine life cycle. *Miner. Eng.* **2023**, *199*, 108121. [CrossRef]
22. Coetzee, L.L.; Theron, S.J.; Martin, G.J.; Van der Merwe, J.D.; Stanek, T.A. Modern gold deportments and its application to industry. *Miner. Eng.* **2011**, *24*, 565–575. [CrossRef]
23. Petruk, W. *Applied Mineralogy in the Mining Industry*; Elsevier: Amsterdam, The Netherlands, 2000.
24. Chryssoulis, S.L.; Cabri, L.J. Significance of Gold Mineralogical Balances in Mineral Processing. *Trans. Inst. Min. Metall. Sect. A Min. Technol.* **1990**, *99*, C1–C10.
25. Nazari, A.M.; Ghahreman, A.; Bell, S. A comparative study of gold refractoriness by the application of QEMSCAN and diagnostic leach process. *Int. J. Miner. Process.* **2017**, *169*, 35–46. [CrossRef]
26. Goodall, W.R.; Scales, P.J.; Ryan, C.G. Applications of PIXE and diagnostic leaching in the characterisation of complex gold ores. *Miner. Eng.* **2005**, *18*, 1010–1019. [CrossRef]
27. Lorenzen, L.; Tumilty, J. Diagnostic Leaching as an analytical tool for evaluating the effect of reagents on the performance of a gold plant. *Miner. Eng.* **1992**, *5*, 503–512. [CrossRef]
28. Lorenzen, L. Some guidelines to the design of a diagnostic leaching experiment. *Miner. Eng.* **1995**, *8*, 247–256. [CrossRef]
29. Celep, O.; Alp, I.; Deveci, H.; Vicil, M. Characterization of refractory behaviour of complex gold/silver ore by diagnostic leaching. *Trans. Nonferrous Met. Soc. China* **2009**, *19*, 707–713. [CrossRef]
30. Lorenzen, L.; Van Deventer, J. The mechanism of leaching of gold from refractory ores. *Miner. Eng.* **1992**, *5*, 1377–1387. [CrossRef]
31. Tumilty, J.; Sweeney, A.; Lorenzen, L. Diagnostic leaching in the development of flowsheets for new ore deposits. In Proceedings of the Metallurgical Society of the Canadian Institute of Mining and Metallurgy; Elsevier: Amsterdam, The Netherlands, 1987; pp. 157–167.
32. Celep, O.; Yazici, E.Y.; Altinkaya, P.; Deveci, H. Characterization of a refractory arsenical silver ore by mineral liberation analysis (MLA) and diagnostic leaching. *Hydrometallurgy* **2019**, *189*, 105106. [CrossRef]
33. Goodall, W.R.; Scales, P.J.; Butcher, A.R. The Use of QEMSCAN and diagnostic leaching in the characterisation of visible gold in complex ores. *Miner. Eng.* **2005**, *18*, 877–886. [CrossRef]
34. Teague, A.; Swaminathan, C.; Van Deventer, J. The behaviour of gold bearing minerals during froth flotation as determined by diagnostic leaching. *Miner. Eng.* **1998**, *11*, 523–533. [CrossRef]
35. Henley, K.; Clarke, N.; Sauter, P. Evaluation of a diagnostic leaching technique for gold in native gold and gold-silver tellurides. *Miner. Eng.* **2001**, *14*, 1–12. [CrossRef]
36. Celep, O.; Alp, I.; Deveci, H. Application of diagnostic leaching technique for refractory gold ores. *DPÜ. Fen Bilim. Enstitüsü* **2008**, *16*, 81–90.
37. Bidari, E.; Aazami, M.; Aghazadeh, V. Process Mineralogical Study of the Arsenical Zone from a Carlin-type Gold Deposit. *Min. Metall. Explor.* **2020**, *37*, 1307–1315. [CrossRef]

38. Stephen, R.S.; Anuar WN, S.; Ismail, S.; Jabit, N.A. Characterisation and diagnostic leaching of gold-bearing mineral ore, East Coast Peninsular Malaysia. *Malays. J. Microsc.* **2021**, *17*, 139–153.
39. Wang, J.; Wang, W.; Dong, K.; Fu, Y.; Xie, F. Research on leaching of carbonaceous gold ore with copper-ammonia-thiosulfate solutions. *Miner. Eng.* **2019**, *137*, 232–240. [CrossRef]
40. Eric, O.J. Diagnostic Leaching for Refractory Gold Ores. Available online: https://www.srk.com/en/publications/diagnostic-leaching-for-refractory-gold-ores (accessed on 1 March 2019).
41. Acarkan, N.; Bulut, G.; Gül, A.; Kangal, O.; Karakaş, F.; Kökkılıç, O.; Önal, G. The effect of collector's type on gold and silver flotation in a complex ore. *Sep. Sci. Technol.* **2010**, *46*, 283–289. [CrossRef]
42. Dunne, R. Flotation of gold and gold-bearing ores. In *Gold Ore Processing*; Elsevier: Amsterdam, The Netherlands, 2016; pp. 315–338. [CrossRef]
43. Forrest, K.; Yan, D.; Dunne, R. Optimisation of gold recovery by selective gold flotation for copper-gold-pyrite ores. *Miner. Eng.* **2001**, *14*, 227–241. [CrossRef]
44. Lu, M.; Sun, Y.; Duan, H.; Xie, D.; Chen, Z.; Wang, J.; Wang, C. A working condition recognition method based on multivariable trend analysis for gold-antimony rougher flotation. *Miner. Eng.* **2020**, *156*, 106493. [CrossRef]
45. Teague, A.; Van Deventer, J.; Swaminathan, C. A conceptual model for gold flotation. *Miner. Eng.* **1999**, *12*, 1001–1019. [CrossRef]
46. O'Connor, C.; Dunne, R. The flotation of gold bearing ores—A review. *Miner. Eng.* **1994**, *7*, 839–849. [CrossRef]
47. Røisi, I.; Aasly, K. The effect of graphite filler in sample preparation for automated mineralogy—A preliminary study. *Mineralproduksjon* **2018**, *8*, A1–A23.

*minerals*

*Article*

# Study on Electrochemical Behavior of Oxidized Pyrite in Alkaline Electrolyte

**Yehao Huang [1,2], Zipei Jia [1], Wen Wang [1], Jia Yao [1], Runbo Gao [1], Laifu Xu [1], Hongtao Zhang [1], Yongsheng Zhang [1] and Xiangyu Song [1,3,*]**

1 School of Chemical Engineering, Zhengzhou University, Zhengzhou 410001, China; huangyehao1688@163.com (Y.H.)
2 Henan Academy of Geology, Zhengzhou 450016, China
3 Zhongyuan Critical Metals Laboratory, Zhengzhou 410001, China
* Correspondence: songxiangyu5268@163.com

**Abstract:** Nowadays, refractory gold ore production around the world accounts for about 30% of total gold production, and the low leaching rate of such gold concentrate seriously limits the efficient utilization of gold resources. The alkaline preoxidation process can improve the leaching rate of this kind of gold deposit and has good development and application prospects. Therefore, it is of great significance to study the oxidation and dissolution behavior of pyrite in an alkaline environment. In this paper, the oxidation process of gold-bearing pyrite in an alkaline electrolyte was studied using electrochemical techniques, and the oxidation products of a pyrite electrode in an alkaline solution were characterized using XPS, SEM, and other analytical methods. The results show that the optimum pH for pyrite electrochemical oxidative dissolution is about 12, and the oxidation potential of pyrite should be above 0.8 V. In the process of alkaline oxidative dissolution of pyrite, part of the S element enters the electrolyte in the forms of $S_x^{2-}$, $S_2O_3^{2-}$, $SO_3^{2-}$, and $SO_4^{2-}$, and a small amount of the S element is adsorbed on the surface of the electrode in the form of $S^0$ and becomes a part of the passive layer. The Fe element is adsorbed on the surface of the electrode in the forms of $Fe(OH)_2$, $Fe_2O_3$, and $Fe_2(SO_4)_3$, which become the main components of the passivation layer. This study provides a theoretical basis and reference data for the chemical preoxidation treatment of gold-bearing sulfide ores.

**Keywords:** pyrite; alkaline oxidation; electrochemical technique; surface passivation; chemical preoxidation

**Citation:** Huang, Y.; Jia, Z.; Wang, W.; Yao, J.; Gao, R.; Xu, L.; Zhang, H.; Zhang, Y.; Song, X. Study on Electrochemical Behavior of Oxidized Pyrite in Alkaline Electrolyte. *Minerals* **2023**, *13*, 1070. https://doi.org/10.3390/min13081070

Academic Editors: Marinela Ivanova Panayotova and Vladko Panayotov

Received: 3 July 2023
Revised: 5 August 2023
Accepted: 7 August 2023
Published: 12 August 2023



## 1. Introduction

Refractory gold ore refers to gold ores with a low leaching rate under conventional leaching conditions, and gold ores with a leaching rate lower than 80% after direct cyanide leaching treatment are generally considered refractory gold ores [1–4]. Nowadays, about 60%~70% of the global gold resource reserves belong to refractory gold sources, and the total gold produced from refractory gold ores accounts for roughly 30% of the gold in the world. Microfine disseminated gold ore is one of the main types of refractory gold deposits [5].

The structures of difficult-to-process gold ores can be divided into three categories: (1) high-sulfur, high-arsenic gold ores (the gold in such ores is dispersed by sulfide ore packages, which are the world's largest reserves of difficult-to-leach gold ores [6–8]); (2) silicate parcels of gold ores (the gold in such ores is dispersed by the vein mineral package [9,10]); (3) carbonaceous gold ore (although the gold in such ore is not wrapped by other minerals, in the leaching process, the organic carbon in the slurry competes for the adsorption of gold-containing complexes, also known as the "gold hijacking phenomenon", which affects the leaching efficiency [11–13]). Gold-bearing sulfide ore is the largest class of difficult-to-manage gold ore reserves; the most common form of gold-bearing sulfide ore is gold-bearing pyrite [5]. Gold is often wrapped by pyrite in the form of fine particles and is not easy to dissociate; in addition, the sulfur element in the ore consumes many gold-leaching chemicals, so that the subsequent leaching rate drops. Therefore, before leaching gold, gold-bearing pyrite pretreatment generally needs to be conducted, through the pretreatment and then opening of the outer layer of fine-grained gold parcels, eliminating the impact of minerals on the leaching chemicals, and improving the efficiency of gold leaching [9,14–16].

The preoxidation treatment methods of refractory gold ore containing sulfur and arsenic can be roughly divided into three categories: roasting oxidation, biological oxidation, and wet chemical oxidation. The roasting oxidation method is the most widely used in industry, but it has the following main drawbacks in the production process: firstly, the roasting process not only has a high cost, but also produces a large amount of sulfur and arsenic oxides, which are discharged into the air and cause serious pollution; secondly, the following cyanide leaching process uses highly toxic cyanide, and the residue of cyanide in the discharged wastewater and tailings also causes toxic pollution issues. The biological oxidation method has the advantages of environmentally friendly production, strong adaptability, and stable and reliable operation of the production process, but it also has the disadvantages of a long oxidation cycle, high requirements for processing equipment, and the need to adjust alkalinity in subsequent cyanide leaching, which directly leads to high production costs. According to different pH values of pulp, chemical oxidation can be roughly divided into alkali and acid pretreatment methods. According to different reaction pressure conditions, it also can be divided into atmospheric pretreatment and hot-pressing pretreatment.

In fact, there are many studies on the chemical oxidation of pyrite, and the studies on oxidation in an acidic environment generally agree that iron is mainly converted to free $Fe^{2+}$ and $Fe^{3+}$, sulfur is mostly converted to $SO_4^{2-}$, and a small portion of the latter is converted to elemental S. Especially in the process of bio-oxidation, the elemental sulfur that is generated on the wall of the container can be clearly seen [17,18]. However, in alkaline environments, the results of studies on the mechanism and products of pyrite oxidation are not consistent. For example, some studies demonstrate the belief that the oxidation products contain elemental sulfur [19], while others demonstrate the belief that elemental sulfur is only a temporary intermediate in the oxidation process of pyrite, and that there is no elemental sulfur in most medium oxidation products [20,21]. In addition, some researchers have suggested that the oxidation product of iron is mainly FeOOH [22]; however, others have suggested that $Fe(OH)_3$ is the final stable oxidation product [23,24]. The oxidation process of pyrite has also shown that S0 is formed on the surface of pyrite electrodes during acidic oxidation, which inhibits the oxidation of the pyrite to a certain extent, but S0 is eventually further oxidized, and the passivation effect is weakened [25,26]. However, alkaline oxidation affects the oxidation rate of pyrite by producing a passivation film, in which the main component of the passivation film is iron hydroxide [27,28]. The acid method and hot-pressing method have the disadvantages of high cost, high equipment requirements (acid corrosion resistance), difficult process control (high-pressure danger), immature waste liquid recycling technology, and environmental pollution due to harsh conditions such as pressurization, heating, and acidity. Therefore, they are less often applied in production. However, because of the advantages of low equipment requirements, low production costs, easy process control, and less environmental pollution, the alkaline preoxidation process at room temperature and pressure has become a hot spot of research in recent years and has good development and application prospects.

Many research data on the alkaline oxidation of sulfide ores show that the oxidation sequence of common gold-bearing sulfide ores (arsenopyrite, chalcopyrite, and pyrite, etc.) in an alkaline oxidation system is as follows: arsenopyrite $\rightarrow$ chalcopyrite $\rightarrow$ pyrite [29–32]. Since pyrite is relatively more difficult to oxidize than the other sulfides, it is of great significance to study the oxidation and dissolution behaviors of pyrite in an alkaline environment. In this paper, the oxidation and passivation of gold-bearing pyrite in alkaline electrolytes were studied using an electrochemical method.

## 2. Materials and Methods

### 2.1. Preparation of Massive Pyrite Electrode

Massive pyrite electrodes are mainly used to research the corrosion and passivation processes generated on surface of pyrite. The electrode requires not only high pyrite mineral purity, but also a large crystal size, no crack defects, and a smooth surface finish after grinding. The above-mentioned blocky pyrite is cut and ground into a square sheet with an area of $10 \times 10$ mm$^2$ and a thickness of 3~5 mm, embedded in a special square electrode casing, the surrounding contact parts are sealed with sealant to ensure that one side of the pyrite electrode is in contact with the electrolyte, and it is ensured that pyrite is in good contact with the electrode copper wire through the electrode casing. To improve the smoothness and freshness of the electrode's surface, it is polished by using polishing powders of 10, 5, and 1 μm in sequence before electrochemical measurement, then the electrode is cleaned with ultrapure water and acetone, and finally, the electrode is blown dry with argon gas and coated with a film in vacuum for standby.

### 2.2. Electrochemical Device and Supporting Electrolyte

In this experiment, a cylindrical glass reactor with a capacity of 200 mL and an electrolyte of NaOH solution was used as the electrolytic cell for the electrochemical reaction test. During the test, a small amount of inert electrolyte, $Na_2SO_4$, was also added to increase the conductivity of the electrolyte and lower the liquid contact potential. Electrochemical measurements were performed using a three-electrode system. The working electrode was the previously prepared block pyrite electrode, the counter electrode was a platinum electrode with an area of 1 cm$^2$, and the reference electrode was the Hg/HgO electrode (filled with 1 M KOH). Unless otherwise stated, all potentials reported in this work are relative to Hg/HgO electrode (0.098 V vs. SHE at room temperature). The Autolab PGSTAT204 (Metrohm, Herisau, Switzerland) electrochemical workstation was used for the electrochemical measurements. The measurement data were automatically collected by a computer. The electrochemical test system (AUTOLAB) used in the tests is shown in Figure 1.

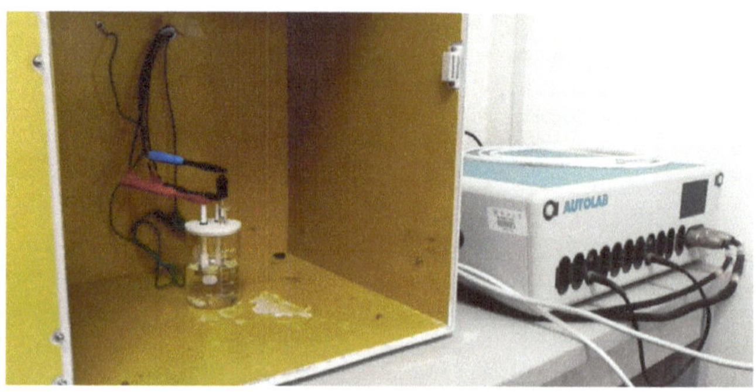


**Figure 1.** The AUTOLAB for the experiment.

### 2.3. Electrochemical Tests

Four different electrochemical testing techniques were used in the experiments, namely cyclic voltammetry (CV), linear scanning voltammetry (LSV), current–time (i-t), and electrochemical impedance spectroscopy (EIS). Before each electrochemical measurement, the working electrode was maintained in the electrolyte for 30 min to ensure stability or reproducibility.

After the working electrode was immersed in electrolyte for 1 h, the open-circuit potential ($E_{OCP}$) was measured using an electrochemical workstation with a sampling time interval of 1 s for 30 min. In the early stage, the cyclic voltammetry curve with a wide range of potential was first carried out, and then the test range of potential was finally determined by the locations of the oxidation peak and reduction peak. LSV analysis was performed from $E_{OCP}$ to an applied potential of 1.4 V at a scan rate of 0.01 V/s. CV analysis was performed between −2 V and 0.5 V at a scan rate of 0.01 V/s. Tafel polarization curve experiments were carried out in an electrolyte with a pH of 12, a voltage measurement range of −0.5 to 0.4 V, and a scan rate of 0.01 V/s. With a holding time of 10 h, the i-t curves were measured at constant potentials of 0.4 V, 0.6 V, 0.8 V, 1.0 V, and 1.2 V, respectively. EIS curves were recorded in the frequency range of 0.05~$10^5$ Hz at potentials of 0.2 V, 0.4 V, 0.6 V, and 0.8 V in the electrolyte with a pH of 12. EIS data were fitted by using the software ZSimpwin 3.60 (2014).

### 2.4. X-ray Photoelectron Spectroscopy (XPS) Analysis

XPS analysis was performed by using a Thermo Escalab 250XI (Thermo Fisher Scientific, America) spectrometer under ultravacuum conditions ($2 \times 10^{-9}$ Torr) with an Al-Kα light source (hv = 1486.6 eV), operating voltage of 15 kV, a power of 25 W, and a temperature of 25 °C. The samples were analyzed using full spectrum scanning as well as fine spectrum scanning during the test, and the binding energies of the characteristic peaks of all elements measured were corrected by using $C_{1s}$(284.6 eV) as a reference.

### 2.5. Sulfur Ion Analysis

On the basis of reviewing and analyzing the literature, a method for analyzing sulfur-containing ions in the electrolyte was developed through analytical verification:

(1) Determination of polysulfide ion concentration

Adjust the electrolyte pH > 9, take 10 mL, add 10% copper chloride solution dropwise until the dark green copper sulfide is completely precipitated, then filter the precipitate and wash it with acetic acid 5 times, dissolve the washed copper sulfide precipitate with 10 mL of nitric acid, then dilute the volume to 100 mL. Finally, the concentration of polysulfide ions is is determined through the converted determination of copper ions concentration.

(2) Determination of thiosulfate ion concentration

Take 10 mg of benchmark $K_2Cr_2O_7$ solution, dissolve in 25 mL of distilled water; add 2 g of potassium iodide and 20 mL of 20% sulfuric acid solution; shake well; place in the dark for 10 min; add 150 mL of distilled water; titrate with the solution containing thiosulfate to be measured; towards the end of the titration, add 2 mL starch indicator solution (10 g/L); continue titration of the solution from blue into a bright green; calculate the thiosulfate concentration.

(3) Determination of sulfate and sulfite ion concentration

Ion chromatography was used for the determination of sulfate and sulfite ion concentrations; the specific steps were as follows: dispense 10 mL of the sample and pretreat it with a hydrogen column, then filter it through a 0.45 μm filter membrane, and then inject it into the autosampler for detection.

## 3. Results

### 3.1. SEM Analysis of Original Pyrite Electrodes

SEM analysis of the surface of the original pyrite electrode without an oxidation reaction was performed, and the results are shown in Figure 2. The surface of the pyrite electrode after grinding, polishing, and cleaning treatment is smooth, with occasional shadow spots, which may be caused by some impurities and lattice defects, and the EDS analysis results show that the elemental content of S is 53.29% and the elemental content of Fe is 46.71%, which is basically consistent with the elemental composition of pure pyrite.

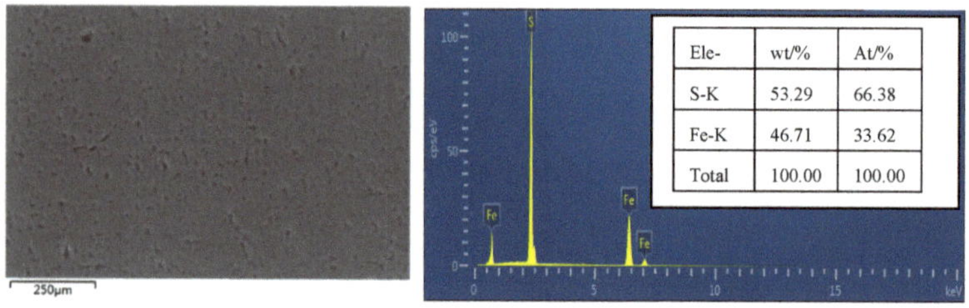


**Figure 2.** SEM and EDS analysis results of pyrite electrode surface.

### 3.2. Open-Circuit Potential ($E_{ocp}$) Measurement

To confirm the stability of the electrochemical test system, the pyrite electrode was immersed in a constantly stirred electrolyte with a temperature of 25 °C and a pH value ranging from 9 to 14 for 1 h, and then the $E_{ocp}$ test was carried out. The test results are shown in Figure 3.

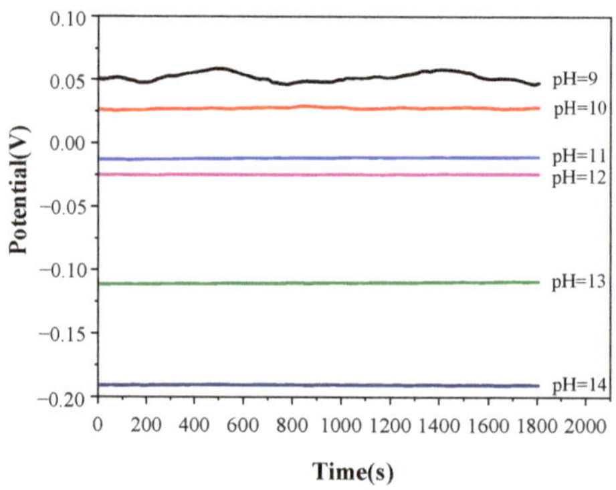


**Figure 3.** $E_{ocp}$ of pyrite electrodes recorded in electrolyte of various pH values for 0.5 h.

According to Figure 3, the $E_{ocp}$ value is relatively unstable when the pH of the electrolyte is 9, which may be due to the weak potential difference fluctuation of the electrolyte caused by agitation, and the potential difference fluctuation cannot be ignored when the concentration of $OH^-$ in the electrolyte is too low. When the pH value of the electrolyte is between 10 and 14, the variation in $E_{ocp}$ with time is less than 2 mV, which indicates that the electrochemical testing system is relatively stable in this range of pH values. Therefore, the variation in the subsequent electrochemical test data is only caused by the alkaline oxidation reaction of the pyrite electrode. Furthermore, $E_{ocp}$ decreases with the increase in the pH value in the electrolyte, which may be caused by a spontaneous oxidation reaction between the pyrite and the electrolyte when the electrochemical test system is in the open-circuit state. With the increase in the concentrations of $S_x^{2-}$ and $S_2O_3^{2-}$ produced by pyrite oxidation in an electrolyte, the oxidation potential moves in a negative direction continuously.

### 3.3. Cyclic Voltammetry (CV) Measurement

According to the literature reports [27,33,34], the potential interval for pyrite redox is generally between −1 and 1 V. Pre-exploratory tests showed that the oxidation potential tends to be more negative in alkaline environments, so the potential interval was chosen to be between −2 and 0.5 V for the CV tests. CV curves are recorded in Figure 4 at a scan rate of 0.01 V/s in various pH value electrolytes, with the pH values of the electrolyte being adjusted to 10, 11, 12, and 13 with NaOH, respectively, a scan voltage of −2 V to 0.5 V, and a temperature of 25 °C. The number of scan turns was five, and the fourth scan turn's results were used for data analysis.

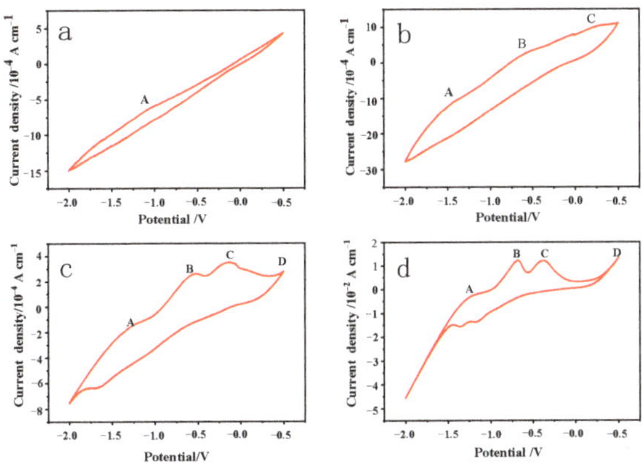


**Figure 4.** Typical voltammetry curve recorded of pyrite electrode in alkaline electrolyte at different pH values. (**a**) pH = 10; (**b**) pH = 11; (**c**) pH = 12; (**d**) pH = 13.

With the increase in the electrolyte pH from 10 to 13, four electrochemical reaction anode peaks, A, B, C, and D, gradually appeared on the CV curves (Figure 4), and the reaction intensity gradually increased, indicating that the oxidation intensity on the surface of the pyrite electrode was positively correlated with the reaction voltage. The oxidation peaks at B, C, and D were more noticeable, indicating that the strength of oxidation on the surface of the pyrite electrode was likewise positively associated with voltage. Furthermore, when the electrolyte pH increased, the reaction intensity gradually increased, indicating that the higher the pH value, the easier the oxidation reaction.

Because the reaction voltage is low and the oxidation intensity is weak, the chemical reaction (1) appears to take place primarily at the anode peak A, according to the available data [35,36]. At this point, the Fe-S bond in $FeS_2$ is broken to form $Fe(OH)_2$ and $S_M$, with $Fe_{1-x}S_2$ and $S_X^{2-}$ as the main components of $S_M$. The pyrite electrode continues to be oxidized as the oxidation voltage increases, and the anodic peak at B can be explained by the reaction shown in Equation (2), which states that $Fe(OH)_2$ is further oxidized into $Fe(OH)_3$, and the main components of $S_L$ are $S^0$ and minor amounts of $Fe_{1-x}S_2$ and $S_X^{2-}$. Furthermore, the anodic peak at C can be explained by the reaction shown in Equation (3), which states that $Fe(OH)_2$ is further oxidized into $Fe(OH)_3$, and the main components of $S_L$ are further converted to $SO_3^{2-}$. Interestingly, when the CV curves at the pH values 13 and 14 are scanned at 0.50 V, a clear sharp oxidation peak D appears, owing to pyrite's high oxidation reaction rate at relatively high voltages. Two chemical reactions occurred on the pyrite electrode at this high voltage: one was the oxidation reaction of the pyrite itself that formed $Fe(OH)_3$ and $SO_4^{2-}$ (Equation (4)), and the other was the process of reoxidation of the surface oxidation product $SO_3^{2-}$ formed at a low voltage to form $SO_4^{2-}$ (Equation (4)).

$$FeS_2 + 2H_2O \rightarrow Fe(OH)_2 + 2S_M + 2H^+ + 2e^- \quad (1)$$

$$FeS_2 + 3H_2O \rightarrow Fe(OH)_3 + 2S_L + 3H^+ + 3e^- \quad (2)$$

$$S_L + 3H_2O \rightarrow SO_3^{2-} + 6H^+ + 6e^- \quad (3)$$

$$FeS_2 + 11H_2O \rightarrow Fe(OH)_3 + 2SO_4^{2-} + 19H^+ + 15e^- \quad (4)$$

$$SO_3^{2-} + H_2O \rightarrow SO_4^{2-} + 2H^+ + 2e^- \quad (5)$$

The reduction peak of the pyrite bulk electrode is not apparent when scanned from 0.50 V to low potential, and the weak reduction peak can be observed only at pH 12. It is inferred that the reduction of the oxidation product $Fe(OH)_3$ to $Fe^{2+}$ occurs between −1.4 V and −1.6 V when the pH is 13 or 14 (as shown in Equation (6)). There is subsequently a smaller reduction peak, which may indicate that pyrite or its products are directly reduced to form $H_2S$ or ferrous sulfides (Equation (7)). The analysis of the electrolyte composition also reveals that a low concentration of $S^{2-}$ existed in the electrolyte solution.

$$Fe(OH)_3 + 3H^+ + e^- \rightarrow Fe^{2+} + 3H_2O \quad (6)$$

$$FeS_2 + 2H^+ + 2e^- \rightarrow FeS + H_2S \quad (7)$$

### 3.4. *Linear Sweep Voltammetric Curve (LSV) Measurements*

According to the literature reports [27,28], the passivation potential interval when pyrite is oxidized is generally between 0.5 and 1 V. Pre-exploratory tests have shown that the oxidation potential tends to be more negative in alkaline environments, so the potential interval was chosen to be between −2 and 0.5 V for the LSV test. The LSV curve test of the pyrite electrode was conducted in an electrolyte at pH 12 and 25 °C. The test voltage was from 0 V to 1.4 V and the scan rate was 0.01 V/s. The test results are shown in Figure 5.

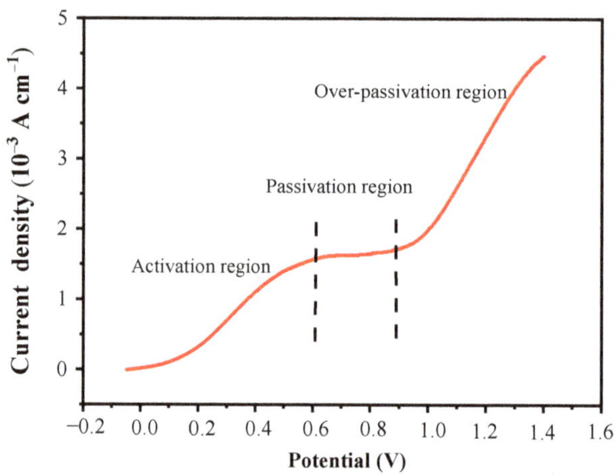


**Figure 5.** LSV curve recorded of pyrite electrode in alkaline solution of pH 12.

In this LSV curve, three distinct regions can be divided: the active region, the passive region, and the over-passive region, as shown in Figure 5. In the active region between 0 and 0.6 V, the current density increases rapidly with the increase in voltage, which is likely due to the oxidative dissolution of pyrite in this voltage range. In the passivation region between 0.6 and 0.9 V, the current density is almost constant, and the response to the voltage increase is not obvious; this may be because the oxidation products of pyrite, such as $Fe(OH)_2$, $Fe(OH)_3$, and so on, deposit on the electrode surface to form a passivation layer, which prevents the current density from continuing to increase with the voltage increase. In the over-passivation region of 0.9~1.4 V, the current density increases rapidly again with the increase in voltage. This is likely because when the voltage increases to 0.9 V, the high voltage provides a stronger driving force for the reaction to break down the passivation layer, so the pyrite covered by the passivation layer can continue to be oxidized.

### 3.5. *i-t Curve Measurements*

The results of CV and LSV demonstrate that the oxidative dissolution of pyrite occurs over a broad range of voltages. To determine the accumulation of products on the surface of pyrite over time and to observe the oxidative dissolution on the surface of pyrite, an i-t curve test was performed at pH 12 and 25 °C for the pyrite electrode at test voltages of 0.4 V, 0.6 V, 0.8 V, and 1.0 V, respectively. The results are shown in Figure 6.

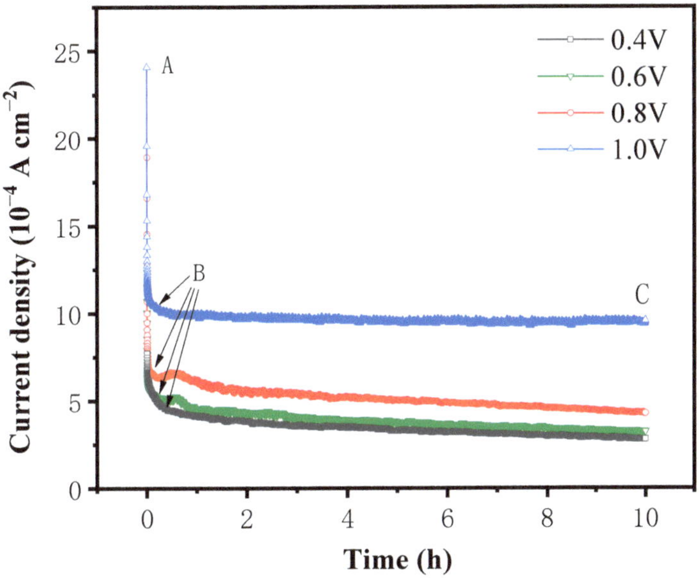


**Figure 6.** The i-t curves of pyrite electrode after alkaline oxidation at different voltages.

As depicted in Figure 6, at all four potentials, the current densities finally tended to be steady at the lower value after dropping acutely within the first several minutes. The variation in current density with reaction time can be roughly divided into two stages, AB and BC. In the AB stage, the initial current density increases with the rise in the applied potential, while the current density decreases rapidly as the reaction time decreases. In the BC stage, the current density tends to be stable with the increase in the reaction time. In addition, with higher system potential, the relatively stable current density is also higher.

As the electrochemical oxidation reaction occurs on the surface of the pyrite electrode under the action of the applied potential, the reaction products rapidly accumulate on the surface of the pyrite and form a passivation layer, which hinders the reaction to a certain extent, resulting in a rapid decrease in current density in the AB stage. In addition, the increase in voltage accelerates the decrease in current density in the AB stage because the higher the voltage, the faster the oxidation and dissolution rates of pyrite, the faster the formation of the passivation layer on the surface of pyrite per unit time, and the greater the degree of hindrance to the reaction.

When the potential was between 0.4 V and 0.8 V, the current density in the BC stage increased slightly with an increase in voltage and decreased with time. This may be because the passivation layer formed in the AB stage failed to completely prevent the oxidation reaction from proceeding, resulting in the accumulation of the passivation layer with the extension of the reaction time and the gradual deepening of the impeded reaction, so that the current density showed a gradual decrease with time. When the potential reaches 1.0 V, the current density in the BC stage basically does not change with time, indicating that the oxidation reaction is faster in the AB stage under high potential conditions, and the passivation layer can quickly reach the thickness that prevents the reaction from proceeding, which in turn makes the reaction rate in the BC stage extremely slow. The hindering effect of the passivation layer on the reaction also remains stable, and the current density also tends to be stable.

### 3.6. Electrochemical Impedance Spectroscopy Analysis

Electrochemical impedance spectroscopy is an important electrochemical measurement technique that obtains information about the properties of the electrode surface by applying a sine wave signal between the electrode and the electrolyte interface and measuring its impedance response. The principle of the technique is based on circuit theory, and the reaction mechanism and charge transfer kinetics of the electrode surface can be inferred by analyzing the change in impedance with frequency. The principle of electrochemical impedance spectroscopy is based on the equivalent circuit model; that is, the system between the electrode and the electrolyte interface is equivalent to the resistance, capacitance, and inductance components in a circuit. When a sine wave signal is applied to the circuit, the impedance response depends on the characteristics of the individual components and their interactions with each other. By analyzing the response signal, information about the surface properties of the electrode can be derived, such as the thickness of the surface film, the charge transfer coefficient, and the ion diffusion coefficient.

An EIS test was conducted at a pH level of 12 and a temperature of 25 °C. The frequency scanning range was 0.01~1 × $10^5$ Hz, and the amplitude was 5 mV. The fitted Nyquist plots by the equivalent circuits at specific potentials of 0.5, 0.6, 0.7, and 0.8 V in the alkaline electrolyte are respectively shown in Figure 7. The fitting parameters of the equivalent circuits simulated from the EIS are shown in Table 1.

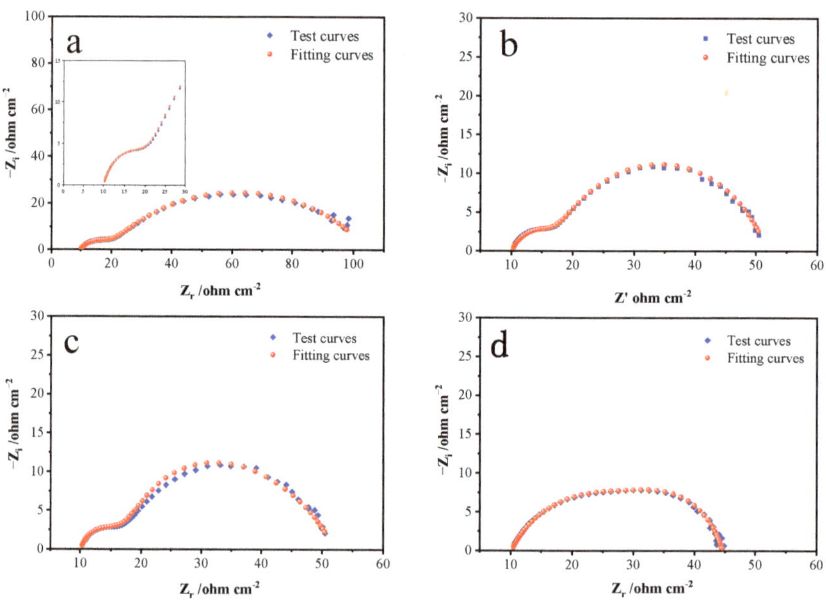


**Figure 7.** The measured and calculated Nyquist plots of pyrite electrodes at different anodic potentials in an alkaline supporting electrolyte ((**a**) 0.2 V, (**b**) 0.4 V, (**c**) 0.6 V, (**d**) 0.8 V).

**Table 1.** Fitting parameters of the equivalent circuits simulated from the EIS in Figure 7.

| U (V) | 0.2 (V) | 0.4 (V) | 0.6 (V) | 0.8 (V) |
|---|---|---|---|---|
| $R_1$ (Ω cm$^{-2}$) | 10.02 | 10.12 | 10.32 | 10.16 |
| $R_2$ (Ω cm$^{-2}$) | 12.07 | 8.49 | 21.35 | 25.56 |
| $R_3$ (Ω cm$^{-2}$) | 82.04 | 34.34 | 17.48 | 9.163 |
| $CPE_1$-T (F cm$^{-2}$) | 0.0021609 | 0.0014452 | 0.001271 | 0.0089737 |
| $CPE_1$-P (F cm$^{-2}$) | 0.67268 | 0.72403 | 0.78395 | 0.89019 |
| $CPE_2$-T (F cm$^{-2}$) | 0.00016282 | 0.00016892 | 0.00023613 | 0.00031208 |
| $CPE_2$-P (F cm$^{-2}$) | 0.69829 | 0.69665 | 0.66611 | 0.64281 |

As can be seen from Figure 7, a small semicircular arc and a large one appear successively in the high-frequency region and the middle-frequency region. The small semicircular arc in the high-frequency region corresponds to capacitance related to the electrode–solution interface, and the large semicircular arc in the middle-frequency region is generated by the Heimholtz double capacitance layers on the pyrite electrode surface. According to Figures 5 and 6, when the system potential is between 0.2 and 0.6 V, the pyrite electrode is in an activation zone during the alkaline oxidative dissolution process, the reaction products have not yet formed a stable passivation layer on the electrode surface, and the electrochemical reaction is active, which means that the passivation layer is not enough to hinder the chemical oxidation reaction and the electrode reaction rate is primarily controlled by the electrochemical reaction. The deviation of the small semicircular arc in Figure 7 from the standard semicircular state is due to the polarization of the electrode surface as the frequency decreases, and the chemical reaction caused by the polarization also results in another larger arc until the $S^0$ and $Fe(OH)_3$ generated during the chemical reaction adsorb on the electrode surface to form a passivation layer, and as the passivation layer thickens, it finally leads the chemical reaction to stop, and the real and imaginary parts are basically no longer changeable. For the same reason, there is no Warburg impedance characteristic slope line with a representation of the diffusion process in the impedance diagram's low- and middle-frequency regions. As the potential continues to rise to 0.8 V, the distinction between the small and large semicircular arcs becomes no longer obvious; this is because the chemical reaction process accelerates at higher potentials, obscuring the surface polarization of the electrode.

Figure 8 presents the equivalent circuit models used to fit the measured EIS data, where R1 represents the solution resistance, R2 represents the interface polarization resistance, capacitance is represented by CPE1, and R3 represents the charge transfer resistance. Based on the test and fitted curves in Figure 7, the fitted data match with the experimental data closely when this circuit is used, which indicates that the equivalent circuit model can reflect the dynamical behaviors of the bulk pyrite electrode more accurately during the alkaline oxidation process.

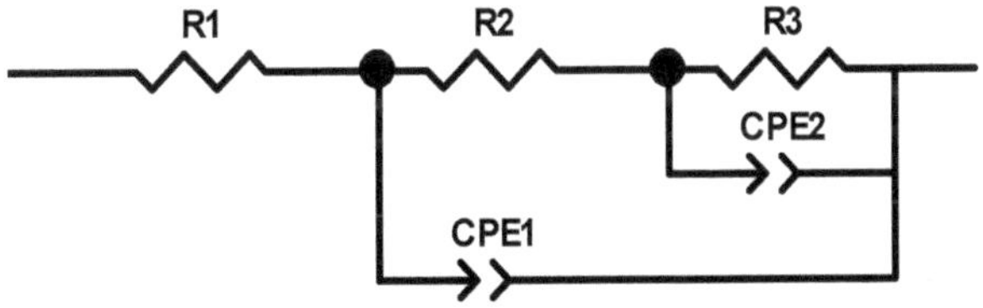


**Figure 8.** Equivalent circuit models used for fitting measured EIS data (R1, solution resistance; R2, polarization resistance generated by the interface capacitor CPE1; R3, charge transfer resistance).

Table 1 lists the parameter values of the equivalent circuit model (Figure 8) at different potentials. The charge transfer resistance $R_3$ decreases from 82.04 $cm^{-2}$ to 9.163 $cm^{-2}$ as the system's reaction voltage rises from 0.2 V to 0.8 V. As the reaction voltage increases, R3 decreases, the charge transfer rate during the reaction increases, and the alkaline oxidative dissolution reaction of pyrite occurs more rapidly. As shown in Figure 7, the rate rises while the radius of the semicircular arc falls. The constant phase element dispersion index ($CPE_2$-T), which reflects how far the corresponding capacitance deviates from the ideal capacitance, is typically used. According to Table 1, the $CPE_2$-T gradually rises as the voltage rises. This may be because as the voltage rises, the electrode surface's reaction rate rises and more $S^0$ and $Fe(OH)_3$ accumulate on the electrode surface, which causes the electrode surface's roughness to rise and, in turn, raises the $CPE_2$-T.

### 3.7. XPS Analysis

The XPS technique was applied to identify the iron and sulfur oxidation products on the bulk pyrite electrode surface after 10 h of the i-t curve test at potentials of 0.4 V,

0.6 V, 0.8 V, and 1.0 V. The measured and fitted plots are shown in Figure 9 and the fitting parameters are present in Table 2.

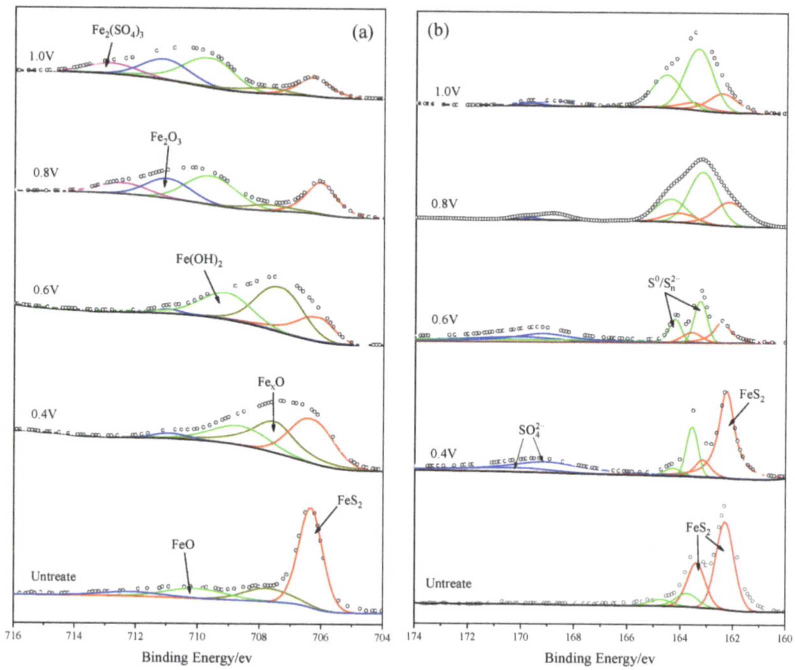

**Figure 9.** Iron and sulfur XPS spectra of unreacted and chronoamperometry-treated pyrite electrodes ((**a**) $Fe_{2P3/2}$ spectrum of pyrite electrodes; (**b**) $S_{2P}$ doublet spectrum of pyrite electrodes).

**Table 2.** Fitting parameters of XPS spectra in Figure 9.

| Spectrum | Original Mineral | 0.4 V | 0.6 V | 0.8 V | 1.0 V | Interpretation |
|---|---|---|---|---|---|---|
| $Fe_{2p3/2}$ | 706.30 | 706.34 | 706.38 | 706.40 | 706.50 | $FeS_2$ |
| | 707.82 | 707.88 | 707.90 | 707.91 | 707.93 | $Fe_XO$ |
| | 710.0 | 709.10 | 709.51 | 709.80 | 709.82 | $FeO/Fe(OH)_2$ |
| | | 710.8 | 710.84 | 710.85 | 710.90 | $Fe_2O_3/FeOOH$ |
| | | | | 713.12 | 713.25 | $Fe_2(SO_4)_3$ |
| $S_{2p3/2}$ | 162.20 | 162.23 | 162.35 | 162.24 | 162.35 | $FeS_2$ |
| | 163.80 | 163.72 | 163.42 | 163.70 | 164.05 | $S^0/S_n{}^{2-}$ |
| | | 168.48 | 168.25 | 168.40 | 168.52 | $SO_4{}^{2-}$ |

The $Fe_{2p3/2}$ spectra of unreacted and anodic-reacted pyrite electrodes are shown in Figure 9a; five characteristic peaks appeared at 706.3 eV, 707.82 eV, 710.0 eV, 710.8 eV, and 713.12 eV for different reaction voltages. The peak with a binding energy of 706.3 eV was considered to correspond to $FeS_2$ [37–39]. The peak at 707.82 eV corresponded to $Fe_XO$ with an iron valence between 0 and +2, which gradually transforms to divalent or trivalent iron as the oxidation potential increases. The peaks at 710.0 eV and 710.8 eV corresponded to $FeO/Fe(OH)_2$ and $Fe_2O_3/FeOOH$, respectively [40–42]. It is possible that both iron oxides and hydroxides existed on the surface of the pyrite electrode because of the alkaline oxidation environment and the ultravacuum environment during XPS detection. Iron hydroxides may be partially dehydrated to become iron oxides under super vacuum conditions. Another peak at 713.12 eV corresponded to $Fe_2(SO_4)_3$ [43], which was difficult to detect at low oxidation potentials because of its solubility. When the oxidation potential increased, a large amount of $Fe_2(SO_4)_3$ products did not diffuse into the solution in time and produced a small amount of deposition on the electrode surface, which was then detected. In conclusion, these stronger peaks at strong oxidation potentials indicated the presence of a large amount of Fe(III) in the forms of $Fe(OH)_3$, FeOOH, $Fe_2O_3$, or $Fe_2(SO_4)_3$ on the surface of the pyrite electrode due to electrochemical polarization, which was consistent with the conclusions of the discussed electrochemical measurements.

Figure 9b shows $S_{2p}$ spectra of the pyrite electrode before and after the oxidation reaction. The dominant peak at the binding energy of 162.2 eV corresponded to $FeS_2$ [37,38,44]. The intensity of the oxidation reaction increased with increasing voltage, and the thickness of the passivation layer, which resulted from the accumulation of adhesion products on the surface of the pyrite electrode, increased rapidly; therefore, the intensity of the $FeS_2$ characteristic peak decreased gradually. The peak at 163.8 eV was supposed to be $S_0$ or $S_X^{2-}$ [39,45–47], and the faint peak at 168.42 eV was considered to be $SO_4^{2-}$ [38,43,48], and these may be the final oxidation products of S in pyrite.

### *3.8. SEM and EDS Analysis*

To observe the passivation layer formed on the surface of the pyrite electrode, the bulk pyrite electrode was subjected to chronoamperometry measurements at 0.8 V with various reaction times of 10 min, 20 min, 30 min, 40 min, 50 min, 60 min, 120 min, 180 min, and 240 min in an alkaline electrolyte of pH 12 and 25 °C. The tested pyrite electrodes were subsequently analyzed using SEM and EDS, and the results are shown in Figures 10 and 11.

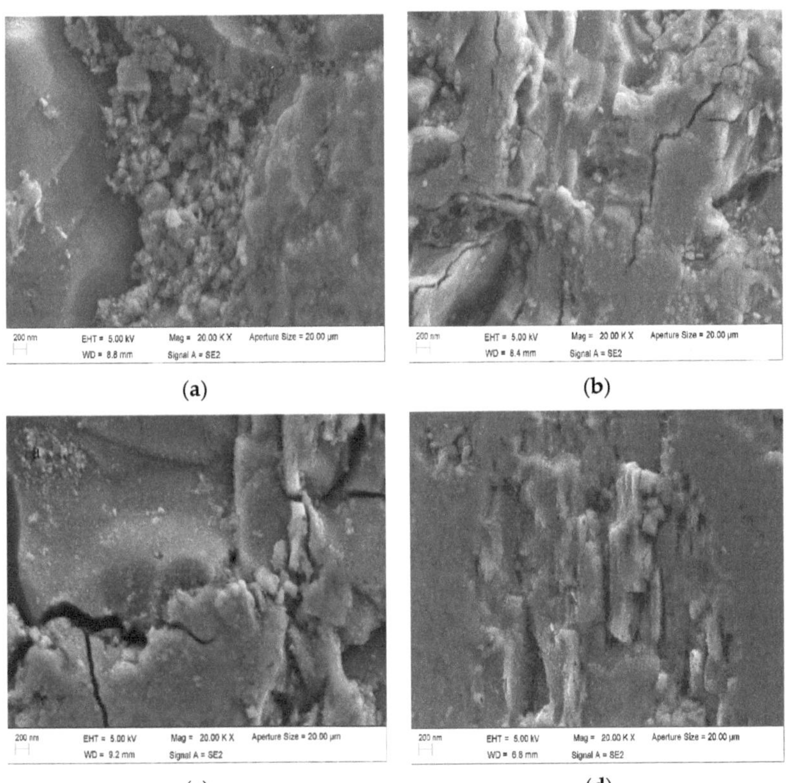


**Figure 10.** SEM images of pyrite electrode surface at different reaction times ((**a**) 10 min; (**b**) 60 min; (**c**) 120 min; (**d**) 240 min).

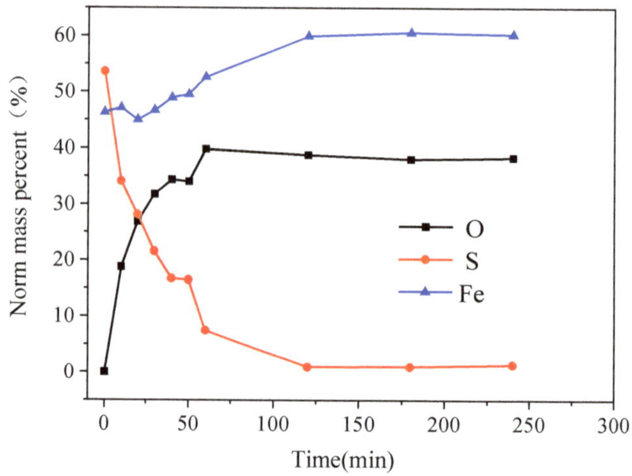


**Figure 11.** Variation in O, S, Fe content of pyrite electrode surface with time.

When the reaction time was 10 min, the oxidation–dissolution reaction occurred primarily on the surface, and a small amount of oxidation product accumulation could be seen on the electrode surface, which was not completely covered. It is evident from Figure 10a that there were only a few slight corrosion pits on the surface of the pyrite electrode and no obvious passivation layer was formed. For the reaction time of 60 min, the passivation layer formed by the oxidation products completely covered the electrode surface and contained microcracks (Figure 10b). For the reaction time of 120 min, the reaction products continued to accumulate on the surface, the passivation layer appeared as blocky aggregates, the blocky crack traces deepened, and the blocky area grew, indicating that the passivation degree increased (Figure 10c). When the pyrite electrode continued to react for 240 min, the surface cracks on the passivation layer were filled with new reaction products, the oxidation reaction of the pyrite electrode basically stopped, and the accumulation of the passivation layer slowed or ceased (Figure 10d).

As shown in Figure 11, the content of O and Fe elements in the oxidation products on the surface of the pyrite electrode increases as the reaction time increases, while the content of S elements decreases. In 0–60 min, the elemental content of S decreased rapidly from 53.66% to 7.55%, the elemental content of O increased rapidly from 0% to 39.81%, and the elemental content of Fe first decreased slightly and then increased gradually. The changes in the O, S, and Fe elemental contents on the surface of the pyrite electrode slowed after 60 min and stabilized after 120 min. During the electrochemical oxidation and decomposition of the pyrite, the S element mainly entered the electrolyte in the form of ions, which did not deposit on the electrode surface easily, so the S element content decreased rapidly. A small portion of Fe elements entered the solution in the form of soluble $Fe(OH)^+$ after being oxidized, so the Fe content decreased initially. However, because the electrolyte was strongly alkaline, $Fe(OH)^+$ continued to react with $OH^-$ as its concentration increased, the passivation layer was gradually formed on the electrode surface as $Fe(OH)_3$, and as the passivation layer gradually thickened over time, its composition tended to remain stable.

### *3.9. Ion Content Change Pattern in Electrolyte*

The oxidation reaction of the pyrite electrode was carried out in an alkaline electrolyte at pH 12 and 25 °C at different voltages and times, and the reacted electrolyte was analyzed for its ionic composition; the results are shown in Figure 12.

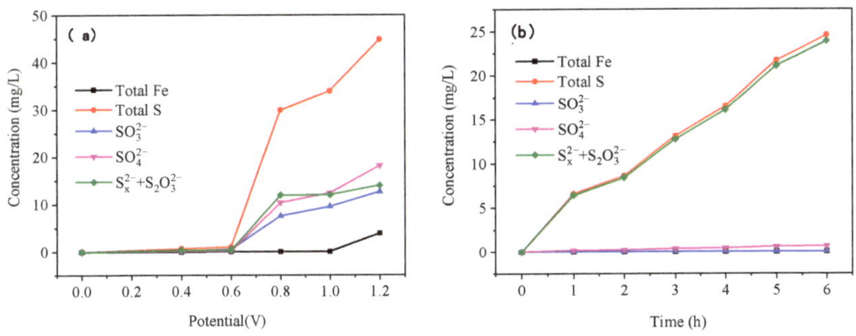


**Figure 12.** Changes in ion content in electrolytes under different reaction conditions. (**a**) The ionic compositions of the electrolytes in which the pyrite electrode reacted for 10 h at different potentials. (**b**) The ionic compositions of the electrolytes in which the pyrite electrode reacted at 0.8 V potential for different oxidation times.

As the reaction voltage increased, the S content in the electrolytes increased significantly, while the Fe content increased only slightly (Figure 12a). This is because the S element enters the electrolyte in the ionic state when the pyrite is oxidized and dissolved, whereas it is difficult for the Fe element to exist in the strong alkaline electrolyte in the form of ions. The contents of $SO_3{}^{2-}$, $SO_4{}^{2-}$, and ($S_x{}^{2-}$ + $S_2O_3{}^{2-}$) ions increased significantly as the reaction voltage rose, with the increase in ($S_x{}^{2-}$ + $S_2O_3{}^{2-}$) being greater than those in $SO_3{}^{2-}$ and $SO_4{}^{2-}$. During the oxidative dissolution of the pyrite, the S element was oxidized and converted to $S_x{}^{2-}$ and $S_2O_3{}^{2-}$ before being further oxidized to $SO_3{}^{2-}$ and $SO_4{}^{2-}$, resulting in a greater increase in $S_x{}^{2-}$ and $S_2O_3{}^{2-}$ than in $SO_3{}^{2-}$ and $SO_4{}^{2-}$. Moreover, since $S_x{}^{2-}$ and $S_2O_3{}^{2-}$ were in the unstable intermediate states, increasing the reaction voltage could accelerate the conversion of ($S_x{}^{2-}$ + $S_2O_3{}^{2-}$) to $SO_3{}^{2-}$ and $SO_4{}^{2-}$, as evidenced by the increase in $SO_3{}^{2-}$ and $SO_4{}^{2-}$ concentration with increasing voltage.

The content of S in the electrolyte increased significantly as the reaction time increased, while the content of Fe remained essentially unchanged (Figure 12b). This is because the Fe ions produced by oxidation in the strong alkaline electrolyte quickly formed iron hydroxide, which was deposited on the surface of the pyrite electrode; therefore, the ionic iron content in the electrolyte was very low. With an increase in reaction time, the content of $SO_3{}^{2-}$ and $SO_4{}^{2-}$ was low and did not change significantly, whereas the content of ($S_x{}^{2-}$ +$S_2O_3{}^{2-}$) was high, indicating that under suitable alkaline oxidation conditions, $S_x{}^{2-}$ and $S_2O_3{}^{2-}$ produced by the pyrite oxidation reaction can exist steadily in the solution, which is the basis for gold leaching by using pyrite's own oxidation products.

## 4. Conclusions

In this paper, the oxidation process of pyrite in an alkaline electrolyte was studied using the electrochemical testing technique, and the oxidation products of a pyrite electrode in an alkaline electrolyte were characterized by combining various analytical testing methods, such as XPS and SEM. The conclusions of this study are as follows.

(1) The electrochemical oxidative dissolution of pyrite occurs preferably in an alkaline environment with a pH of around 12, and an appropriate oxidation potential is above 0.8 V. These parameters are still of reference significance in the preoxidation treatment process of sulfide ores with chemical oxidants.

(2) During the alkaline oxidative dissolution of pyrite, some elemental S enters the electrolyte as $S_x{}^{2-}$, $S_2O_3{}^{2-}$, $SO_3{}^{2-}$, and $SO_4{}^{2-}$ ions, and a small amount of the element S is adsorbed on the electrode surface as $S^0$ to form the passivation layer. As the main component of the passivation layer, Fe is essentially adsorbed on the electrode surface as $Fe(OH)_2$, $Fe_2O_3$, and $Fe_2(SO_4)_3$.

(3) When the oxidation potential was between 0 and 0.6 V, the increase in reaction voltage had no significant effect on the change in ion concentration in the electrolyte. Low potential polarization could result in the formation of a sulfur-rich passivation film on the surface of the pyrite electrode, which caused the reaction to slow down or stop. When the voltage was greater than 0.6 V, as the potential increased, the inert oxidation products dissolved, the oxidation process continued, and the concentrations of soluble ions in the electrolyte began to rise rapidly. As the reaction time prolonged, the S element was first converted into $S_x^{2-}$ and $S_2O_3^{2-}$, and then further oxidized into $SO_3^{2-}$ and $SO_4^{2-}$, whereas $S_x^{2-}$ and $S_2O_3^{2-}$ could remain stable in the solution for a longer time. Therefore, the content of $S_x^{2-}$ and $S_2O_3^{2-}$ in the electrolyte increased rapidly with the prolongation in the reaction time, whereas the content of $SO_3^{2-}$ and $SO_4^{2-}$ changed little.

(4) Comprehensively, the results of this paper—the change rule of the pyrite surface composition and the ionic composition in the solution during alkaline electrochemical oxidation—are expressed in Table 3.

**Table 3.** Changing law of surface composition of pyrite alkaline electrochemical oxidation and ionic composition in solution.

| Classifications | Original | Early Stages of Oxidation | Intermediate Stage of Oxidation | Later Stage of Oxidation |
|---|---|---|---|---|
| Main components of pyrite surface | $FeS_2$ | $Fe_xO$<br>$Fe(OH)_2$ | FeOOH<br>$Fe_2O_3$<br>$S^0$<br>$Fe_2(SO_4)_3$ | FeOOH<br>$Fe_2O_3$<br>$Fe_2(SO4)_3$ |
| Major iron–sulfur ion components in electrolyte solutions | none | $S_x^{2-}$<br>$S_2O_3^{2-}$<br>$Fe^{2+}$<br>$Fe(OH)^+$ | $S_x^{2-}$<br>$S_2O_3^{2-}$<br>$SO_3^{2-}$<br>$Fe^{2+}$<br>$Fe^{3+}$ | $S_2O_3^{2-}$<br>$SO_3^{2-}$<br>$SO_4^{2-}$<br>$Fe^{3+}$ |

In conclusion, $S_x^{2-}$ and $S_2O_3^{2-}$ generated by the alkaline oxidation of pyrite can stably exist in the solution, which provides a new idea for the one-step gold extraction process of "oxidative pretreatment-gold leaching" for microfine-grained leaching type difficult-to-treat gold-bearing sulfide ores, i.e., to leach gold by making full use of the sulfur in the gold-bearing sulfide minerals to directly generate polysulfides and thiosulfates with oxidizing and coordinating abilities in the alkaline system. This technology has the advantages of being environmentally friendly and having a short process and a low cost, and has good prospects for industrial application. The authors will continue conducting research on the simultaneous dissolution of gold and sulfur to finish the whole technological process of one-step gold extraction, so that this work can be put into practice as soon as possible.

**Author Contributions:** Y.H.: data curation, methodology, formal analysis, writing—original draft. Z.J.: methodology and writing—review and editing. W.W.: writing—review and editing. J.Y.: writing—review and editing. R.G.: writing—review and editing. L.X.: writing—review and editing. H.Z.: writing—review and editing. Y.Z.: supervision. X.S.: conceptualization, resources, writing—review and editing, funding acquisition. All authors have read and agreed to the published version of the manuscript.

**Funding:** This work was supported by the National Natural Science Foundation of China [grant numbers 51874259 and 52074244].

**Data Availability Statement:** Not applicable.

**Conflicts of Interest:** The authors declare no conflict of interest.

## References

1. Amankwah, R.K.; Pickles, C.A. Microwave roasting of a carbonaceous sulphidic gold concentrate. *Miner. Eng.* **2009**, *22*, 1095–1101. [CrossRef]
2. Brierley, C.L. Biohydrometallurgical prospects. *Hydrometallurgy* **2010**, *104*, 324–328. [CrossRef]

3. Surimbayev, B.N.; Kanaly, Y.S.; Bolotova, L.S.; Shalgymbayev, S.T. Behaviour of Sulfur, Arsenic and Organic Carbon in a Gravity Concentration of Gold from Refractory Ore. *News Natl. Acad. Sci. Repub. Kazakhstan-Ser. Chem. Technol.* **2020**, *5*, 116–125. [CrossRef]
4. Yang, H.-Y.; Gong, E.-P.; Yang, L. Process mineralogy of low-grade double refractory gold ore and influencing factor on gold. *J. Northeast. Univ. (Nat. Sci.)* **2008**, *29*, 1742–1745.
5. Lei, Z.; Xueyi, G.; Qinghua, T.; Shuiping, Z.; Dong, L.; Hong, Q. Research progress and industrial application of pretreatment methods for refractory gold ores. *Gold* **2021**, *42*, 60–68.
6. Liu, Z.; Ye, X.; Wei, D. The Experimental Research on the Separation for Floatation of the Sulfur and Arsenic in a Refractory Gold Ore. In Proceedings of the 1st International Conference on Energy and Environmental Protection (ICEEP 2012), Hohhot, China, 23–24 June 2012.
7. Ma, Y.; Luo, X.; Yin, W.; Yao, J.; Liu, M.; Hong, J. Separation Tests of Low-grade Gold Ore with High Sulfur and Arsenic. In Proceedings of the International Forum on Powder Technology and Application, China Powder Technol Assoc, Anshan, China, 27–29 October 2012.
8. Wang, Q.; Hu, X.; Zi, F.; Qin, X.; Nie, Y.; Zhang, Y. Extraction of gold from refractory gold ore using bromate and ferric chloride solution. *Miner. Eng.* **2019**, *136*, 89–98. [CrossRef]
9. Zhang, L.; Guo, X.-y.; Tian, Q.-h.; Li, D.; Zhong, S.-p.; Qin, H. Improved thiourea leaching of gold with additives from calcine by mechanical activation and its mechanism. *Miner. Eng.* **2022**, *178*, 107403. [CrossRef]
10. Zhu, C.; Ge, B.; Lu, Z.; Zhu, W. Biooxidation Pretreatment Cyanide Leaching of a Flotation Gold Concentrate. In Proceedings of the International the Conference on Chemical Engineering, Metallurgical Engineering and Metallic Materials (CMMM 2012), Kunming, China, 12–13 October 2012.
11. Jin, J.; Han, Y.; Li, H.; Huai, Y.; Peng, Y.; Gu, X.; Yang, W. Mineral phase and structure changes during roasting of fine-grained carbonaceous gold ores and their effects on gold leaching efficiency. *Chin. J. Chem. Eng.* **2019**, *27*, 1184–1190. [CrossRef]
12. Owusu, C.; Mensah, S.; Ackah, K.; Amankwah, R.K. Reducing preg-robbing in carbonaceous gold ores using passivative or blanking agents. *Miner. Eng.* **2021**, *170*, 106990. [CrossRef]
13. Qian, L.; Hong Ying, Y.; Lin Lin, T.; Li Li, Q.; Jing, P.; Zeng Liang, Z.; Hai Liang, L. Effect of *Phanerochaete chrysosporiumon* Graphite Degradation during Pre-Treatment of Simulated Gold Preg-Robbing Ore. *Adv. Mater. Res.* **2015**, *1130*, 169–174. [CrossRef]
14. Tamayo, J.A.; Melendez, A.M.; Calderon, J.A. An integrated approach to evaluating the effect of associated minerals on copper ammoniacal thiosulfate leaching of a gold-bearing sulfide concentrate. *Hydrometallurgy* **2019**, *184*, 9–21. [CrossRef]
15. Sitando, O.; Senanayake, G.; Dai, X.; Nikoloski, A.N.; Breuer, P. A review of factors affecting gold leaching in non-ammoniacal thiosulfate solutions including degradation and in-situ generation of thiosulfate. *Hydrometallurgy* **2018**, *178*, 151–175. [CrossRef]
16. Xu, B.; Yang, Y.; Jiang, T.; Li, Q.; Zhang, X.; Wang, D. Improved thiosulfate leaching of a refractory gold concentrate calcine with additives. *Hydrometallurgy* **2015**, *152*, 214–222. [CrossRef]
17. Boon, M.; Heijnen, J.J. Chemical oxidation kinetics of pyrite in bioleaching processes. *Hydrometallurgy* **1998**, *48*, 27–41. [CrossRef]
18. Descostes, M.; Vitorge, P.; Beaucaire, C. Pyrite dissolution in acidic media. *Geochim. Cosmochim. Acta* **2004**, *68*, 4559–4569. [CrossRef]
19. Tu, Z.; Guo, C.; Zhang, T.; Lu, G.; Wan, J.; Liao, C.; Dang, Z. Investigation of intermediate sulfur species during pyrite oxidation in the presence and absence of *Acidithiobacillus ferrooxidans*. *Hydrometallurgy* **2017**, *167*, 58–65. [CrossRef]
20. Liu, H.; Gu, G.; Xu, Y. Surface properties of pyrite in the course of bioleaching by pure culture of *Acidithiobacillus ferrooxidans* and a mixed culture of *Acidithiobacillus ferrooxidans* and *Acidithiobacillus thiooxidans*. *Hydrometallurgy* **2011**, *108*, 143–148. [CrossRef]
21. Rodriguez, Y.; Ballester, A.; Blazquez, M.L.; Gonzalez, F.; Munoz, J.A. New information on the sphalerite bioleaching mechanism at low and high temperature. *Hydrometallurgy* **2003**, *71*, 57–66. [CrossRef]
22. Todd, E.C.; Sherman, D.M.; Purton, J.A. Surface oxidation of pyrite under ambient atmospheric and aqueous (pH = 2 to 10) conditions: Electronic structure and mineralogy from X-ray absorption spectroscopy. *Geochim. Cosmochim. Acta* **2003**, *67*, 881–893. [CrossRef]
23. Giannetti, B.F.; Bonilla, S.H.; Zinola, C.F.; Raboczkay, T. A study of the main oxidation products of natural pyrite by voltammetric and photoelectrochemical responses. *Hydrometallurgy* **2001**, *60*, 41–53. [CrossRef]
24. Schaufuss, A.G.; Nesbitt, H.W.; Kartio, I.; Laajalehto, K.; Bancroft, G.M.; Szargan, R. Reactivity of surface chemical states on fractured pyrite. *Surf. Sci.* **1998**, *411*, 321–328. [CrossRef]
25. Lowson, R.T. Aqueous oxidation of pyrite by molecular-oxygen. *Chem. Rev.* **1982**, *82*, 461–497. [CrossRef]
26. Bryson, L.J.; Crundwell, F.K. The anodic dissolution of pyrite ($FeS_2$) in hydrochloric acid solutions. *Hydrometallurgy* **2014**, *143*, 42–53. [CrossRef]
27. Antonijevic, M.M.; Dimitrijevic, M.D.; Serbula, S.M.; Dimitrijevic, V.L.J.; Bogdanovic, G.D.; Milic, S.M. Influence of inorganic anions on electrochemical behaviour of pyrite. *Electrochim. Acta* **2005**, *50*, 4160–4167. [CrossRef]
28. Lara, R.H.; Viridiana Garcia-Meza, J.; Cruz, R.; Valdez-Perez, D.; Gonzalez, I. Influence of the sulfur species reactivity on biofilm conformation during pyrite colonization by *Acidithiobacillus thiooxidans*. *Appl. Microbiol. Biotechnol.* **2012**, *95*, 799–809. [CrossRef]
29. Ciminelli, V. *Oxidation of Pyrite in Alkaline Solutions and Heterogeneous Equilibria of Sulfur-And Arsenic-Containing Minerals in Cyanide Solutions*; The Pennsylvania State University ProQuest Dissertations Publishing: State College, PA, USA, 1987.
30. Ciminelli, V.; Osseo-Asare, K. Kinetics of Pyrite Oxidation in Sodium Carbonate Solutions. *Metall. Mater. Trans.* **1995**, *26B*, 209–218. [CrossRef]

31. Khoso, S.A.; Hu, Y.-h.; Lu, F.; Gao, Y.; Liu, R.-q.; Sun, W. Xanthate interaction and flotation separation of $H_2O_2$-treated chalcopyrite and pyrite. *Trans. Nonferr. Met. Soc. China* **2019**, *29*, 2604–2614. [CrossRef]
32. Tao, D.P.; Li, Y.Q.; Richardson, P.E.; Yoon, R.H. The incipient oxidation of pyrite. *Colloids Surf. A Physicochem. Eng. Asp.* **1994**, *93*, 229–239. [CrossRef]
33. Lin, H.K.; Say, W.C. Study of pyrite oxidation by cyclic voltammetric, impedance spectroscopic and potential step techniques. *J. Appl. Electrochem.* **1999**, *29*, 987–994. [CrossRef]
34. Almeida, C.; Giannetti, B.F. The electrochemical behavior of pyrite-pyrrhotite mixtures. *J. Electroanal. Chem.* **2003**, *553*, 27–34. [CrossRef]
35. Jiang, W.R.; Tu, Z.H.; Zhou, S. A Brief Overview on the Mechanism and Kinetic Influencing Factors of the Pyrite Surface Oxidation. *Met. Mine* **2021**, *50*, 88–102. [CrossRef]
36. Xu, G.; Deng, F.; Fan, W.; Shi, Z.; Ma, R.; Wang, C. Pre-oxidation of refractory gold concentrate by electrochemical methods in alkaline electrolyte. *Mater. Today Commun.* **2022**, *31*, 103397. [CrossRef]
37. Naveau, A.; Monteil-Rivera, F.; Guillon, E.; Dumonceau, J. XPS and XAS studies of copper(II) sorbed onto a synthetic pyrite surface. *J. Colloid Interface Sci.* **2006**, *303*, 25–31. [CrossRef] [PubMed]
38. Nesbitt, H.W.; Bancroft, G.M.; Pratt, A.R.; Scaini, M.J. Sulfur and iron surface states on fractured pyrite surfaces. *Am. Mineral.* **1998**, *83*, 1067–1076. [CrossRef]
39. Wittstock, G.; Kartio, I.; Hirsch, D.; Kunze, S.; Szargan, R. Oxidation of galena in acetate buffer investigated by atomic force microscopy and photoelectron spectroscopy. *Langmuir* **1996**, *12*, 5709–5721. [CrossRef]
40. Ejtemaei, M.; Nguyen, A.V. Characterisation of sphalerite and pyrite surfaces activated by copper sulphate. *Miner. Eng.* **2017**, *100*, 223–232. [CrossRef]
41. Yang, X.; Mu, Y.; Peng, Y. Comparing lead and copper activation on pyrite with different degrees of surface oxidation. *Miner. Eng.* **2021**, *168*, 106926. [CrossRef]
42. Zhang, T.; Wang, Y.; Hu, Y.; Wang, Z.; Chen, J.; Niu, X.; Li, Y.; Gong, X. HO center dot selective cleavage Fe-S bond for $FeS_2$ electrolysis in alkaline solution. *Electrochim. Acta* **2019**, *306*, 327–338. [CrossRef]
43. Tu, Z.; Wan, J.; Guo, C.; Fan, C.; Zhang, T.; Lu, G.; Reinfelder, J.R.; Dang, Z. Electrochemical oxidation of pyrite in pH 2 electrolyte. *Electrochim. Acta* **2017**, *239*, 25–35. [CrossRef]
44. Derycke, V.; Kongolo, M.; Benzaazoua, M.; Mallet, M.; Barres, O.; De Donato, P.; Bussiere, B.; Mermillod-Blondin, R. Surface chemical characterization of different pyrite size fractions for flotation purposes. *Int. J. Miner. Process.* **2013**, *118*, 1–14. [CrossRef]
45. Buckley, A.N.; Woods, R. The surface oxidation of pyrite. *Appl. Surf. Sci.* **1987**, *27*, 437–452. [CrossRef]
46. Hampton, M.A.; Plackowski, C.; Nguyen, A.V. Physical and Chemical Analysis of Elemental Sulfur Formation during Galena Surface Oxidation. *Langmuir* **2011**, *27*, 4190–4201. [CrossRef] [PubMed]
47. Niu, X.; Chen, J.; Li, Y.; Xia, L.; Li, L.; Sun, H.; Ruan, R. Correlation of surface oxidation with xanthate adsorption and pyrite flotation. *Appl. Surf. Sci.* **2019**, *495*, 143411. [CrossRef]
48. Cai, Y.F.; Paan, Y.G.; Xue, J.Y.; Su, G.Z.; Li, X. Comparative XPS study between experimentally and naturally weathered pyrites. *Geochim. Cosmochim. Acta* **2009**, *73*, A184. [CrossRef]

minerals

MDPI

*Article*

# Extraction of Gallium from Brown Corundum Dust by Roasting—Acid Leaching Process

**Juhua Zhang [1,2,3,*], Yuwei Chang [1,2], Cong Gao [1,2], Xujie Hui [1,2] and Ari Jokilaakso [3]**

1 The State Key Laboratory of Refractories and Metallurgy, Wuhan University of Science and Technology, Wuhan 430081, China; lalawzdan@163.com (Y.C.); gao1967107188@163.com (C.G.); hxj@wust.edu.cn (X.H.)
2 Key Laboratory for Ferrous Metallurgy and Resource Utilization of Ministry of Education, Wuhan University of Science and Technology, Wuhan 430081, China
3 Department of Chemical and Metallurgical Engineering, School of Chemical Engineering, Aalto University, Kemistintie 1F, P.O. Box 16100, FI-00076 Aalto, Finland; ari.jokilaakso@aalto.fi
* Correspondence: zhangjuhua@wust.edu.cn

**Abstract:** Brown corundum dust is a solid waste produced during the preparation of brown corundum with bauxite as the raw material. The dust has a relatively high gallium content; therefore, it is of great value to recover the gallium from this kind of dust. In this paper, a range of analysis and characterization methods, including XRD, XRF, SEM-EDS, and EPMA, were used to determine the occurrence of gallium. It was found that gallium was mainly present in the potassium-rich phase, wrapped by amorphous silicate and the corundum phase. Roasting activation followed by an acid leaching process was proposed to extract gallium from brown corundum dust. An investigation was carried out on the effects of roasting temperature, roasting time, and additive dosage on the recovery of gallium and the evolution of the phase composition of the dust. The results show that the roasting activation of sodium carbonate was better than that of calcium oxide. After roasting at 1073 K for 40 min with a sodium carbonate dosage of 0.5 (mass ratio of sodium carbonate to dust), the phase composition changed completely to mainly consist of sodium silicate, sodium aluminosilicate, and potassium aluminosilicate. In that case, around 93% of Ga could be recovered from the roasted dust through $H_2SO_4$ (4.6 mol/L) leaching for 90 min. The leaching process was described well by the kinetic equation of $k_3t = 1/(1-\alpha)^{1/3} - 1$, with an apparent activation energy of 16.81 kJ/mol, suggesting that the leaching rate was limited by the transfer of leaching agent across the contacting interface of the dust particles.

**Keywords:** amorphous silicate; roasting activation; sodium carbonate; calcium oxide; leaching kinetics

**Citation:** Zhang, J.; Chang, Y.; Gao, C.; Hui, X.; Jokilaakso, A. Extraction of Gallium from Brown Corundum Dust by Roasting—Acid Leaching Process. *Minerals* **2023**, *13*, 900. https://doi.org/10.3390/min13070900

Academic Editors: Marinela Ivanova Panayotova and Vladko Panayotov

Received: 14 June 2023
Revised: 28 June 2023
Accepted: 29 June 2023
Published: 1 July 2023



## 1. Introduction

Gallium is a critical element widely applied in the chemical industry, in polymerization, dehydration, and catalysis, for instance [1,2]. Gallium and gallium-based compounds are commonly used as an important semiconductor component material due to their superior electronic and optical properties, and gallium is also required in the preparation of solar cells [3–5]. In addition, gallium plays a pivotal role in some medical treatments [6,7]. For instance, gallium nitrate has a good therapeutic effect in the treatment of urothelial tumors, radioactive gallium can be used to detect and locate malignant tumor cells, and alloys of gallium with indium, silver, and tin are utilized as tooth filling material.

The content of gallium in the Earth's crust is about 15 ppm, and the reserves are relatively high compared with molybdenum, tungsten, antimony, mercury, and arsenic [8]. However, the gallium mineral concentration is not enough to become the main mineral of an ore body, so it is difficult for gallium to form ore deposits [9], and it usually enters other minerals. Therefore, at present, gallium is usually recovered and extracted from secondary resources or obtained as a by-product during the aluminum oxide production process [10,11]. Commonly, the secondary resources for extracting gallium include red mud [12,13], fly ash [14,15], electronic waste [16,17], phosphorus flue dust [18,19], and zinc refinery residues [20,21].

Brown corundum is produced through electric arc furnace smelting with bauxite, anthracite, and iron filings as raw materials. It contains 95%–97% $Al_2O_3$ and a small amount of Fe, Si, Ti, etc., and is brownish black in color. It is commonly used as a refractory castable, sandblasting material, abrasive material, filtration medium, or functional filler. In the brown corundum production process, a lot of flue dust is generated, in which the content of gallium oxide is as high as 0.13%–0.16% [22]. Therefore, recovering gallium from this kind of brown corundum dust not only improves the gallium resource utilization efficiency but also creates economic benefits for the brown corundum industry. Up to now, there have been only a few studies on the extraction of gallium from brown corundum dust. Ding and his team [23] recovered gallium through microwave-assisted acid leaching and achieved a recovery of 82.56% when they conducted an experiment under the conditions of 25 wt.% $H_2SO_4$, temperature of 90 °C, leaching time of 50 min, solid-liquid ratio of 1:5 (g/mL), and stirring speed of 200 rpm, with an ultrasonic power of 900 W. Wen et al. [24] used a two-stage alkaline leaching, including 120 min of concentrated alkali leaching and 30 min of dilute alkali leaching, to recover gallium from corundum dust, and a recovery of 93% was achieved. In addition, that research team also treated the corundum dust with a mixture of sulfuric acid and hydrogen fluoride. Under leaching conditions of a solid–liquid ratio of 1:5 (g/mL), 80 °C, 4 h, 1.5 mol/L $H_2SO_4$, and 6.4 mol/L HF, the gallium leaching rate was able to reach up to 91% [25]. However, attention should be paid to the reasonable disposal and treatment of the fluorine-containing wastewater produced by this method, and the high corrosiveness of hydrofluoric acid should also be considered.

In this paper, based on the chemical and phase composition and the morphology of the corundum dust, a process of roasting activation followed by acid leaching was used to extract gallium. We studied the effects of the roasting parameters on the recovery of Ga and the phase evolution of the roasted dust, including temperature time, additive dosage, and leaching factors consisting of acid concentration, leaching temperature, leaching time, and stirring rate. The mechanism of gallium extraction and the kinetics of the acid leaching process were also investigated.

## 2. Experimental

### 2.1. Brown Corundum Dust

The raw material used in the experiments was collected from a brown corundum plant in Guizhou province, China. The chemical composition of the dust was analyzed by an X-ray Fluorescence Spectrometer (XRF, Rigaku/ZSXPrimus IV, Thermo Fisher Scientific, Waltham, MA, USA), and the results are listed in Table 1, where the contents of the main elements, including Si, K, Al, S, Fe, and the target element of Ga, were further detected by inductively coupled plasma optical emission spectrometry (ICP-OES, Agilent 5110, Agilent Technologies Inc., Santa Clara, CA, USA) and inductively coupled plasma mass spectrometry (ICP-MS, Agilent 7800, Agilent Technologies Inc., Santa Clara, CA, USA). Additionally, the carbon and nitrogen contents were determined using an oxygen–nitrogen–hydrogen analyzer (ONH836, Leco, San Jose, CA, USA) and a carbon and sulfur analyzer (HX-HW88, Nanjing Huaxin Analytical Instrument Manufacturing Co. Ltd., Nanjing, China), respectively.

**Table 1.** Chemical composition of brown corundum dust (wt.%).

| $SiO_2$ | $K_2O$ | $Al_2O_3$ | $SO_3$ | $Fe_2O_3$ | MgO | MnO | $P_2O_5$ | $Na_2O$ | $TiO_2$ | PbO | F |
|---|---|---|---|---|---|---|---|---|---|---|---|
| 40.82 | 17.7 | 23.39 | 3.085 | 3.542 | 0.870 | 0.757 | 0.712 | 0.533 | 0.491 | 0.415 | 0.345 |
| ZnO | CaO | CuO | $Ga_2O_3$ | Cl | $Cr_2O_3$ | $SnO_2$ | $ZrO_2$ | $Rb_2O$ | $As_2O_3$ | SrO | $Bi_2O$ |
| 0.153 | 0.150 | 0.147 | 0.114 | 0.084 | 0.032 | 0.028 | 0.027 | 0.026 | 0.022 | 0.012 | 0.012 |
| $V_2O_5$ | $ThO_2$ | $Nb_2O_5$ | NiO | $GeO_2$ | $La_2O_3$ | $Y_2O_3$ | $Co_2O_4$ | LOI | C | N | |
| 0.007 | 0.006 | 0.005 | 0.004 | 0.004 | 0.004 | 0.003 | 0.002 | 5.940 | 0.350 | 0.081 | |

As shown in Table 1, the main components in the dust were $SiO_2$, $K_2O$, and $Al_2O_3$, and these three components accounted for around 82% in total. The total iron content calculated by $Fe_2O_3$ was 3.542%, of which 0.22% of Fe existed in the form of metallic iron, which was determined by chemical titration. The content of Ga was 0.085%, far beyond the content range of 0.003%–0.0087% of a typical bauxite [26].

The phase composition of the brown corundum dust was analyzed by X-ray diffraction (XRD, X'Pert Pro MPD, PAnalytical B.V., Almelo, Netherlands; HighScore (Plus) V3.X database, PDF 2021), and the pattern is shown in Figure 1. The main crystal phases in the raw material were corundum ($Al_2O_3$) and $\beta$-potassium sulfate ($K_8S_4O_{16}$, reference code 96-120-0012). Elements of K and S mainly existed in the form of potassium sulfate. As shown in Table 1, the content of silicon was the highest, but no characteristic peaks of silicon-containing phases were found, indicating that the silicon elements were mainly present as amorphous silicate. The peaks referring to gallium-containing minerals did not appear in this pattern. This was due to the low content of gallium in the raw material and its dispersed distribution in the amorphous silicate phase.

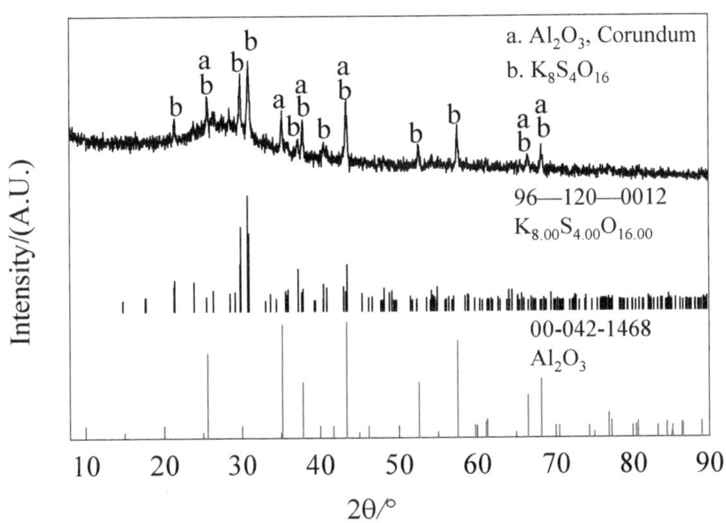


**Figure 1.** X-ray diffraction pattern of brown corundum dust.

A small amount of dust, acrylic powder (Anfu Weiyi Metallographic Equipment Co., Ltd., Jian, China), and curing agent (Anfu Weiyi Metallographic Equipment Co., Ltd., Jian, China) were placed in a beaker and then mixed evenly. The mass ratio of the added acrylic powder and curing agent was 10:8. The mixture was transferred to a mold and allowed to stand for more than half an hour until the sample was completely cured. The mounted sample was polished with a polishing machine until there were no obvious scratches on the polished surface when viewed under a light. The polished surface was used for SEM and EDS (scanning electron microscope–energy dispersive spectrometer, Nova 400 NanoSEM,

FEI Company, Hillsboro, OR, USA; HKL Channel 5 EBSD, Oxford Instruments, Oxford, UK; INCA IE 350 PentaFET X-3 EDS, Oxford Instruments, Oxford, UK) analysis to determine the morphology and distribution of the main elements of the brown corundum dust. The results are shown in Figure 2 and Table 2.

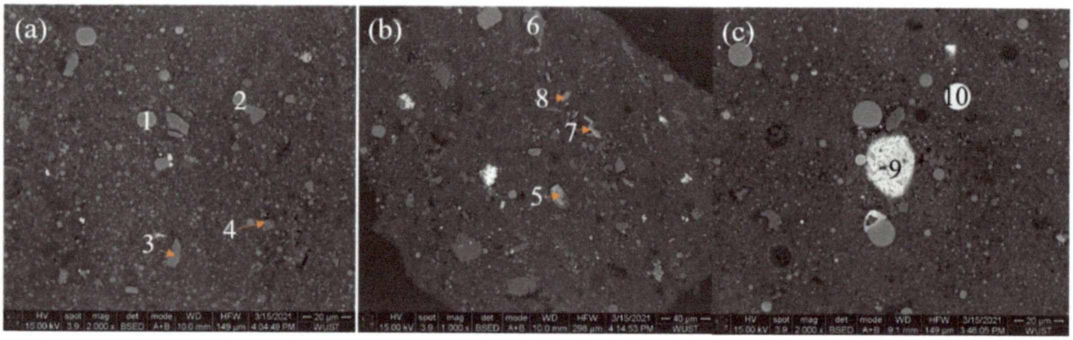


**Figure 2.** Backscattered electron image (BSE) of brown corundum dust: (**a**) ×2000; (**b**) ×1000; (**c**) ×2000.

**Table 2.** Chemical composition of each point in Figure 2.

| Point | Element/wt.% | | | | | | | | |
|---|---|---|---|---|---|---|---|---|---|
| | O | Si | K | Al | Fe | Ti | Mn | Ca | S |
| 1 | 29.59 | 30.03 | 18.19 | 15.08 | 5.80 | 1.31 | - | - | - |
| 2 | 33.45 | 26.13 | 17.62 | 18.81 | 2.13 | 1.01 | 0.85 | - | - |
| 3 | 8.33 | 89.50 | 1.45 | - | - | - | - | 0.72 | - |
| 4 | 42.64 | - | - | 57.36 | - | - | - | - | - |
| 5 | 45.03 | 1.82 | 17.83 | - | - | - | - | 15.81 | 19.50 |
| 6 | 36.95 | 1.15 | 0.54 | 56.15 | 1.17 | 0.75 | - | 3.28 | - |
| 7 | 50.06 | 1.67 | 17.24 | - | - | - | - | 13.05 | 17.98 |
| 8 | 46.44 | 8.70 | 15.75 | 4.41 | - | - | - | 10.68 | 14.02 |
| 9 | 28.43 | 0.78 | 0.51 | 12.59 | 55.80 | 0.55 | 1.33 | - | - |
| 10 | 26.98 | 1.28 | 0.34 | 1.55 | 69.86 | - | - | - | - |

According to the chemical and phase composition, as well as the EDS analysis results listed in Table 2, points 1 and 2 in Figure 2 are spherical potassium aluminosilicate. Point 3 refers to monatomic silicon. Points 4 and 6 correspond to corundum with irregular shape. Points 5, 7, and 8 are flaky calcium sulfate and potassium sulfate. Points 9 and 10 have high iron content and are composed of iron oxides. The black area covering the particles is the embedding resin. To sum up, the potassium mainly existed in potassium sulfate, the silicon was present in the form of amorphous silicate, and the occurrence of aluminum included corundum and aluminosilicate. No gallium was detected by EDS due to its low content and relatively dispersed distribution. Hence, an EPMA (Electron Probe X-ray micro analyzer, EPMA-8050G, Shimadzu-Kratos, Kyoto, Japan) was used to verify the existence of gallium, and the results are illustrated in Figure 3.

As shown in Figure 3, gallium occurred densely with sodium and potassium together, whereas the areas where they existed also contained high amounts of aluminum, oxygen, and silicon. According to the color of the area where gallium occurred, the gallium content ranged from 0.118% to 0.196%, which is higher than the average gallium content of 0.085% listed in Table 1, indicating that gallium was enriched to some extent. It is speculated that gallium occurred in the form of a sodium–gallium compound and potassium–gallium compound, wrapped by the amorphous silicate and corundum.

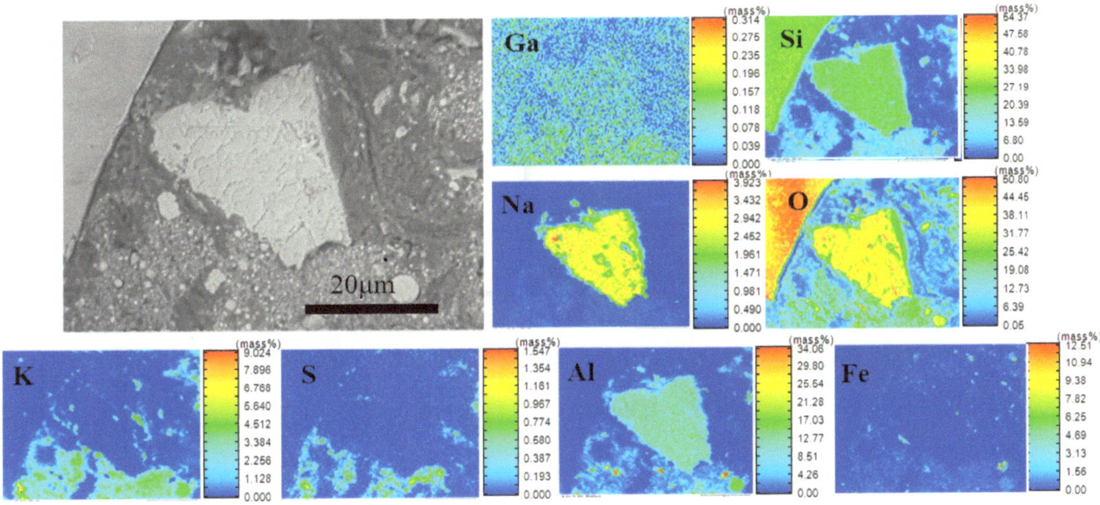


**Figure 3.** Elemental distribution of brown corundum dust detected by EPMA.

*2.2. Reagents and Apparatus*

The reagents used for the experiments, including anhydrous $Na_2CO_3$ (analytical reagent, abbreviated as AR), sulfuric acid (AR), hydrochloric acid (AR), titanium trichloride (AR), thiourea (AR), toluene, and methyl isobutyl ketone (AR), were purchased from Sinopharm Chemical Reagent Co. of China. Gallium standard solution (GSBG62026-90, 10% HCl, 1000 μg/mL) was used to measure the concentration of gallium in the leaching solution by means of a spectrophotometer. All of the leaching experiments and spectrophotometer detection were conducted using deionized water with an RO effluent conductivity of 2–10 μS/cm. The instruments included a vertical tube furnace (BLMT-1700 °C, Boleymant Test Electric Furnace Co., Ltd., Wuhan, China), as shown in Figure 4a, a heat-collecting thermostatic water bath equipped with magnetic stirrer (DF-101S, Gongyi Yuhua Instruments Co., Ltd., Gongyi, China), as shown in Figure 4b, an electronic balance (FA2004, Shanghai Shangping Instruments Co., Ltd., Shanghai, China), a vacuum drying oven (DZF-6050, Gongyi Yuhua Instruments Co., Ltd.), and a sample mill (XQM-100, Wuhan Prospecting Plant, Wuhan, China).

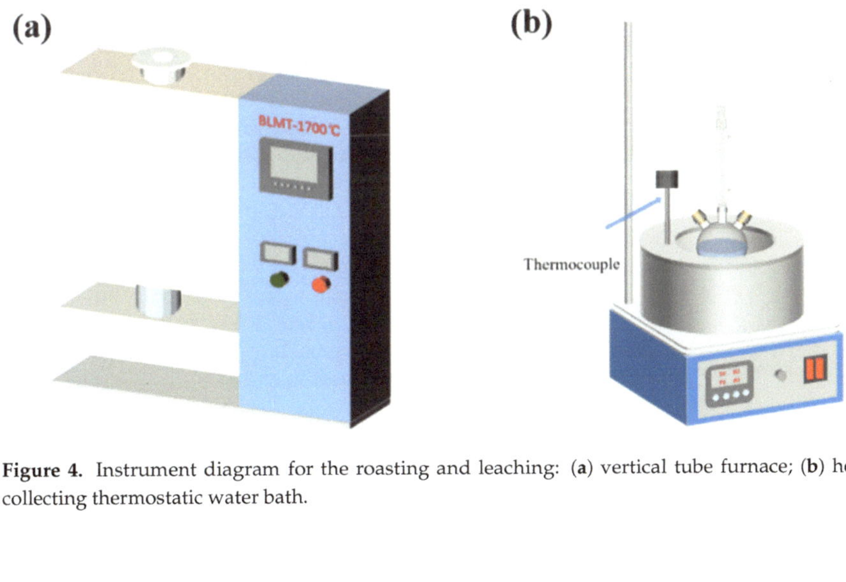


**Figure 4.** Instrument diagram for the roasting and leaching: (**a**) vertical tube furnace; (**b**) heat-collecting thermostatic water bath.

### 2.3. Operational Procedure

The brown corundum dust was mixed with the roasting additive at a determined mass ratio. The mixture was put in a corundum crucible (ϕ75 mm × H27 mm), and then the crucible was placed in the thermostatic area of the vertical tube furnace calibrated by a standard couple. The bottom end of the tube was sealed during roasting and a hole was left in the upper lid of the furnace tube to allow carbon dioxide and sulfur dioxide to flow out. After roasting for a period, the crucible was gradually pulled to the cold end of the furnace and taken out to be cooled in air. Because a certain degree of sintering happened to the sample during the roasting, the roasted sample was ground in a pulverizer (YP-200-1, Nanchang Yongping Laboratory Equipment Co., Ltd., Nanchang, China) for 15 s to ensure that the particles passed a 200-mesh sieve (74 μm). Then the ground sample was leached with diluted sulfuric acid in a three-neck flask, which was heated by the thermostatic water bath. When the leaching was complete, the slurry was filtered. The filtrate and leaching residue were collected to determine the recovery of gallium. The concentration of gallium was measured by the rhodamine B spectrophotometry method (Chinese GB/T 20127.5-2006).

## 3. Results and Discussion

### 3.1. Effect of Roasting

#### 3.1.1. Roasting Temperature

Since the gallium component was embedded in the amorphous silicate and corundum in the dust, CaO and $Na_2CO_3$ were used as roasting additives, with the aim of reacting with the silicates and improving the acid solubility of the gallium-containing components. The effect of roasting temperature on the recovery of gallium is shown in Figure 5. The recovery of gallium first increased and then decreased sharply with the increase in roasting temperature from 823 K to 1323 K, with the maximum at 1123 K. It is known that the melting point of $Na_2CO_3$ is around 1124 K and most sodium salts soften or melt below 1273 K. Hence, when the temperature was over 1123 K, the liquid sodium salt phase was produced. In fact, we found that the samples were obviously sintered, and the sintering phenomenon became more serious as the temperature increased from 1123 K to 1223 K. These lower-melting substances, especially the silicates, surrounding other higher-melting phases that contain gallium, made gallium difficult to extract by acid leaching. Additionally, these lower-melting substances could have filled pores within and between the sample particles, hindering the internal transfer of the leaching agent. Hence, the sintering was not beneficial to the subsequent leaching process. As displayed by the curve, the recovery of gallium declined rapidly from more than 98% to less than 55% in this temperature range. Compared with sodium carbonate, calcium oxide exhibited lower activation efficiency, and the corresponding recovery of gallium was lowered by 5%–12% at temperatures ranging from 823 K to 1123 K. However, unlike the trend in gallium recovery in sodium carbonate roasting, the recovery of gallium did not decrease but rather tended to level off when the roasting temperature was raised from 1123 K to 1323 K, and a maximum of 82.83% recovery was achieved. This was attributed to the higher melting point of calcium salts. In addition, no obvious sintering occurred during roasting within the chosen temperature range.

Figure 6a,b display the XRD patterns of samples roasted at 823 K, 1123 K, and 1323 K. As shown in Figure 6a, after roasting at 823 K for 2 h, the corundum and potassium sulfate were still present, and peaks for the sodium carbonate additive appeared, indicating that sodium carbonate was stable at this temperature and that the reactions between the additive and the dust were not sufficient. When roasted at 1123 K, most of the amorphous silicates and corundum in the original brown corundum dust reacted with sodium carbonate and formed new silicates ($Na_2SiO_3$ and $Mg_{1.1}Fe_{0.9}Ca_4Si_4O_{14}$) and aluminosilicates ($NaAlSiO_4$, $K_{0.985}AlSiO_4$). The characteristic peaks of $Na_2CO_3$ were still found, showing that there may have been an excess of the roasting additive. When increasing the temperature to 1323 K, the sodium carbonate disappeared and more sodium elements participated in the formation of silicates, such as $Na_2SiO_3$, $NaAlSi_3O_8$ (albite), and $Na_{0.68}Fe_{0.68}Si_{0.32}O_2$. Compared with the XRD patterns for 1123 K, no peaks related to the potassium components were observed.

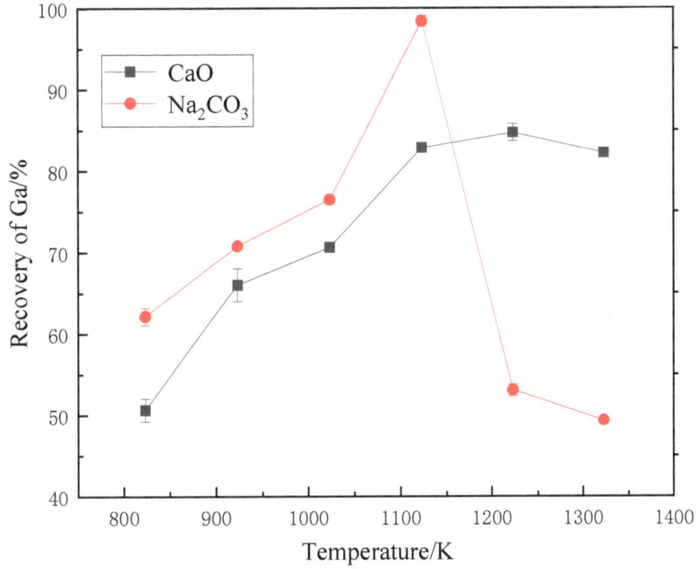


**Figure 5.** Effect of roasting temperature (roasting time 2 h, additive mass/dust mass = 1.0, acid leaching time 3 h, leaching temperature 353 K, L/S (mL/g) = 10, stirring rate 800 rpm, $C(H_2SO_4)$ = 25 vol.%).

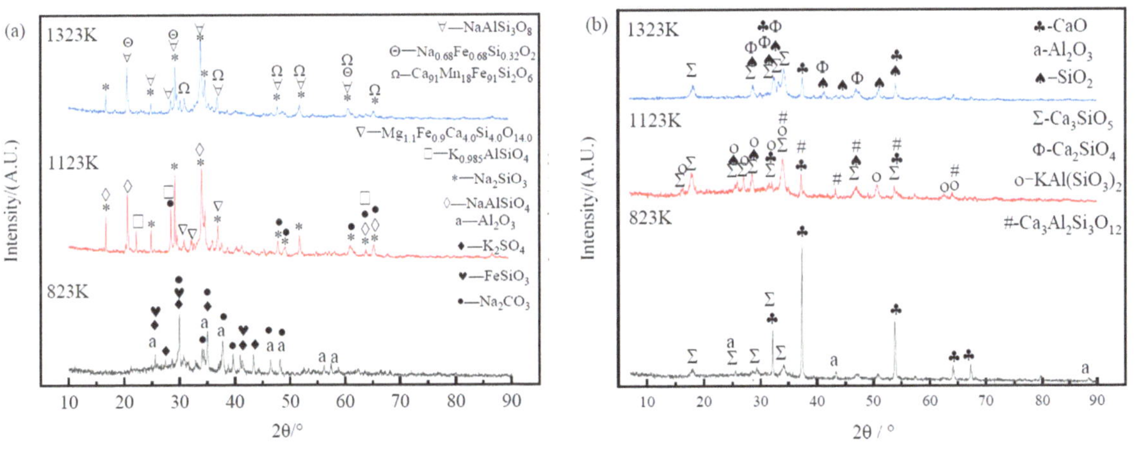


**Figure 6.** XRD patterns of samples roasted at different temperatures with (**a**) $Na_2CO_3$ and (**b**) CaO as additives.

When CaO was used as the roasting activation additive, after roasting at 823 K for 2 h some corundum still existed, as shown in Figure 6b. The characteristic peak intensity of calcium oxide was strong, suggesting that there was plenty of unreacted calcium oxide in the roasted sample. However, obviously, reactions between the additive and the dust had started and $Ca_3SiO_5$ was produced. After raising the temperature to 1123 K, the peak strength of calcium oxide decreased significantly, indicating that more calcium oxide took part in the reactions. Apart from $Ca_3SiO_5$, aluminosilicate ($KAl(SiO_3)_2$, $Ca_3Al_2Si_3O_{12}$) and $SiO_2$ were generated. When the roasting temperature reached 1323 K, as in the sodium carbonate roasting, no peaks in the potassium-containing phase were found. The phase composition of the roasted sample changed at this temperature and included $Ca_3SiO_5$, $Ca_2SiO_4$, $SiO_2$, and excess CaO. However, as illustrated in Figure 5, the recovery of gallium was not improved further when the temperature rose from 1123 K to 1323 K.

### 3.1.2. Additive Dosage

As mentioned above, an excess of additive was found in the roasted sample when the mass ratio of additive to dust was 1.0. Figure 7 shows the change in gallium recovery against the additive dosage. The recovery of gallium increased with an increasing amount of $Na_2CO_3$, and it tended to level off when the mass ratio was over 0.5. This was lower than the theoretical stoichiometric value (0.72) for transforming all the amorphous silicate into sodium silicate ($Na_2SiO_3$) but within the stoichiometric value range from $NaAlSiO_4$ to $Na_2SiO_3$. When CaO was used as additive, the recovery of gallium increased when increasing the additive dosage and the highest recovery occurred at a mass ratio of 0.8, which was close to the theoretical value of 0.76 for CaO and $SiO_2$, forming $Ca_2SiO_4$. This is in agreement with the XRD analysis, namely, that $Ca_2SiO_4$ was more stable than $Ca_3SiO_5$. When the mass ratio was increased to 1.0, the extraction percentage of gallium decreased. This may have resulted from the decrease in the leaching kinetic rate caused by the solid layer of $Ca_2SO_4$ produced by the reaction between the excess CaO and $H_2SO_4$.

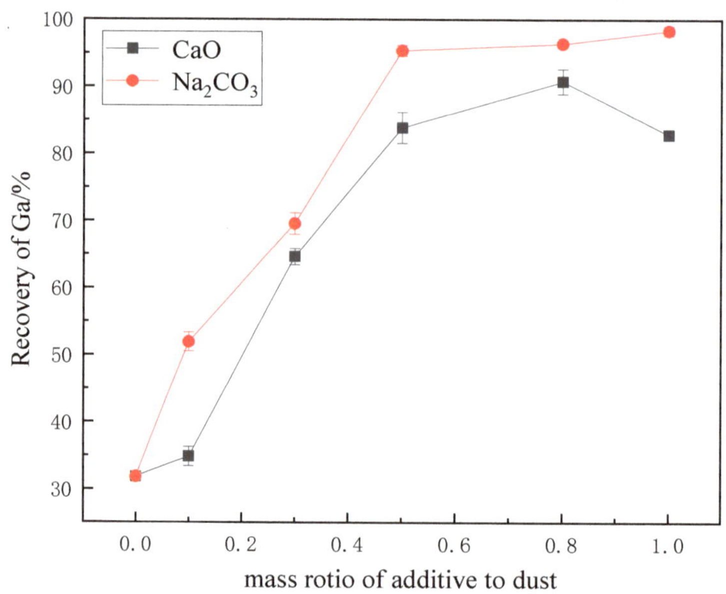


**Figure 7.** Effect of additive dosage (roasting time 2 h, roasting temperature 1123 K, acid leaching time 3 h, acid leaching temperature 353 K, L/S (mL/g) = 10, stirring rate 800 rpm, C($H_2SO_4$) = 25 vol.%).

Figure 8 shows the XRD patterns of roasted products with different dosages of additives. When the mass ratio was 0.1, there was some unreacted corundum ($Al_2O_3$) in the roasted dust, and $Na_2Si_2O_5$, $SiO_2$, and some potassium aluminosilicate salts ($KAlSiO_4$, $KAlSi_2O_6$) were produced. As the dosage of $Na_2CO_3$ was increased to 0.5, the characteristic peaks of $Al_2O_3$ and $SiO_5$ disappeared and the occurrence of sodium silicate changed from $Na_2Si_2O_5$ to $Na_2SiO_3$ because more sodium participated in the reactions. Some potassium elements were replaced with the sodium, and then sodium aluminosilicate ($NaAlSiO_4$) was formed. When the mass ratio of additive to dust equaled 1.0, clearly there was an excess of sodium carbonate and its characteristic peaks appeared in the roasted dust. The phase composition of the dust was thoroughly changed. The original amorphous silicate and corundum were completely invisible and new silicates ($Na_2SiO_3$ and $Mg_{1.1}Fe_{0.9}Ca_{4.0}Si_{4.0}O_{14.0}$) and aluminosilicates ($NaAlSiO_4$ and $K_{0.985}AlSiO_4$) were generated.

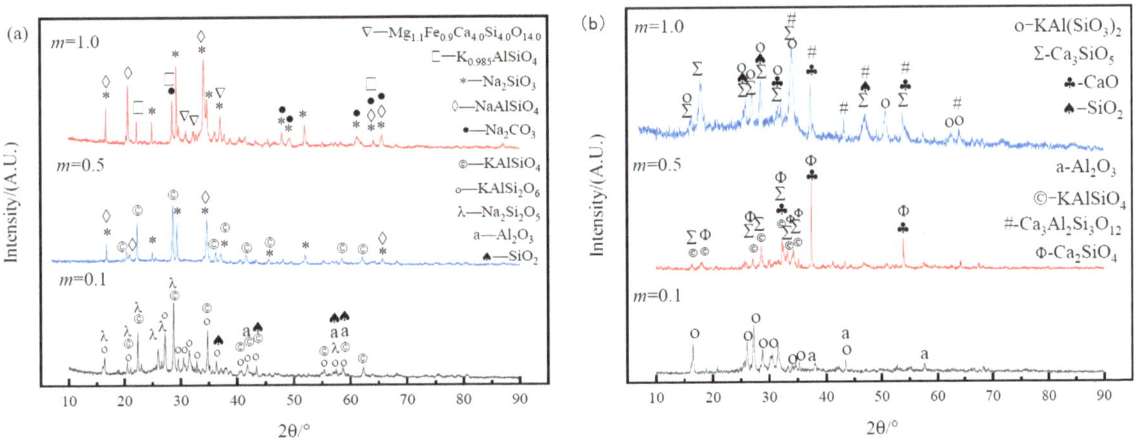


**Figure 8.** XRD patterns of samples roasted with different dosages of (**a**) $Na_2CO_3$ and (**b**) CaO.

Similarly, when the CaO additive was insufficient, there was no calcium silicate found, as shown in Figure 8b, and the main silicate phase was potassium aluminosilicate ($KAl(SiO_3)_2$). When increasing the mass ratio to 0.5, the extent of the reaction between the dust and activation additive became greater; dicalcium silicate ($Ca_2SiO_3$) and tricalcium silicate ($Ca_3SiO_5$) coexisted. By further raising the dosage of CaO, the dicalcium silicate was converted into tricalcium silicate. Additionally, calcium aluminosilicate ($Ca_3Al_2Si_3O_{12}$) appeared. Although the characteristic peaks of calcium oxide appeared in both patterns of additive mass ratios of 0.5 and 1.0, the reason for the mass ratio of 0.5 was attributed to reaction kinetic limitation; consequently, the reactions between the additive and the dust were insufficient. The remaining calcium oxide in the sample with a mass ratio of 1.0 predominantly resulted from dosing more of the additive than was required for the reactions.

The effect of additive dosage on activation roasting was further studied by SEM–EDS, and the results for roasted dust with a mass ratio ($Na_2CO_3$/dust) of 0.1 and 1.0 are shown in Figure 9 and Table 3.

**Table 3.** Chemical compositions of different points obtained by EDS analysis in Figure 9 (wt.%).

| Point | O | Na | Mg | C | Si | K | Al | Fe | Ti | Ca | Zn |
|---|---|---|---|---|---|---|---|---|---|---|---|
| 1 | 50.5 | 41.7 | - | - | 4.2 | 1.8 | 1.8 | - | - | - | - |
| 2 | 48.1 | 26.6 | - | 25.3 | - | - | - | - | - | - | - |
| 3 | 34.2 | 1.1 | - | - | 33.4 | - | 9.0 | - | - | 20.1 | 2.2 |
| 4 | 32.1 | - | 66.6 | - | 0.3 | - | - | 1.0 | - | - | - |
| 5 | 36.8 | 17.7 | - | - | 22.6 | 5.3 | 14.3 | 2.8 | 0.5 | - | - |
| 6 | 35.4 | 2.4 | - | - | 23.2 | 22.8 | 14.8 | 1.4 | - | - | - |
| 7 | 36.9 | - | - | - | 25.5 | - | 15.3 | - | - | 22.3 | - |
| 8 | 20.2 | - | 4.6 | - | 5.7 | - | 3.3 | 66.2 | - | - | - |
| 9 | 38.9 | 12.8 | - | - | 21.8 | 5.5 | 17.6 | 3.4 | - | - | - |
| 10 | 37.1 | - | - | - | 15 | 10.7 | 37.2 | - | - | - | - |
| 11 | 46.3 | 9.5 | 12.3 | - | - | - | - | 1.2 | - | 30.7 | - |
| 12 | 38.9 | - | 61.1 | - | - | - | - | - | - | - | - |
| 13 | 31 | 3.9 | - | - | 27.5 | 23.3 | 14.3 | - | - | - | - |
| 14 | 43.8 | 1.1 | - | - | 54.6 | 0.5 | - | - | - | - | - |

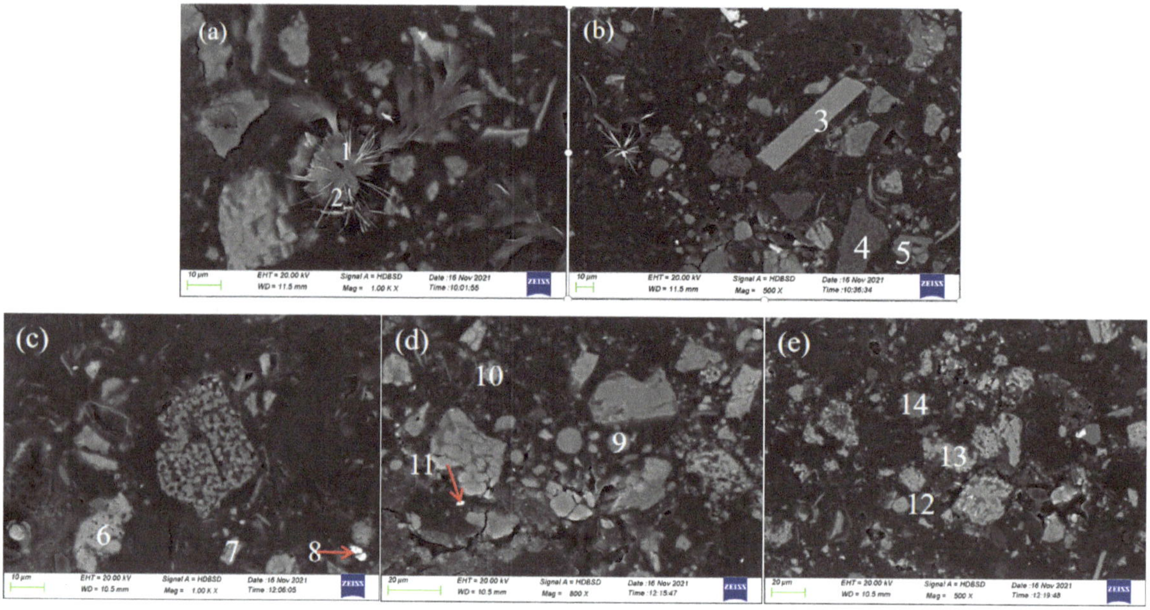


**Figure 9.** BSE images for the dust roasted with different dosages of $Na_2CO_3$: (**a**) $m$ = 1, ×1000; (**b**) $m$ = 1, ×500; (**c**) $m$ = 0.1, ×1000; (**d**) $m$ = 0.1, ×800; (**e**) $m$ = 0.1, ×800.

By comparing the images in Figure 9, it can be seen that there was a surplus of sodium carbonate when the additive was added at a mass ratio of 1.0. Multiple feather-like and needle-like substances appeared in the roasted sample. Based on the chemical composition in Table 3, we speculated that these were sodium oxide (point 1) and sodium carbonate (point 2). This is in agreement with the XRD analysis shown in Figure 8a. Although the thermodynamic calculation results show that sodium carbonate was stable at this roasting temperature and at a pressure of 1 atm, the decomposition of sodium carbonate occurred due to the far lower partial pressure of the produced $CO_2$ gas, which promoted the decomposition reaction [27]. Besides, the thermodynamic calculations were based on a simplification of the reactants contained in the dust. In fact, the components involved in the reaction were complex in the real reaction system and some substance may have accelerated the decomposition of sodium carbonate. Point 3 is calcium aluminosilicate, point 4 contains magnesium oxide, and points 5 and 9 consist of sodium aluminosilicate. Points 6, 10, and 13 in Figure 9c–e are potassium aluminosilicates. Point 7 is composed of calcium aluminosilicate. Point 8 indicates iron oxide. Points 11 and 12 are mainly composed of magnesium oxide, calcium oxide, and sodium oxide. Point 14 is silicon oxide. These results are consistent with the XRD patterns. The simple oxides (MgO, CaO, FeO) were present in the roasted dust with low dosages of $Na_2CO_3$, meaning that the amorphous silicate was not completely activated with roasting, and these oxides did not transfer to the silicates.

The BSE–EDS analysis for the dust roasted with CaO as the additive are shown in Figure 10 and Table 4. Although the maximum recovery of gallium occurred at a CaO dosage of 0.8, as displayed in Figure 7, the surplus CaO additive was detected in the roasted samples (points 3, 4, and 6 in Figure 10). Calcium aluminosilicate (point 1), potassium aluminosilicate (points 8, 16), calcium silicate (point 12), and silicon dioxide (points 2, 7, and 9) were generated after roasting activation, which is consistent with the XRD analytical results in Figure 8b. The metallic iron and ferrous iron were oxidized as ferric oxide (point 13). Points 5 and 11 in Figure 10a,c had high contents of silicon and sulfur, suggesting that the potassium sulfates were not completely decomposed and that some of them transformed into aluminosilicates.

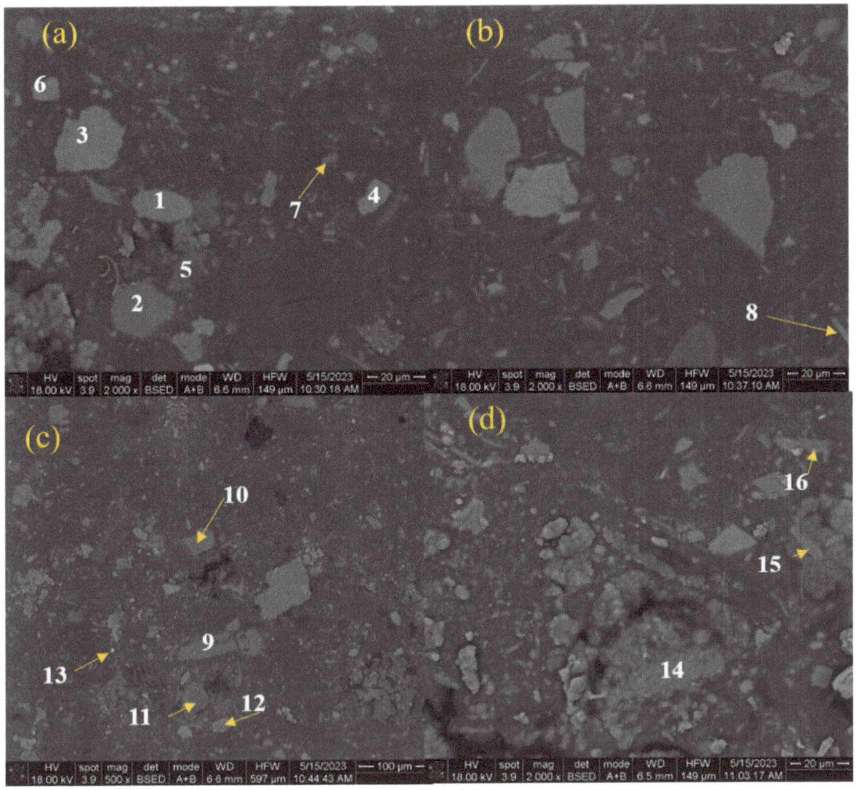


**Figure 10.** BSE images for different areas (**a**–**d**) of dust roasted with CaO dosage (mass ratio of CaO to dust) of 0.8.

**Table 4.** Chemical compositions of different points obtained by EDS analysis in Figure 10 (wt.%).

| Point | O | Na | Mg | Si | K | Al | Fe | Ti | Ca | S |
|---|---|---|---|---|---|---|---|---|---|---|
| 1 | 38.67 | - | - | 33.23 | - | 8.15 | - | - | 19.95 | - |
| 2 | 42.23 | - | - | 57.77 | - | - | - | - | - | - |
| 3 | 46.28 | - | - | - | - | - | - | - | 53.72 | - |
| 4 | 44.65 | - | - | - | - | - | - | - | 55.35 | - |
| 5 | 49.04 | - | 2.79 | 16.75 | 2.10 | 4.57 | 1.53 | - | - | 23.22 |
| 6 | 45.69 | - | - | - | - | - | - | - | 54.31 | - |
| 7 | 45.96 | - | 0.99 | 53.04 | - | - | - | - | - | - |
| 8 | 35.60 | - | 1.14 | 28.91 | 10.87 | 17.86 | 4.73 | 0.90 | - | - |
| 9 | 42.98 | 1.23 | - | 51.06 | 1.03 | - | - | - | - | 3.69 |
| 10 | 39.96 | - | 29.95 | 30.09 | - | - | - | - | - | - |
| 11 | 41.54 | 12.00 | 1.37 | 8.99 | 2.48 | 2.55 | - | - | - | 31.07 |
| 12 | 43.13 | - | 4.30 | 32.48 | -- | - | - | - | 14.29 | 5.80 |
| 13 | 23.67 | - | - | 9.21 | - | 6.20 | 60.93 | - | - | - |
| 14 | 31.14 | - | - | 2.05 | - | 66.82 | - | - | - | - |
| 15 | 38.77 | - | - | 2.28 | - | 54.89 | 1.87 | - | - | 2.20 |
| 16 | 38.88 | - | 4.01 | 21.64 | 10.00 | 14.43 | 3.65 | - | 0.90 | 6.48 |

3.1.3. Roasting Time

Figure 11 shows that when the additive was sodium carbonate, there was no significant change in gallium recovery within the time range from 40 min to 120 min at temperatures over 1098 K; the recoveries were higher than 93%. In particular, at the higher temperature of 1123 K, the activation reactions reached equilibrium within 20 min. For the CaO additive, the gallium extraction percentage increased with the extended time and a recovery of 90.8% was reached after 120 min. The continuously increasing trend in recovery demonstrates that calcium oxide reacted more slowly with the dust than sodium carbonate did. This further proves that the remaining unreacted calcium oxide existed in the dust after roasting for 120 min with the additive mass ratio of 0.5. Overall, in comparison with calcium oxide roasting, a higher recovery of gallium was obtained with sodium carbonate roasting at a lower temperature, a lower dosage, and within a shorter roasting time. Hence, sodium carbonate performed better than calcium oxide in the activation of raw brown corundum dust.

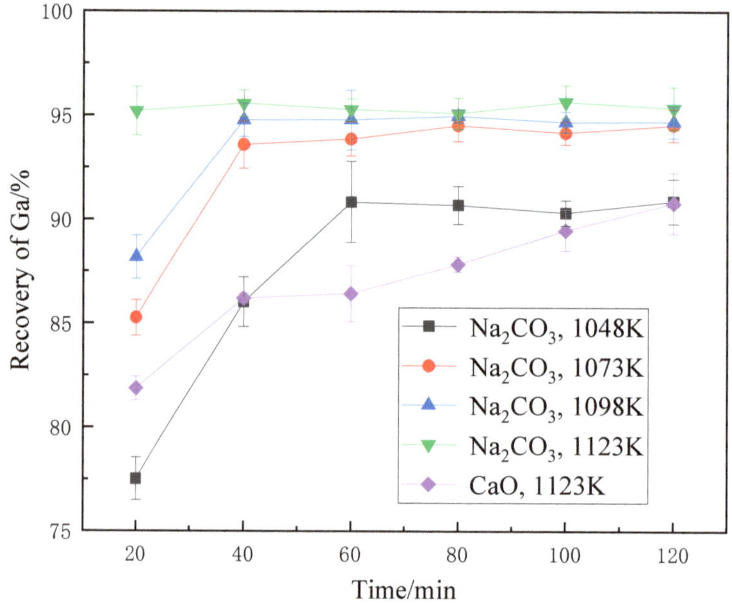


**Figure 11.** Variation in recovery of gallium in terms of roasting time ($Na_2CO_3$ dosage of 0.5, CaO dosage of 0.8, acid leaching time 3 h, acid leaching temperature 353 K, L/S (mL/g) = 10, stirring rate 800 rpm, $C(H_2SO_4)$ = 25 vol.%).

Figure 12 presents the BSE images of the brown corundum dust roasted at 1123 K and 1073 K. The EDS analysis results for different points marked in Figure 12 are shown in Table 5. The unreacted sodium carbonate additive (point 9), with a feathery shape, still appeared in Figure 12b after roasting for 120 min with the additive dosage of 0.5. Evidently, the original potassium sulfate, corundum, and amorphous silicate were converted into sodium silicate (points 6 and 8), sodium aluminosilicate (point 3), and potassium aluminosilicate (point 5) after reacting with the sodium carbonate. The magnesium oxide and calcium oxide (points 1 and 7) were relatively stable and only a certain amount of them was transformed into silicates (point 4), although free silicon oxide (point 2) was present at the same time. Furthermore, other elements, such as iron, titanium, manganese, and phosphorus, mainly occurred in aluminosilicates. Compared with the sample roasted at 1123 K, the sample roasted at 1073 K for 40 min was also effectively activated. Many of the potassium aluminosilicates (points 10, 11, 15, and 17) were included in Figure 12c,d, and some of the magnesium oxide was converted into the magnesium silicate (point 13). However, the corundum was not completely reacted, and the phase with high aluminum and oxygen content was discovered in the roasted dust (Figure 12d, point 16). In general, after roasting at 1073 K for 40 min, the reactions between the additive and the dust proceeded to a high degree, which explains why the recovery of gallium was as high as 93.11%.

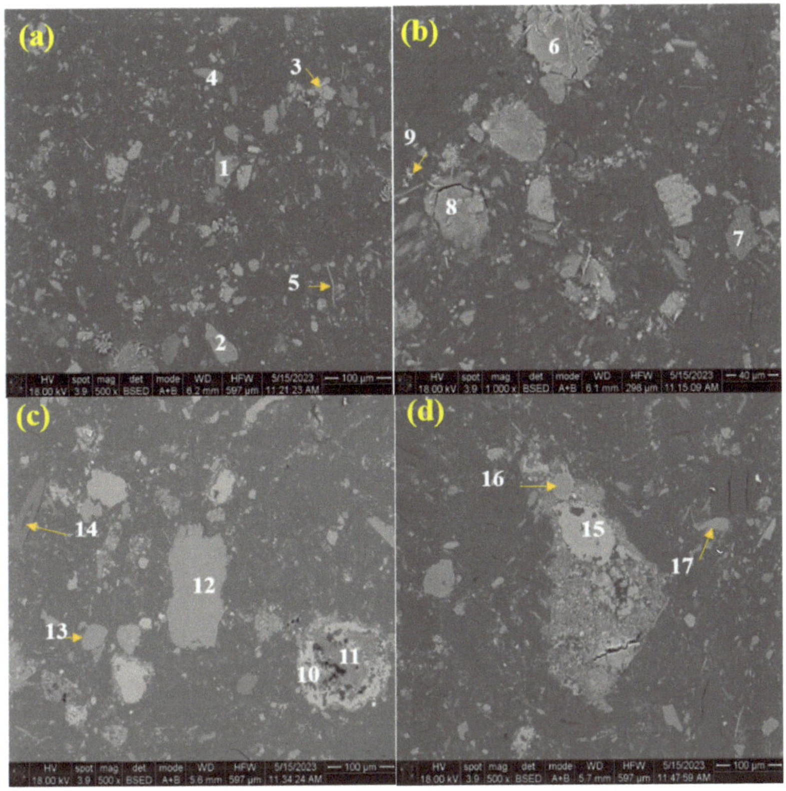


**Figure 12.** BSE images for the dust roasted with $Na_2CO_3$ dosage of 0.5: (**a**,**b**) roasted at 1123 K for 120 min; (**c**,**d**) roasted at 1073 K for 40 min.

**Table 5.** Chemical compositions of different points in Figure 12 (wt.%).

| Point | O | Na | Mg | Si | K | Al | Fe | Ti | Ca | S | P | Mn | C |
|---|---|---|---|---|---|---|---|---|---|---|---|---|---|
| 1 | 47.32 | - | 17.94 | - | - | - | - | - | 34.74 | - | - | - | - |
| 2 | 44.46 | - | - | 55.54 | - | - | - | - | - | - | - | - | - |
| 3 | 41.70 | 19.42 | - | 15.31 | 11.06 | 6.08 | 1.71 | 0.59 | - | - | 4.12 | - | - |
| 4 | 41.26 | - | 20.21 | 38.53 | - | - | - | - | - | - | - | - | - |
| 5 | 35.58 | - | 16.99 | 25.84 | 10.56 | 9.34 | 2.71 | - | - | - | - | - | - |
| 6 | 40.07 | 31.59 | - | 22.94 | 5.41 | - | - | - | - | - | - | - | - |
| 7 | 30.31 | - | 68.14 | - | - | - | 1.55 | - | - | - | - | - | - |
| 8 | 38.04 | 21.27 | - | 34.72 | 4.35 | - | - | 1.62 | - | - | - | - | - |
| 9 | 42.99 | 21.71 | - | 1.04 | 0.60 | - | - | - | - | 3.27 | - | - | 30.39 |
| 10 | 28.56 | - | - | 32.24 | 25.81 | 8.70 | 2.06 | - | 1.24 | - | - | 1.39 | - |
| 11 | 33.89 | - | - | 25.62 | 28.31 | 12.19 | - | - | - | - | - | - | - |
| 12 | 46.95 | - | 18.03 | - | - | - | - | - | 35.02 | - | - | - | - |
| 13 | 36.39 | - | 30.60 | 30.96 | - | - | 2.05 | - | - | - | - | - | - |
| 14 | 48.37 | - | - | 51.63 | - | - | - | - | - | - | - | - | - |
| 15 | 30.47 | - | 0.60 | 28.30 | 27.67 | 9.81 | 2.04 | - | 1.12 | - | - | - | - |
| 16 | 33.50 | - | 1.88 | 6.81 | 4.45 | 50.03 | 1.41 | 0.80 | - | - | - | 1.12 | - |
| 17 | 34.40 | - | 0.76 | 29.42 | 11.08 | 18.63 | 4.57 | 1.13 | - | - | - | - | - |

### 3.2. *Effect of Acid Leaching*

#### 3.2.1. Leaching Parameters

After roasting activation with sodium carbonate, the phase composition of the dust was restructured. The amorphous silicate and corundum were transformed into sodium silicate and aluminosilicates. The potassium sulfate was decomposed and potassium participated in the formation of potassium aluminosilicate. The gallium-containing components encapsulated in these original phases could then be extracted by leaching. In our previous study, different leaching methods, including acid leaching, alkaline leaching, and water leaching, were compared; the results showed that acid leaching performed with the highest efficiency in gallium extraction, whereas the dissolution percentage of silicon was kept as low as 0.31% [27].

Leaching experiments were carried out upon the brown corundum dust roasted at 1073 K for 40 min with a mass ratio of additive to dust of 0.5 to investigate the influence of the acid leaching parameters on the extraction efficiency of gallium.

As shown in Figure 13, the leaching time, temperature, and concentration of sulfuric acid all had a great effect on the recovery of gallium. When the leaching time was extended from 30 to 90 min, the recovery increased sharply by around 14%, and at least 90 min leaching were required to obtain a recovery higher than 93%. As for the leaching temperature, clearly, as displayed in Figure 13b, a higher leaching temperature was beneficial for extracting gallium from the roasted dust and the recovery increased from 68.78% to 93.11% as the temperature increased from 298 K to 373 K. Although the phase composition of the dust was changed after roasting, it still required a relatively high concentration of sulfuric acid to obtain satisfactory recovery of gallium. Due to the high content of silicon in the sample, only about 30% of Ga was extracted when 1 mol/L $H_2SO_4$ was used for leaching. A significant growth trend in the recovery appeared as the acid concentration increased, until the concentration reached 4.6 mol/L. Meanwhile, a suitable stirring speed had to be maintained to make the leaching agent and roasted dust particles come into contact thoroughly and to overcome the resistance of external diffusion. Figure 13d shows that a recovery of 93.11% was achieved at a stirring rate of 800 rpm using the laboratory leaching apparatus.

#### 3.2.2. Characterization of Samples before and after Leaching

In order to further determine the mechanism of sulfuric acid leaching, the samples were characterized by XRD and SEM–EDS before and after leaching.

As seen in Figure 14, after acid leaching, sulfate ($K_2Ca_2(SO_4)_3$) was generated. Aluminosilicates, including sodium aluminosilicate and potassium aluminosilicate, did not react with acid completely, and some of them remained in the residue. Amorphous silicon-containing components were produced during the acid leaching process, which was proven by the wide and flat peak appearing within 2 θ ranging from 20° to 30°.

Figure 15 presents the EDS mapping analysis of the leaching residue. The potassium concentration was low and appeared in a dispersed manner. Previously, we found that around 70% of potassium was dissolved into the acid solution [27]. The remaining potassium mainly co-existed with elements of silicon and aluminum. Based on the analysis of the XRD pattern of the leaching residue, it was deduced that the remaining potassium was present in the form of potassium aluminosilicate. Additionally, some unreacted sodium aluminosilicate and calcium aluminosilicate were left in the leaching residue. This indicates that the aluminosilicates were less likely than sodium silicate to react with sulfuric acid. Silicon dioxide was produced by the reactions of sodium silicate and the aluminosilicate with sulfuric acid and remained in the residue due to its low acid solubility. Additionally, potassium calcium sulfate and magnesium silicate were found in the residue. These results are in good agreement with the XRD analysis.

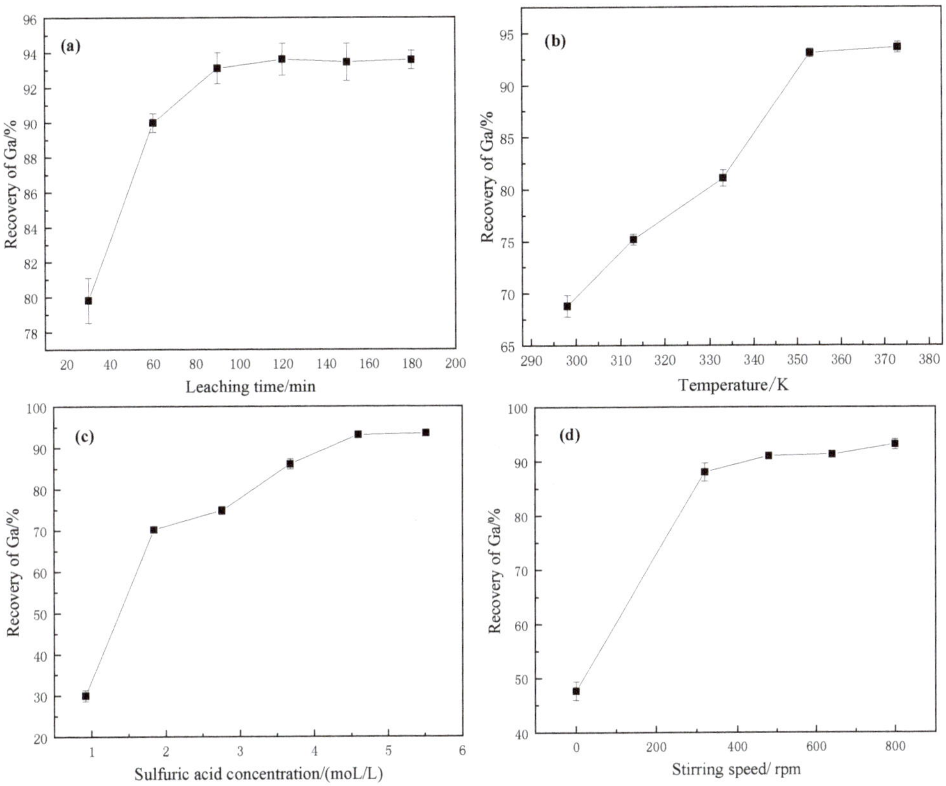


**Figure 13.** Effect of leaching parameters on the recovery of gallium: (**a**) leaching time (353 K, 800 rpm, $C(H_2SO_4)$ = 4.6 mol/L, L/S (mL/g) = 10); (**b**) leaching temperature (90 min, 800 rpm, $C(H_2SO_4)$ = 4.6 mol/L, L/S (mL/g) = 10 g/mL); (**c**) sulfuric acid concentration (90 min, 353 K, 800 rpm, L/S (mL/g) = 10); (**d**) stirring rate (90 min, 353 K, $C(H_2SO_4)$ = 4.6 mol/L, L/S (mL/g) = 10).

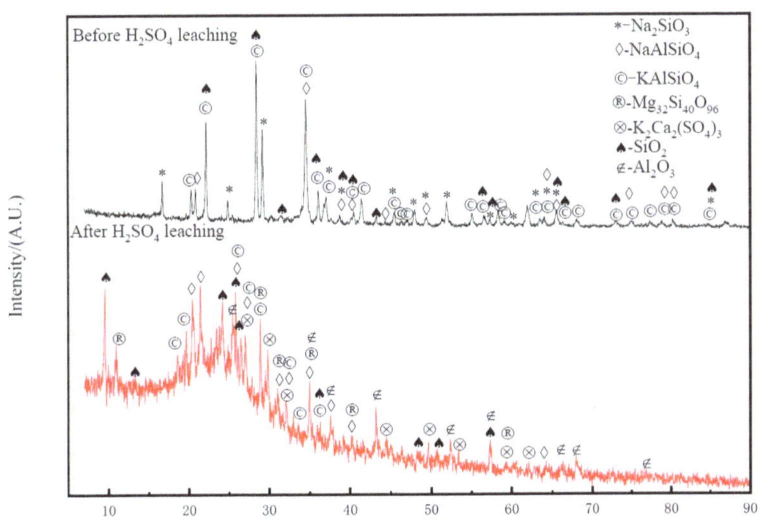


**Figure 14.** XRD patterns for roasted dust and leaching residue.

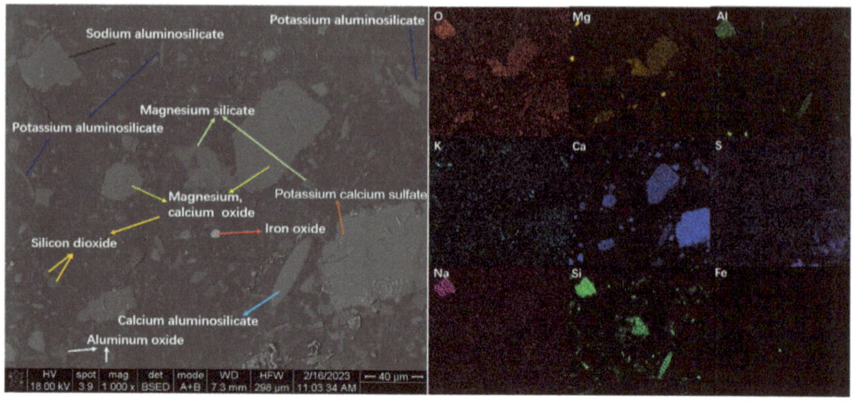


**Figure 15.** BSE and EDS mapping images of leaching residue.

The morphology of the dust before and after leaching is displayed in Figure 16. As shown in Figure 16a, the roasted dust particles were extremely uneven and clearly agglomerated. Based on the chemical composition analysis, it is known that these particles were mainly composed of silicates and aluminosilicates with different sodium and potassium contents. Other elements, such as Mg and Fe, also transferred into the aluminosilicates. After acid leaching, the particles were still agglomerated. Numerous sheet-like sulfates (points 8, 10, and 11) and irregular-shaped silicon dioxides (point 9) were distributed on the surface of the agglomerated particles.

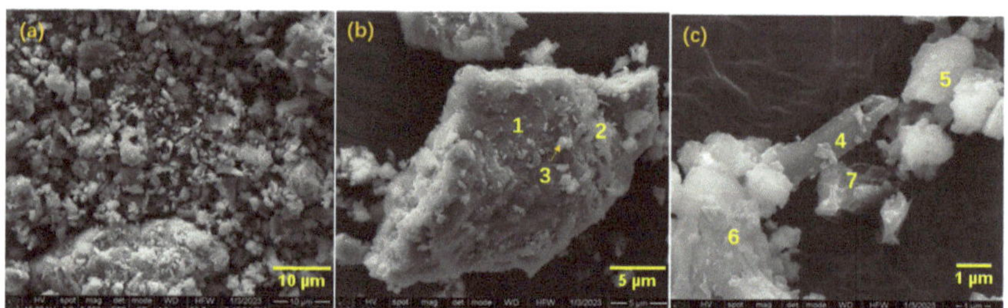


Point 1

| Element | O | Na | Al | Si | K |
|---|---|---|---|---|---|
| wt., % | 38.79 | 24.91 | 3.86 | 25.83 | 6.60 |

Point 2

| Element | O | Na | Al | Si | K |
|---|---|---|---|---|---|
| wt., % | 45.14 | 24.80 | 5.17 | 18.73 | 6.16 |

Point 3

| Element | O | Na | Al | Si | K |
|---|---|---|---|---|---|
| wt., % | 38.04 | 15.51 | 9.36 | 21.57 | 15.51 |

Point 4

| Element | O | Na | Si | K |
|---|---|---|---|---|
| wt., % | 49.05 | 25.23 | 22.70 | 3.02 |

Point 5

| Element | O | Na | Al | Si | K | Fe |
|---|---|---|---|---|---|---|
| wt., % | 40.87 | 6.60 | 12.82 | 17.44 | 20.42 | 1.85 |

Point 6

| Element | O | Na | Mg | Al | Si | K | Fe |
|---|---|---|---|---|---|---|---|
| wt., % | 37.45 | 19.32 | 5.76 | 9.36 | 22.34 | 2.13 | 3.65 |

Point 7

| Element | O | Na | Si | K |
|---|---|---|---|---|
| wt., % | 45.90 | 25.93 | 20.24 | 7.94 |

Point 8

| Element | O | Al | Si | S | K | Fe |
|---|---|---|---|---|---|---|
| wt.,% | 23.77 | 5.16 | 11.48 | 42.77 | 10.71 | 6.11 |

Point 9

| Element | O | Si | S |
|---|---|---|---|
| wt., % | 50.71 | 47.45 | 1.85 |

Point 10

| Element | O | Al | Si | S | K |
|---|---|---|---|---|---|
| wt., % | 56.77 | 6.83 | 3.92 | 23.62 | 8.86 |

Point 11

| Element | O | Na | Al | Si | S | K |
|---|---|---|---|---|---|---|
| wt., % | 42.62 | 2.21 | 5.32 | 9.56 | 32.07 | 8.22 |

Point 12

| Element | O | Na | Al | Si | S | K |
|---|---|---|---|---|---|---|
| wt., % | 50.82 | 3.03 | 2.40 | 22.58 | 18.57 | 2.60 |

**Figure 16.** SEM images of the dust (**a**–**c**) before leaching and after (**d**,**e**) leaching.

### 3.2.3. Leaching Kinetics

The extraction of gallium from the roasted dust by means of sulfate acid leaching was a liquid–solid multiphase reaction process. During the leaching process, the following reactions occurred.

$$Na_2SiO_3 + 2H^+ = 2Na^+ + H_2SiO_3 \quad (1)$$

$$NaAlSiO_4 + 4H^+ = Na^+ + Al^{3+} + H_2SiO_3 + H_2O \quad (2)$$

$$KAlSiO_4 + 4H^+ = K^+ + Al^{3+} + H_2SiO_3 + H_2O \quad (3)$$

After the silicates and aluminosilicates reacted with sulfuric acid, gallium was then released and transferred into the solution. Our previous study [27] found that the leaching rates of K and Al were 71.33% and 48.42%, respectively, when 98.38% of Ga was recovered with the acid leaching, suggesting that only part of the potassium aluminosilicate was dissolved by sulfuric acid leaching and, at the same time, some aluminum present as the corundum did not participate in the leaching reaction.

The common liquid–solid reaction models include the unreacted nuclear shrinkage model and the uniform model. The results of leaching experiments showed that this leaching process did not conform to the characteristics of the uniform model, and therefore the unreacted nuclear shrinkage model was chosen for studying the leaching process. This model has been widely used in the field of hydrometallurgy to simulate the leaching reaction process. Based on the relationships between the stirring speed and the extraction efficiency of gallium, as shown in Figure 13d, it was deduced that the leaching rate is not controlled by external diffusion when the agitation rate is kept over 480 rpm. Three kinetic equations were selected to determine the limiting step in this leaching process.

When the reaction rate is controlled by internal diffusion (Ginstling–Braunstein):

$$k_1t = 1 - 2\alpha/3 - (1 - \alpha)^{2/3} \quad (4)$$

When the reaction rate is controlled by a chemical reaction (interface):

$$k_2t = 1 - (1 - \alpha)^{1/3} \quad (5)$$

When the leaching process is controlled by the transfer across the contacting interface, the reaction rate is related to the contacting area of the sphere [28]. The kinetics of the leaching process conform to:

$$k_3t = 1/(1 - \alpha)^{1/3} - 1 \quad (6)$$

where $\alpha$ is the recovery of gallium (%); $k_1$, $k_2$, and $k_3$ are the rate constants; and $t$ represents the leaching time (min).

In order to determine the controlling steps of the sulfuric acid leaching process and the corresponding kinetic parameters, the variation in gallium recovery with leaching time at 298 K, 313 K, 333 K, and 353 K were obtained, as illustrated in Figure 17. The data were fitted with the three equations above, and the linear correlation coefficient ($R^2$) for fitting the lines is given in Table 6.

**Table 6.** Correlation coefficients ($R^2$) for lines fitted with the three kinetic models.

| **Temperature/K** | **Correlation Coefficient ($R^2$)** | | |
|---|---|---|---|
| | $k_1t = 1 - 2\alpha/3 - (1 - \alpha)^{2/3}$ | $k_2t = 1 - (1 - \alpha)^{1/3}$ | $k_3t = 1/(1 - \alpha)^{1/3} - 1$ |
| 298 | 0.96335 | 0.99182 | 0.99683 |
| 313 | 0.98612 | 0.9834 | 0.99712 |
| 333 | 0.99562 | 0.97385 | 0.99768 |
| 353 | 0.99676 | 0.96936 | 0.9979 |

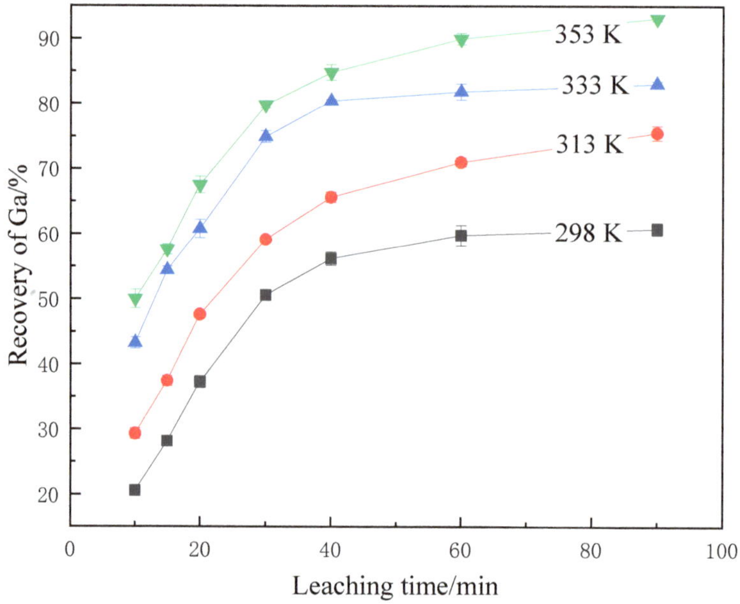


**Figure 17.** Variation of gallium recovery at different leaching temperatures with leaching time.

As listed in Table 6, the linear correlation coefficients for lines fitted with $k_3t = 1/(1-\alpha)^{1/3} - 1$ were closest to 1.0, indicating that this kinetic equation describes the leaching process best. The rate constants ($k_3$) at different temperatures were obtained based on the slopes of the straight lines in Figure 18. ln$k$ against $1/T$ was plotted as displayed in Figure 19. According to the Arrhenius equation (ln$k = -E_a/(RT + \ln A)$), the apparent activation energy ($E_a$) and the pre-exponential factor ($A$) were calculated as 16.81 kJ/mol and 7.34 min$^{-1}$ based on the slope and the intercept of the straight line.

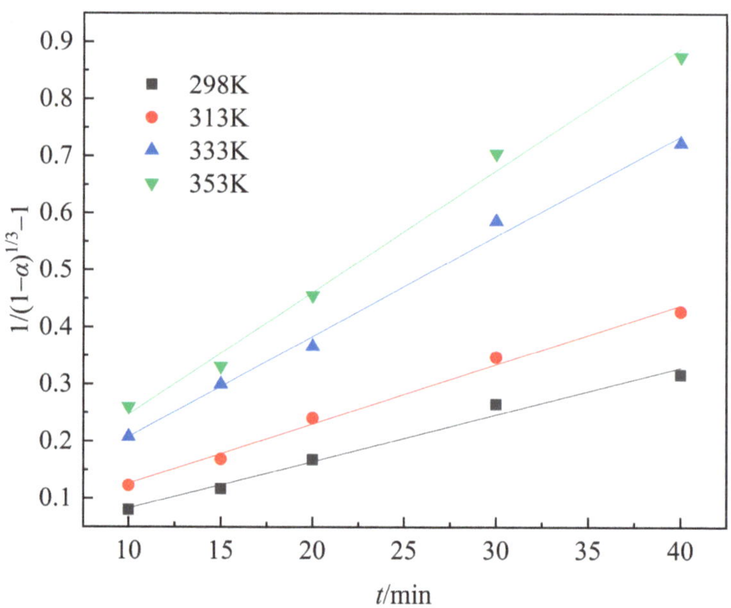


**Figure 18.** Plot of $1/(1-\alpha)^{1/3} - 1$ versus $t$ for data obtained at different temperatures.

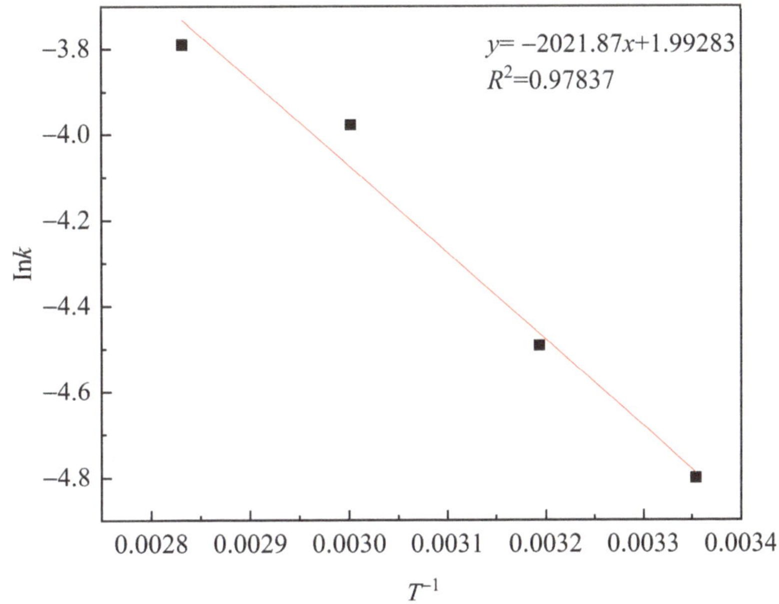


**Figure 19.** Plot of ln*k* against $1/T$.

## 4. Conclusions

In this paper, a process route involving roasting activation and sulfuric acid leaching was first proposed to treat brown corundum dust and then successfully extract gallium from this kind of dust.

The main phases contained in brown corundum dust are potassium sulfate, corundum, and amorphous silicate. Gallium is dispersed in the potassium-rich phase, which is surrounded by corundum and amorphous silicate. The distribution characteristics of gallium explain why direct water leaching and conventional acid leaching cannot achieve satisfactory gallium recovery [27]. Other enhanced acid leaching methods and concentrated alkali leaching [23–25] may destroy the corundum and amorphous silicate phases and recover gallium from the dust. The roasting activation proposed in this study effectively changed the phase composition and distribution. After roasting, the corundum and amorphous silicate were converted into acid-soluble phases.

Both sodium carbonate roasting and calcium oxide roasting can activate brown corundum dust and thus make gallium extraction feasible by sulfuric acid leaching. In comparison with calcium oxide roasting, sodium carbonate roasting performs better in activation efficiency through transforming the corundum and amorphous silicate into sodium silicate and aluminosilicate salts. Further, it requires a lower roasting temperature and a lower dosage. After the brown corundum dust was roasted at 1073 K for 40 min with a sodium carbonate dosage of 0.5, over 93 % of gallium was recovered by 4.6 mol/L $H_2SO_4$ leaching at 353 K for 90 min.

During the leaching process, aluminosilicates react with sulfuric acid and produce silicic acid, which remains in the leaching residue. Hence, compared with direct concentrated alkali leaching, the silicon dissolution of this proposed process route is as low as 0.31% [24,27]. This is another advantage of this process. Since silicon dioxide accounts for 40.82 wt.% of the raw dust, if a high percentage of silicon enters the gallium-containing solution, it will have a major adverse effect on the subsequent purification and sedimentation separation of the solution.

Based on the characterization results of the leaching residue, it was found that almost all of the sodium silicate reacted with sulfuric acid, whereas aluminosilicates were less likely to react with sulfuric acid, and some of them remained in the leaching residues. Silicon dioxide and sulfate were distributed on the surface of the leaching tailing particles but did not form a dense product layer. The leaching kinetics agreed well with the equation of $kt = 1/(1-\alpha)^{1/3} - 1$ with an apparent activation energy of 16.81 kJ/mol. This suggests that the leaching process is controlled by transfer across the contacting interface of the reactive particles and is closely related to the specific surface area of the diffusion interface. The measures to overcome this diffusion resistance and increase the leaching kinetic rate merit further investigation.

**Author Contributions:** Conceptualization, J.Z.; data curation, Y.C. and X.H.; formal analysis, J.Z. and Y.C.; funding acquisition, J.Z.; investigation, Y.C. and C.G.; project administration, J.Z.; software, X.H.; supervision, J.Z.; writing—original draft, J.Z. and Y.C.; writing—review and editing, A.J. All authors have read and agreed to the published version of the manuscript.

**Funding:** This research was funded by the Hubei Provincial Key Technologies R & D Program (2022BCA058), the China Scholarship Council (202008420104), and the National Natural Science Foundation of China (51804230).

**Data Availability Statement:** They will be available on request.

**Acknowledgments:** We would like to thank Guohong Zhang and Wei Yuan at the Analytical & Testing Center of Wuhan University of Science and Technology for their help with EPMA analysis.

**Conflicts of Interest:** The authors declare no conflict of interest.

## References

1. Tang, S.Y.; Tabor, C.; Kalantar-Zadeh, K.; Dickey, M.D. Gallium liquid metal: The devil's elixir. *Annu. Rev. Mater. Res.* **2021**, *51*, 381–408. [CrossRef]
2. Harvey, A.; Backes, C.; Gholamvand, Z.; Hanlon, D.; Mcateer, D.; Nerl, H.; Mcguire, E.; Seral-Ascaso, A.; Ramasse, Q.; Mcevoy, N.; et al. Preparation of gallium sulfide nanosheets by liquid exfoliation and their application as hydrogen evolution catalysts. *Chem. Mater.* **2015**, *27*, 3483–3493. [CrossRef]
3. Chelvanathan, P.; Hossain, M.I.; Amin, N. Performance analysis of copper–indium–gallium–diselenide (CIGS) solar cells with various buffer layers by SCAPS. *Curr. Appl. Phys.* **2010**, *10*, S387–S391. [CrossRef]
4. Hedayati, M.; Olyaee, S.; Ghorashi, S.M.B. The effect of adsorbent layer thickness and gallium concentration on the efficiency of a dual-junction copper indium gallium diselenide solar cell. *J. Electron. Mater.* **2020**, *49*, 1454–1461. [CrossRef]
5. Nishinaka, K.; Terakado, O.; Tani, H.; Hirasawa, M. Pyrometallurgical recovery of gallium from GaN semiconductor through chlorination process utilizing ammonium chloride. *Mater. Trans.* **2017**, *58*, 688–691. [CrossRef]
6. Sigel, A.; Sigel, H.; Freisinger, E.; Sigel, R.K.O. (Eds.) *Metallo-Drugs: Development and Action of Anticancer Agents*; De Gruyter: Berlin, Germany, 2018; ISBN 9783110470734.
7. Lee, B.F.; Chiu, N.T.; Chang, J.K.; Liu, G.C.; Yu, H.S. Technetium-99m(V)-DMSA and gallium-67 in the assessment of bone and joint infection. *J. Nucl. Med.* **1998**, *39*, 2128–2131.
8. Zhao, C.; Qin, S.; Yang, Y.; Li, Y.; Lin, M. Concentration of gallium in the permo-carboniferous coals of China. *Energy Explor. Exploit.* **2009**, *27*, 333–343. [CrossRef]
9. Naumov, A.V. Status and prospects of world gallium production and the gallium market. *Metallurgist* **2013**, *57*, 367–371. [CrossRef]
10. Redlinger, M.; Eggert, R.; Woodhouse, M. Evaluating the availability of gallium, indium, and tellurium from recycled photovoltaic modules. *Sol. Energy Mater. Sol. Cells* **2015**, *138*, 58–71. [CrossRef]
11. Xu, K.; Deng, T.; Liu, J.; Peng, W. Study on the recovery of gallium from phosphorus flue dust by leaching with spent sulfuric acid solution and precipitation. *Hydrometallurgy* **2007**, *86*, 172–177. [CrossRef]
12. Lu, F.; Xiao, T.; Lin, J.; Li, A.; Long, Q.; Huang, F.; Xiao, L.; Li, X.; Wang, J.; Xiao, Q.; et al. Recovery of gallium from Bayer red mud through acidic-leaching-ion-exchange process under normal atmospheric pressure. *Hydrometallurgy* **2018**, *175*, 124–132. [CrossRef]
13. Xue, B.; Wei, B.; Ruan, L.; Li, F.; Jiang, Y.; Tian, W.; Su, B.; Zhou, L. The influencing factor study on the extraction of gallium from red mud. *Hydrometallurgy* **2019**, *186*, 91–97. [CrossRef]
14. Huang, J.; Wang, Y.; Zhou, G.; Gu, Y. Exploring a promising technology for the extraction of gallium from coal fly ash. *Int. J. Coal Prep. Util.* **2022**, *42*, 1712–1723. [CrossRef]
15. Huang, J.; Wang, Y.; Zhou, G.; Gu, Y. Investigation on the effect of roasting and leaching parameters on recovery of gallium from solid waste coal fly ash. *Metals* **2019**, *9*, 1251. [CrossRef]
16. Zhou, J.; Zhu, N.; Liu, H.; Wu, P.; Zhang, X.; Zhong, Z. Recovery of gallium from waste light emitting diodes by oxalic acidic leaching. *Resour. Conserv. Recycl.* **2019**, *146*, 366–372. [CrossRef]

Carbonaceous gold-bearing ores (CGOs) (Figure 2) can be classified as refractory because of the large amount of carbon.

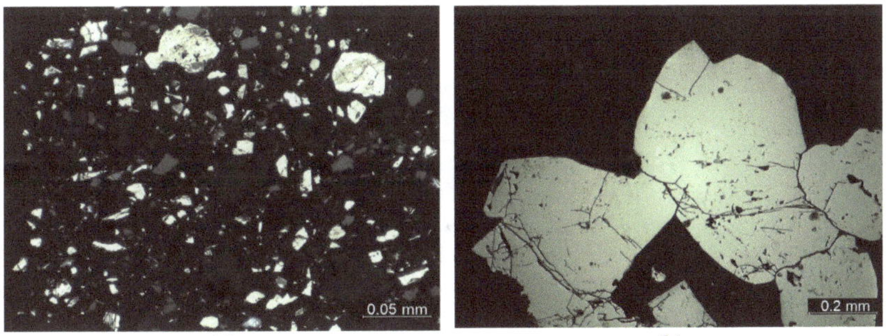


**(a)** **(b)**

**Figure 2.** Microphotograph of the carbonaceous gold-bearing ore initial sample: (**a**) general view and (**b**) a pyrite crystal.

Chemical composition of the sample: $SiO_2$—56.11%, $Al_2O_3$—23.2%, $Fe_2O_3$—4.12%, $K_2O$—3.03%, CaO—1.91%, MgO—1.13%, $TiO_2$—0.63%, $Na_2O$—0.584%, $P_2O_5$—0.199%, MnO—0.061%, As—0.45%, Au—5.99 g/t, Ag—0.29 g/t, $C_{tot}$—2.65% (including $C_{org}$—1.52), $S_{tot}$—1.59%. Pyritohedron habitus pyrite and needle arsenopyrite are gold-bearing. In both sulfides, gold is present in microscopic and submicroscopic forms. To increase the extraction of noble metals in the processing of this type of ore, it is imperative to reduce the negative impact of carbon at the stage of flotation by using activation techniques at the successive stages of the technological treatment of ores.

## 2.2. Research Methodology

### 2.2.1. Grinding and Sieve Analysis

To study the effect of mechanical activation and to develop the criteria characterizing the efficiency of overcoming the energy barrier in the process of minerals' comminution processes, experimental studies of grinding GSO in a ball mill were carried out. The diameter of the mill's work volume was 118 mm and the length was 125 mm. The particle size of the initial material is −2000 + 0 microns. The initial material weight for each crushing cycle was 250 g and the weight of the ball loading was 2.94 kg. The size of 0.071 mm was accepted as the control sieve. Grinding was carried out in an aqueous medium (L/S = 1:1 by mass). After each grinding, a sieve analysis was carried out using a standard set of sieves. The study of grinding process intensification was carried out with the use of grinding intensifier reagents. The effect of potassium permanganate, sodium carbonate, and aqueous alkaline solution of sodium silicate on the yield of control class and distribution of components in size classes of milled products was studied. The consumption rate of each reagent in a series of experiments was 100 g/t. The specific surface of the ground products was measured using a laser particle analyzer Microsizer 201C (VA Instalt, St. Petersburg, Russia).

The analysis of the form of mineral assemblages was carried out using an Axio Imager A2m optical polarization microscope (Zeiss, Jena, Germany). The luster contrast of ore minerals allowed the software analysis of images (DG Analyzer 1.5 program). The image obtained with a microscope was converted into black and white gamma. Inclusions of ore minerals represented white areas. The program calculated geometric parameters of these areas and downloaded the result as values of equivalent diameters of each area. Filling of the enclosed areas was used to eliminate false detections. The representativeness of the results was ensured by processing five images of different areas of the sample and joint mathematical processing of selected objects (Figure 3).

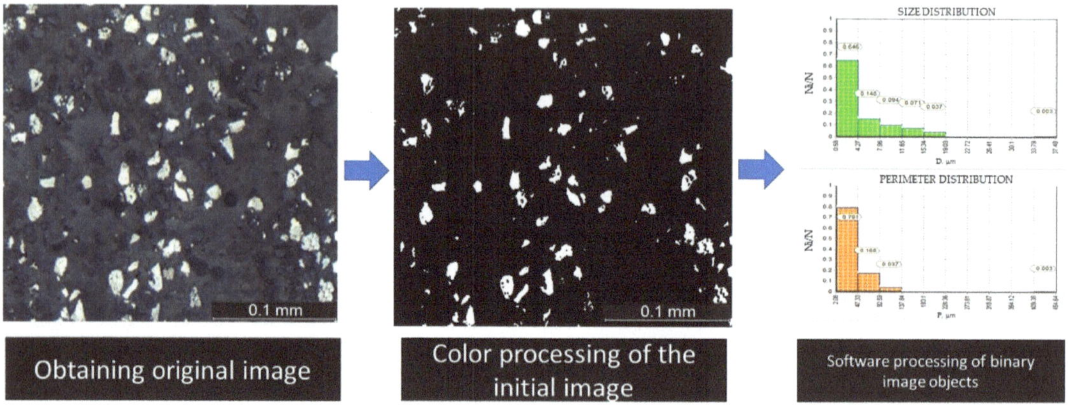


**Figure 3.** Algorithm of optical and microscopic studies of the grinding operation products.

### 2.2.2. Flotation Experiments

Flotation experiments were carried out with the use of a pneumomechanical flotation machine Flotation Bench Test Machine (Laarmann Group B.V., Roermond, The Netherlands) with a chamber volume of 1.5 l. Experiments to obtain flotation concentrates included the following: grinding the sample to a size of 60% of the −71 μm class and carrying out flotation. The reagent suite of flotation included the following: media regulator (sodium carbonate) with a dosage of 400 g/t; carbon depressor (dextrine) with a dosage of 100 g/t; activator of sulfide minerals (copper sulfate) with a dosage of 300 g/t; collectors (butyl potassium xanthate and sodium-dibutyl dithiophosphate) with a dosage of 175 and 25 g/t, respectively; and frother with a dosage of 50 g/t.

The elemental composition of the sample was studied using an EDX 7000 X-ray fluorescence energy dispersive analyzer (Shimadzu Corporation, Kyoto, Japan) according to calibration curves. The organic carbon content of the sample was investigated using a TOC-L organic carbon analyzer (Shimadzu Corporation, Kyoto, Japan). The elemental composition and organic carbon content were measured at least three times to minimize the error and reproducibility of the results.

### 2.2.3. Thermogravimetric Analysis Methodology and Calculation of Activation Energy

Differential thermal analysis and thermogravimetric analyses (DTA-TG) were carried out at three heating rates of 5, 10, and 20 K/min in a nitrogen atmosphere using an SDT Q600 thermal analyzer. Representative samples with a mass of 5 g were taken for investigation. The resulting samples were rubbed down to a particle size of −20 μm and heated to 1250 K. After the measurements, DTA-TG curves were obtained. The TGA curve data were used to calculate the activation energy by systematising the mass loss of the samples as a function of temperature. From these data, the conversion rates were calculated. The degree of transformation was calculated using the following formula:

$$x = \frac{m_0 - m_i}{m_0 - m_f}, \quad (2)$$

where $m_0$—mass of the sample at the initial moment of time, g; $m_i$—current mass, g; and $m_f$—sample mass at the end of analysis, g.

The activation energy was calculated by the Ozawa–Flynn–Wall method and the Friedman method. To calculate the activation energy by the Friedmann method, the logarithm of the conversion rate $\ln\beta(dx/dT)$ was plotted as a function of the inverse temperature $1000/T$. The tangent of the obtained dependence allows the calculation of the activation energy. The equation is provided in Table 1.

The following formula is used to calculate the activation energy according to the Friedman method:

$$E_a = -0.95 \cdot \alpha \cdot R \tag{3}$$

where $R$—universal gas constant and $\alpha$—tangent of the slope of the straight line for the chosen degree of conversion at the three heating rates.

The interval cut off on the ordinate axis represents the pre-exponential factor ln$A$.

In order to calculate the activation energy by the OFW method, the logarithm of the heating rate $\log_{10}\beta$ at three values was plotted as a function of the inverse temperature $1000/T$. The tangent of the resulting slope allows the value of the activation energy to be calculated. The equation is provided in Table 1.

The following formula is used to calculate the activation energy according to the OFW method:

$$E_a = -2.19 \cdot \alpha \cdot R \tag{4}$$

where $R$—universal gas constant and $\alpha$—tangent of the slope of the straight line for the chosen degree of conversion at the three heating rates.

## 3. Result and Discussion

### *3.1. Studies to Substantiate the Energy Parameter of Fracture Efficiency and Assess the Effect of Intensifiers on the Grinding Process*

To formulate a parameter of the efficiency of overcoming the energy barrier of internal traction forces, a generalized energy–size ratio was taken as the basis:

$$\frac{dE}{dl} = \frac{-k}{l^n} \tag{5}$$

where $E$—energy spent on grinding of a certain mass of material, J/kg; $l$—grain size, (m); $k$—empirical coefficient of proportionality, characterizing the durability of the material; and its dimensionality depends on the value of the exponent n. When $n$ = 1, the ratio is an expression of the Kirpitchev–Kick law; when $n$ = 1.5, the ratio is an expression of the law of F. Bond; and when $n$ = 2, the ratio is an expression of the law of P. Rittinger [27].

By integrating the expression (5) for the case $n \neq 1$ in limits from $l_i$—finite particle size, m, to $l_0$—values of the initial particle size, m, we obtain the following:

$$E = \frac{k}{n-1}\left(\frac{1}{l_i{}^{n-1}} - \frac{1}{l_0{}^{n-1}}\right) \tag{6}$$

Based on the Gibbs–Helmholtz equation for the surface free energy, the increase in the formed surface will be proportional to the energy expended for the destruction of a certain mass of material [28]. The energies expended to reduce the coarseness and form a new surface are equivalent:

$$\sigma \Delta S_i = \frac{k}{n-1}\left(\frac{1}{l_i{}^{n-1}} - \frac{1}{l_0{}^{n-1}}\right) \tag{7}$$

where $\sigma$—the value of specific surface free energy, (J/m$^2$); $\Delta S$—the increase in specific surface area; m$^2$/kg.

Let us introduce $K_\sigma$ = $k/\sigma$, the parameter of the energy barrier overcoming of the internal bonding forces' efficiency during the fracture of mineral aggregates. This relationship relates the specific free surface energy of minerals ($\sigma$) to the empirical material fracture resistance coefficient ($k$). The dimensionality of the index at $n$ = 2 in the SI system is [m$^{2.5}$/kg]. From the point of view of evaluating the effect of intensifying grinding methods, the higher the value of this coefficient, the more effective the effect of the method. For the case of ball milling, we can take $n$ = 1.5 according to F. Bond's law. Substituting the parameter $P_{80}$ instead of $l$, we obtain the expression for $K_\sigma$:

$$K_\sigma = \frac{0.5\Delta S_i}{\left(\frac{1}{\sqrt{P_{80i}}} - \frac{1}{\sqrt{P_{800}}}\right)} \quad (8)$$

Using the proposed parameter, it is possible to quantitatively estimate different methods of grinding process intensification while maximizing the yield of the control class.

In order to confirm the possibility of using this parameter to assess the efficiency of comminution under real grinding conditions, experimental studies were carried out. The results of the analysis of the distribution of the main components in the initial sample by sieve classes are shown in Figure 4.

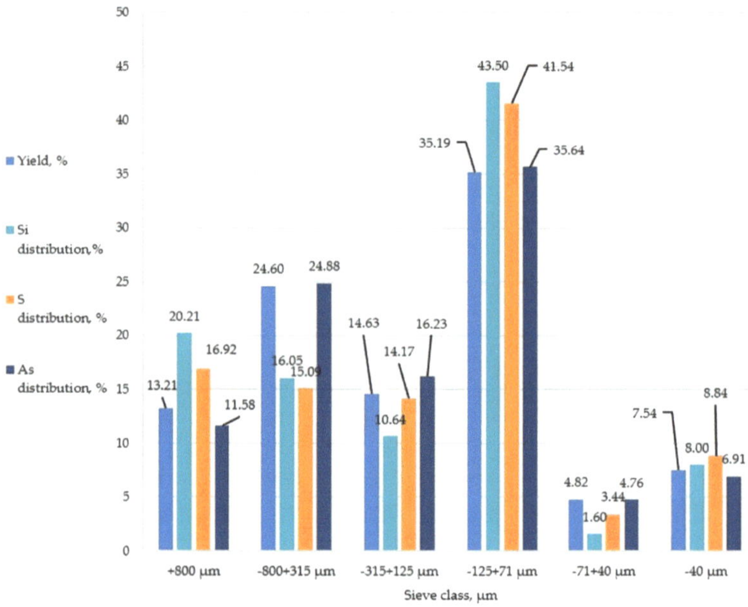


**Figure 4.** Distribution of the sample's main components by sieve classes.

Based on the analysis of the obtained data, it was found that the highest values of component distribution are in the −125 + 71 μm class, which is also characterized by the highest yield value. This is probably because of the presence of the richest clusters of sulfide minerals in this class. The second class with the highest values of silicon and sulfur distribution is the +800 μm class, which is probably associated with the high content of closed aggregates.

Based on previous studies, it was found that the sample is characterized by high hardness, so it was established that the required time for grinding is 30 minutes [34]. Graphical interpretation of particle size distribution and distribution of components into particle size classes depending on the size of the grinded product is shown in Figure 5.

On the basis of data analysis (Figure 5), it was found that the highest values of partial distributions of components associated with sulfide minerals, with the parameter $P_{80}$ value being less than 132.29 μm, correspond to the class −40 μm. For the product with $P_{80}$ = 205.04 μm, the maximum values of sulfur and arsenic distributions correspond to classes −125 + 71 μm and −250 + 125 μm, respectively. Such redistribution of components is probably related to the peculiarities of sulfide minerals and the host rock minerals' liberation.

Optical and microscopic investigations of grinded products were carried out to analyze changes in geometric characteristics of sulfide mineral inclusions. Graphic interpretation of the obtained results is shown in Figure 6.

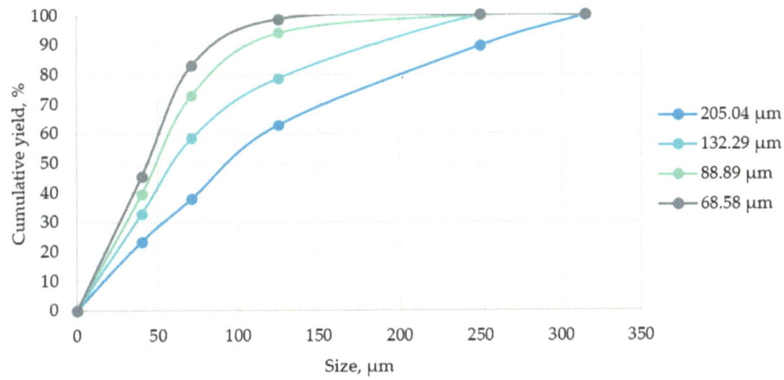


(**a**)

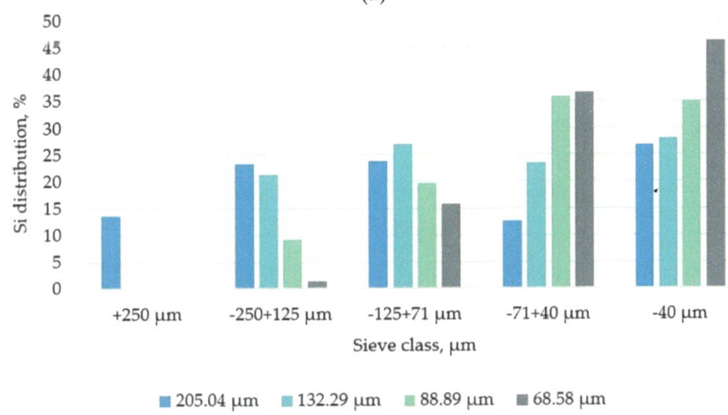


(**b**)

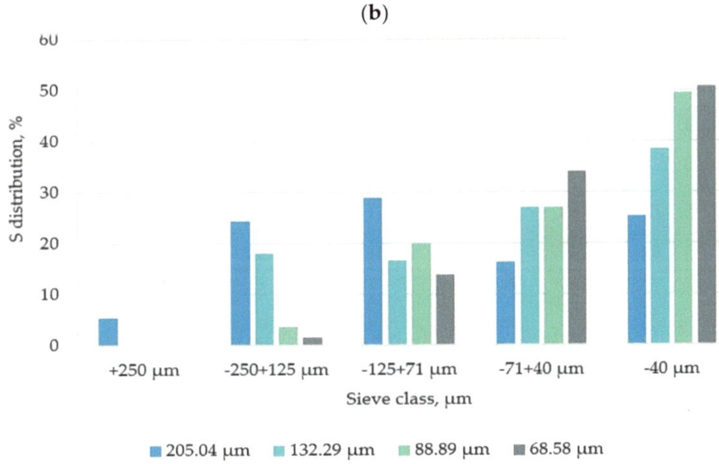


(**c**)

**Figure 5.** *Cont.*

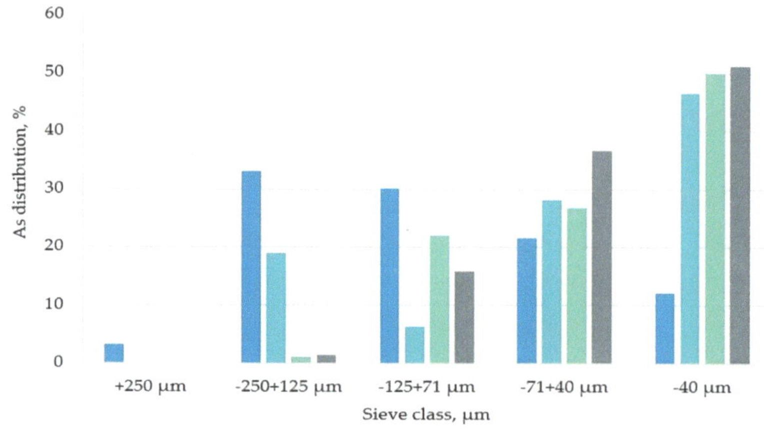


(**d**)

**Figure 5.** Characteristics of the granulometric and component composition depending on the parameter $P_{80}$: (**a**) cumulative yield values, %; (**b**) Si distribution values; (**c**) S distribution values; and (**d**) As distribution values.

In Figure 6, there are clearly expressed peaks of grain size of −20 + 10 μm at the value of parameter $P_{80}$ = 205.04 μm. This grain size class of sulfide minerals is dominant by the relative content in all product size classes. For values $P_{80}$ = 132.29 μm and 88.89 μm, there are also peaks of sulfide mineral grains of the size class −20 + 10 μm. The highest relative grain content is characteristic of the −40 μm class. At $P_{80}$ = 68.58 μm, there is no clearly expressed peak, and there is a gradual increase in the sulfide mineral grain content up to its maximum at −20 + 10 μm. When grinding the sample from the value of $P_{80}$ = 205.04 μm to 132.29 μm, an increase in the absolute content of −70 + 60 μm grains was observed. When grinding the sample from the value of $P_{80}$ = 132.29 μm to 88.89 μm, an increase in the absolute content of grains of adjacent grain size classes was observed: −60 + 50 μm and −80 + 70 μm.

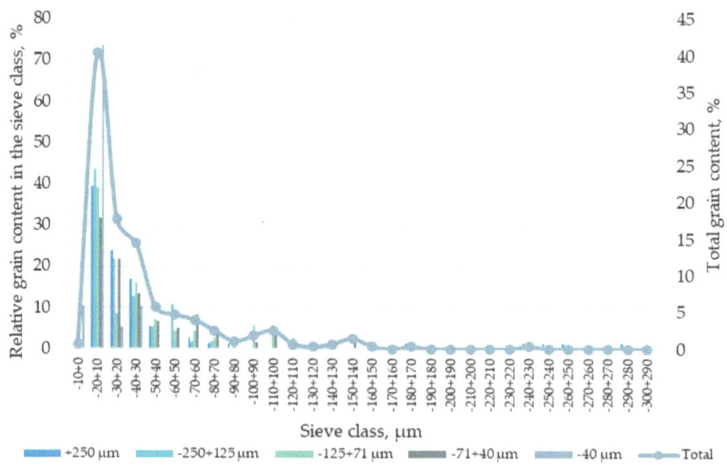


(**a**)

**Figure 6.** *Cont.*

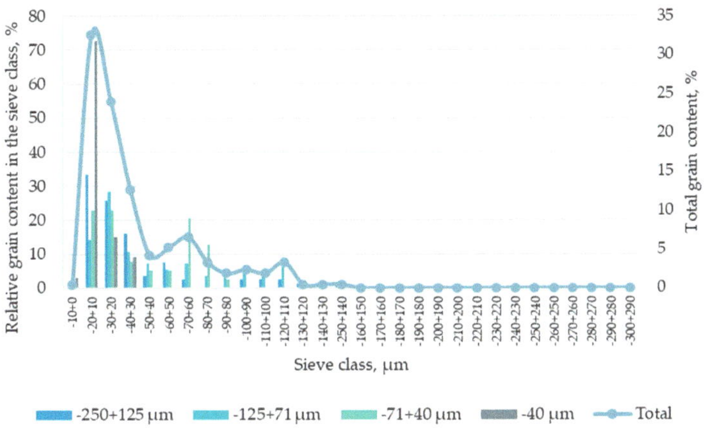


(b)

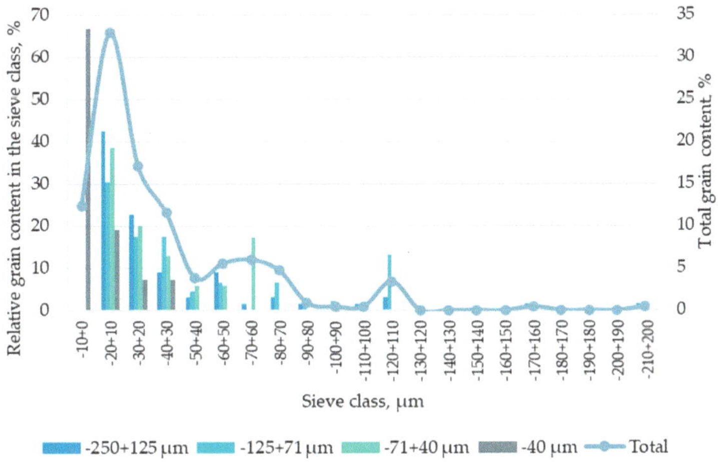


(c)

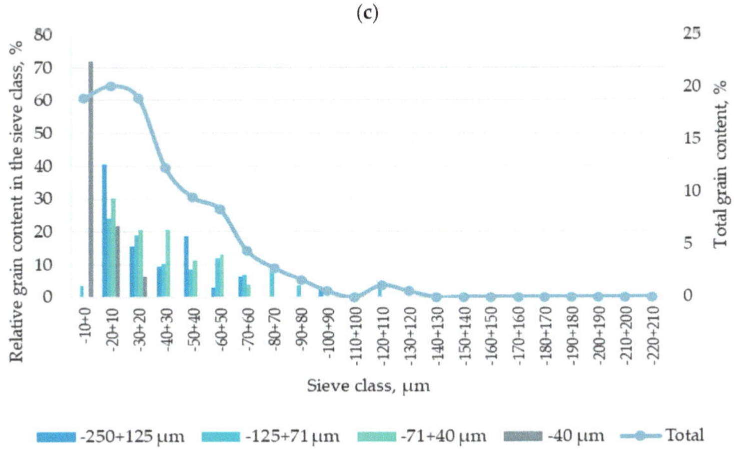


(d)

**Figure 6.** Distribution of sulfide aggregates by size classes: (**a**) $P_{80}$ = 205.04 µm, (**b**) $P_{80}$ = 132.29 µm, (**c**) $P_{80}$ = 88.89 µm, and (**d**) $P_{80}$ = 68.58 µm.

Redistribution of −20 + 10 μm grains at $P_{80}$ = 132.29 μm between sample size classes and a significant decrease in the relative grain content in the size classes above 40 μm is probably associated with the large mineral aggregates' destruction and liberation of ore minerals at this size. The peak of absolute content of ore mineral grains of −70 + 60 μm in the size class is possibly connected to the increase in the share of sulfide grains in the size class −71 + 40 μm. Subsequent grinding leads to an increase in the content of −10 μm grains and their redistribution into the general size class of the −40 μm sample, which is probably due to the grinding of already revealed sulfide minerals.

The maximum values of absolute grain sizes of −20 + 10 μm and their high relative content for all size classes of the sample are probably due to the fact that this size coincides with the prevailing size of ore minerals' dissemination. The smaller peaks of absolute sulfide mineral grain contents of −70 + 60 μm and −120 + 110 μm in size with a subsequent decrease in the value of their excesses during milling confirm the conclusion about the dissolution of ore minerals' large aggregates and their lower grinding rate, which leads to their accumulation in the above classes. Further transition of ore minerals into fine size classes with the destruction of large aggregates is undesirable, as it will have a negative effect on further enrichment operations.

Thus, the value of $P_{80}$ = 132.29 μm, which corresponds to the yield of the −71 μm class = 58.35%, was accepted for further research. Under grinding conditions, which allowed to achieve the given coarseness, a series of experiments to study the effect of grinding intensifiers on the redistribution of components by size classes and change in the specific surface were carried out.

Interpretation of the analysis of components' distribution by size classes in the grinding products when using different intensifier results is shown in Figure 7.

The analysis of the data presented in Figure 6 shows that the sodium silicate application resulted in the growth of the −40 μm class, which was 12.77% compared with the grinding without reagents. The application of sodium silicate resulted in the largest increases in sulfur and arsenic of 17.96%, 17.72%, and 17.68% in the −40 μm class. When sodium carbonate was used, the increase in the −40 μm class was 13.76% as compared with milling without the use of reagents. The increments of distribution values in the −40 μm class of silicon, sulfur, and arsenic were 0.16%, 3.95%, and 6.83%, respectively. The application of potassium permanganate resulted in an increase in the −40 μm class of 11.40% compared with the milling without the reagents. The increments in distribution values of silicon, sulfur, and arsenic in the −40 μm class were 1.34%, 2.01%, and 15.86%, respectively.

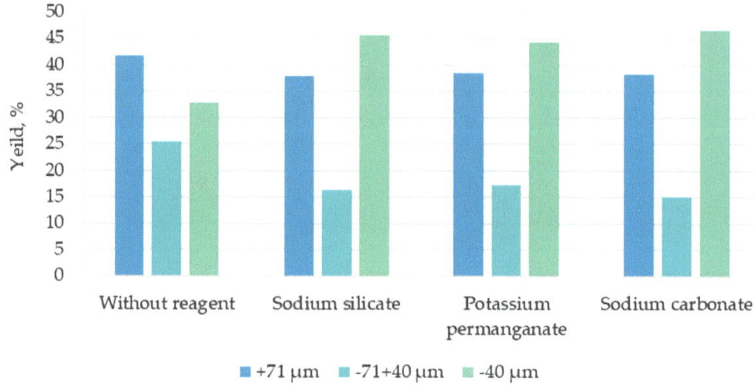


(a)

**Figure 7.** *Cont.*

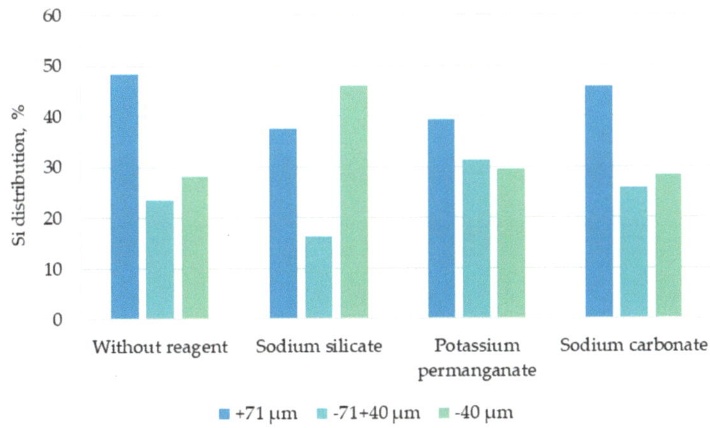


(**b**)

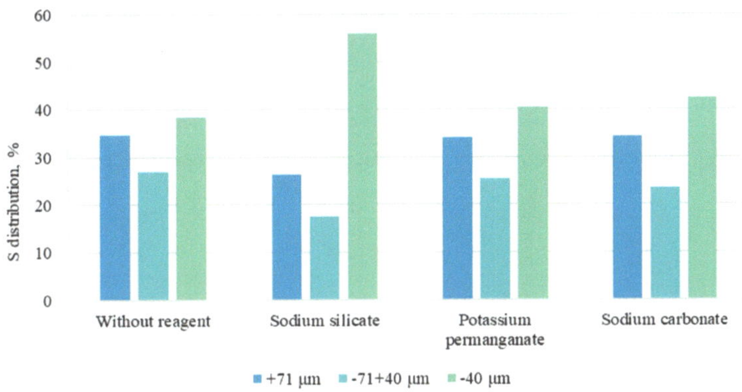


(**c**)

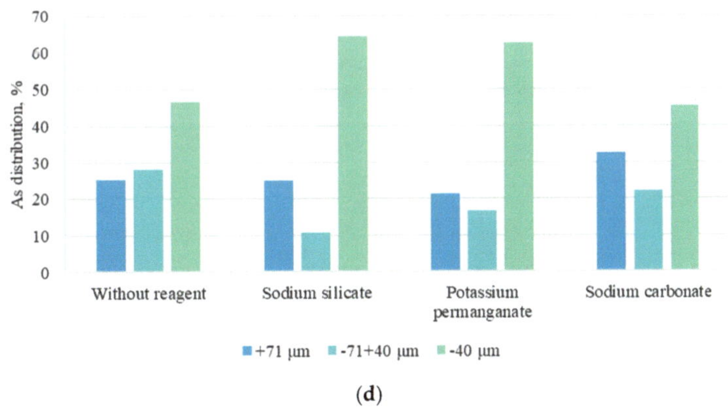


(**d**)

**Figure 7.** Values of the sieve classes' yields and distribution of silica, sulfur, and arsenic by sieve classes depending on the use of different intensifiers for grinding: (**a**) yield values, (**b**) Si distribution values, (**c**) S distribution values, and (**d**) As distribution values.

To determine the $K_\sigma$ value, a series of experimental measurements of the specific surface area for grinding products of different particle sizes were carried out. The $K_\sigma$ value was determined as a value of the angular coefficient of the straight line constructed with coordinates $1/\sqrt{P_{80}}$, $\Delta S_s$. Graphical interpretation of the obtained results is shown in Figure 8.

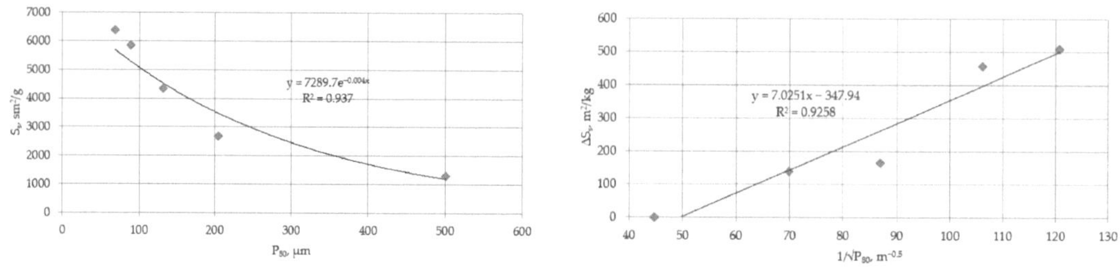


**Figure 8.** Diagram of the dependence of the values of the derivative function of the increment of the specific surface area on the parameter $P_{80}$.

Based on the analysis of the data obtained, $K_\sigma$ = 3.513 $m^{2.5}/kg$. The calculation of $K_\sigma$ for cases of application of unstrengthening reagents during grinding was carried out according to Formula (8). The results of the calculations are summarized in Table 2.

**Table 2.** Values of $P_{80}$, $S_s$, $\Delta S_s$, and $K_\sigma$ for grinding cases with the use of grinding intensifiers.

| No. | Experiment's Conditions | $P_{80}$, µm | $S_s$, $sm^2/g$ | $\Delta S_s$, $sm^2/g$ | $K_\sigma$, $m^{2.5}/kg$ |
|---|---|---|---|---|---|
| 1 | Initial probe | 501.45 | 1295.54 | - | - |
| 2 | Grinding without reagent | 132.29 | 4342.78 | 3047.24 | 3.513 |
| 3 | Grinding with sodium silicate | 127.31 | 5847.80 | 4552.26 | 5.176 |
| 4 | Grinding with potassium permanganate | 128.17 | 4553.50 | 3257.96 | 3.730 |
| 5 | Grinding with sodium carbonate | 127.91 | 4415.76 | 3120.22 | 3.565 |

Based on the data analysis in the table, it was found that the highest value of $K_\sigma$ corresponds to grinding with the use of sodium silicate. The increase in the parameter was 1.663 $m^{2.5}/kg$. When potassium permanganate and sodium carbonate were used, an increase in the value of the energy parameter, $K_\sigma$, was also recorded. However, the parameters increased by 0.217 $m^{2.5}/kg$ and 0.052 $m^{2.5}/kg$, respectively.

For the comparative assessment of the applied intensification methods, the specific parameter $K'_\sigma$ can be used:

$$K\prime_\sigma = \frac{K_{\sigma i} - K_{\sigma 0}}{K_{\sigma 0}} \tag{9}$$

The zero value $K_{\sigma 0}$ corresponds to the case of grinding without reagent.

### 3.2. *Activation Energy Calculation for Crushed Ore and Flotation Concentrates as an Example*

As a result of this research, the possibility of quantifying the activation energy on the example of comminuted ore and flotation concentrates was substantiated:

- quantification of activation energy on comminuted ore;
- quantifying the energy for flotation concentrates;
- assessment of the application of impacts based on thermogravimetric analysis using the developed criterion.

#### 3.2.1. Quantification Based on Thermogravimetric Analysis by Calculating the Activation Energy on Comminuted Ore

A thermogravimetric analysis was carried out on comminuted carbonaceous ore samples at three heating rates (5, 10, and 20 K/min). The results are shown in Figure 9.

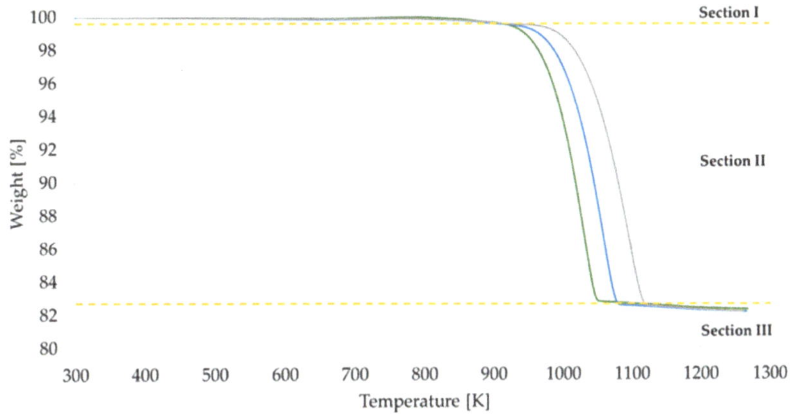


**Figure 9.** Results of thermogravimetric investigation of comminuted samples at three heating rates (green 5 K/min, blue 10 K/min, grey 20 K/min, yellow—splitting into sections).

The analysis of the data presented in Figure 1 shows that, for these samples, three weight-loss sections are observed for the 5 K/min heating rate: the first one—low mass loss (less than 1%) up to 934 K, the second one—intensive mass loss (about 16%) in the temperature range 934–1048 K, and the third one—low mass loss (less than 1%) above 1048 K. For heating rates of 10 K/min and 20 K/min, the boundary between the second and third region lies at 1076 K and 1115 K, respectively. The values of the degree of transformation are <0.05 for the first section, 0.05–0.95 for the second section, and >0.95 for the third section for all heating rates. The main mass loss occurs in section 2, for which the calculations of the activation energy were carried out.

Graphical data for calculating the activation energy using the Friedman method and the OFW method for conversion rates from 0.05 to 0.95 are shown in Figure 10.

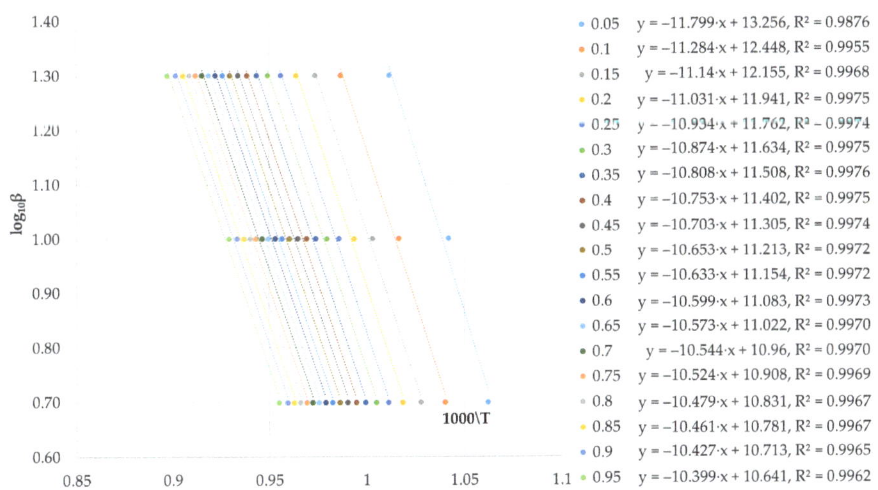


(**a**)

**Figure 10.** *Cont.*

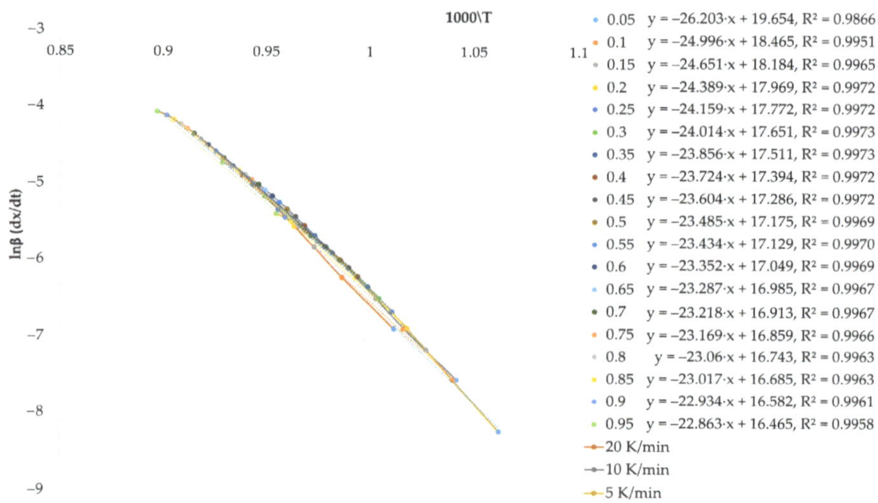


(**b**)

**Figure 10.** Calculation results for the activation energy for three heating rates for comminuted ore samples: (**a**) using the OFW method and (**b**) using the Friedman method.

The activation energy and the pre-exponential factor of the two methods were calculated from the dataset (Table 3).

**Table 3.** Activation energy calculation results for comminuted carbon sulphide ore samples for section 2.

| No. | x | $E_a$, kJ/mol (OFW Method) | $E_a$, kJ/mol (Fiedman Method) | ln*A* |
|---|---|---|---|---|
| 1 | 0.05 | 214.79 | 207.08 | 19.654 |
| 2 | 0.10 | 205.43 | 197.54 | 18.465 |
| 3 | 0.15 | 202.80 | 194.81 | 18.184 |
| 4 | 0.20 | 200.81 | 192.75 | 17.969 |
| 5 | 0.25 | 199.05 | 190.93 | 17.772 |
| 6 | 0.30 | 197.96 | 189.79 | 17.651 |
| 7 | 0.35 | 196.75 | 188.53 | 17.511 |
| 8 | 0.40 | 195.75 | 187.49 | 17.394 |
| 9 | 0.45 | 194.84 | 186.54 | 17.286 |
| 10 | 0.50 | 193.93 | 185.60 | 17.175 |
| 11 | 0.55 | 193.57 | 185.20 | 17.129 |
| 12 | 0.60 | 192.94 | 184.55 | 17.049 |
| 13 | 0.65 | 192.47 | 184.04 | 16.985 |
| 14 | 0.70 | 191.95 | 183.50 | 16.913 |
| 15 | 0.75 | 191.59 | 183.11 | 16.859 |
| 16 | 0.80 | 190.76 | 182.24 | 16.743 |
| 17 | 0.85 | 190.44 | 181.90 | 16.685 |
| 18 | 0.90 | 189.82 | 181.25 | 16.582 |
| 19 | 0.95 | 189.30 | 180.69 | 16.465 |

Analysis of the data presented in Table 3 shows that, for the initial samples in section 2, where an intensive loss of mass occurs, the value of activation energy decreases with the increasing degree of conversion. The average value of activation energy for ground ore samples is 187.77 kJ/mol according to the Friedman method and 196.05 kJ/mol according to the OFW method. The discrepancy between the average values of activation energy calculated by the two methods as well as for the values of activation energy obtained at different degrees of conversion does not exceed 5%, which shows a high convergence of the results.

### 3.2.2. Quantification of the Application of Effects Based on Thermogravimetric Analysis by Calculating the Activation Energy using Sulphide Flotation Concentrates

Flotation concentrate samples were investigated using thermogravimetric analysis at three heating rates (5, 10, and 20 K/min). The results are shown in Figure 11.

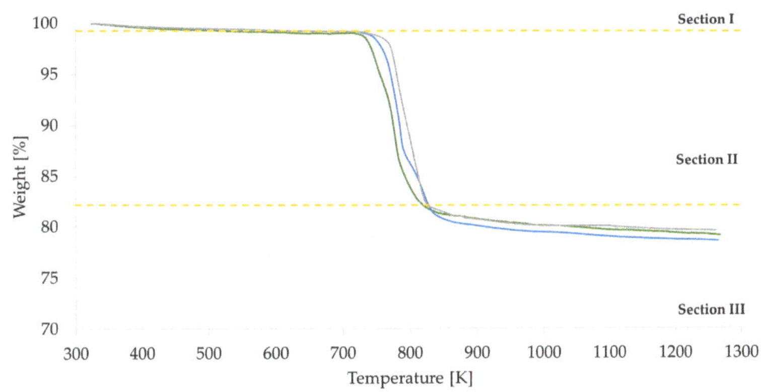


**Figure 11.** Results of the thermogravimetric investigation of flotation sulphide concentrates at three heating rates (green 5 K/min, blue 10 K/min, grey 20 K/min, yellow—separation into sections).

The interpretation of the results in Figure 12 allows to determine that, for flotation concentrates, it is possible to divide a thermogravimetric curve into three areas: the first—low mass loss (less than 1% up to a temperature of 738 K); the second—intensive mass loss in a range of temperatures from 738 K to 820 K; and the third—low mass loss at temperatures over 820 K. The values of the degree of transformation are <0.10 for the first section, 0.10–0.70 for the second section, and ≥0.70 for the third section for all heating rates. The values of activation energy for the heating rates were calculated, and the graphical data for calculating the activation energy by the Friedman method and the OFW method for conversion degrees from 0.05 to 0.95 are presented in Figure 12.

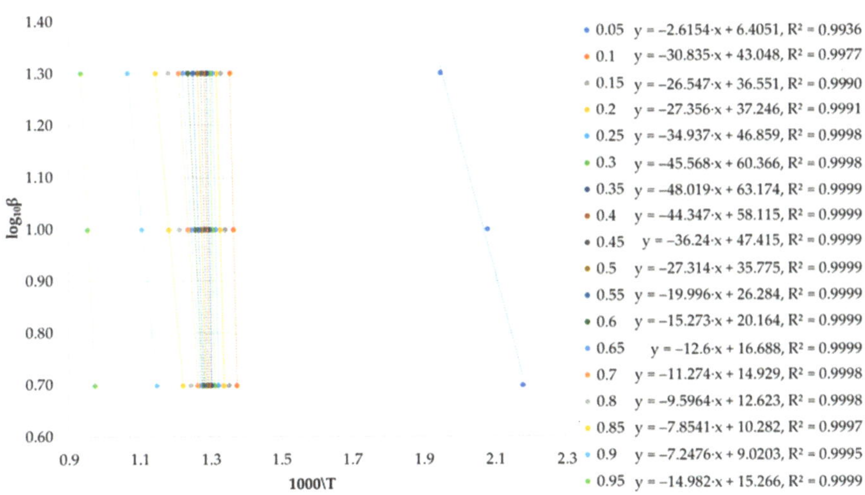


(a)

**Figure 12.** *Cont.*

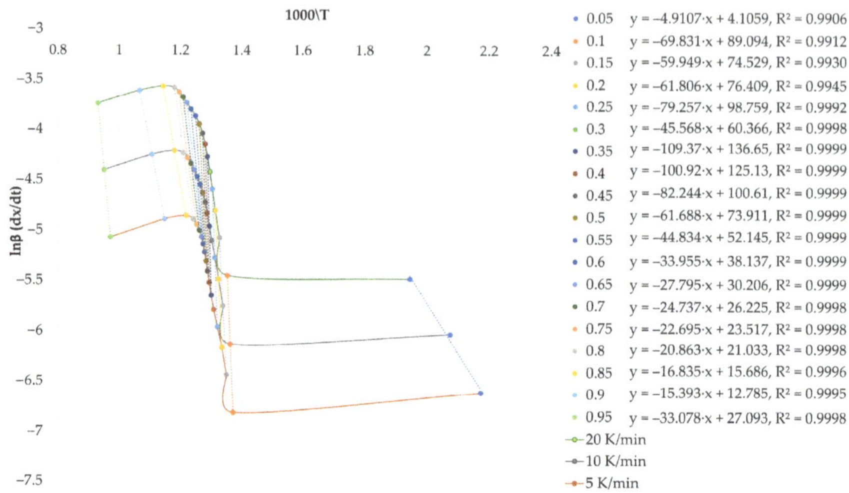


(**b**)

**Figure 12.** Calculation results for activation energy for three heating rates for flotation concentrate samples: (**a**) using the OFW method and (**b**) using the Friedman method.

The activation energy and the pre-exponential factor of the two methods were calculated from the dataset (Table 4).

**Table 4.** Activation energy calculation results for crushed carbon sulphide ore samples for sections 2 and 3.

| No. | x | $E_a$, kJ/mol (OFW Method) | $E_a$, kJ/mol (Fiedman Method) | ln$A$ |
|---|---|---|---|---|
| | | **Section 2** | | |
| 1 | 0.10 | 561.33 | 551.88 | 43.05 |
| 2 | 0.15 | 483.27 | 473.78 | 36.55 |
| 3 | 0.20 | 498.00 | 488.46 | 37.25 |
| 4 | 0.25 | 636.00 | 626.37 | 46.86 |
| 5 | 0.30 | 829.54 | 819.80 | 60.37 |
| 6 | 0.35 | 874.16 | 864.37 | 63.17 |
| 7 | 0.40 | 807.32 | 797.54 | 58.11 |
| 8 | 0.45 | 659.73 | 649.98 | 47.41 |
| 9 | 0.50 | 497.24 | 487.53 | 35.78 |
| 10 | 0.55 | 364.02 | 354.33 | 26.28 |
| 11 | 0.60 | 278.04 | 268.35 | 20.16 |
| 12 | 0.65 | 229.37 | 219.66 | 16.69 |
| | | **Section 3** | | |
| 13 | 0.70 | 205.23 | 195.50 | 14.93 |
| 14 | 0.75 | 189.13 | 179.36 | 13.73 |
| 15 | 0.80 | 174.70 | 164.88 | 12.62 |
| 16 | 0.85 | 142.98 | 133.05 | 10.28 |
| 17 | 0.90 | 131.94 | 121.65 | 9.02 |
| 18 | 0.95 | 272.74 | 261.42 | 15.27 |

The interpretation of the data obtained in Table 2 shows that the average value of the activation energy for section 2 was 559.83 kJ/mol using the OFW method and 550.17 kJ/mol using the Friedman method. For section 3, the average activation energy was 178.01 kJ/mol and 168.19 kJ/mol using the OFW method and the Friedman method, respectively.

The discrepancy between the average activation energies calculated by the two methods as well as for the activation energies calculated at different conversion rates does not exceed 5%, which shows a high convergence of the results.

#### 3.2.3. Evaluation of the Application of Impacts Based on Thermogravimetric Data Using the Developed Criterion

For quantitative estimation of the degree of physical–energetic influence on the samples, a complex criterion was developed; its calculation was carried out on the basis of activation energy data, calculated by the differential and integral kinetic methods, for degrees of transformation corresponding to the stage of intensive mass loss, i.e., chemical reaction of destruction at various heating rates.

$$K_\alpha = A - \left[ \sqrt{\left(1 + \frac{n_a \sum_{i=1}^{n_a} E^a_{\alpha i}}{n_b \sum_{i=1}^{n_b} E^b_{\alpha i}}\right)^2} \right]^{-1} \tag{10}$$

where $K_a$—comprehensive criterion for assessing the effectiveness of physical and energy impacts; $\sum_{i=1}^{n_a} E^a_{\alpha i}$—the value of activation energy after the physical and energetic action, calculated using the differential method for degrees of transformation with order number $i$ from the beginning of the chemical destruction reaction, kJ/mol; $\sum_{i=1}^{n_b} E^b_{\alpha i}$—the value of activation energy before the physical and energetic impact, calculated using the differential method for degrees of transformation with row number i from the beginning of the chemical destruction reaction, kJ/mol; $n_b$ and $n_a$—the number of degree of transformation values evaluated before and after the physical and energetic impact, respectively; and $A$—a coefficient depending on the type of action (activation/destruction), which is chosen on the basis of the following ratios:

$$\text{at } \frac{\sum_{i=1}^{n_a} E^a_{\alpha i}}{n_a} > \frac{\sum_{i=1}^{n_b} E^b_{\alpha i}}{n_b} \quad A = 1$$

$$\text{at } \frac{\sum_{i=1}^{n_a} E^a_{\alpha i}}{n_a} < \frac{\sum_{i=1}^{n_b} E^b_{\alpha i}}{n_b} \quad A = 0$$

$$\text{at } \frac{\sum_{i=1}^{n_a} E^a_{\alpha i}}{n_a} = \frac{\sum_{i=1}^{n_b} E^b_{\alpha i}}{n_b} \quad A = 0.5$$

The range of the criterion is $-1$ to 1, where $K_a$ = 1 corresponds to the maximum theoretically possible activating effect, $K_a$ = $-1$ corresponds to the maximum theoretically possible destructive effect, and $K_a$ = 0 corresponds to no effect.

This criterion not only allows to estimate the degree of effect from the position of an average evaluation over the whole selected range of degrees of reaction, but also makes it possible to consider the required number of degrees of reaction n, e.g., when it is necessary to evaluate the reaction over narrow temperature intervals.

The data presented in Tables 3 and 4, as well as data for the corresponding raw materials obtained earlier in [34], were used to calculate the criteria values.

Criteria were calculated for the data obtained, which allowed a quantitative assessment of the destruction and activation processes:

- for samples of comminuted ore, the criterion is $K_a$ = $-0.73$, which corresponds to the value of the proposed criterion for the destruction effect;
- for samples of flotation concentrates, the criterion is $K_a$ = 0.54, which corresponds to the value of the proposed criterion for the activation effect.

According to the term introduced by Arrhenius, the activation energy indicates the minimum amount of energy that must be applied to the system in order for a reaction to occur. The destructive effect of comminution on samples based on the calculated criterion shows that, by reducing the particle size and increasing the specific surface area of the particles, comminution helps to reduce the value of the energy needed to change the system. This is in accordance with Section 3.1, which calculates the $K_\sigma$ coefficient at the comminution stage using sulphide ore reagents.

From the perspective of the activation effect of flotation, this is due to the reduced recovery of the carbonaceous component in the concentrate and the extraction of ore mineralization (sulfide minerals) in the flotation concentrate during the flotation process, increasing the energy of the system, which increases the energy needed to communicate the reaction and further processing to the system. The resulting criterion values will allow the quantification of the degree of impact on the ore for different physical and energy impact methods.

Analysis of the presented results allows to substantiate a possibility of application of the thermogravimetric investigation of samples for a quantitative estimation of the influence of ore and enrichment products by the calculation of activation energy and the suggested criterion.

To assess the overall effect of applied intensification solutions on the whole grinding beneficiation stage, the module of the $K_a$ and $K'_\sigma$ multiplication can be used (Figure 13).

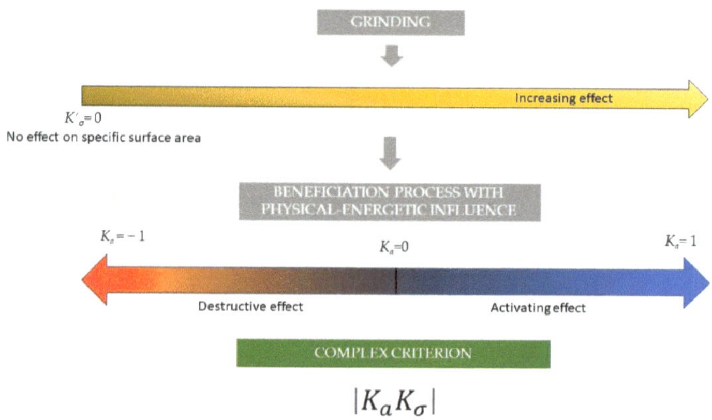


**Figure 13.** Algorithm of intensification solution assessment.

Increasing the value of $K'_\sigma$ corresponds to more specific surface area available after the grinding, and $K_a$ shows how this affects the performance of the intensification on the beneficiation stage. The increasing module of the multiplication shows the growing efficiency of the applied intensification method complex. The obtained value can be used to characterize the impact of proposed intensification technologies on the whole beneficiation process and, in the case of various proposed solutions, choose the most prospective one.

## 4. Conclusions

This paper presents the results of experimental and theoretical studies to substantiate and develop criteria that characterize the efficiency of overcoming the energy barrier of internal cohesion forces during grinding of complex ores of compound material composition, as well as the substantiation of methods of evaluation of decomposition and activation during flotation. Sulfide and carbonaceous gold-bearing ores were selected as objects of research on the basis of the presence of noble and rare metals and their fine dissemination.

As a result of the studies, a criterion for evaluating the efficiency of overcoming the energy barrier of internal forces ($K_\sigma$) was proposed as the ratio of the empirical coefficient of proportionality of the specific fracture energy and the relative value of reduction of the characteristic grain size to the value of free surface energy. The highest value of $K_\sigma$ is established for the case of sodium silicate application.

The highest values of the distribution of components in the −40 micrometer class as well as the highest value of specific surface when using sodium silicate are probably caused by the adsorptive decrease in the strength of silicate mineral aggregates due to the selective sorption of sodium silicate molecules on them. Adsorption of these compounds leads to a decrease in free energy of the newly formed mineral surface and, as a result, decreases the strength of internal bonds and facilitates deformation. Because ore minerals are embedded in an array of silicate minerals, the destruction of the latter also leads to their transition into fine sieve classes. The mechanism of action of potassium permanganate consists of the interaction with minerals prone to oxidation. Oxidation products are sorbed on the surface of minerals and create points of local decrease in the surface energy, which leads to an increase in the rate of grinding of these minerals and their selective transition into finer size classes. It is likely that high ratios of arsenic growth in the −40 micrometer class to sulfur and silicon growth when potassium permanganate is used are caused by this fact.

As a result of the research carried out, the possibility of quantitative estimation of the influence of physical and energy impacts on ore and products based on the interpretation of the results of thermogravimetric research by the calculation of values of activation energy was substantiated. The calculations were executed using two methods: the Friedman method and the OFW method. The results obtained have high convergence (error less than 5%). The developed criterion enables to quantitatively estimate the degree of impact on ore and concentration and define the character of impact by the coefficient value. Thus, the coefficient value for ore activation is 0.54, while for destruction, the coefficient is −0.73.

**Author Contributions:** T.A.—conceived and designed the experiments, and analyzed the data; N.N. and A.R.—implementation and processing of the analysis results; A.A. and V.K.—performed the experiments. All authors have read and agreed to the published version of the manuscript.

**Funding:** This work was carried out within the grant of the Russian Science Foundation (Project N 23-47-00109).

**Data Availability Statement:** Not applicable.

**Conflicts of Interest:** The authors declare no conflict of interest.

## References

1. Chanturiya, V.A. Scientific substantiation and development of innovative processes for the extraction of zirconium and rare earth elements in the deep and comprehensive treatment of eudialyte concentrate. *J. Min. Inst.* **2022**, *256*, 505–516. [CrossRef]
2. Litvinenko, V.S. Digital economy as a factor in the technological development of the mineral sector. *Nat. Resour. Res.* **2020**, *29*, 1521–1541. [CrossRef]
3. Litvinenko, V.S.; Sergeev, I.B. Innovations as a Factor in the Development of the Natural Resources Sector. *Stud. Russ. Econ. Dev.* **2019**, *30*, 637–645. [CrossRef]
4. Panayotova, M.; Panayotov, V. Studies on zinc recovery from technogenic waste. *E3S Web Conf.* **2021**, *295*, 03008. [CrossRef]
5. Sohn, H.Y. Nonferrous Metals: Production and History. In *Reference Module in Materials Science and Materials Engineering;* Elsevier: Amsterdam, The Netherlands, 2017; ISBN 9780128035818.
6. Espinoza, R.D.; Rojo, J. Towards sustainable mining (Part I): Valuing investment opportunities in the mining sector. *Resour. Policy* **2017**, *52*, 7–18. [CrossRef]
7. Litvinenko, V.S.; Petrov, E.I.; Vasilevskaya, D.V.; Yakovenko, A.V.; Naumov, I.A.; Ratnikov, M.A. Assessment of the Role of the State in the Management of Mineral Resources. *J. Min. Inst.* **2022**, *259*, 95–111. [CrossRef]
8. Romasheva, N.V.; Babenko, M.A.; Nikolaichuk, L.A. Sustainable development of the Russian Arctic region: Environmental problems and ways to solve them. *MIAB. Min. Inf. Anal. Bull.* **2022**, *10*, 78–87. [CrossRef]
9. Panayotova, M.; Panayotov, V. Application of membrane processes in mining and mineral processing. *E3S Web Conf.* **2021**, *280*, 08016. [CrossRef]
10. Duryagina, A.M.; Talovina, I.V.; Lieberwirth, H.; Ilalova, R.K. Morphometric Parameters of Sulphide Ores as a Basis for Selective Ore Dressing. *J. Min. Inst.* **2022**, *256*, 527–538. [CrossRef]
11. Koteleva, N.; Kuznetsov, V.; Vasilyeva, N. A Simulator for Educating the Digital Technologies Skills in Industry. Part One. Dynamic Simulation of Technological Processes. *Appl. Sci.* **2021**, *11*, 10885. [CrossRef]
12. Aleksandrova, T.N.; Aleksandrov, A.V.; Nikolaeva, N.V.; Romashev, A.O. Noble and Rare Metals in Caustobioliths and Prospects of Their Recovery. *J. Min. Sci.* **2015**, *51*, 1254–1261. [CrossRef]

13. Litvinenko, V.S.; Tsvetkov, P.S.; Molodtsov, K.V. The Social and Market Mechanism of Sustainable Development of Public Companies in the Mineral Resource Sector. *Eurasian Min.* **2020**, *2020*, 36–41. [CrossRef]
14. Dzhevaga, N.; Lobacheva, O. Reduction in Technogenic Burden on the Environment by Flotation Recovery of Rare Earth Elements from Diluted Industrial Solutions. *Appl. Sci.* **2021**, *11*, 7452. [CrossRef]
15. Lieberwirth, H.; Kühnel, L. Particle size effects on selectivity in confined bed comminution. *Minerals* **2021**, *11*, 342. [CrossRef]
16. Popov, O.; Talovina, I.; Lieberwirth, H.; Duryagina, A. Quantitative Microstructural Analysis and X-Ray Computed Tomography of Ores and Rocks—Comparison of Results. *Minerals* **2020**, *10*, 129. [CrossRef]
17. Kanda, Y.; Kotake, N. Comminution energy and evaluation in fine grinding. In *Handbook of Powder Technology*; Elsevier: Amsterdam, The Netherlands, 2007; Volume 12, pp. 529–550.
18. Austin, L.G.; Klimpel, R.R. The theory of grinding operations. *Ind. Eng. Chem.* **1964**, *56*, 18–29. [CrossRef]
19. Lieberwirth, H.; Hillmann, P.; Hesse, M. Dynamics in double roll crushers. *Miner. Eng.* **2017**, *103–104*, 60–66. [CrossRef]
20. Hesse, M.; Popov, O.; Lieberwirth, H. Increasing efficiency by selective comminution. *Miner. Eng.* **2017**, *103–104*, 112–126. [CrossRef]
21. Boyko, V.F.; Litvinova, N.M.; Melnikova, T.N. Selection of surfactant consumption during ore grinding of the Mnogovershinnoe deposit. *Obogashchenie Rud.* **2004**, *6*, 3–5.
22. Boyko, V.F.; Melnikova, T.N. *Natural-Technogenic Free-Dispersed Systems. Origin. Evolution*; Khabarovsk State Technical University: Khabarovsk, Russia, 2005; ISBN 5-7389-0366-8.
23. Litvintsev, V.S.; Melnikova, T.N.; Yatlukova, N.G.; Litvinova, N.M. Mechanical activation in ore dressing processes. *Gorny Zhurnal* **2006**, *6*, 95–96.
24. Melnikova, T.N.; Yatlukova, N.G.; Litvinova, N.M. To a question of optimization of grinding of ores. *Obogashchenie Rud.* **2006**, *4*, 5–7.
25. Khopunov, E.A. The role of surface-active substances in ore destruction under volumetric deformation. *Izvestia Vuzov./Min. J.* **2017**, *2*, 93–99.
26. Golini, D.; Jacobs, S.D. Transition between brittle and ductile mode in loose abrasive grinding. In *Advanced Optical Manufacturing and Testing*; SPIE: Bellingham, DC, USA, 1990; Volume 1333, pp. 80–91.
27. Tikhonov, O.N. Energy-averaged particle mixture coarseness and its use in Bond, Rittinger and Kick-Kirpichev formulas. *Obogashchenie Rud.* **2008**, *4*, 13–18.
28. Ip, S.W.; Toguri, J.M. The equivalency of surface tension, surface energy and surface free energy. *J. Mater. Sci.* **1994**, *29*, 688–692. [CrossRef]
29. Menzinger, M.; Wolfgang, R. The meaning and use of the Arrhenius activation energy. *Angew. Chem. Int. Ed. Engl.* **1969**, *8*, 438–444. [CrossRef]
30. Kissinger, H.E. Reaction Kinetics in Differential Thermal Analysis. *Anal. Chem.* **1957**, *29*, 1702–1706. [CrossRef]
31. Ozawa, T. Kinetic analysis of derivative curves in thermal analysis. *J. Therm. Anal.* **1970**, *2*, 301–324. [CrossRef]
32. Flynn, J.H.; Wall, L.A. A Quick, Direct Method for the Determination of Activation Energy from Thermogravimetric Data. *Polym. Lett.* **1966**, *4*, 323–328. [CrossRef]
33. Friedman, H.L. Kinetics of thermal degradation of char-forming plastics from thermogravimetry. Application to a phenolic plastic. *J. Polym. Sci. Part C Polym. Symp.* **1964**, *6*, 183–195. [CrossRef]
34. Alexandrova, T.N.; Nikolaeva, N.V.; Romashev, A.O.; Afanasova, A.V. *Low-Dimensional Structures of Noble and Rare Metals in Carbonaceous Rocks and Methods of Their Recovery Using Energy Impacts Based on Big Data*; Publishing and Printing Association of Higher Education Institutions: Saint-Petersburg, Russia, 2021; 154p, ISBN 978-5-91155-133-9.

minerals

MDPI

*Article*

# Mineral Phase Evolution during Oxidation Roasting Pretreatment of Typical Carlin Gold Ore and Effects on Gold Leaching Efficiency

**Licheng Ma [1,*], Xiang Liu [1,2], Lei Wang [3] and Jungang Qi [1]**

1 State Key Laboratory of Multiphase Complex Systems, Institute of Process Engineering, Chinese Academy of Sciences, No. 1 North 2nd Street, Zhongguancun, Haidian District, Beijing 100190, China
2 School of Chemical Engineering, University of Chinese Academy of Sciences, Beijing 100049, China
3 GRINM Resources and Environment Tech. Co., Ltd., Beijing 100088, China
* Correspondence: malicheng@ipe.ac.cn; Tel.: +86-10-82544923

**Abstract:** Arsenious and sulphur-bearing micro-disseminated gold ore is a kind of typical refractory Carlin-Type. The gold in Carlin-Type gold ore grains is distributed finely, existing as invisible or submicroscopic gold, encapsulated in arsenopyrite and pyrite. The technical difficulty of treatment Carlin-Type gold ore lies in how to release the fine gold wrapped in pyrite and arsenopyrite. In this study, the oxidation roasting pre-treatment technique was used to treat the Carlin-Type gold ore. This included a two-stage roasting process: the arsenic was removed in the first roasting process, and the sulphur was removed in the second roasting process. The thermodynamic of the roasting process was analyzed, and the mineral phase evolution of the roasting process was investigated by using XRD, SEM and EDS. Finally, the influence of sodium cyanide dosage and leaching time on leaching efficiency was investigated. The results suggest that for the first roasting temperature at 550 °C, and the second roasting at temperature 700 °C with air flow 2.5 L/min, the sodium cyanide dosage is 1.75 kg/t and leaching time is 27 h; a good leaching efficiency is obtained with 83.85%.

**Keywords:** oxidation roasting pretreatment; Carlin-Type gold deposit arsenopyrite; pyrite; cyanide leaching

**Citation:** Ma, L.; Liu, X.; Wang, L.; Qi, J. Mineral Phase Evolution during Oxidation Roasting Pretreatment of Typical Carlin Gold Ore and Effects on Gold Leaching Efficiency. *Minerals* **2023**, *13*, 558. https://doi.org/10.3390/min13040558

Academic Editors: Marinela Ivanova Panayotova and Vladko Panayotov

Received: 9 March 2023
Revised: 10 April 2023
Accepted: 10 April 2023
Published: 16 April 2023



## 1. Introduction

The natural resources of gold mine are decreased and the high–quality gold ores are gradually depleted. The refractory gold ores have become the main source of gold product. The feature of refractory gold ore is that the gold grain size is extremely fine, and exists in invisible or submicroscopic gold, which is encapsulated in arsenopyrite, sulfide, carbonate or siliceous rock [1–3]. Hence, it is difficult to extract gold from ores using cyanide leaching, even after ultra-fine grinding [4]. In order to render gold amenable to the subsequent cyanide leaching, oxidation pretreatment is considered to be an effective method for treatment of refractory gold ore, and breaks up the sulphide to oxides or sulphates before cyanidation [5–9].

A number of projects have been conducted to improve the extraction of gold from fine-grained refractory gold ores by using oxidation roasting pretreatment [10–12]. Oxidation roasting pretreatment is the most common pretreatment approach to treatment refractory gold ores [13–17]. To achieve good recoveries in subsequent leaching, the majority of the sulfur and organic carbon has to be oxidized to $SO_2$ and $CO_2$; the arsenic of ore has to be transformed to the volatile trivalent compound. Some of $SO_2$ and $SO_3$ and volatile trivalent compound can be captured with lime or dolomite of ores forms the stable compounds and ends up in the calcine [18].

In this study, the gold in arsenious and sulfur-bearing Carlin-Type gold ore is distributed extremely finely, existing as invisible or submicroscopic gold, encapsulated in arsenopyrite and pyrite. The technical difficulty of treatment Carlin-Type gold ore lies in how to release the fine gold encapsulated in pyrite and arsenopyrite. In order to achieve a good gold recovery, the thermodynamic analysis was used to verify the feasibility and reliability of the oxidation roasting in this paper. The mineral phase evolution of calcining was investigated by using XRD, SEM and EDS. Finally, a central composite design was used for investigating the operating variables corresponding to the leaching efficiency of gold.

## 2. Material and Methods

### *2.1. Raw Materials*

An arsenious and sulfur-bearing Carlin-Type gold ore was obtained from Guizhou. The ore sample was prepared by crushing and grinding. Inductively coupled plasma optical emission spectrometry (ICP-OES, PerkinElmer Optima, USA) was used to determine the chemical composition of the sample after dissolution, whereas gold was analyzed with the fire assay method followed by atomic absorption spectroscopy (AAS, WFX-130A, Beijing, China). The main phases were identified via mineral liberation analyzer (MLA 650, FEI Company, Czech Republic), scanning electron microscope (SEM Zeiss Sigma, Germany) equipped with energy-dispersive X-ray spectroscopy (EDS, Oxford, UK), electron probe scanning and X-ray diffraction (Rigaku SmartLab, Japan).

The main elements of the gold ore are shown in Table 1. It shows that the arsenious and sulfur-bearing Carlin-Type gold ore contains 2.50 g/t of Au, 9.15% of S and 0.79% of As. The SEM-EDS image of arsenious and sulfur-bearing Carlin-Type gold ore is shown in Figure 1. The metal minerals are mainly pyrite and arsenopyrite, and gangue minerals are mainly quartz, sericite, dolomite, calcite, ferrodolomite, kaoline, etc. The occurrence of gold is determined via diagnostic leaching. The occurrence of gold in different phases of Carlin-Type gold ore is shown in Figure 2. The dissemination characteristics and size fraction of main minerals were investigated with a mineral liberation analyzer (MLA). The gold in ore is extremely fine, and exists as invisible or submicroscopic gold, and is encapsulated in pyrite, arsenopyrite and silicate minerals. The particles of pyrite and arsenopyrite are also fine, 76.5% of pyrite have particle size −300 µm, and 83.36% of arsenopyrite have particles size −38µm. It is interesting that the distribution of gold in pyrite is only 6.08%, and the distribution of gold in arsenpyrite is 93.92%. The large amount of gold in arsenpyrite is the reason it is difficult to treat. The carbon content in the ore is 1.70%. The occurrence of carbon in different phases of Carlin-Type gold ore shows that 93.53% of carbon was presented in inorganic carbon; only 4.41% of carbon was presented in organic carbon and the rest was graphitic carbon. The content of organic and graphitic carbon in the ore was below 1%. Hence, there is almost no preg-robbing organic carbon.

**Table 1.** The main elements of the Carlin-Type gold ore.

| Element | Content (%) | Mineral | Content (%) |
|---|---|---|---|
| Au (g/t) | 2.50 | Pyrite | 22.93 |
| Ag (g/t) | 0.82 | Arsenopyrite | 1.27 |
| S | 9.15 | Sericite | 32.00 |
| As | 0.79 | Quartz | 25.09 |
| Fe | 9.55 | Dolomite | 5.95 |
| Mg | 1.75 | Calcite | 2.68 |
| Ca | 3.9 | Ferrodolomite | 3.06 |
| Ti | 1.5 | Kaoline | 2.53 |
| K | 2.46 | Chlorite | 1.51 |
| Na | 0.03 | Rutile | 1.33 |
| P | 0.33 | Others | 1.65 |
| C | 1.70 | | |
| Al | 3.12 | | |
| $SiO_2$ | 40.74 | | |

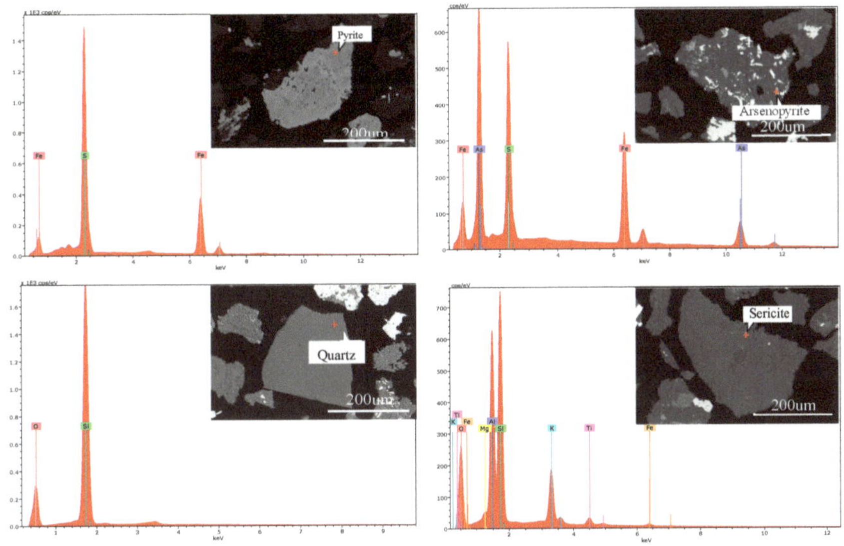


**Figure 1.** The SEM-EDS images of arsenious and sulfur-bearing Carlin-Type gold ore.

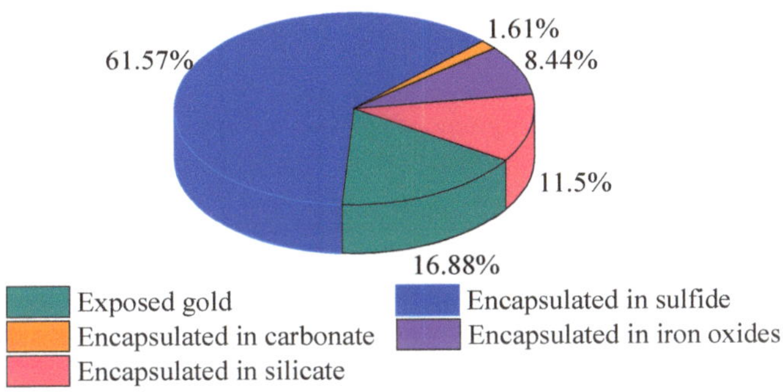


**Figure 2.** The occurrence of gold in different phases of Carlin-Type gold ore.

*2.2. Roasting and Cyanide Leaching*

The ground ore, with 85% passing size of 74 μm, was dried. Air oxidation roasting experiments were carried out in controlled atmosphere chamber furnace (static furnace). A 20 g sample was weighed and placed in a quartz crucible; a desired air flow was set which was controlled by adjusting the flowmeter. The sample was heated to a predetermined temperature with a temperature increase speed of 20 °C/min. When the desired temperature was reached, the temperature was maintained for a certain time. Subsequently, the calcine was cooled to room temperature and gold was leached under the following conditions: the calcine was ground to 92.23% passing size of 38 um, concentration of slurry 25%, pH of leaching solution 11–12 and leaching temperature 25 °C. After the leaching test, the slurry was filtrated via a vacuum filtration device. The residue was dried and the gold content was analyzed via the fire assay method followed by atomic absorption spectroscopy.

## 3. Results and Discussion

### *3.1. The Oxidation Roasting Pretreatment*

#### 3.1.1. The Thermodynamics Fundamental of Roasting

The oxidation roasting of arsenious and sulfur-bearing Carlin-Type gold ore is a complex chemical reaction process. The arsenopyrite and pyrite in the gold ore is oxidized to iron oxides, arsenic oxides, sulfur oxides, sulfate and arsenate in the oxidation roasting process so that the finely disseminated gold particles can be dissolved via cyanide during leaching. The main equilibrium reaction can be expressed as follows:

$$2FeAsS(s) = FeS_2(s) + 2As(g) \tag{1}$$

$$4FeS_2(s) + 11O_2(g) = 2Fe_2O_3(s) + 8SO_2(g) \tag{2}$$

$$2As(g) + O_2(g) = 2AsO(g) \tag{3}$$

$$2Fe_2O_3(s) + 4AsO(g) + 3O_2(g) = 4FeAsO_4(s) \tag{4}$$

To understand the phase evolution of arsenopyrite and pyrite of gold ore in the roasting process, a thermodynamic model was calculated by using the Equilibrium Composition module of HSC Chemistry 6.0. The amount of $O_2$ (g) and $N_2$ (g) in the calculation was normalized to be 1.0 mol and 3.71 mol, respectively. The equilibrium amounts of species and temperature for the reaction of FeAsS and $O_2$, and $FeS_2$ and $O_2$ at different content of $O_2$ are shown in Figures 3 and 4, respectively. It can be seen from Figures 3 and 4 that the content of oxygen is an important factor for affecting FeAsS and $FeS_2$ oxidation roasting. At temperature 200–600 °C, the $As_2O_3$ is easily formed at lower amounts of oxygen, and the FeAsS is decomposed to $FeS_2$, $As_2O_3$, $SO_2$ and iron oxide is lower. When the amount of oxygen is increased, the main compositions are changed to $FeAsO_4$, $SO_3$ and $SO_2$. Compared to FeAsS, $FeS_2$ is easily oxidized to $SO_2$, $Fe_3O_4$ and $Fe_2O_3$ at lower amounts of oxygen. When the amount of oxygen is increased, the $FeS_2$ is oxidized to $SO_2$, $SO_3$ and $Fe_2O_3$. From the analysis, the arsenic and sulfur can be removed by using control roasting conditions, e.g., amount of oxygen and roasting temperature.

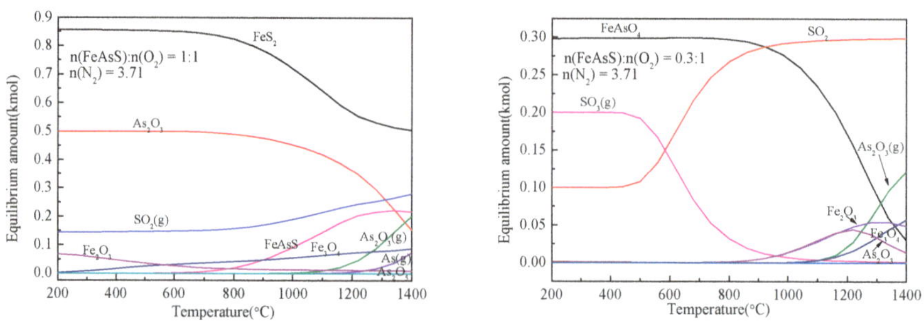


**Figure 3.** Equilibrium amounts of species as a function of temperature for reaction of FeAsS and $O_2$.

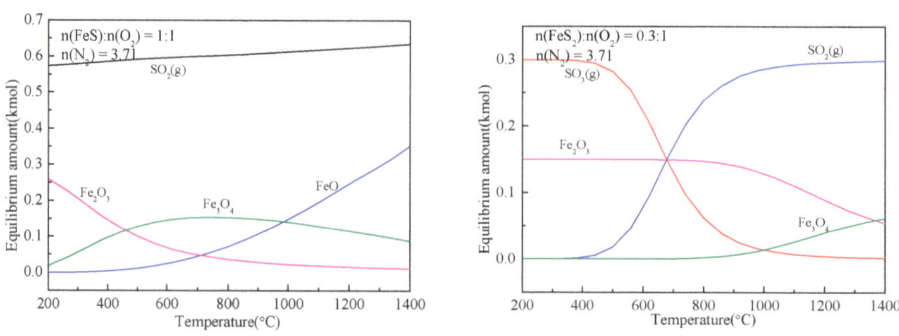


**Figure 4.** Equilibrium amounts of species as a function of temperature for reaction of $FeS_2$ and $O_2$.

3.1.2. Effect of Roasting Condition on Removal Efficiency of As and S, and Leaching Efficiency of Gold

The gold of refractory gold ore is encapsulated in arsenopyrite and sulfide. The content of gold in pyrite is only 6.08%, and the distribution of gold in arsenpyrite is 93.92%; it is necessary to remove As and S from the sulfide refractory gold ore using the air oxidation roasting for releasing the fine gold encapsulated in pyrite and arsenopyrite. Figure 5 shows effects of the temperature (a) and air flow rate (b) on the removal efficiency of As and S, and leaching efficiency of gold in single stage roasting. It can be seen from Figure 5a (air flow rate 2.5 L/min, roasting 1 h) that the removal efficiency of As is less than 30%, and decreases on increasing the roasting temperature; the removal efficiency of S increases on increasing the roasting temperature. The leaching efficiency of gold is less than 80%, and increases on increasing the roasting temperature from 500 °C to 700 °C. The effects of air flow rate on the removal efficiency of As and S, and leaching efficiency of gold are shown in Figure 5b (roasting temperature 700 °C, roasting 1 h), which shows that the removal efficiency of As decreases on increasing the airflow rate; however, the removal efficiency of S shows the opposite trend. The leaching efficiency of gold decreases on increasing the air flow rate. It also can be seen that the removal of As has an important effect on leaching of gold.

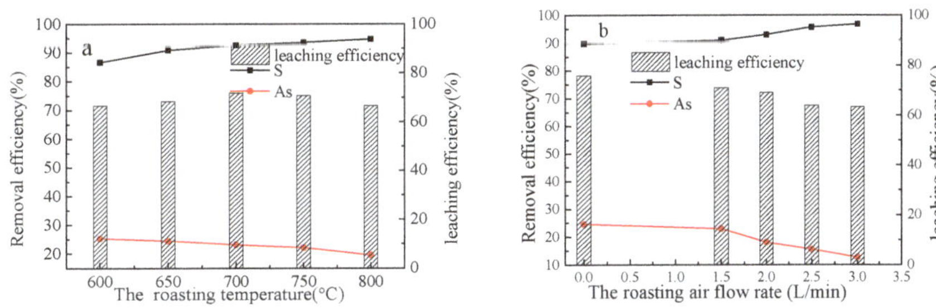


**Figure 5.** Effects of the temperature (**a**) and air flow rate (**b**) on the removal efficiency of As and S, and leaching efficiency of gold in single stage roasting.

In order to obtain the best removal efficiency of As and S, and leaching efficiency of gold, a series of two stage roasting experiments was used to investigate the effect of roasting condition on removal efficiency of As and S, and leaching efficiency of gold, which investigated the first roasting temperature, the second roasting temperature and the second roasting air flow. The effects of roasting condition on removal efficiency of As and S, and leaching efficiency of gold are shown in Figure 6. The effects of first roasting temperature on the removal efficiency of As and S are shown in Figure 6a (the sample was roasted in air (air flow 0 L/min) at different temperatures (450–600 °C) for 1 h). It can be seen from Figure 6a that the optimum removal efficiency of As is 46.91% at temperature 550 °C with only 38.41% S removal. To further remove S from the first stage roasting calcine (obtained at temperature 550 °C, air flow rate 0 L/min, roasting time 1 h), the higher temperature (600 °C–800 °C) and higher air flow rate (1.5 L/min -3.0 L/min) were used to roast the first stage roasting calcine. It can be seen from Figure 6b that the removal efficiency of S increased on increasing temperature, the leaching efficiency also increased on increasing the second stage roasting temperature below 750 °C. The leaching efficiency of gold decreased when the temperature exceeded 750 °C; this is because the sample was sintered under over roasting conditions and as a result the gold was not accessible to cyanide. The effects of the air flow on the removal efficiency of S, and leaching efficiency of gold in second stage roasting are shown in Figure 6c. It can be seen that the removal efficiency of S increases on increasing of air flow; the leaching efficiency of gold also increases with air flow. Above all, the As and S can be removed by controlling the operating parameters and condition of roasting process, which can give a good leaching efficiency of gold.

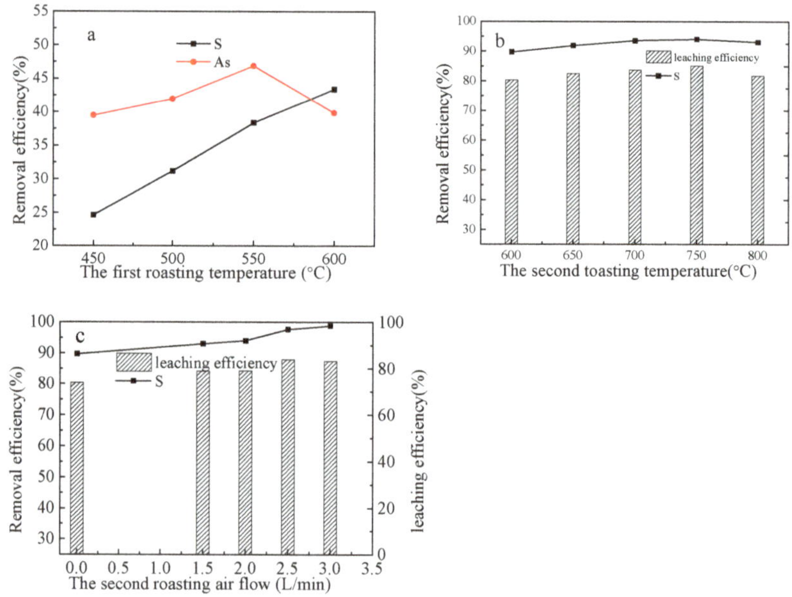


**Figure 6.** Effects of the first roasting temperature (**a**), the second roasting temperature (**b**) and air flow rate of second stage roasting (**c**) on the removal efficiency of As and S, and leaching efficiency of gold in the two stage roasting.

### 3.2. *Mineral Phase Evolution of Carlin-Type Gold Ore*

The arsenic was removed in the first roasting process at a temperature of 550 °C, and the sulfur was removed in the second roasting process at 700 °C with air flow 2.5 L/min. The mineral composition and microstructure of calcine was determined via X-ray diffraction, the electron probe microanalysis technique (EPMA) and SEM. The X-ray diffraction pattern of arsenious and sulfur-bearing refractory gold ore and calcine are shown in Figure 7. It can be seen from Figure 7 that the main components of calcine were quartz and hematite. Comparing with raw material, the diffraction peak of pyrite and arsenopyrite disappeared completely in calcine and is replaced by the hematite peak. This indicates that the pyrite and arsenopyrite were almost completely oxidized to hematite in calcine.

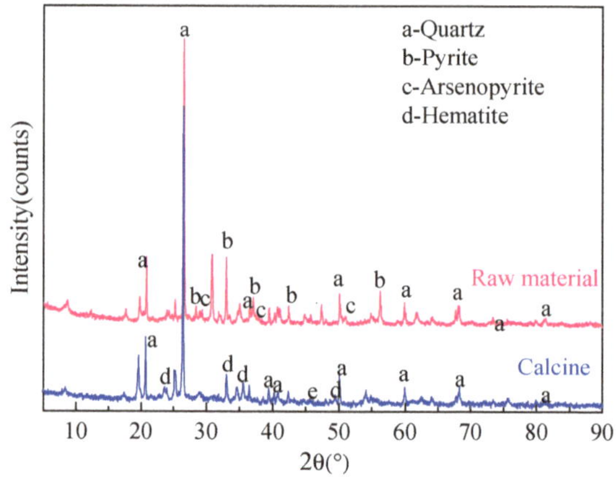


**Figure 7.** X-ray diffraction pattern of arsenious and sulfur-bearing refractory gold ore and calcine.

The surface micro structure and elemental composition of raw ores and hematite in calcine were observed via map analyses with the electron probe microanalysis technique (EPMA). The electron probe microanalysis of hematite in calcine is shown in Figures 8 and 9. It can be seen from Figures 8 and 9 that the pyrite contains some arsenic in raw ore. The gold in ore is extremely fine, and exists as invisible or submicroscopic gold, which does not reveal the presence of any discrete grains of gold. In these gold ores, gold is chemically bound within the arsenopyrite in a non-metallic form, owing to its suitable atomic spacing and crystal chemistry with respect to gold (Swash, 1988). As shown in Figures 8 and 9, comparing with raw ore, most arsenic and sulfur was removed in roasting process, and hematite was the main product of the calcine. The microstructure of hematite is shown in Figure 10. It can be seen from Figure 10 that the hematite possesses a porous structure resulting from the escape of the arsenic or sulphur atoms from the sulphide grains. Those micro-pores could provide a diffusion channel for CN−, which is conducive to improving the leaching efficiency of gold.

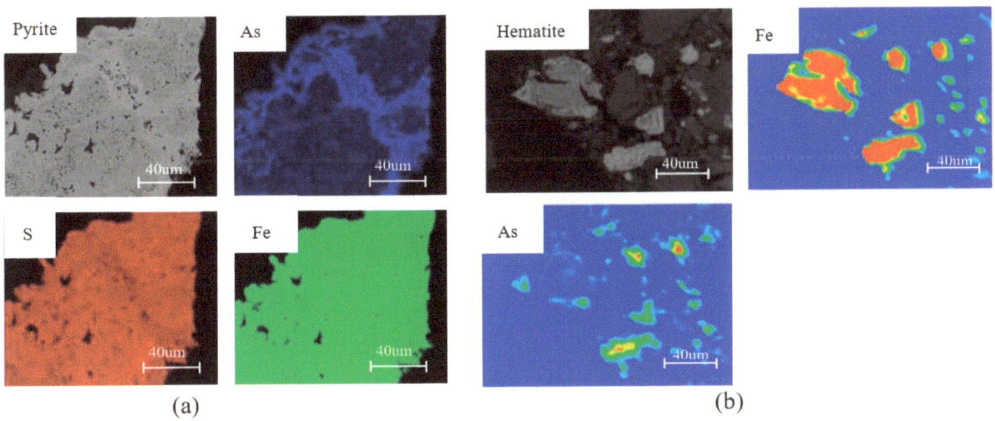


**Figure 8.** The electron probe microanalysis of pyrite (**a**) and hematite in calcine (**b**).

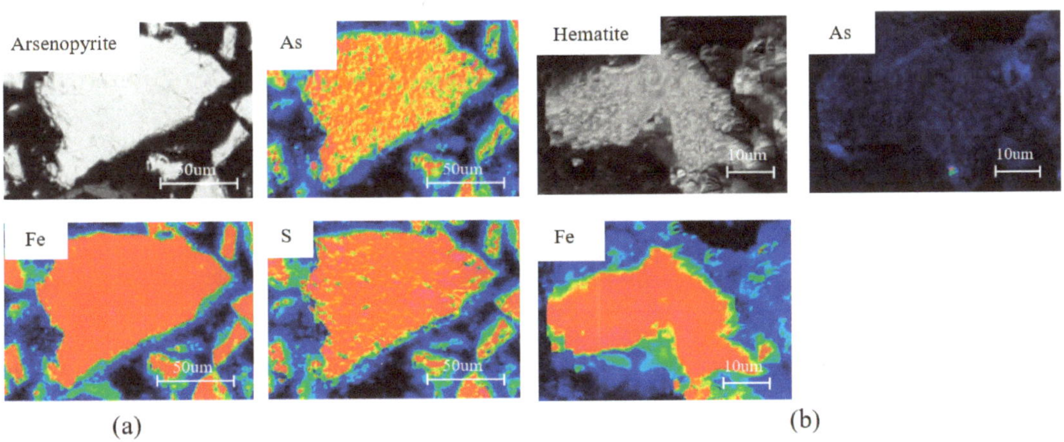


**Figure 9.** The electron probe microanalysis of arsenopyrite (**a**) and hematite in calcine (**b**).

The mineral phase evolution during roasting is followed by a change in the occurrence state of gold. The change in the occurrence state of gold directly affects the gold leaching efficiency. The occurrence of gold is determined via diagnostic leaching. The occurrence of gold in different phases of the arsenious and sulfur-bearing Carlin-Type gold ore and the calcine is shown in Figure 11. It can be seen from Figure 11 that only 16.88% of gold is exposed in raw ore, with up to 61.6% of the gold encapsulated in sulfides. After roasting, the calcin contains up to 56.4% exposed gold, iron oxide-encapsulated gold up to 31.50% and sulfide-encapsulated gold almost reduced to 0 %. This will facilitate the subsequent gold leaching process.

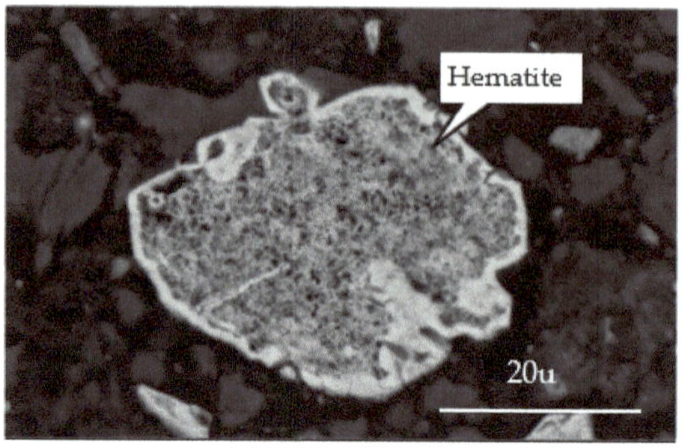


**Figure 10.** The microstructure of hematite in calcine.

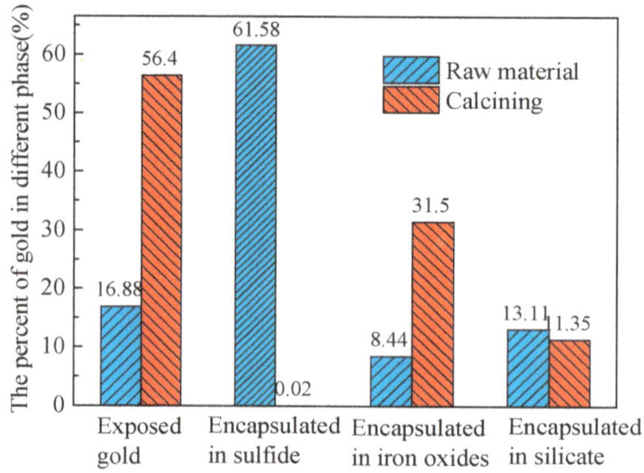


**Figure 11.** The occurrence of gold in different phases of Carlin-Type gold ore and calcine.

### 3.3. *The Effects on Gold Leaching Efficiency*

As mentioned above, the arsenic was removed in the first stage roasting process at temperature 550 °C, air flow rate 0 L/min and roasting time 1 h. The sulfur was removed in the second roasting process at 700 °C, air flow 2.5 L/min with roasting time 1 h. Hence, the calcine was prepared at the above conditions, and the influence of sodium cyanide dosage and leaching time on leaching efficiency was investigated. Figure 12a shows the leaching efficiency of gold at different sodium cyanide dosages with a constant leaching time of 24 h. It can be seen from Figure 12a that the leaching efficiency of gold increases with increasing sodium cyanide dosage. When the sodium cyanide dosage is more than 1.75 kg/t, the leaching efficiency of gold does not significantly increase. Figure 12b shows the leaching efficiency of gold at different leaching times with a constant sodium cyanide dosage of 1.75 kg/t. It can be seen from Figure 12a that the leaching efficiency of gold increases with increasing leaching time. When leaching time is more than 27 h, the leaching efficiency of gold does not significantly increase. Based on the above analysis, the results suggest that for the first roasting temperature at 550 °C, and the second roasting at temperature 700 °C with air flow 2.5 L/min, the sodium cyanide dosage is 1.75 kg/t and leaching time is 27 h; a good leaching efficiency is obtained with 83.85%.

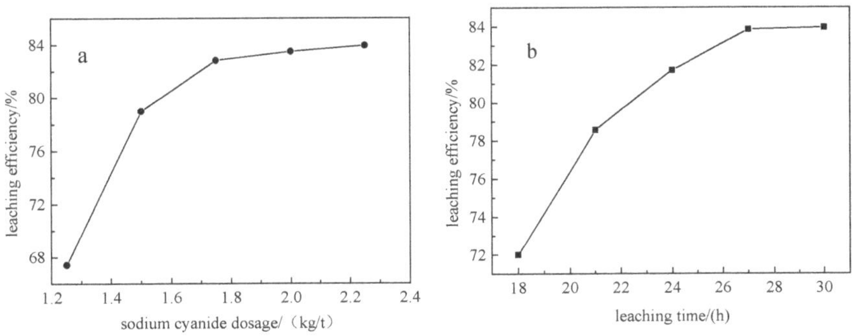


**Figure 12.** The effect of sodium cyanide dosage (**a**) and leaching time (**b**) on the leaching efficiency.

## 4. Conclusions

This study investigated the two stage oxidation roasting technique, which was used to treat the arsenious and sulfur-bearing Carlin-Type gold ore with the purpose of improving leaching efficiency of gold. In the roasting process, the arsenic was removed in the first roasting, and the sulfur was removed in second roasting. The thermodynamic of the roasting process was analyzed, and the mineral phase evolution of calcining was investigated by using XRD, SEM and EDS. After roasting, the pyrite and arsenopyrite were almost completely oxidized to hematite. The exposed gold is up to 56.40%, iron oxide-encapsulated gold is up to 31.50% and sulfide-encapsulated gold is reduced almost to 0%. The porous structure of the hematite facilitates the subsequent gold leaching process. Finally, a small scale (20 g) batch test with first roasting temperature at 550 °C for 1 h without air flow, and second roasting at temperature 700 °C for 1 h with air flow 2.5 L/min, followed by 27 h cyanidation with 1.75 kg/t cyanide addition gave 83.5% gold extraction.

**Author Contributions:** Investigation, Writing—original draft, Data curation, Visualization, Writing—review and editing, L.M. and L.W., Conceptualization, Methodology, L.M. and X.L., Resources, Validation, Funding acquisition, L.M. and L.W., Investigation, Data, J.Q., Review, L.M. and J.Q. All authors have read and agreed to the published version of the manuscript.

**Funding:** This research was supported by Beijing Natural Science Foundation (2232067) and National Key R&D Program during the 14th Five-year Plan Period (2021YFC2900900).

**Data Availability Statement:** Data are contained within the article.

**Conflicts of Interest:** The authors declare no conflict of interest.

## References

1. Cline, J.S. Timing of gold and arsenic sulfide mineral deposition at the Getchell Carlin-type gold deposit, north-central Nevada. *Econ. Geol.* **2001**, *96*, 75–89. [CrossRef]
2. Rui-Zhong, H.; Wen-Chao, S.; Xian-Wu, B.; Guang-Zhi, T.; Hofstra, A.H. Geology and geochemistry of Carlin-type gold deposits in China. *Miner. Depos.* **2002**, *37*, 378–392. [CrossRef]
3. Su, W.; Xia, B.; Zhang, H.; Zhang, X.; Hu, R. Visible gold in arsenian pyrite at the Shuiyindong Carlin-type gold deposit, Guizhou, China: Implications for the environment and processes of ore formation. *Ore Geol. Rev.* **2008**, *33*, 667–679. [CrossRef]
4. Corrans, I.J.; Angove, J.E. Ultra fine milling for the recovery of refractory gold. *Miner. Eng.* **1991**, *4*, 763–776. [CrossRef]
5. Komnitsas, C.; Pooley, F.D. Mineralogical characteristics and treatment of refractory gold ores. *Miner. Eng.* **1989**, *2*, 449–457. [CrossRef]
6. Piervandi, Z. Pretreatment of refractory gold minerals by ozonation before the cyanidation process: A review. *J. Environ. Chem. Eng.* **2023**, *11*, 109013. [CrossRef]
7. Li, J.; Yang, H.; Zhao, R.; Tong, L.; Chen, Q. Mineralogical characteristics and recovery process optimization analysis of a refractory gold ore with gold particles mainly encapsulated in pyrite and Arsenopyrite. *Geochemistry* **2023**, *83*, 125941. [CrossRef]
8. Majumdar, S.; Singh, S.; Sahoo, P.R. Characterization of organic matter and its implications for pyrite hosted refractory gold mineralization along the South Purulia Shear Zone, eastern India. *Ore Geol. Rev.* **2020**, *124*, 103584. [CrossRef]
9. Qin, H.; Guo, X.; Tian, Q.; Yu, D.; Zhang, L. Recovery of gold from sulfide refractory gold ore: Oxidation roasting pretreatment and gold extraction. *Miner. Eng.* **2021**, *164*, 106822. [CrossRef]

10. Amankwah, R.K.; Pickles, C.A. Microwave roasting of a carbonaceous sulphidic gold concentrate. *Miner. Eng.* **2009**, *22*, 1095–1101. [CrossRef]
11. Fraser, K.S.; Walton, R.H.; Wells, J.A. Processing of refractory gold ores. *Miner. Eng.* **1991**, *4*, 1029–1041. [CrossRef]
12. Nanthakumar, B.; Pickles, C.A.; Kelebek, S. Microwave pretreatment of a double refractory gold ore. *Miner. Eng.* **2007**, *20*, 1109–1119. [CrossRef]
13. Swash, P.M. A mineralogical investigation of refractory gold ores and their beneficiation with special reference to arsenical ores. *J. South Afr. Inst. Min. Metall.* **1988**, *88*, 173–180.
14. Arriagada, F.J.; Osseo-Asare, K.A. Gold extraction from refractory ores: Roasting behaviour of pyrite & arsenopyrite in Precious Metals. In *Min Extn & Processing*; Kudryk, V., Liang, W.W., Eds.; Metallurgical Society of AIME: Englewood, CO, USA, 1984; pp. 367–385.
15. Marsden, J.; House, I. Oxidative Pre-treatment. In *Chemistry of Gold Extraction*; SME: Chandler, AZ, USA, 2006; Chapter 5; pp. 147–231.
16. Fernández, R.R.; Collins, A.; Marczak, E. Gold recovery from high arsenic containing ores at Newmont's roasters. *Miner. Metall. Process.* **2010**, *27*, 60–64. [CrossRef]
17. Hammerschmidt, J.; Güntner, J.; Kerstiens, B.; Charitos, A. Roasting of gold ore in the circulating fluidized-bed technology. *Dev. Miner. Process.* **2005**, *15*, 433–453.
18. Thomas, K.G.; Cole, A.P. Roasting developments–especially oxygenated roasting. In *Gold Ore Processing*; Elsevier: Amsterdam, The Netherlands, 2016; pp. 373–392.

minerals

MDPI

*Article*

# Combined Physicochemical and Energy Methods to Improve the Recovery of Rare Earth Elements from Eudialyte Concentrate

**Valentine A. Chanturiya, Vladimir G. Minenko, Andrey L. Samusev, Maria V. Ryazantseva * and Elizaveta V. Koporulina**

N.V. Melnikov's Institute of Comprehensive Exploitation of Mineral Resources Russian Academy of Sciences, Kryukovsky Tupik 4, 111020 Moscow, Russia
* Correspondence: ryzanceva@mail.ru; Tel.: +7-(495)360-58-68

**Abstract:** The parameters for efficient nitric acid leaching were experimentally determined, which ensured the recoveries of Zr and REEs from eudialyte concentrate up to 87.0%–91.7% and 76.0%–81.1%, respectively. The possibility was shown of intensifying the leaching process through preliminary energy treatments to ensure the intensive breakdown of mineral complexes and grains; as a result, the recovery of Zr and REEs increased by more than 10%. A process was developed for the selective recovery of up to 91.5% of zirconium and up to 71.2% of REEs in the form of carbonate compounds from the pregnant solution of nitric acid leaching by chemical precipitation as well as up to 81.1% REEs and up to 91.7% zirconium on hypercrosslinked polystyrene sorbents.

**Keywords:** eudialyte; acid leaching; energy impacts; zirconium; rare earth elements

**Citation:** Chanturiya, V.A.; Minenko, V.G.; Samusev, A.L.; Ryazantseva, M.V.; Koporulina, E.V. Combined Physicochemical and Energy Methods to Improve the Recovery of Rare Earth Elements from Eudialyte Concentrate. *Minerals* **2023**, *13*, 414. https://doi.org/10.3390/min13030414

Academic Editors: Marinela Ivanova Panayotova and Vladko Panayotov

Received: 30 January 2023
Revised: 2 March 2023
Accepted: 10 March 2023
Published: 15 March 2023



## 1. Introduction

REEs are indispensable elements in the modern automotive, aerospace, metallurgical, chemical, electronic, energy and nuclear industries. It is impossible to produce high-tech communications, modern magnets, special steels, batteries, and computer chips without use of the REEs. In some sources, rare earth elements are called "metals of a new technological order [1]".

However, the reserves of rich mineral resources (bastnasite, monazite, and xenotime) are limited, which is why there is a need for the development of new technology for processing the secondary REE resources. These resources include mineral feeds with a low content of REEs (apatites, eudialyte ores, and ocean bottom sediments) and waste (electronic scrap, phosphors, lighting bulbs, coal fly ash, etc.) [2–11].

A promising source of REE in Russia is the eudialyte ores (Lovozero deposit, Kola Peninsula). The resources of eudialyte ores of the Lovozero massif are huge ($ZrO_2$: 300 million ton;, REE oxides: 60–70 million tons). The content of the valuable components in eudialyte ores is quite low (0.5 g/t), but the content of the most valuable heavy REE (HREE) is sufficiently high (20%). Eudialyte is a potential source of Hf, and it usually contains very little thorium and uranium. This is why the development of technology for the recovery and separation of REE from the eudialyte is of current interest, taking into account the depletion of rich REE resources [12,13].

Various methods for the processing of eudialyte concentrates have been proposed in the literature [14–17]. The main problems of hydrometallurgical eudialyte processing is the formation silicic acid. It undergoes a polycondensation process with the formation of polymers of various compositions and structures up to stable gels. The latter leads to a drop in the recovery of valuable components. The various methods of silicic acid polymerization (silicate gel formation) are reported in the literature. Various solutions the silica gel formation problem have been proposed: silica gel formation can be avoided by applying a combination of mechanochemical and extraction with nitric acid/tri-n-butylphosphate [18], the introduction of $F^-$ ions during leaching [19], and a two-stage hydrometallurgical process [17], including preliminary acid treatment of the heated concentrate followed by aqueous leaching. In this paper, the use of energy effects on mineral pulp (ultrasound and electrochemical processing) is suggested. The experimental results show an increase in leaching efficiency and the opening of the silicate matrix [20,21].

Another significant problem with eudialyte processing is the recovery and separation of the valuable components (Zr and REE) from the pregnant solution of the eudialyte concentrate through leaching. However, the pregnant solution of the eudialyte concentrate lichate is quite complex in terms of the high concentration of the leaching agents, the low concentration of the valuable components, and the high concentration of the accompanying elements, most significantly alkaline earth. Extractive and sorption technologies are used for the selective recovery of the REEs from the leaching solutions. Among the human-made selective sorption materials, sorbents based on hypercrosslinked polystyrene are important [22]. The chemical inertness of hypercrosslinked polystyrene allows its use in aggressive media such as concentrated solutions of $HNO_3$, HF, and $H_2SO_4$. The important advantage of this sorbent is the stability of the volume of the sorbent layer in the column during contact with electrolyte solutions of different concentrations; additionally, the method is simple and can be easily automated. It is also important that all used sorbents are produced on an industrial scale [23]. This paper presents the results of the application of sorbents based on hypercrosslinked polystyrene for the processing of a real pregnant solution of eudialyte acid leaching.

## 2. Materials and Methods

Eudialyte concentrate from the Lovozero field was used in leaching experiments (Table 1).

**Table 1.** Chemical composition of eudialyte concentrate (wt %).

| Component | **$Na_2O$** | **MgO** | **$Al_2O_3$** | **$SiO_2$** | **$K_2O$** | **CaO** | **$TiO_2$** | MnO | $Fe_2O_3$ |
|---|---|---|---|---|---|---|---|---|---|
| Content | 10.14 | 0.18 | 6.90 | 44.96 | 1.37 | 5.42 | 3.29 | 1.66 | 5.20 |
| Component | **Cl** | **$SO_3$** | **$HfO_2$** | **$Nb_2O_5$** | **$Rb_2O$** | **SrO** | **Ba** | U | Th |
| Content | 0.97 | 0.20 | 0.14 | 0.81 | 0.01 | 2.63 | 0.09 | <0.01 | <0.02 |
| Component | **$ZrO_2$** | **$Ta_2O_5$** | **As** | **$CeO_2$** | **$Y_2O_3$** | **$Nd_2O_3$** | **$La_2O_3$** | Other REE oxides | LOI |
| Content | 11.50 | 0.05 | 0.03 | 0.68 | 0.46 | 0.32 | 0.29 | 0.75 | 1.97 |

Note: LOI, ignition loss.

Leaching conditions were varied depending on the aims of the experiments. These conditions are carefully detailed in [24,25] and are presented in Table 2.

**Table 2.** Leaching conditions.

| **Autoclave Name** | **Sample Mass, G** | **Acid Concentration, M** | **Temperature Range, °C** | **Residence Time, min** | **S:L Ratio** | **Agitation Speed, $min^{-1}$** |
|---|---|---|---|---|---|---|
| TOP 120 series (TOP INDUSTRY) | 60 g | 2.4–9.6 | 80–140 °C | 60–300 | 1:10 | 550 |
| IPKON RAS Equipment * | | | | | | |

* autoclave equipped with the fitting for US and EC treatment of mineral pulp during leaching.

Preliminary ultrasonic (US) treatment in water (5, 10, 20, and 30 min) and in a 7.2 M acid solution (10 min) was carried out with a dispersion machine (MEF-15, MELFIZ, manufacturer—MELFIZ, Moscow, Russia) at a power of 600 W and an operating frequency of 22 kHz at S:L = 1:10. Electrochemical (EC) treatment conditions were as follows: anode current density of 200 $A/m^2$ and volume current density of 1.5 A/L.

After the leaching, the insoluble mineral residue was separated from the pregnant solution with centrifugation (UC-1536 E, ULAB, China) then it was washed with distilled

$H_2O$ and dried. The chemical composition of the solid residues was analyzed by ICP-MS (ICPE-9000, Shimadzu, Kyoto, Japan) after digestion with $HNO_3$ solution and XRF (ARL ADVANT'X, Thermo Scientific, Waltham, MA, USA). The chemical composition of the liquid phases was analyzed using ISP-AES (Elan-6100, PerkinElmer, Waltham, MA, USA) and MS.

The surface transformation of the eudialyte grains was examined with XPS (Versa Probe II spectrometer, ULVAC-PHI, Chigasaki, Japan) and DRIFT (IR-Affinity, Shimadzu, Kyoto, Japan).

The micromorphology and chemical composition of the eudialyte grains were determined using an analytical scanning electron microscope (ASEM) LEO 1420VP (Carl Zeiss, Cambridge, UK), equipped with an energy-dispersive X-ray INCA 350 spectrometer (Oxford Instruments Ltd., Oxford, UK) at an accelerating voltage of 20 kV in high-vacuum mode. The samples were fixed on a carbon conductive tape and coated with a thin layer of carbon.

Chemical precipitation of Zr and REE was performed in two steps. Zirconium hydroxides were recovered during the 1st step, when the leach solution was alkalized with $CaCO_3$ to pH 3.8–4.0. The REEs were recovered during the second stage in the form of hydroxides and hydrocarbonates, with further solution alkalization with caustic soda ($Na_2CO_3$) to pH 6.0–6.4. Precipitates were separated from the solution. The final precipitates were washed with distilled $H_2O$ and dried to constant weight.

Highly crosslinked polystyrene sorbents DOWEX 1–X–8 (Acros Organics), PCA-433 (−1.2 + 0.3 mm), and Puromet MTS 9500, Purolite C 160 (Purolite) sized −63 + 43 μm were used in the study. The specifications of the sorbents used are presented in Table 3.

**Table 3.** Ion exchange resins used in the study.

| Resin | Dowex 1 × 8 Strong Base Anion Exchange Resin Strongly Basic | Purolite PCA433 Strong Base Anion Exchange Resin | Purolite C160 Strong Acidic Exchange Resin | Puromet MTS 9500 Chelating Resin |
|---|---|---|---|---|
| Polymeric matrix | Styren divinylbenzene copolymer | | | |
| Functional group | Trimethyl ammonium | Quaternary ammonium type I | Sulfonic/aminophosphonic | Aminophosphonic |
| Exchange capacity | 1.2 eq/L | 1.3 eq/L | 2.4 eq/L | 1.3 eq/L Ca |
| Other properties | Gel polymer Crosslinking 8% Operating pH 0–14 | Gel polymer Operating pH 0–14 | Macroporous Operating pH 0–14 | Macroporous chelating Operating pH 0–14 |

The preparation of polystyrene sorbents for the experiment was as follows: the sorbent sample was kept in a 50% $HNO_3$ solution for 24 h, then washed with distilled $H_2O$ until the neutral pH of the wash was reached.

In the sorption experiment, a pregnant acid leach solution was used (line 1 in Tables 4 and 5) as well as a refined REE solution obtained by dissolution (50% $HNO_3$) of REE precipitate isolated from the initial pregnant solution during its neutralization with calcium carbonate, according to the method described above (line 2 in Tables 4 and 5).

The sorption experiment involved the following protocol: a 30 mL glass column was filled with water and sorbent granules, then a solution (1 mL/min) was fed through the column until the concentration of La at the column outlet was equal to the concentration at the column inlet. La was used as indicative element to control the column saturation with REEs as one of the representatives of the REE group.

After the column with the initial solution was equilibrated, the electrolyte was displaced by a top-to-bottom flow of the mobile phase. In both cases, the eluate was collected in 2.0 mL fractions, the content of metals in which was measured via ISP-MS, AES, and MS.

**Table 4.** Composition of the pregnant leach solution and refined REE solution.

| | Na | Mg | Al | K | Ca | Ti | Si | Sc | Cr | Ni |
|---|---|---|---|---|---|---|---|---|---|---|
| | | | | | | mg/L | | | | |
| 1 | 2272.0 | 16.1 | 520.7 | 115.7 | 1121.0 | 233.7 | 0.1 | 0.1 | 0.5 | 1.6 |
| 2 | 187.65 | 5.7 | 288.4 | 43.9 | 263.0 | 89.6 | 0.1 | 0.1 | 0.4 | 1.5 |
| | **Zn** | **Cu** | **Sr** | **Mn** | **Fe** | **Zr** | **Ba** | **Hf** | **Nb** | **ΣREE** |
| 1 | 13.6 | 0.8 | 580.0 | 527.0 | 580.0 | 2657.0 | 36.9 | 29.8 | 0.7 | 596.0 |
| 2 | 10.2 | 0.7 | 34.0 | 87.7 | 398.0 | n/d | n/d | 29.8 | 0.6 | 571.6 |

Note: n/d—not detected, below detection limits.

**Table 5.** REE concentration.

| | Sc | Y | La | Ce | Pr | Nd | Sm | Eu | Gd | Tb | Dy | Ho | Er | Tm | Yb | Lu |
|---|---|---|---|---|---|---|---|---|---|---|---|---|---|---|---|---|
| | | | | | | | | | mg/L | | | | | | | |
| 1 | 0.4 | 138.4 | 76.7 | 18.6 | 22.5 | 66.1 | 20.1 | 6.7 | 23.5 | 4.1 | 23.2 | 4.8 | 12.9 | 2.0 | 11.6 | 1.6 |
| 2 | 0.3 | 135.4 | 68.4 | 17.4 | 21.8 | 63.3 | 19.7 | 6.5 | 22.7 | 4.0 | 23.2 | 4.7 | 12.8 | 1.9 | 11.0 | 1.5 |

## 3. Results

### *3.1. Effect of Acids on the Leaching Efficiency of Eudialyte Concentrate*

SPECIAL FEATURES OF EUDIALYTE SURFACE WHEN EXPOSED TO ACIDS. The microscopic investigations revealed the morphological features of eudialyte particles after chemical interaction with acid solutions. We found that a main feature of the eudialyte particles after interaction with $H_2SO_4$ and HCl solutions was the appearance of deep, vertical cracks (Figure 1b,c). It can be assumed that, in these acids, the mineral normal dissolution rate is comparable to and/or exceeds the tangential (layer-b-layer) rate, which contributes to a more active dissolution of the entire volume of the particle and not just its surface. A specific feature of the interaction of eudialyte with a $HNO_3$ solution is the layer-by-layer mechanism of mineral dissolution (Figure 1a).

a b c

**Figure 1.** Different types of eudialyte grains transformation at tangential (layer-by-layer) dissolution in $HNO_3$ (**a**) at the normal and tangential dissolution in $H_2SO_4$ (**b**) due to the normal mechanism of dissolution in HCl (**c**).

Using the ASEM technique, it was determined that the degree of eudialyte grain destruction varied during the leaching process, while the grains of potassium feldspar, loparite, and aegirine were morphologically unchanged; the particles of lamprophyllite demonstrated the initial stage of dissolution.

Some secondary phases were found among the solid materials after the leaching of concentrate with $H_2SO_4$: the intergrowth of elongated gypsum and celestine crystals formed due to the interaction of $Ca^{2+}$ and $Sr^{2+}$ ions passing into the solution upon dissolution of lamprophyllite and the anionic group $[SO_4]^{2-}$ (Figure 2).

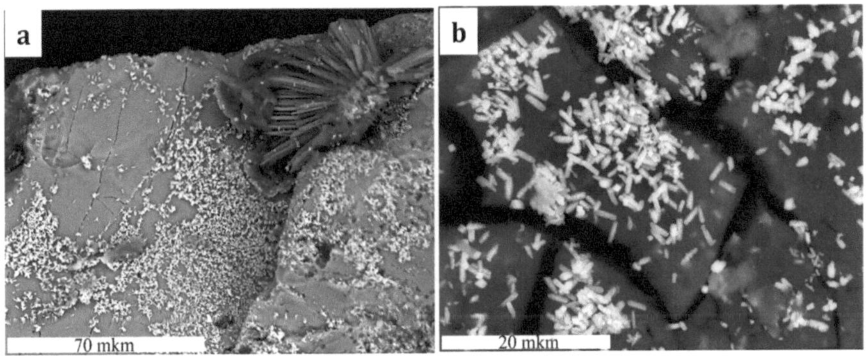


**Figure 2.** Secondary phases (celestite $SrSO_4$ (**a**) and gypsum $CaSO_4{\bullet}2H_2O$ (**b**)) in the residues of the leachate.

Most mineral particles after interaction with the $HNO_3$ solution were considerably changed and covered with a dense layer of redeposited leaching products. The exception was individual grains of aegirine, which retained their initial needle-prismatic habitus. In addition, the formation of secondary phases was not excluded. Apparently, they were represented by a thin mixture of silicate gels with various degrees of hydration, with captured micron particles of acid-resistant minerals (egirine, loparite, feldspars), hydrolysis products of aluminum salts, as well as various ions of rare, alkaline, and alkaline earth elements sorbed from solution and complex ionic groups, presumably $[Nb(H_2O)_6](OH)_2\}^{3+}$, $[Zr(NO_3)_6]_2$, $[Zr(OH)_2]^{2+}$, $[NaSiO_3]^{-}$, etc.

The influence of various inorganic acids (hydrochloric, sulfuric, and nitric) on the eudialyte surface state was also studied by IR spectroscopy. The IR spectra of the sample both initially and after the interaction with HCl, $H_2SO_4$, and $HNO_3$ are given in Figure 3. The analysis of the obtained results led us to the following conclusion: the most considerable effect on the structural and chemical properties of the mineral was exerted by the HCl solution. Figure 3 shows that the spectra profile in the region from 800 $cm^{-1}$ to 1350 $cm^{-1}$ significantly changed compared with that of the initial (untreated) sample. For the sample after the interaction with HCl solution. A change in the band symmetry with an obvious broadening was observed: the shoulder at 1135 $cm^{-1}$ transformed into a distinct band at 1145 $cm^{-1}$; the peaks at 980 $cm^{-1}$ and 1025 $cm^{-1}$ disappeared.

A lowering of the mentioned band peak area was also observed. In the initial sample, the peak area value was about 585 rel. units, while for the sample treated with HCl, it was about 132 rel. units, i.e., a change by a factor of 4.4.

The action of the aqueous solutions of $H_2SO_4$ and $HNO_3$ on eudialyte modified its surface to a lesser extent. The spectra in the region of Si-O-Si bond-stretching vibrations maintained the symmetry; the area under the spectra in the discussed range dropped from 585 rel. units (initial sample) to 355 rel. units and 350 rel. units, respectively.

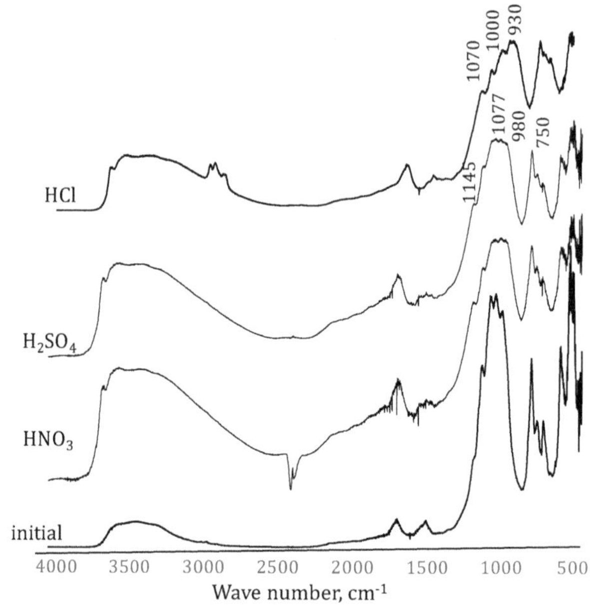


**Figure 3.** IR spectra of the initial eudialyte and of the sample after the interaction with HCl, $H_2SO_4$, and $HNO_3$.

The XPS results generally agreed with the IR spectroscopy data. The results of the XPS study of the elemental composition (at.%) of eudialyte particles after the interaction with the solutions of $HNO_3$, $H_2SO_4$, and HCl are presented in Table 6. The surface of the treated samples was depleted in metal content as a result of the intensive removal of cations from the surface into the leaching solution. The surface concentration of aluminum and sodium dropped by ~98%; calcium by 80%; magnesium, manganese, and potassium by 75%; titanium by 70%; and iron by 70%. The surface concentration of Si enhanced by 1.3–1.7 times.

**Table 6.** Atomic concentration (at.%) of eudialyte surface according to XPS data.

| ACID | Atomic Concentration, at.% | | | | | | | | | | | | | | | | ΣMe/Si |
|---|---|---|---|---|---|---|---|---|---|---|---|---|---|---|---|---|---|
| | C | N | O | S | F | Na | Mg | Ca | Al | Si | K | Ti | Mn | Fe | Sr | Zr | |
| No treatment | 4.7 | 0.6 | 60.0 | n.d. | 0.6 | 12.2 | 0.5 | 0.5 | 2.8 | 14.8 | 0.4 | 0.9 | 0.5 | 0.7 | 0.4 | 1.2 | 1.36 |
| $NHO_3$ | 4.8 | 1.9 | 66.2 | n.d. | 0.1 | 3.4 | 0.2 | 0.2 | 1.4 | 19.2 | 0.2 | 0.6 | 0.2 | 0.5 | 0.2 | 1.2 | 0.42 |
| $H_2SO_4$ | 2.5 | 0.3 | 69.2 | 2.2 | <0.1 | 0.9 | 0.1 | 0.1 | 0.1 | 23.1 | 0.1 | 0.3 | 0.2 | 0.2 | 0.5 | 0.1 | 0.11 |
| HCl | 3.9 | 0.2 | 70.2 | n.d. | 0.1 | 0.2 | <0.1 | <0.1 | <0.1 | 24.6 | n.d. | 0.1 | <0.1 | 0.1 | n.d.. | 0.4 | 0.05 |

Note: n.d.—not detected.

Through this process, the ratio of Me to Si significantly decreased, as shown in Table 6. The largest decrease in the Me/Si ratio was observed for the sample after the interaction with HCl (from 1.36 to 0.05), which confirmed that HCl was the most efficient agent for the leaching.

LEACH CAKES. The highest yield of metals from the eudialyte concentrate was observed when HCl was used as a leaching agent (Table 7). The ratio of ΣMe/Si in the cakes dropped by ~8 times (from ~1.7 to 0.2). When using the solutions of $HNO_3$ and $H_2SO_4$ as the solvent, the obtained ratio of ΣMe to Si in the cakes was about 0.3.

**Table 7.** XRF results of the concentrate before the treatment and after leaching with various acids.

| ACID | Chemical Composition, at.% | | | | | | | | | | | | | | | | |
|---|---|---|---|---|---|---|---|---|---|---|---|---|---|---|---|---|---|
| | $SO_3$ | $Na_2O$ | MgO | CaO | $Al_2O_3$ | $K_2O$ | $TiO_2$ | MnO | $Fe_2O_3$ | SrO | $ZrO_2$ | $CeO_2$ | $La_2O_3$ | $Nd_2O_3$ | $Y_2O_3$ | ΣMeO | $SiO_2$ | Me/Si |
| Without treatment | 0.20 | 10.1 | 0.18 | 5.42 | 6.9 | 1.37 | 3.29 | 1.66 | 5.2 | 2.6 | 11.5 | 0.67 | 0.29 | 0.32 | 0.46 | 50.16 | 44.96 | 1.14 |
| $NHO_3$ | <0.02 | 1.6 | 0.08 | 1.60 | 1.6 | 0.99 | 2.39 | 0.45 | 3.0 | 1.6 | 5.9 | 0.35 | 0.18 | 0.14 | 0.12 | 20.02 | 64.61 | 0.30 |
| $H_2SO_4$ | 2.39 | 2.0 | 0.07 | 2.54 | 1.7 | 1.06 | 2.96 | 0.51 | 3.2 | 2.7 | 4.1 | 0.60 | 0.30 | 0.22 | 0.17 | 24.52 | 63.62 | 0.35 |
| HCl | 0.02 | 0.8 | 0.04 | 1.11 | 1.1 | 0.72 | 2.05 | 0.27 | 2.4 | 1.0 | 3.7 | 0.27 | 0.12 | 0.10 | 0.08 | 13.78 | 70.57 | 0.20 |

It should be noted that in the cakes, in comparison with the initial concentrate, there was a significant increase in the concentration of gel, for example, $SiO_2$ from 44.96% to 64.61% ($HNO_3$), 63.62% ($H_2SO_4$), and 70.57% (HCl), which apparently caused minimal gelation when using HCl as the leaching agent (Table 7).

GEL. The effect of the acids on the content (vol.%) of the gel-like silicate was investigated. As shown, the use of $H_2SO_4$ resulted in a considerable volume of gel. The percentage of gel-like silicate was ~63%, 49%, and 48% for $H_2SO_4$, $HNO_3$, and HCl leaching agents, respectively. The centrifugation of the gel-like silicate reduce the volume of the gel by 1.5 times.

Regarding the occurrence form of the valuable components in the silicate gel, it was found that during sulfuric acid leaching, the loss of REEs with the gel was due to the REE sulfate phases precipitating from the pregnant solution (Figure 4) as well as the precipitation of CaTi $(SO_4)_3$, whose crystals precipitate in a cubic or pseudocubic form (Figure 5).

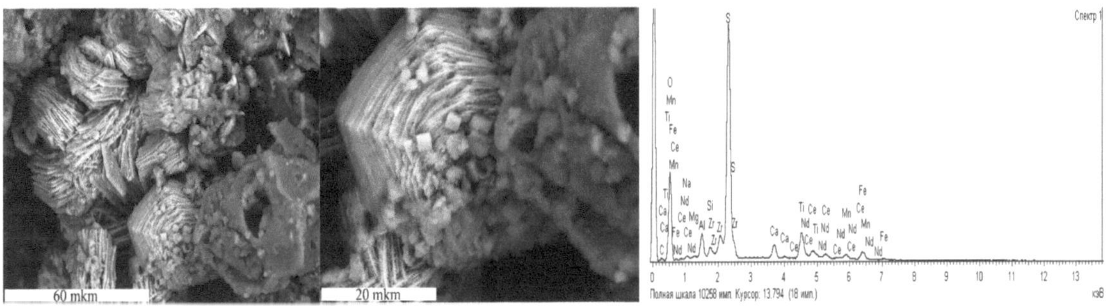


**Figure 4.** Secondary crystalline phases captured into gel-like silicate after leaching with $H_2SO_4$ solution: intergrowths and single crystals of sulfate of rare earth element $REE_2(SO_4)_3$.

**Figure 5.** Cubic or pseudocubic crystals of $CaTi(SO_4)_3$.

DISTRIBUTION OF REEs TO THE LEACHING PRODUCTS. The usage of $H_2SO_4$ provided the most effective leaching of the concentrate: the Zr recovery was about 92%, the Σ REE recovery was about 82%, and the loss of mass was about 78%. In the cases of $HNO_3$

and HCl, the recovery of Zr was 76% and 84%, respectively, and the Σ REE recovery was 79% and 83%, respectively (Table 8).

**Table 8.** Recovery of Zr and REE and losses with silicate gel under various leaching conditions and their concentrations in the pregnant solutions after the gel-like phase separation.

| ACID | REE Recovery, % | | Zr Recovery, % | | Zr Loss with Gel, % | ΣREE Loss with the Gel,% | $C_{Zr}$ in Pregnant Solution, g/dm³ | $C_{\Sigma REE}$ in Pregnant Solution, mg/dm³ |
|---|---|---|---|---|---|---|---|---|
| | Gel | Solution | Gel | Solution | | | | |
| $H_2SO_4$ | 70.22 | 12.34 | 26.90 | 64.65 | 30.0 | 85.0 | 4.50 | 221.6 |
| HCl | 18.76 | 64.74 | 8.34 | 75.57 | 10.0 | 23.0 | 5.00 | 999.3 |
| $HNO_3$ | 44.32 | 35.38 | 41.04 | 35.83 | 54.0 | 56.0 | 2.40 | 494.2 |
| $HNO_3$(US treatment) | 29.39 | 54.52 | 29.39 | 57.87 | 32.25 | 35.03 | 3.10 | 637.1 |

However, sulfuric acid leaching provided the most significant losses of valuable components with gel-like silicate: the losses of Zr and Σ REE were about 30% and 85%, respectively, while in the $HNO_3$ and HCl solutions, the losses of Zr and REE were about 54% and 56%, and 10% and 23%, respectively (Table 8). At the same time, it should be noted that the US treatment of the mineral pulp produced an increase in the recovery and a drop in the loss of valuable components. The increase in the REE recovery was about 4% and the recovery of Zr rose to nearly 10% (Table 8) compared with those of the experiment without US treatment (Table 8). The reduction in the loss of Zr and REEs with the gel -like silicate was ~21% (from 54.0% to 32.25%).

Based on the above, in purified pregnant solutions (after removing the silicate gel), the maximum concentration of Zr and REE cations was observed with HCl leaching, at ~5.04 and 1.0 g/dm$^3$, respectively (Table 8).

However, despite the good indicators of the leaching with the usage of HCl, this technology has no future in industrial processes for many practical reasons. This is why $HNO_3$ was chosen as the leaching agent in further studies.

### 3.2. *Effect of Preliminary EC and US Treatments on the Nitric Acid Leaching*

The utilization of various kinds of energy [20,26] is the alternative way of the hydrometallurgical process intensification. In this study, the effect of energy (US and EC) treatments and their combination on the efficiency of eudialyte concentrate leaching was examined.

EFFECT OF EC AND US TREATMENTS ON EUDIALYTE SURFACE EC treatment led to a significant reduction in the areas of particles covered by gel-like silica (Figure 6a,b) and to an increase in the areas composed of parallel teeth. The trigonal symmetry of the parallel-oriented teeth inherited the symmetry of the three- and nine-membered rings of silicon–oxygen tetrahedra located along the c axis of the eudialyte crystal structure. This indicated that leaching processes proceeded most intensively along this direction. The qualitative elemental composition of these regions was close to that of the original eudialyte, but the content of the rare earth elements significantly decreased and became below the detection limit of the energy-dispersive spectrometer.

The eudialyte particles after US treatment during the leaching demonstrated intensive surface fracturing until complete breakdown and the absence of silicate gel deposition (Figure 6c).

The results clearly demonstrated that the mineral surface layer was noticeably depleted in Al, Na, and Mg (% at.). The surface concentration of Al, Na, and Mg decreased by 2.0–4.5, 3.0–5.5, and 2.5–5.0 times, respectively. These data support the conclusion that the concentrations of Al, Na, and Mg in a sequence of studied samples consistently decreased. This shows that US treatment and its combination with EC treatment is more effective for breaking down eudialyte silicate matrices.

**Figure 6.** Surface of eudialyte grain (**a**) and area with serrate morphology (**b**) after leaching with EC treatment. Eudialyte grains after acid leaching with the application of US treatment (**c**).

Table 9 presents XPS results obtained for the initial (untreated) eudialyte sample and for the samples after the various treatments.

**Table 9.** XPS results of the eudialyte surface layer.

| TREATMENT | Atomic Concentration, at.% | | | | | | | | | | | | | |
|---|---|---|---|---|---|---|---|---|---|---|---|---|---|---|
| | C | N | O | F | *Na* | *Mg* | *Al* | Si | K | Ti | Mn | Fe | Sr | Zr |
| 1. None | 4.7 | 0.6 | 60.0 | 0.6 | *12.2* | *0.5* | *2.8* | **14.8** | 0.4 | 0.9 | 0.5 | 0.7 | 0.4 | 1.2 |
| 2. $HNO_3$ | 4.8 | 1.9 | 66.2 | 0.1 | *3.4* | *0.2* | *1.4* | **19.2** | 0.2 | 0.6 | 0.2 | 0.5 | 0.2 | 1.2 |
| 3. $HNO_3$ + EC treat.—200 $A/m^2$ | 6.2 | 1.9 | 63.8 | 0.2 | *4.1* | *0.2* | *0.9* | **18.3** | 0.2 | 1.3 | 0.5 | 0.5 | 0.3 | 1.2 |
| 4. $HNO_3$ + EC treat. 200 $A/m^2$ + US 20 kHz | 6.1 | 0.9 | 66.6 | 0.1 | *2.2* | *0.2* | *0.6* | **21.0** | 0.1 | 0.6 | 0.2 | 0.3 | 0.2 | 1.0 |
| 5. $HNO_3$ + US 20 kHz | 4.9 | 1.2 | 67.7 | 0.2 | *2.4* | *0.1* | *0.7* | **20.2** | 0.1 | 0.7 | 0.2 | 0.5 | 0.1 | 1.1 |

The XPS and IR spectroscopy [24,25] data indicated that the eudialyte decomposition mechanism during the interaction with the leaching agent could be explained by the following processes: The first is associated with the selective extraction of alumina, sodium, and manganese. The second is associated with the breakdown of the silicate structure. At the same time, the concentration (at.%) of metal cations reduced, and the destruction of the silicate matrix was accompanied by the transformation of the silicate matrix into a gel-like phase. US and EC treatments significantly promoted the breakdown of the silicate matrix and improved the efficiency of eudialyte leaching.

### 3.3. *Validation of the Optimized Process Modes and Process Flows for Nitric Acid Leaching of Eudialyte Concentrate*

Special methods have been developed for identifying the dominant factors affecting various processes, which are called screening experiments. A mathematical model of a screening experiment with the preparation of a Plackett–Berman design was used in the study for the separation of the factors that insignificantly influenced the recovery of the valuable components with a minimum number of experiments, while ensuring the reliability of the results obtained (Table 10). The value of Zr recovery from the pregnant solution during the leaching of eudialyte concentrate was taken as a response. The influencing parameters ($x_n$) were: ultrasonic treatment ($x_1$), mineral pulp temperature in the leaching process ($x_2$), the concentration of nitric acid ($x_3$), electrochemical treatment ($x_4$), and leaching duration ($x_5$) (Table 10).

**Table 10.** Screening experiment plan.

| Factor | UStreat. | T, °C | $C_{HNO3}$, g/dm$^3$ | EC Treat. | t, h | Fictive Factors | | | | $y$,% |
|---|---|---|---|---|---|---|---|---|---|---|
| Base level | | 60 | 450 | | 2 | | | | | |
| Variability interval | | 20 | 100 | | 1 | | | | | |
| Lower level (−) | no | 40 | 350 | no | 1 | | | | | |
| Higher level (+) | yes | 80 | 550 | yes | 3 | | | | | |
| code | $x_1$ | $x_2$ | $x_3$ | $x_4$ | $x_5$ | $x_6$ | $x_7$ | $x_8$ | $x_9$ | $y$ |
| 1 | + | + | + | + | − | − | − | + | − | 64.0 |
| 2 | − | − | + | + | + | − | − | − | + | 54.0 |
| 3 | + | + | − | + | + | + | − | − | − | 93.0 |
| 4 | − | + | + | − | + | + | + | − | − | 55.0 |
| 5 | − | − | + | + | − | + | + | + | − | 25.0 |
| 6 | − | + | − | + | + | − | + | + | + | 46.0 |
| 7 | + | − | + | − | + | + | − | + | + | 56.0 |
| 8 | + | − | − | + | − | + | + | − | + | 36.0 |
| 9 | + | − | − | − | + | − | + | + | − | 65.0 |
| 10 | − | + | − | − | − | + | − | + | + | 54.0 |
| 11 | + | + | + | − | − | − | + | − | + | 68.0 |
| 12 | − | − | − | − | − | − | − | − | − | 12.5 |
| $b_i$ | ***11.29*** | ***10.96*** | 1.29 | 0.62 | ***9.12*** | 0.79 | −3.21 | −0.71 | −0.045 | |

The most valuable coefficients were determined: $S_y^2 = 29.91$, $S_{b_i}^2 = 2.8612$, and $\Delta_{b_i} = 4.702$, from the results of the 12 experiments [27–30].

The analysis of the obtained results proved that the major parameters influencing recovery were US treatment, temperature of the pulp, and duration of the process.

The experimental studies of the influencing parameters were performed for the selection of the optimal parameters. The analysis of the obtained results showed that the correlation between the Zr recovery and US treatment followed a simple equation. The most significant values were achieved during the stable US treatment of the pulp. This experiment condition improved the Zr recovery by 2.8 times. This increase in the Zr recovery under the US treatment of the pulp could be explained by the fracturing of the eudialyte grains [27]. US treatment inhibited the formation of gel-like silica on the grains surfaces.

The results in Table 11 illustrate the influence of the leaching stage number (stage duration was 1 h) on Zr and REEs recovery. Two stages of the process increased the Zr recovery by approximately 10% and the REE recovery by 2%. In this case, the volume of gel-like silicate increased by 11% (from ~55% to ~66%), which led to increases in the Zr and REEs loss by ~10% and ~6%, respectively (Table 11, lines 1 and 3). A further increase in the number of stages from two to three led to the growth in the Zr recovery by 4.5% and in the REEs recovery by 2% (Table 11, lines 3–4). The total volume of gel-like silicate increased to 70% and, as a consequence, the loss of Zr and REE of with three stages of leaching was 60%.

The decreases in REEs and Zr losses with by the lowering of the duration of the first and second leaching stages with an increase in the duration of the third stage (while the total process duration was unchanged) were investigated. Reducing the durations of the first and second stages and the lengthening of the third-stage during provided gains in the Zr recovery by approximately 5% (Table 11, lines 4 and 5) and in REE recovery by 10% (Table 11, lines 4 and 5).

**Table 11.** Effect of the number and duration of leaching stages on the recovery of valuable components.

| N | Acid/Treatment | n | t, min | ε,% | | ν,% | |
|---|---|---|---|---|---|---|---|
| | | | | Zr | REE | Zr | REE |
| 1 | $HNO_3$ | 1 | 60 | 76.9 | 79.5 | 41 | 44.2 |
| 2 | $HNO_3$ (US treatment) | 1 | 60 | 87.3 | 83.5 | 32.3 | 35.0 |
| 3 | $HNO_3$ | 2 | 60/60 | 86.5 | 81.4 | 51.9 | 50.8 |
| 4 | $HNO_3$ | 3 | 60/60/60 | 91.1 | 83.2 | 60.3 | 59.9 |
| 5 | $HNO_3$ | 3 | 40/40/100 | 96.2 | 92.8 | 41.5 | 44.7 |
| 6 | $HNO_3$ (1st-stage US treatment) | 3 | 40/40/100 | 97.1 | 94.5 | 31.4 | 33.8 |
| 7 | $HNO_3$ ($H_2SO_4$ at 3rd stage) | 3 | 40/40/100 | 95.8 | 91 | 41.5 | 44.7 |

Note: n—number of the stages; t—stage duration; ε—recovery (total); ν—loss of valuable component with gel.

At the same time, the volume of the gel-like silica insignificantly increased, while the loss of the valuable components was the same as that of the one-stage leaching process (Table 11, lines 1 and 5). The application of US treatment during the first leaching stage increased the total Zr and REEs recovery by 1% and approximately 1.5%, respectively (Table 11, lines 5 and 6), and reduced the volume of gel-like silica, which led to a drop in the loss of valuable components by 1.3 times.

The use of $H_2SO_4$ acid during the third leaching stage had a little impact on the leaching efficiency: Zr and REEs recovery slightly dropped (Table 11, line 7).

Thus, a three-stage leaching process is recommended with the following conditions: duration of the first stage, 40 min, second stage, 40 min; and third stage, 100 mi;, nitric acid concentration, 450 g/L; mineral pulp temperature, 80 °C; S:L ratio, 1:20; continuous US treatment of the mineral pulp in the first stage of leaching. The application of the mentioned procedure enabled a Zr recovery of 97.1% and a REE recovery of 94.5%

The technological scheme of eudialyte concentrate nitric acid leaching is depicted in Figure 7.

### *3.4. Recovery of Valuable Components from Pregnant Eudialyte Acid Leach Solution*

#### 3.4.1. Recovery by Chemical Precipitation

The process of neutralizing the pregnant solution to pH 4.0 using calcium carbonate is accompanied by the chemical precipitation of zirconium and metal impurities [31,32]. The precipitate forms hydroxides (zirconium aluminum, and iron), carbonates, and hydroxides (REEs). The real solution has high mineralization, which is why the actual processes and chemical reactions may be more complicated than the theoretical reactions and require additional study.

Figure 8 a shows the dependence of the recoveries of Zr, Al, and Fe into a solid residue as a function of pH at 20 °C.

The efficiency of element precipitation increase with the rise in pH. At pH 4.0, the recovery of Zr and Al was about of 100% and that of Fe was 40%. At the same time, recovery of sodium, manganese, strontium, titan and potassium was 15–35% and REEs loss was ~9%. The sharp increase in REE loss up to 40.0 % (due to co-precipitation with zirconium) was observed during the further pH rising. At that time, the $CaCO_3$ consumption raised from 375 to 600 $g/dm^3$ due to the growth in solubility.

The separation of Zr and REEs must take place at pH 4.0–4.2 in order to achieve entire recovery of Zr, Al, and Fe. Further chemical precipitation of REE compounds should be carried out using $Na_2CO_3$ and NaOH. The obtained results indicate the highest REEs recovery (96%) was achieved at a $Na_2CO_3$ consumption of 6 $g/dm^3$ (pH 6.0). With NaOH application, the highest REEs recovery was 59%.

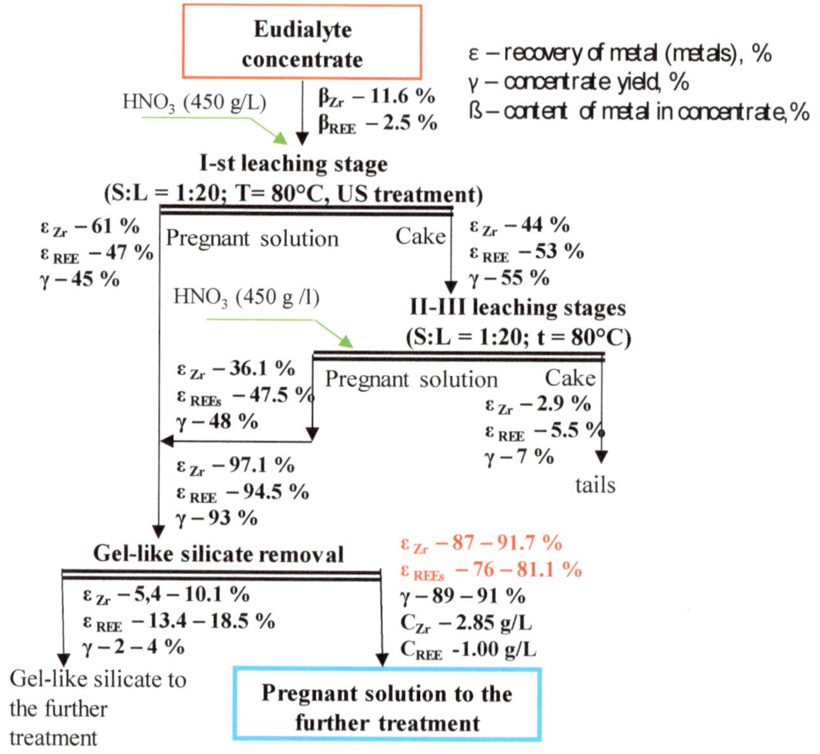


**Figure 7.** The developed scheme of eudialyte concentrate leaching.

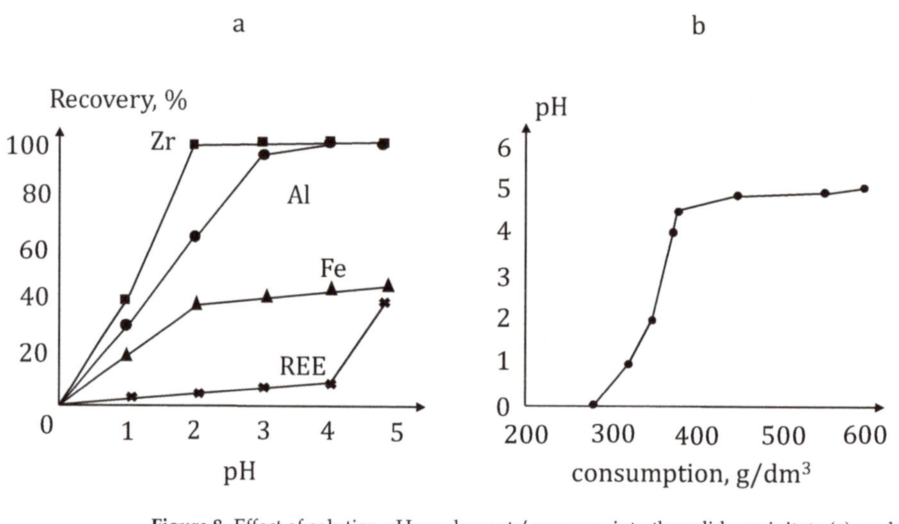


**Figure 8.** Effect of solution pH on elements' recovery into the solid precipitate (**a**), and the correlation of $CaCO_3$ consumption and pH solution at 20 °C (**b**).

Small rounded particles were found among the main mass of solid precipitate. They contained rare earth elements of the cerium group (La, Ce, and Nd), yttrium, and synthetic hydroxyl analogs of bastnasite-group minerals: hydroxylbastnesite (Ce, La, Y)$CO_3$(OH, F) or hydroxylsynchysite CaREE$(CO_3)_2$OH. At the same time, we could not exclude the deposition of aqueous rare earth carbonates calcinsite $(Ce, La)_2(CO_3)_3{\cdot}4H_2O$, lanthanite $(Ce, La, Nd)_2(CO_3)_3{\cdot}8H_2O$, or calcioanquilite $CeCa(CO_3)_2(OH){\cdot}H_2O$. Because of the micron size of these particles, it was impossible to correctly determine their composition.

It was determined that the Zr recovery was independent of the initial concentration; it was 99.6%–99.9%. However, the increase in the initial REE concentration from 150 to 1200 mg/dm$^3$ led to increased loss with the residue from 0.3% to 8.5% due to the coprecipitation of rare earth elements with impurities [33,34].

#### 3.4.2. Sorption Method for the Recovery of REEs and Zirconium from Pregnant Eudialyte Acid Leach Solution

Three types of sorbents were used in this study: a strongly basic anion exchange resin (Dowex 1 × −8 and Purolite PCA 433), a strongly basic cation resin (Purolite C 160), and a chelate resin (Puromet MTS 9500). As expected, the two grades of highly basic anion resins produced very similar results. Although they differed somewhat due to the fact that Purolite PCA 433 carries a quaternary ammonium base as a functional group, which is known to have more pronounced basic properties than its tertiary forms. Figures 9 and 10 show the concentration curves obtained by the elution of a pregnant acid leach solution of eudialyte through the strongly basic polystyrene anion resins Purolite PCA 433 and Dowex 1 × −8.

The analysis of the obtained experimental data and the calculation of the material balance when using Purolite PCA 433 showed a sorption of 63.5% of Σ REE into the pregnant solution of the following elements: cerium (79.8%), lanthanum (67.5%), neodymium (99.5%), and praseodymium (81.1%) when eluate fractions 22 to 30 (0.88 to 1.20 of the column volume) were selected.

It was found that when eluting with water, the recovery of zirconium, lanthanum, cerium, and praseodymium from the resin was about 30%, 68%, 80%, and 20%, respectively. Thus, it was found that approximately 70% zirconium, 32% lanthanum, 20% cerium, and 80% praseodymium remained associated with the sorbent and could not be recovered in the reverse experiment (Table 12, column 1).

In order to desorb these elements, the sorption column was washed with 10 column volumes of the following solutions: 0.1 N and 0.5 N hydrochloric acid; 0.1 N, 0.5 N, and 8 N nitric acid; and 0.1 N sodium hydroxide at a flow rate of 1.0 mL/min. The results obtained are presented in Table 12, from which it can be seen that when washed with 0.1 N $HNO_3$, the recovery of zirconium increased by 5.7% (from 29.9% to 35.6%); those of lanthanum, cerium, and praseodymium increased by 2.1%–2.3%.

Elution with 1 N $HNO_3$ was found to perform better and made it possible to increase the recovery of zirconium from 29.9% to 51.9%, i.e., by 22%. At the same time, the recovery of lanthanum increased from 67.5% to 72.5%, that of cerium from 79.8% to 85.4%, and that of preseodymium from 81.1% to 83.5%, which led to the total recovery of REEs increasing to 73.7%.

Under the same conditions, when passing the refined REE solution through the strongly basic anion resin Purolite PCA 433, the recovery of REEs increased to 75.3% (Σ REE content) in the initial solution (75.8% Ce, 80.6% La, 64.5% Pr, and 46.5% Nd) and to 59.5% Zr. When passing the initial pregnant solution through a strongly basic anion exchange resin Dowex 1 × −8 in a single pass, 61.3% Σ REE was extracted into the bulk selective fraction: cerium (84.1%), lanthanum (87.9%), neodymium (44.1%), and praseodymium (28.5%), when selecting eluate fractions 22 to 30 (0.88 to 1.20 of the column volume). As can be seen from the data presented in the Figure 9, it is important that when using DOWEX X 8, in contrast to PCA433, zirconium is recovered along with heavy REEs and alkaline earth metals in the first three-quarters of the column volume.

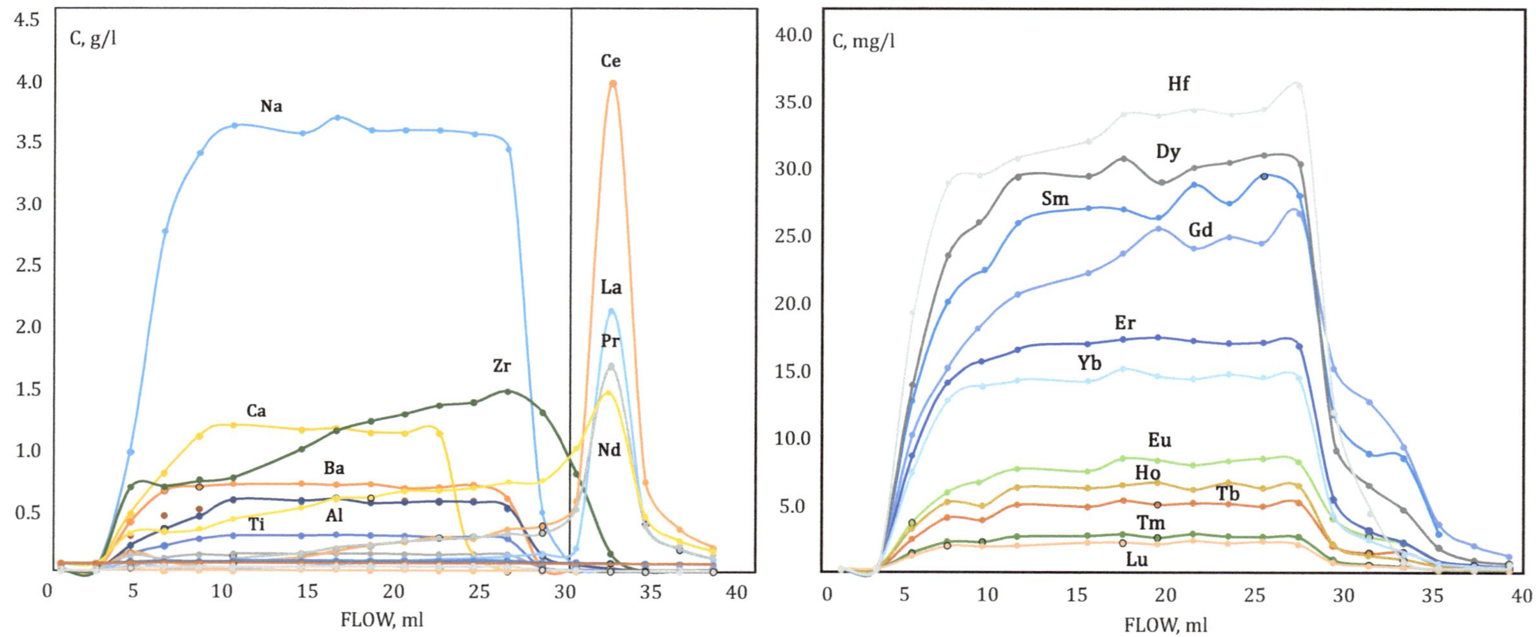


**Figure 9.** Pregnant solution elution curves through Dowex 1 × −8 sorbent.

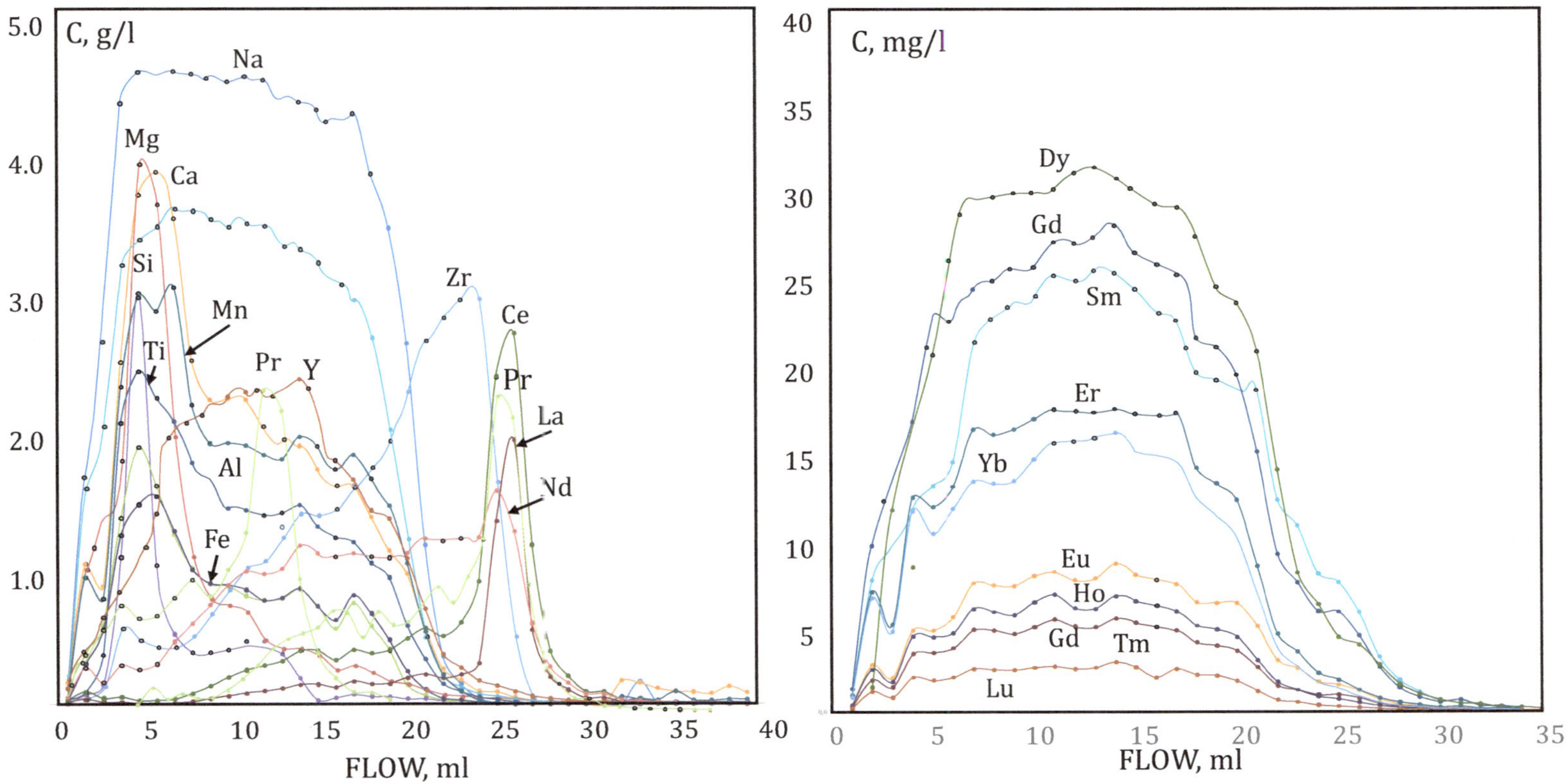


**Figure 10.** Refined REE solution elut:on curves through Purolite PCA-433.

**Table 12.** REE and Zr recovery with elution with various eluents.

| | Recovery in $H_2O_{distill.}$, % | Recovery in 0.1 N $HNO_3$, % | Recovery in 1.0 N $HNO_3$, % |
|---|---|---|---|
| Zr | 29.9 | 35.6 | 51.9 |
| La | 67.5 | 69.7 | 72.5 |
| Ce | 79.8 | 82.2 | 85.4 |
| Pr | 81.1 | 83.2 | 84.6 |
| Nd | 99.5 | 99.5 | 99.5 |

Passing a refined REE solution through a sorbent improved the process performance and made it possible to recover 70% of Σ REE in the solution in the bulk selective fraction due to an increase in the recovery of cerium by 1.1% (from 86.9 to 97.9), lanthanum by 9.1% (from 75.0 to 84.1), praseodymium by 9.3% (from 57.1% to 66.4%), and neodymium by 8.8% (from 35.3% to 44.1%), with additional recoveries of 28.5% of europium, 18.1% of samarium, and 35% of holmium and lutetium.

The analysis of the experimental data obtained for the strongly acidic cation exchanger Purolite C 160 and the chelate resin Puromet MTS 9500 revealed the priority of the sorption of total REEs. The recovery on the cation exchanger C 160 was 95% REEs, 83% Zr, 72% Sr, 80% Fe, 85% Mn, 83% Ca, 55% Al, and 67% Na. For the Puromet MTS 9500 sorbent, the recovery of total REEs and Zr into the resin was 99%; that of the associated elements ranged from 20% to 95% (Na 21%, Al 95%, Ca 28%, Mn 38%, Fe 95%, and Sr 17%).

Because the use of Puromet MTS 9500 made it possible to achieve higher recoveries of valuable components into the resin, while reducing the sorption of accompanying impurities, further research was aimed at finding effective desorbing agents (eluents) for the selective recovery of REEs and other sorbed elements.

During desorption, the feed method and feed rate of eluents were similar to those used in the process of feeding solutions to the sorbent; the volume of the passed desorption solution was 5–10 times resin volume (column volume). The list of desorption solutions is given in Table 13.

**Table 13.** Eluents and molarities.

| Compound | Concentration, mol/L | | | | | | | |
|---|---|---|---|---|---|---|---|---|
| $H_2SO_4$ | 0.1 | 0.25 | 0.5 | 1.0 | 2.0 | 5.0 | 10.0 | |
| $HNO_3$ | 0.25 | 0.5 | 1.0 | 2.0 | 3.0 | 5.0 | 6.0 | 8.0 |
| HCl | 0.25 | 0.5 | 1.0 | 2.0 | 4.0 | 6.0 | | |
| $NH_4HF_2$ | 0.25 | 0.5 | 1.0 | 2.0 | 3.0 | 5.0 | | |
| EDTA | 0.125 (pH 5.8) | 0.125 (pH 4.5) | 0.125 (pH 4.0) | | | | | |
| Citric acid | 0.25 | 0.4 | 0.8 | 1.5 | 2.0 | | | |
| Oxalic acid | 0.20 | 0.4 | 0.8 | 1.5 | 2.0 | | | |

From our analysis of the experimental data (Figure 11) and calculation of the material balances, we drew the following histograms regarding the performance of the eluents used. Citric acid and EDTA showed poor performance in the process of desorption of the elements adsorbed by Puromet MTS 9500. The use of a 0.25 M solution of ammonium bifluoride ($NH_4HF_2$) made it possible to recover 98.9% of the zirconium, without any loss of REEs.

The use of 0.2 M oxalic acid as an eluate made it possible to recover 97.2% of the Na, while the concentration of rare earth elements in the eluate was below the detection threshold with the 0.05 ppm method, in other words, without the loss of REEs.

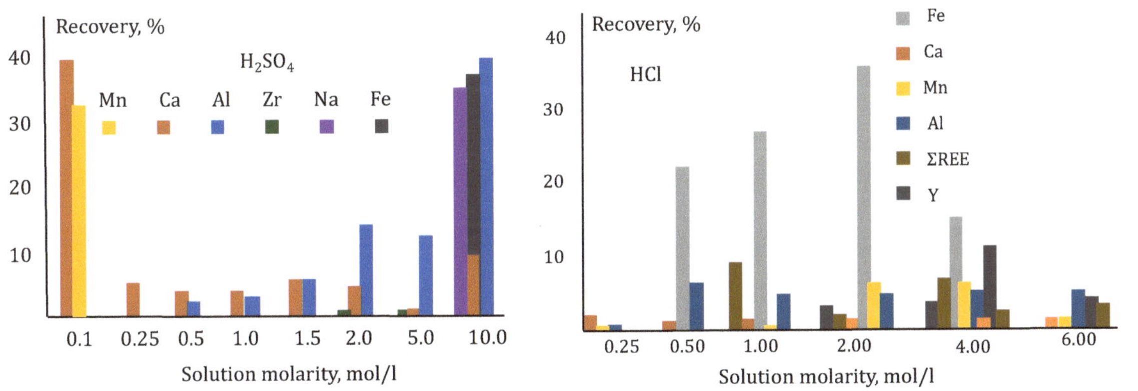


**Figure 11.** Recovery of the metals in sulfuric and hydrochloric acid (150 mL; %).

Sulfuric acid ($H_2SO_4$) at a concentration of 0.1 mol/L was an effective elution solution for Ca and Mn: when passing five column volumes (150 mL), 34% and 33% passed into the solution, respectively. Under the same conditions, sulfuric acid at a concentration of 10 mol/L caused the transition into the solution of 35% Na, 39% Ca, 37% Fe, and 38% Al. At the same time, the loss of REEs with the eluate of a 0.1 M $H_2$ $SO_4$ solution was 0.5% adsorbed on resin and 1.7% in the eluate of a 10 M solution (Figure 11).

Hydrochloric acid at concentrations of 0.5–2.0 mol/L, when passing fives column volumes (150 mL), caused the transition to the elution solution of 22.54% to 36.3% Fe; at the same time, 2.31% to 9.29% of the REEs passed into the eluate (Figure 11b).

Finally, concentrated (5 and 8 M) solutions of nitric acid ($HNO_3$) were found to be effective eluting agents for REEs. The 8 M solution performed better: when passing five column volumes, a recovery of 83.8% REEs was achieved; for 5 M, the recovery was 9.8% REEs. At the same time, 13.5% Al and 8.5% Ca passed into the 10 M nitric acid. The composition and concentration of REEs in the final product of REE desorption are shown in Table 14.

**Table 14.** REE concentration in the effluent (8 M $HNO_3$).

| La | Ce | Pr | Nd | Sm | Eu | Gd | Tb | Dy | Ho | Er | Tm | Yb | Lu | ΣREE |
|---|---|---|---|---|---|---|---|---|---|---|---|---|---|---|
| | | | | | | | mg/L | | | | | | | |
| 90.54 | 218.03 | 26.88 | 78.33 | 24.20 | 7.99 | 27.85 | 5.04 | 28.14 | 5.87 | 15.77 | 2.39 | 13.57 | 1.92 | 715.4 |

Based on the data obtained, two processes for the selective recovery of Zr and REEs from pregnant solutions of eudialyte concentrate leaching were proposed: sorption and precipitation–sorption (Figure 12).

The sorption extraction process involved feeding the pregnant solution in its native form to the Puromet MTS 9500 sorbent and the subsequent recovery from the sorbent by stepwise elution with mineral acid solutions (10–15 column volumes each, which ensured 100% removal of impurities), first of sodium; second of calcium, manganese, iron, and aluminum; third of REEs; and, finally, of zirconium into a solution of ammonium bifluoride.

The precipitation–sorption process involves sequential alkalization of the initial pregnant solution with calcium carbonate in order to isolate the zirconium product (pH 4) and REE carbonates with further introduction of sodium carbonate to pH 6.1. In order to remove impurities (Fe, Al, Ca, etc.), the resulting REE carbonates are proposed to be dissolved in 50% nitric acid and then REEs can be recovered into sorbents, followed by the selection of impurities and valuable REEs by elution with mineral acid solutions.

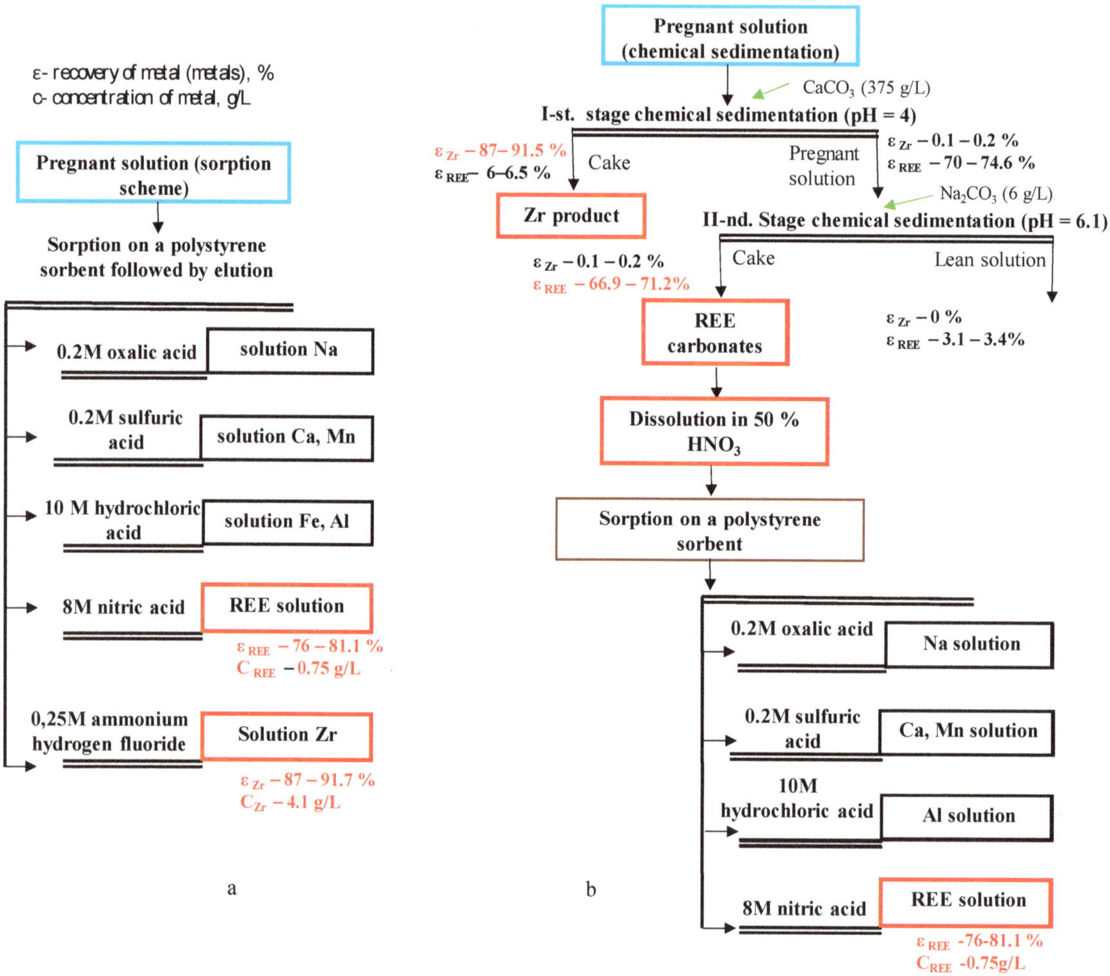


**Figure 12.** Schemes of the REE recovery from pregnant leaching solution through sorption (**a**) and precipita-tion-sorption schemes (**b**).

## 4. Conclusions

The mechanism of the effect of various acids on the structural and chemical properties and dissolution of eudialyte concentrate was revealed. It was determined that during the interaction of eudialyte with HCl and $H_2SO_4$, the mineral surface is described by the appearance of a series of deep, parallel-oriented fractures because of the predominance of the normal dissolution rate. When interacting with $HNO_3$, the tangential (layered) mechanism of dissolution dominates.

In the nitrogen–sulfuric–hydrochloric acid sequence, a consistent decrease in the content (at. %) on the eudialyte surface was identified for Al by 90%; Na by 98.4%; for Ca, Mg, and Mn by 75%–80%; for Ti by 88%, and for Zr by approximately 92%. The phase composition of the silicate gel for all acids was characterized by Zr compounds, REEs, and nanofragments of the main concentrate minerals. According to the silicate gel formation efficiency and, accordingly, Zr and REE loss, the acids could be ranked as follows: sulfuric acid, nitric acid, and hydrochloric acid.

The most promising leaching results were obtained with HCl and $HNO_3$, which provided comparable qualitative and quantitative process indicators. Because the application of HCl is difficult because of many practical reasons, it cannot be recommended for industrial applications of the eudialyte concentrate leaching process.

The influence of combined energy treatment (thermal, electrochemical, and ultrasonic treatment) was determined. It consisted of dispersing the colloidal silicate gel, preventing the formation of its precipitates on the surface of mineral particles, and the formation of numerous defects and microcracks on the surface of eudialyte grains up to breakage during the ultrasonic treatment. It led to increases in the intensity and rate of the leaching process.

The greatest intensifying effect on the process of concentrate acid leaching was exerted by ultrasonic treatment of the mineral pulp. It enabled the highest Zr and REE recovery at a coarser feed size, lower slurry temperatures, and shorter leaching duration, as well as a reduction in REE loss to the gel by a factor of 2–2.5.

A highly efficient leaching process was developed for eudialyte concentrate, the practical implementation of which ensures a Zr recovery of 97.1% and a REE recovery of 94.5%, with significant reductions in silicate gel-like formation and valuable components loss by 21.0%, as well as increases in Zr and REE recovery by 10 % and 4%, respectively.

Our study demonstrated the high performance of Zr and REEs chemical precipitation. It was established that an increase in the pH of the initial pregnant solution during the first stage of precipitation with $CaCO_3$ to 4.0 provided 99.95% Zr recovery into the solid phase and the associated recoveries of impurities: 100% of Al and 43.8% of Fe. The total recovery of REEs into the solid phase was as high as 74.6%, and the total loss of REEs with the product of the precipitation was 9%. A further increase in the pH of the solution using $Na_2CO_3$ to 6.1 provided an REE recovery into the solid phase of 87.5%, with a total recovery from solution of 95.5%.

It was found that the use of the Puromet chelate resin MTS 9500 as a sorbent provided the highest (99%) recovery of REEs and Zr from the productive solutions. As a result of the subsequent stepwise gradient elution of the saturated sorbent, the following results were obtained: a Zr solution with a concentration of 4.1 g/L (Zr recovery into the solution was as high as 91.7%) and a REE solution with a concentration of 0.75 g/L (REE recovery 81.1%).

A process for the selective recovery of Zr and REEs from the pregnant leaching solutions of eudialyte concentrate based on a combination of chemical precipitation, sorption, and elution methods was proposed. Two process flows for treating pregnant solutions of eudialyte acid leaching were developed: sorption and precipitation–sorption.

**Author Contributions:** General scientific supervision of work V.A.C.; formulation of the research goals and aims, data analysis, and paper preparation; performing experiments (leaching), methodology development, IR, SEM, XPS results, data analysis, and preparation of the initial draft V.G.M., A.L.S., M.V.R. and E.V.K. All authors have read and agreed to the published version of the manuscript.

**Funding:** The research was carried out with the financial support of a project of the Russian Federation represented by the Ministry of Science and Higher Education of the Russian Federation No. 13.1902.21.0018 (agreement 075-15-2020-802).

**Data Availability Statement:** Not applicable.

**Acknowledgments:** Research was performed using the equipment of the NRC "Kurchatov Institute" Shared Knowledge Center.

**Conflicts of Interest:** The authors declare no conflict of interest.

## References

1. Vasilevskaya, D.V.; Saparov, S.M. *Mineral Resource Complex in the Period of Uncertainty and New Challenges: Results, Trends, Prospects*; Rusajns: Moscow, Russia, 2022; 130p.
2. Abaka-Wood, G.B.; Johnson, B.; Addai-Mensah, J.; Skinner, W. Recovery of Rare Earth Elements Minerals in Complex Low-Grade Saprolite Ore by Froth Flotation. *Minerals* **2022**, *12*, 1138. [CrossRef]
3. Jyothi, R.K.; Thenepalli, T.; Ahn, J.W.; Parhi, P.K.; Chung, K.W.; Lee, J.Y. Review of rare earth elements recovery from secondary resources for clean energy technologies: Grand opportunities to create wealth from waste. *J. Clean. Prod.* **2020**, *267*, 122048. [CrossRef]

4. Gaustad, G.; Williams, E.; Leader, A. Rare earth metals from secondary sources: Review of potential supply from waste and byproducts. *Resour. Conserv. Recycl.* **2021**, *167*, 105213. [CrossRef]
5. Abaka-Wood, G.B.; Addai-Mensah, J.; Skinner, W. The concentration of rare earth elements from coal fly ash. *J. South. Afr. Inst. Min. Metall.* **2022**, *122*, 21–28. [CrossRef]
6. Patil, A.B.; Tarik, M.; Struis, R.P.; Ludwig, C. Exploiting end-of-life lamps fluorescent powder e-waste as a secondary resource for critical rare earth metals. *Resour. Conserv. Recycl.* **2021**, *164*, 105153. [CrossRef]
7. Ait Brahim, J.; Ait Hak, S.; Achiou, B.; Boulif, R.; Beniazza, R.; Benhida, R. Kinetics and mechanisms of leaching of rare earth elements from secondary resources. *Miner. Eng.* **2022**, *177*, 107351. [CrossRef]
8. Binnemans, K.; Jones, P.T.; Blanpain, B.; Van Gerven, T.; Pontikes, Y. Towards zero-waste valorisation of rare-earth-containing industrial process residues: A critical review. *J. Clean. Prod.* **2015**, *99*, 17–38. [CrossRef]
9. Wang, N.; Sun, X.; Zhao, Q.; Yang, Y.; Wang, P. Leachability and adverse effects of coal fly ash: A review. *J. Hazard. Mater.* **2020**, *396*, 122725. [CrossRef]
10. Blissett, R.S.; Smalley, N.; Rowson, N.A. An investigation into six coal fly ashes from the United Kingdom and Poland to evaluate rare earth element content. *Fuel* **2014**, *119*, 236–239. [CrossRef]
11. Seredin, V.V. Rare earth element-bearing coals from the Russian Far East deposits. *Int. J. Coal Geol.* **1996**, *30*, 101–129. [CrossRef]
12. Savel'eva, I.L. The rare-earth metals industry of Russia: Present status, resource conditions of development. *Geogr. Nat. Resour.* **2011**, *32*, 65–71. [CrossRef]
13. Kuleshevich, L.V.; Dmitrieva, A.V. Rare-earth mineralization in Karelia's alkaline and moderately alkaline complexes, associated metasomatic rocks and ores. *Gorn. Zhurnal* **2019**, *3*, 45–50. [CrossRef]
14. Zakharov, V.I.; Skiba, G.S.; Solovyov, A.V.; Lebedev, V.N.; Mayorov, D.V. Some aspects of acid treatment of eudialyte. *TSvetnye Met.* **2011**, *11*, 25–29.
15. Lebedev, V.N. Sulfuric acid technology of eudialyte concentrate. *Russ. J. Appl. Chem.* **2003**, *76*, 1559–1563. [CrossRef]
16. Lebedev, V.N.; Shchur, T.E.; Maiorov, D.V.; Popova, L.A.; Serkova, R.P. Features of the acid decomposition of eudialyte and some rare metal concentrates the Kola Peninsula. *Russ. J. Appl. Chem.* **2003**, *76*, 1191–1196. [CrossRef]
17. Davris, P.; Stopic, S.; Balomenos, E.; Panias, D.; Paspaliaris, I.; Friedrich, B. Leaching of rare earth elements from eudialyte concentrate by suppressing silica gel formation. *Miner. Eng.* **2017**, *108*, 115–122. [CrossRef]
18. Chizhevskaya, S.V.; Chekmarev, A.M.; Klimenko, O.M.; Povetkina, M.V.; Sinegribova, O.A.; Cox, M. Non—Traditional methods of treating high—Silicon containing are elements. In Proceedings of the Hydrometallurgy'94, Cambridge, UK, 11–15 July 1994.
19. Dibrov, I.A.; Chirkst, D.E.; Litvinova, T.E. Experimental Study of Zirconium(IV) Extraction from Fluoride-Containing Acid Solutions. *Russ. J. Appl. Chem.* **2002**, *75*, 195–199. [CrossRef]
20. Chanturiya, V.A.; Chanturiya, E.L.; Minenko, V.G.; Samusev, A.L. Acid leaching process intensification for eudialyte concentrate based on energy effects. *Obogashchenie Rud* **2019**, 29–36. [CrossRef]
21. Chanturia, V.A.; Minenko, V.G.; Samusev, A.L.; Ryazantseva, M.V.; Koporulina, E.V. Influence Exerted by Ultrasound Processing on Efficiency of Leaching, Structural, Chemical, and Morphological Properties of Mineral Components in Eudialyte Concentrate. *J. Min. Sci.* **2018**, *54*, 285–291. [CrossRef]
22. Inamuddin; Luqman, M. (Eds.) *Ion Exchange Technology I: Theory and Materials*; Springer: Dordrecht, The Netherlands, 2012. [CrossRef]
23. Davankov, V.A.; Tsyurupa, M.P. *Hypercrosslinked Polymeric Networks and Adsorbing Materials, Synthesis, Structure, Properties and Application*; Elsevier: Amsterdam, The Netherlands, 2010; 648p.
24. Chanturia, V.A.; Minenko, V.G.; Koporulina, E.V.; Ryazantseva, M.V.; Samusev, A.L. Influence of Acids on Extraction Efficiency of Zirconium and Rare Earth Metals in Eudialyte Concentrate Leaching. *J. Min. Sci.* **2020**, *55*, 984–994. [CrossRef]
25. Chanturiya, V.A.; Minenko, V.G.; Samusev, A.L.; Chanturia, E.L.; Koporulina, E.V.; Bunin, I.; Ryazantseva, M.V. The Effect of Energy Impacts on the Acid Leaching of Eudialyte Concentrate. *Miner. Process. Extr. Metall. Rev.* **2020**, *42*, 484–495. [CrossRef]
26. Khokhlova, O.V. Povyshenie Effektivnosti Shchelochno-Kislotnogo Sposoba Kompleksnogo Vyshchelachivaniya Evdialitovogo Koncentrata. Ph.D. Thesis, University of Science and Technology (MISIS), Moscow, Russia, 30 May 2018.
27. Chanturia, V.A.; Minenko, V.G.; Samusev, A.L.; Chanturia, E.L.; Koporulina, E.V. The Mechanism of Influence Exerted by Integrated Energy Impacts on Intensified Leaching of Zirconium and Rare Earth Elements from Eudialyte Concentrate. *J. Min. Sci.* **2017**, *53*, 890–896. [CrossRef]
28. Durakovic, B. Design of Experiments Application, Concepts, Examples: State of the Art. *Period. Eng. Nat. Sci.* **2017**, *5*, 421–439. [CrossRef]
29. Jankovic, A.; Chaudhary, G.; Goia, F. Designing the Design of Experiments (DOE)—An Investigation on the Influence of Different Factorial Designs on the Characterization of Complex Systems. *Energy Build.* **2021**, *250*, 111298. [CrossRef]
30. Lee, B.C.Y.; Mahtab, M.S.; Neo, T.H.; Farooqi, I.H.; Khursheed, A. A Comprehensive Review of Design of Experiment (DOE) for Water and Wastewater Treatment Application—Key Concepts, Methodology and Contextualized Application. *J. Water Process Eng.* **2022**, *47*, 102673. [CrossRef]
31. Obrestad, T. Method for Producing Anhydrous Calcium Nitrate Powder. RU-2680036-C2 2680036C2, 14 February 2019.
32. Pozin, M.E. *Tekhnologiya Mineralnykh Soley*; Khimia: Leningrad, Russia, 1974; Part II.

33. Soltani, F.; Abdollahy, M.; Petersen, J.; Ram, R.; Becker, M.; Javad Koleini, S.M.; Moradkhani, D. Leaching and Recovery of Phosphate and Rare Earth Elements from an Iron-Rich Fluorapatite Concentrate: Part I: Direct Baking of the Concentrate. *Hydrometallurgy* **2018**, *177*, 66–78. [CrossRef]
34. Chen, S.; Huang, X.; Feng, Z.; Xu, Y.; Wang, M.; Xia, C.; Zhao, L. Behavior of Rare Earth, Iron, and Phosphorus during Purification of Rare Earth Sulfate Leach Solution Using Magnesium Oxide. *Hydrometallurgy* **2020**, *196*, 105377. [CrossRef]

minerals

MDPI

*Article*

# Substantiation of New Reagent Compositions for the Effective Extraction of Rhenium in the Processing of Complex Molybdenum Ores

**Valentine A. Chanturiya *, Tamara N. Matveeva *, Viktoriya V. Getman, Anna Yu. Karkeshkina and Nadezhda K. Gromova**

N.V. Melnikov's Institute of Comprehensive Exploitation of Mineral Resources Russian Academy of Sciences, Kryukovsky Tupik, 4, 111020 Moscow, Russia
* Correspondence: vchan@mail.ru (V.A.C.); tmatveeva@mail.ru (T.N.M.)

**Abstract:** Modern trends in technological improvement for producing high-quality rhenium-containing molybdenum concentrates are aimed at developing environmentally friendly solutions com-pared to traditional ones, involving a reduction in consumption or complete replacement of toxic substances and the use of plant reagents and microorganisms and make it possible to eliminate energy-intensive processes of steaming and roasting. It is known that up to 25%–30% of Mo, Cu and Re goes to rough flotation tailings and is lost in wastes. Those losses are a consequence of the ore composition variability, chalcopyrite–molybdenite ratio, an increase in the proportion between primary copper sulfide and finely disseminated molybdenite and the recovery of floating pyrite. High rates of valuable metal recovery into bulk concentrate can be achieved by using new compositions of flotation reagents that selectively change the hydrophobic properties of the target minerals. The application of new reagent compositions, including novel chemicals—dithiopyrylmethane (DTM), composite reagent (CR) and conventional butyl xanthate (ButX)—was theoretically and experimentally substantiated for the effective rhenium extraction in the processing of hard-to-beneficiate complex molybdenum ores. It is identified that DTM forms a complex DTM–Re compound and chemically adsorbed on rhenium-containing molybdenite providing an increase in Re recovery into the bulk Cu–Mo–Re concentrate by 17%, reducing by two times the loss of rhenium with flotation tailings, and the subsequent 97.6% Re extraction of the molybdenum concentrate by autoclave leaching.

**Keywords:** molybdenum ores; molybdenite; flotation; reagents; dithiopyrylmethane; rhenium; reagent composition

**Citation:** Chanturiya, V.A.; Matveeva, T.N.; Getman, V.V.; Karkeshkina, A.Y.; Gromova, N.K. Substantiation of New Reagent Compositions for the Effective Extraction of Rhenium in the Processing of Complex Molybdenum Ores. *Minerals* **2023**, *13*, 372. https://doi.org/10.3390/min13030372

Academic Editors: Marinela Ivanova Panayotova and Vladko Panayotov

Received: 16 January 2023
Revised: 2 March 2023
Accepted: 4 March 2023
Published: 7 March 2023



## 1. Introduction

The main source of rhenium products is molybdenum concentrates and by-products obtained during the enrichment of porphyry copper–molybdenum ores [1,2]. Rhenium does not have its own ore deposits. It is isomorphically included in some sulfide minerals (chalcopyrite, molybdenite) [3–5]. The average content of molybdenum in the ore is 0.003%–0.01%, rhenium 0.02–0.5 ppm [6,7]. In the ore beneficiation process, rhenium is concentrated in the bulk flotation concentrate, and in selective flotation up to 80 wt.%, Re goes to molybdenum concentrate, and up to 20 wt.% follows to copper concentrates. The total extraction of rhenium in molybdenum (copper–molybdenum) ores treatment is relatively low—up to 45 wt.%. This is due to the large metal losses with flotation tailings. Molybdenum concentrates obtained during the enrichment of porphyry copper ores contain 0.02%–0.17% rhenium [8]. The largest Re producer is the Chilean company Molymet [9,10]. Re world production over the past 5 years is about 40–50 t [11].

In terms of Re reserves, Russia is in the third place after Chile and the United States [12].

A vital task in the rare metal production is to reduce the losses of valuable metals with dressing tailings and metallurgical treatment that often reach up to 50%–70% only for rhenium. In this regard, increasing the extraction of rhenium into copper–molybdenum and molybdenum concentrates, additional extraction of rhenium processing tailings is the most important task to increase the production of this strategic metal.

Copper–molybdenum ores belong to very hard-to-beneficiate mineral resource due to low content of valuable metals, the variability of the mineral composition, uneven and relatively large inclusions of molybdenite and pyrite, fine inclusions of copper minerals and a high content of flotation active pyrite. That is why the high amounts of Mo, Cu and Re go to the tailings just on the first stage of bulk sulfide flotation [13]. This circumstance necessitates the use of multi-staged enrichment schemes. After a relatively coarse first-stage ore grinding to 45%–55% less than 0.074 mm, a bulk Cu–Mo concentrate and tailings are obtained. Then, Cu–Mo concentrate after regrinding to 90%–95% less than 0.074 mm goes to the separation into the copper concentrate (21%–32% Cu with recovery of 75%–95%) and molybdenum concentrate (40%–57% Mo with a recovery from 45%–55% to 80%–86%) [4,5]. Copper and molybdenum rhenium-containing concentrates and by-products are exposed to oxidative roasting while sublimating rhenium, sintering molybdenum concentrate with soda or sodium sulfide, followed by leaching and precipitation. Mo–Re concentrate may be subjected to nitric acid autoclave leaching without roasting, and rhenium is extracted on selective sorbents from the leaching solutions [14–17].

In bulk copper–molybdenum flotation, xanthates or their mixtures with different hydrocarbon radical lengths, dithiophosphates and thionocarbamates are used as collectors [10,18]. Most of the dressing plants use several collectors or compositions of individual collectors in bulk, cleaning and scavenger flotation operations [19,20]. Non-polar reagents (diesel fuel, kerosene, engine oil, etc.) together with xanthates are used as molybdenite collectors.

Summarizing the above, we can conclude that the effective Re, Mo and Cu recovery rates into flotation concentrates can be achieved by using new types of flotation reagents that selectively change the hydrophobic and flotation properties of the target minerals [21–25]. The improved rhenium extraction into copper–molybdenum and molybdenum concentrates will decrease its high losses in flotation tailings during Cu–Mo processing of hard-to-beneficiate ores.

This paper aimed at studying the novel reagents and their compositions to improve rhenium extraction in flotation processing of hard-to-beneficiate copper–molybdenum ores.

## 2. Materials and Methods

Monomineral samples of molybdenite ($MoS_2$) 4 by 8 mm in size, polished sections of chalcopyrite and pyrite 10 by 10 mm in size, and ground fractions of molybdenite, chalcopyrite and pyrite with a particle size of 0.063–0.1 mm were used.

Scanning electron microscopy (LEO 1420VP, ZEISS Group, Oberkochen, Germany) equipped with an energy dispersive microanalyzer (Oxford INCA ENERGY 350, Oxford Instruments, Abingdon, UK) and laser scanning microscopy (KEYENCE VK-9700, KEYENCE Corporation of America, Itasca, IL, USA) were used to study the morphology and element composition of adsorbed reagent phases (thin films) on the mineral surface. SEM analysis makes it possible to identify the presence of mineral and organic phases on the mineral surface in the initial state and after contact with reagents by the presence and intensity of absorption peaks [26–28].

Scanning electron LEO 1420 VP INCA 350 microscope operates by the line-by-line moving (scanning) of a focused electron beam along the sample surface. The image is formed sequentially by points and is the result of the electron beam interaction with the sample. As a result of this interaction, a group of radiations is formed, including reflected, secondary electrons and characteristic X-ray radiation. Low-energy secondary electrons are mainly responsible for the formation of a pattern that reflects the shape of the sample surface. The intensity of reflected electrons signal directly relates to the density (average atomic number) of the target at the point of the primary electron beam incidence. The quantum energy of characteristic X-ray radiation is strictly determined by the type of electron transition in a particular atom. The number of quanta of this energy in the summary spectrum of X-ray radiation is proportional to the number of atoms of this type (element). Qualitative elemental analysis is carried out by an INCA-300 X-ray energy-dispersive spectrometer (Oxford Instruments, Abingdon, UK). Analysis is possible both at a point 2–5 microns, and over an arbitrarily specified area. Range of registered elements: Be4–U92. Electron microscopic observation and analysis of thin-film objects (especially organic ones with low average atomic numbers) on a massive substrate has a number of methodological limitations. Based on interaction of the primary electron beam with the sample, the generated zone of the characteristic X-ray radiation for light elements is more than 3–7 μm, depending on the accelerating voltage of the electron microscope. Since the thickness of the reagent's coating is often much less than these values, the intensity of the peaks of its constituent components (carbon, oxygen, nitrogen, and phosphorus) in the summary energy-dispersive spectrum of the sulfide substrate and the film will be low. High-energy reflected electrons weakly interact with a thin layer of reagent. Therefore, the image contrast in reflected electrons is extremely weak. Low-energy secondary electrons interact with the surface layer much stronger. As a result, the areas covered with reagents accumulate an electronic charge that leads to the appearance of bright zone in secondary electrons images.

Additionally, qualitative and quantitative verification of the adsorbed collecting reagents on sulfide minerals was performed by UV spectrophotometric method. The optical densities of the reagent solutions before and after their contact with ground mineral fraction were measured on a Shimadzu UV-1800 spectrophotometer. The spectrophotometer allows receiving data in the form of a graphical spectrum. The error of determination is ±0.002 Abs (at 0.5 Abs), ±0.004 Abs (at 1.0 Abs), ±0.006 Abs (at 2.0 Abs). The solution was prepared as follows: a sample of a mineral weighing 1 g was placed in a glass (volume 50 mL), a reagent solution (0, 5, 10 and 15 mg/L) was added and stirred for 5 min, then filtered. The filtrate was put in a cuvette, and the UV spectrum (190–360 nm) was obtained. The adsorption of the reagent on the mineral was determined by the difference between the initial and residual concentration of the reagent in the liquid phase in accordance with the Bugger–Lambert–Behr law.

Mineral flotation on pure minerals is carried out in a small 20–50 mL cell and on a small sample weight 1–5 g in order to evaluate the flotation activity of the mineral. Ore flotation is performed in 0.75 L cells.

The sample of the copper–molybdenum rhenium-containing porphyry ore from one of the Russian deposits was selected for ore flotation tests. The initial content of copper, molybdenum and rhenium in the sample according to the chemical analysis was 0.54% Cu, 0.0055% Mo and 0.13 ppm Re, respectively. Chemical analyses were performed in the certified Stuart Geochemical and Essay, Ltd. by the analytical method of inductively coupled plasma mass spectrometry after ore sample decomposition in a mixture of concentrated nitric $HNO_3$ and hydrochloric HCl acids at a volume ratio 1:3. Each probe was duplicated. Definition range was 0.001–50 ppm Re, 0.05–10,000 ppm Mo, 0.2–10,000 ppm Cu. Cu content in the concentrate was analyzed by oxidative decomposition in three acids followed by inductively coupled plasma atomic emission analysis (definition range 0.005%–40% Cu). The original ore sample with a size of −10 + 0 mm was crushed in a cone crusher to a class of −2 mm. Then, by quartering, the sample was divided into portions weighing 250 g and crushed to 70% of the class −0.071 mm in a laboratory ball mill at the ratio solid–water–ball 1:0.5:6. Flotation tests were performed according to the direct bulk flotation scheme obtaining the concentrate and tailings in a laboratory mechanical flotation machine MEKHANOBR (0.75 L).

Butyl xanthate (ButX), kerosene, a novel composite reagent (CR), which is a finely emulsified kerosene solution with a dialkyldithiophosphinate (S–P-containing agent) [29,30], and dithiopyrilmethane (DTM) that belongs to the pyrazole group chemicals, were used as collectors [31,32]. For the depressing rock minerals, liquid glass was used 300 g/t, frother-pine oil 50 g/t. New reagent compositions were tested at variable dosages.

## 3. Results and Discussion

### *3.1. Research of a Novel Composite Reagent CR*

Electron-donor sulfur and phosphorus atoms in the CR structure predetermine its ability to complex formation and selective adsorption, while alkyl radicals provide a high collecting activity with respect to the extracted sulfides. In contrast to S–P–O-containing reagents that are well-known and used in flotation practice of non-ferrous and noble metal ores with a high pyrite content, CR reagent has a high selectivity to non-ferrous minerals [29,30]. Laser scanning microscopy (ASEM) and scanning electron microscopy were used to obtain surface images of polished mineral sections. Energy-dispersive spectra of the surface covered by adsorbed CR phase and reagent-free before and after contact with the reagent's solutions were measured and analyzed. On molybdenite, newly formed phases of adsorbed CR appeared after its contact with CR solution. They formed the dark plates of regular shape, evenly distributed on the mineral surface (Figure 1c). Phosphorus, carbon peaks in the energy dispersive spectrum of newly formed phases indicated the CR fixation on the molybdenite surface (Figure 1d).

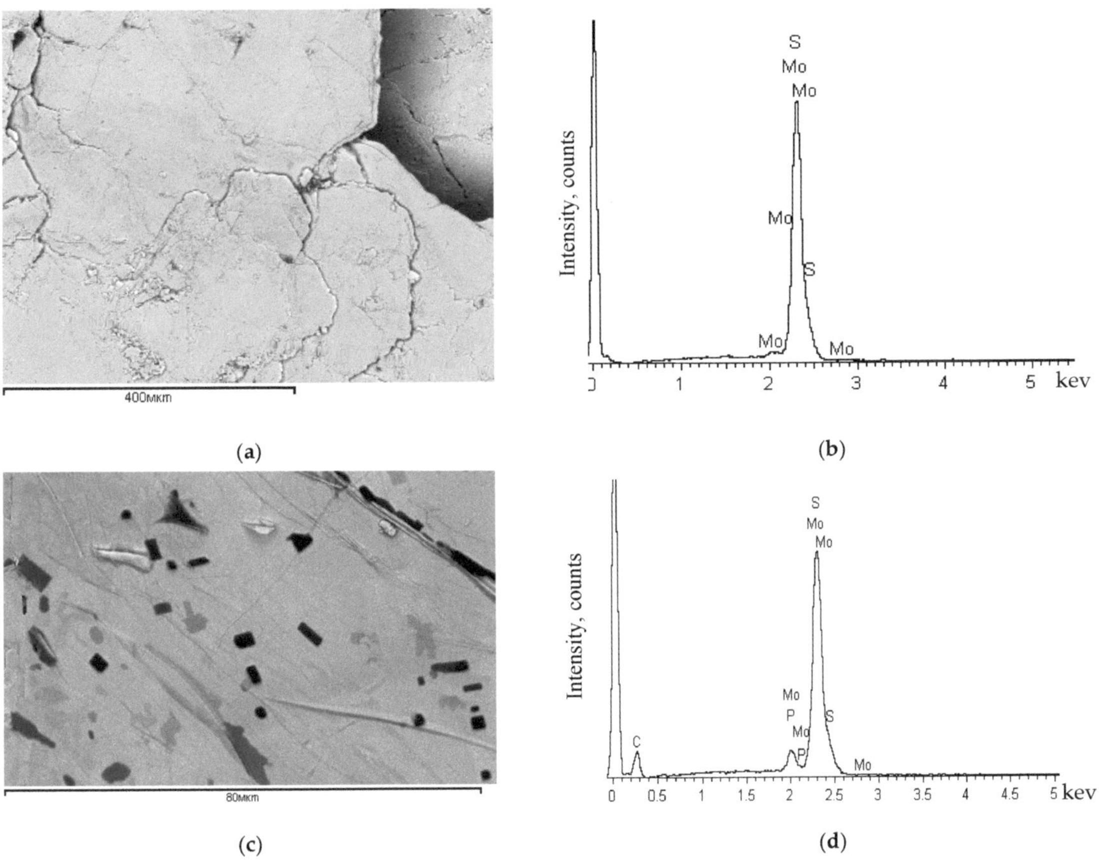


**Figure 1.** SEM image of molybdenite surface before (**a**) and after (**b**) contact with the reagent's solutions and energy-dispersive spectrum of the surface areas before (**c**) and after (**d**) contact with the reagent's solutions.

The energy spectrum of a clean mineral surface before its contact with the reagent's solution (Figure 1a) showed only S and Mo peaks that belong to the $MoS_2$ mineral, but no other organic elements were detected like in the spectrum of the mineral that was in contact with the reagent's solution (Figure 1b). A similar determination was observed on the chalcopyrite sample. Figure 2a demonstrates the surface image before adding the reagent, and the energy dispersion spectrum had Cu, Fe and S peaks of chalcopyrite $CuFeS_2$ (Figure 2b). After chalcopyrite contacted the CR solution, the needle-like spots of the reagent, rather evenly distributed over the surface (Figure 2c,d), appeared on the mineral surface, and C and P peaks in the energy-dispersive spectrum of those spots proved the CR adsorption (Figure 2e).

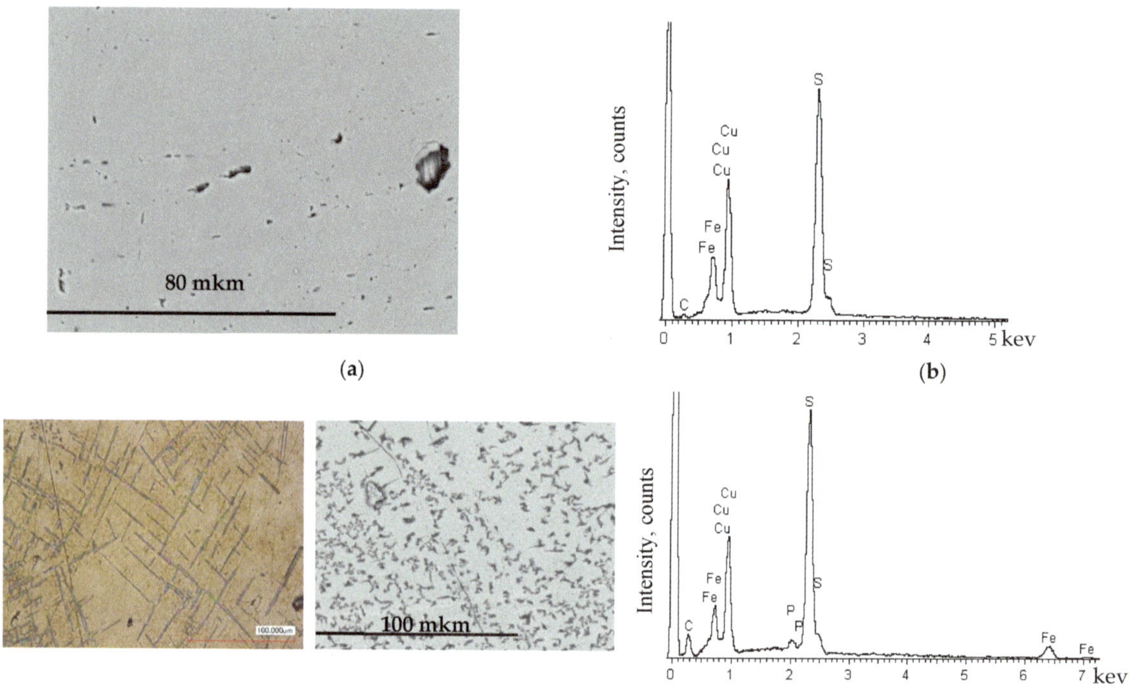


**Figure 2.** SEM image of the chalcopyrite surface (**a**) and energy-dispersive spectrum (**b**) before its contact with the CR; surface image of the chalcopyrite by laser scanning microscopy (**c**), SEM image (**d**); and energy-dispersive spectrum (**e**) after contact with the CR.

Repeated washing by distilled water did not change the surface images that demonstrated a reserved reagent covering, thus proving the stable CR fixation on molybdenite and chalcopyrite. This fact indicated strong chemical interaction of the reagent with the mineral surface. There was no CR phase determined on the pyrite sample, indicating that the CR reagent did not adsorb on the pyrite surface.

UV spectroscopy was applied to determine the adsorbed CR quantitatively on ground fractions of pyrite, chalcopyrite and molybdenite. The CR solutions had characteristic absorption peaks in the UV spectrum at a wavelength 232 nm (Figure 3a) that made it possible to measure its concentration before and after contact with minerals. Measuring the difference between the initial and residual concentrations of the reagent, in accordance with the Bugger–Lambert–Beer law, the CR adsorption after 5 min contact with the mineral suspension was calculated. It was found that at 10 mg/L concentrations, CR demonstrated 60 µg/g adsorption on chalcopyrite and 160 µg/g on molybdenite. The maximal absorption activity 108 mg/g on chalcopyrite and 170 μg/g on molybdenite CR showed at the concentration 15 mg/L. It was identified that CR did not interact with pyrite at the concentrations 0–20 mg/L. The sorption activity of CR with respect to chalcopyrite and molybdenite can be explained by its interaction with copper and molybdenum atoms forming stable complex compounds on the mineral surface.

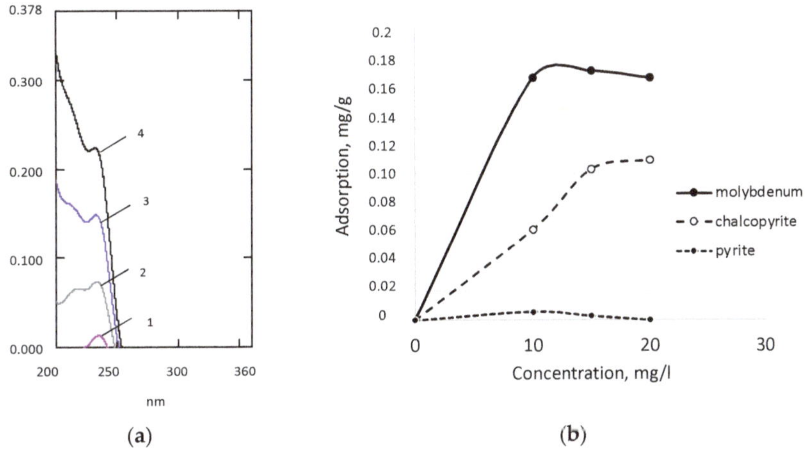


**Figure 3.** UV Spectra of the composite reagent CR at variable concentrations: 5 (1), 10 (2), 15 (3) and 20 (4) mg/L (**a**); CR adsorption curves via CR concentration (**b**).

Molybdenite, chalcopyrite and pyrite floatability by CR reagent was studied in comparison with conventional kerosene monomineral flotation at a variable collector concentration (Figure 4). Monomineral flotation was carried out on the ground fractions of molybdenite, chalcopyrite and pyrite with a particle size of 0.063–0.1 mm.

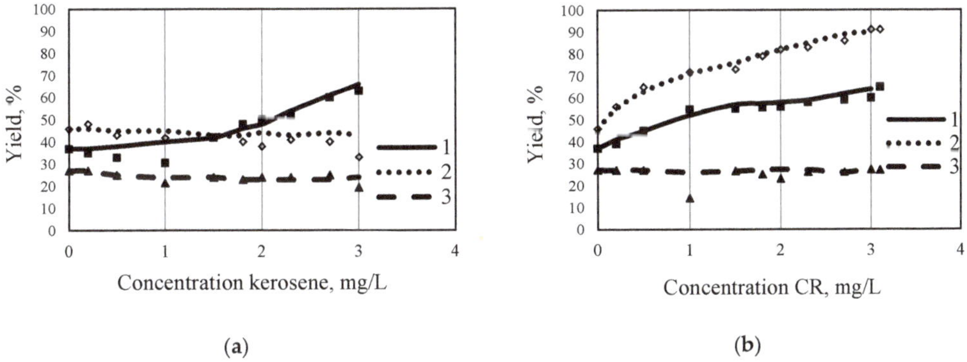


**Figure 4.** Floatability curves of molybdenite (1), chalcopyrite (2) and pyrite (3) via kerosene (**a**) and CR (**b**) concentration.

Floatability curves analysis (Figure 4b) showed that CR composite reagent exhibits selective collecting properties towards Cu and Mo sulfides and did not float pyrite: at 1 mg/L concentration, 53% molybdenite and 70% chalcopyrite floated into the concentrate. At the CR concentration of 3 mg/L the floatability of molybdenite and chalcopyrite increased to 64% and 90%, respectively, while kerosene helped to raise only molybdenite recovery to 62%. An important advantage of the composite reagent CR is that CR did not increase the floatability of pyrite that might give a positive effect on the quality of both the bulk Cu–Mo–Re concentrate and individual Mo–Re and Cu–Re concentrates in ore flotation.

*3.2. Study of the Reagent Dithiopyrilmethane*

The reactive center in the heterocyclic DTM molecule is the N–C=S group, which has a high polarity and complex-forming ability towards non-ferrous, noble and rare metals [31,32]. UV spectrophotometry tests showed the quantitative dependence of DTM adsorption on natural molybdenite as a function of reagent concentration (Figure 5).

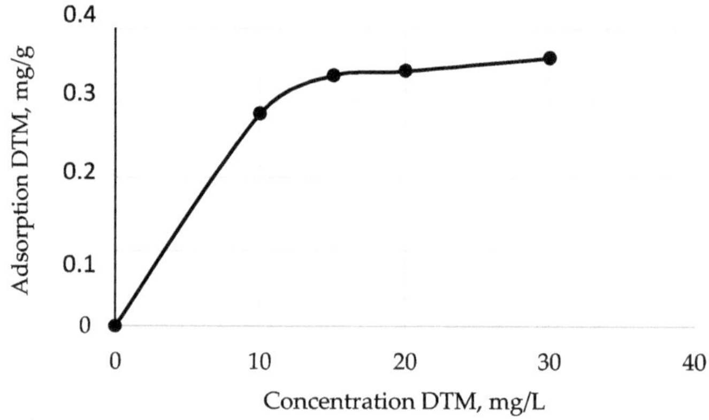


**Figure 5.** DTM adsorption isotherm on molybdenite.

The DTM adsorption on molybdenite sharply increased to 0.34 mg/g when the reagent concentration grew to 15 mg/L; the maximum DTM adsorption was 0.36 mg/g at 20–30 mg/L. The results of scanning electron microscopy confirmed that dithiopyrilmethane fixed on the molybdenite surface.

A SEM image of the mineral surface after its contact with a DTM solution revealed the appearance of dark-colored irregularly shaped neoplasms in the energy-dispersive spectra of which there are both Mo and S peaks (Figure 6), characteristic of molybdenite, and C and N peaks of organic substance appeared, indicating the DTM adsorption on molybdenite.

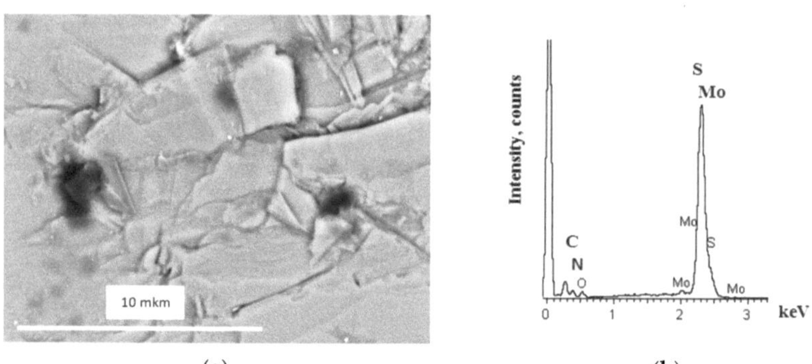


**Figure 6.** SEM image of molybdenite surface after its contact with a 0.1% DTM solution—(**a**); and energy-dispersive spectrum of dark spots/areas with adsorbed reagent—(**b**).

The preservation of newly formed reagent phase on molybdenite after repeated washing confirmed strong DTM adsorption on the mineral surface that may be justified by the chemical interaction between the reagent N–C=S group and Re-containing molybdenite.

minerals

*Article*

# Acid Resistance and Ion-Exchange Capacity of Natural Mixtures of Heulandite and Chabazite

**Vladimer Tsitsishvili [1], Marinela Panayotova [2,*], Nato Mirdzveli [3], Nagima Dzhakipbekova [4], Vladko Panayotov [5], Nanuli Dolaberidze [3] and Manana Nijaradze [3]**

1 Department of Chemistry and Chemical Technologies, Georgian National Academy of Sciences, Tbilisi 0108, Georgia
2 Department of Chemistry, University of Mining and Geology "St. Ivan Rilski", 1700 Sofia, Bulgaria
3 Petre Melikishvili Institute of Physical and Organic Chemistry, I. Javakhishvili Tbilisi State University, Tbilisi 1086, Georgia
4 Department of Chemistry and Pharmaceutical Industry, Mukhtar Auezov South Kazakhstan University, Shymkent City 160012, Kazakhstan
5 Engineering Sciences Unit, Bulgarian Academy of Sciences, 1000 Sofia, Bulgaria
* Correspondence: marichim@mgu.bg; Tel.: +359-28060305

**Abstract:** The recovery and immobilization of metals from wastewater often occurs in an acidic environment that destroys the structure of adsorbents such as zeolites, which are porous crystalline aluminosilicates. The influence of hydrochloric acid solutions on the structure and properties of two natural mixtures of heulandite (HEU) and chabazite (CHA)—tuff from the Georgian Dzegvi-Tedzami deposit (HEU/CHA $\approx$ 8) and rock from the Kazakhstani deposit Chankanay (HEU/CHA $\approx$ 1)—was studied by the X-ray energy dispersion spectra and diffraction patterns, as well as by adsorption of water, benzene, and nitrogen methods. It was found that acid-mediated dealumination, decationization, dissolution, and changes in systems of micro- and mesopores depend on the nature and chemical composition of the initial zeolites. It is concluded that, under the influence of acid, (i) zeolite micropores become accessible to relatively large molecules and ions, and the surface area of the adsorbent increases; (ii) the volume of mesopores decreases, and pores with a diameter of less than 4 nm become predominant; (iii) in terms of the degree of dealumination and dissolution rate, Kazakhstani zeolite is more acid-resistant than Georgian heulandite; and (iv) Kazakhstani zeolite retains a high ion-exchange capacity in an acidic environment, while Georgian heulandite, treated with dilute hydrochloric acid solutions, uptakes relatively high amounts of valuable silver, copper, and zinc.

**Keywords:** heavy metals; wastewater; heulandite-clinoptilolite; chabazite; bactericidal metal ions exchanged zeolites

**Citation:** Tsitsishvili, V.; Panayotova, M.; Mirdzveli, N.; Dzhakipbekova, N.; Panayotov, V.; Dolaberidze, N.; Nijaradze, M. Acid Resistance and Ion-Exchange Capacity of Natural Mixtures of Heulandite and Chabazite. *Minerals* **2023**, *13*, 364. https://doi.org/10.3390/min13030364

Academic Editor: Jorge César Masini

Received: 1 February 2023
Revised: 1 March 2023
Accepted: 2 March 2023
Published: 4 March 2023



## 1. Introduction

The use of zeolites for the recovery or immobilization of heavy metals [1,2], such as cadmium [3–10], chromium [3–5,8,11–14], nickel [3,5,8,15], copper [3,5–9,11,15], zinc [3,5–9,15], lead [5–9,15–19], manganese [4,20,21], cobalt [9,11], iron, and others [14,22–25], from wastewater is an important tool for protecting the environment, and a possible way to recover valuable metals from waste streams.

Zeolites are porous crystalline aluminosilicates, $M_nSi_xAl_nO_{2(n+x)}\cdot mH_2O$ ($M^+ = Na^+$, $K^+, \ldots \frac{1}{2}Ca^{2+}, \frac{1}{2}Mg^{2+}, \ldots$), built from alternating $SiO_4$ and $AlO_4^-$ tetrahedrons forming open framework uniform structures with cages and channels, which have entrance windows with usual sizes up to 1 nm. Currently, 248 zeolite framework types describing natural and synthetic zeolites are known, in accordance with the data of the International Zeolite Association (IZA) [26]. The use of natural and synthetic zeolites is based on a combination of molecular sieve, sorption, ion-exchange, and catalytic properties, among which an important role is played by the ability of zeolites to maintain ion-exchange reactions between external ions and metal ions, compensating for the negative charge of the aluminosilicate crystal framework.

Extraction of metals from wastewater often occurs in an acidic environment, causing dealumination of zeolites [27], resulting in a decrease in their ion exchange capacity, and therefore it is advisable to use acid-resistant zeolites.

In general, synthetic zeolites are more suitable for sequestering heavy metals in wastewater treatment [1–3,5,7–9,11,12,16,19,22,25], followed by modified zeolites [1,2,13,15,20,21,24], whereas natural ones, albeit very interesting and appealing from an environmental and economic perspective, exhibit the least sorption for a good number of heavy metals [1–4,6,14,17,18,20,23,24]. It is noted [2] that the high uniformity of the distribution in pore size, hydrophobicity/hydrophilicity, and the existence of a sole compensation cation in synthetic zeolites warrants their superior capacity for cation exchange in comparison to natural zeolites.

However, the modification of natural zeolites through various chemical treatments can improve their properties [27]. Thus, monocationic natural zeolites obtained by ion exchange modification have a sufficiently high ion exchange capacity [1,2,20,21,24]. It has been shown that treatment with bases and/or acids leads to the washing out of impurities that block pores, resulting in an increase in the cation exchange capacity [28].

The effect of acid treatment on the properties and structure of natural zeolites has been studied by different authors. Rožić et al. studied the dealumination and decationization of Croatian clinoptilolite tuff as a result of its treatment with 0.1 N HCl solution [29]. Akimkhan investigated the eventual changes in the Si/Al ratio in natural clinoptilolite due to its relatively long (24 h) treatment with HCl solutions (3 and 12 M) [30]. Wang and coworkers studied the influence of a $HNO_3$ solution treatment on the structure and properties of natural zeolite clinoptilolite [31]. Silva and coauthors studied the effect of a 24 h acid treatment (nitric, sulfuric, and hydrochloric—1.25 M, 1.43, and 0.97 M, respectively) on the properties of Australian zeolite clinoptilolite [32]. They concluded that the treatment with HCl led to the most pronounced changes in the pore size and surface area of the zeolite. However, most of the published studies used clinoptilolite, and a prolonged treatment time and/or higher concentrations of acids. Although natural chabazite also showed a high sorption capacity with respect to heavy metals [33,34], to the best of our knowledge, there are no studies on the effect of an acid treatment on the properties and structure of natural chabazite, as well as comparing both zeolites with respect to the effect of the acid treatment.

The aim of this study was to examine acid-mediated changes in the chemical composition, structure, and properties of heulandite-bearing tuff from the Rkoni plot of the Tedzami-Dzegvi deposit (Eastern Georgia) and zeolitic rock from the Chankanay deposit (Kazakhstan, Almaty region), both chosen for the fabrication of new zeolite filter materials for removing heavy metals and water disinfection. Our goal was not only to determine the acid resistance of the studied zeolites, but also to establish the acid treatment conditions that improve the performance properties, in particular, the ion exchange capacity. It is well known that $Ag^+$-, $Cu^+$-, and $Zn^+$-loaded zeolites (by ion exchange) possess bactericidal properties [35–38].

## 2. Materials and Methods

Heulandite-chabazite-containing tuff from the Rkoni plot of the Tedzami-Dzegvi deposit (Eastern Georgia) and zeolitic rock of the Chankanay deposit (Kazakhstan, Almaty region), containing heulandite and chabazite, were ground using a standard crusher, fractionated to the particle size of 1–1.4 mm or 14–16 mesh, washed with distilled water to remove clay impurities, and dried at a temperature of 95–100 °C. The main characteristics of rocks, according to the results of their study [39], are given in Table 1.

**Table 1.** Features of the studied Georgian (Rkoni) and Kazakhstani (Chankanay) zeolite-containing rocks.

| Sample from | Rkoni | Chankanay |
|---|---|---|
| Zeolite phase content (%) | 90 | 70 |
| Empirical formula | $(Na_{0.25}K_{0.06}Ca_{0.19}Mg_{0.15})$ $[AlSi_{3.6}O_{9.2}]\cdot 3H_2O$ | $(Na_{0.115}K_{0.079}Ca_{0.228}Mg_{0.175})$ $[AlSi_{2.96}O_{7.92}]$ $3H_2O$ |
| HEU/CHA [1] | 8 | 1 |
| Framework impurity | $Ti_{0.005}$ | $Ti_{0.009}$ |
| Impurities per Al atom | $Fe_{0.2}$, $Ca_{0.14}$ | $(SiO_2)_{\approx 1}$, $Fe_{0.33}$, $Ca_{0.17}$ |

[1] HEU—content of heulandite ($M_8(H_2O)_{24}[Al_8Si_{28}O_{72}]$), CHA—content of chabazite ($M_{12}(H_2O)_{40}[Al_{12}Si_{24}O_{72}]$).

According to these data, the Georgian sample can be identified as a sodium heulandite (Si/Al = 3.6) with a small admixture of chabazite, while the Kazakhstani sample is a mixture of heulandite and chabazite with a large admixture of quartz.

Acid treatment of samples with 0.5, 1.0, and 2.0 N HCl solutions was carried out by mixing 10 g of original zeolitic tuff with 100 mL of the solution in a shaking water bath (OLS26 Aqua Pro, Grant Instruments, US, linear mode) at 75 °C. This temperature was chosen as optimal due to the reproducibility of the results (control of temperature constancy throughout the acid treatment), as well as for comparison with the results of published works [29,40]. To achieve the maximum effect, acid treatment was carried out in three stages: the first stage lasted 1 h, the second stage lasted 2 h, and the third stage lasted 3 h. Each stage was followed by washing with distilled water until no $Cl^-$ ions were detected in the washing water, by using a $AgNO_3$ solution. Acid treatment with dilute solutions (0.032, 0.16, and 0.32 N) was carried out at room temperature and for 2 h.

Ion exchange reactions were carried out in 1 N solutions of silver nitrate, copper, and zinc chlorides at a solid:liquid ratio of 1:10 at room temperature, with stirring for 6 h; the prepared samples were dried at 100–110 °C. Silver(I) nitrate $AgNO_3$, zinc(II) chloride $ZnCl_2$, and copper(II) chloride dihydrate $CuCl_2\cdot 2H_2O$ (p.a., obtained from Merck KGaA—Darmstadt, Germany) were used without any additional purification.

Chemical composition of samples was calculated from the X-ray energy dispersive spectra (XR-EDS) obtained from a high-performance scanning electron microscope JSM-6490LV (JEOL, Japan), equipped with an INCA Energy 350 XRED analyzer (Oxford Instruments). Powder X-ray diffraction (XRD) patterns were obtained from a diffractometer D8 Endeavor (Bruker, Germany) using the Cu-$K_\alpha$ line ($\lambda$ = 0.154056 nm); the powder XRD patterns were recorded by scanning at a rate of 1°/min in the $2\Theta$ range of 5° to 100°, with a 0.02° step. Sorption of water and benzene vapors was measured in static conditions at room temperature and constant pressure. Nitrogen adsorption/desorption isotherms were measured at 77 K using the ASAP 2020 Plus analyzer (Micromeritics, Norcross, GA, USA) on samples vacuum degassed at 350 °C, and data analysis was carried out using the Brunauer–Emmett–Teller (BET) and Barrett-Joyner-Halenda (BJH) models.

## 3. Results and Discussion

### *3.1. Chemical Composition of the Acid-Treated Samples*

The chemical composition of the acid-treated samples indicates their dealumination and decationization. After the first stage of acid treatment (1 h), the degree of dealumination reached 87–92% of the final value obtained, as a result of the second stage of treatment (2 h); after the third stage (3 h), the chemical composition of the resulting product did not change. All results below refer to samples obtained from the three-stage process.

#### 3.1.1. Dealumination

The results for the chemical composition of the studied materials (XR-EDS analysis) are given in Table 2, in terms of averaged empirical formulas of dehydrated zeolites, the Si/Al molar ratio, and degree of dealumination (DD), which is defined as the ratio of the aluminum content of the treated sample to the aluminum content of the original sample.

**Table 2.** Chemical composition of unit cell, Si/Al molar ratio, and degree of dealumination (DD) in original and acid-treated Georgian (Rkoni) and Kazakhstani (Chankanay) zeolites.

| [HCl] (N) [1] | Rkoni | | | Chankanay | | |
|---|---|---|---|---|---|---|
| | Empirical Formula | Si/Al | DD | Empirical Formula | Si/Al | DD |
| 0 | $(Na_{1.95}K_{0.47}Ca_{1.49}Mg_{1.17})[Al_{7.83}Si_{28.2}O_{72}]$ | 3.6 | 1.00 | $(Na_{1.05}K_{0.72}Ca_{2.07}Mg_{1.6})[Al_{9.1}Si_{26.9}O_{72}]$ | 2.96 | 1.00 |
| 0.5 | $(Na_{0.62}K_{0.67}Ca_{0.71}Mg_{0.47})[Al_{4.58}Si_{31.4}O_{72}]$ | 6.85 | 0.58 | $(Na_{0.98}K_{0.82}Ca_{1.07}Mg_{1.25})[Al_{8.96}Si_{27.0}O_{72}]$ | 3.02 | 0.98 |
| 1.0 | $(Na_{0.47}K_{0.4}Ca_{0.65}Mg_{0.46})[Al_{4.26}Si_{31.7}O_{72}]$ | 7.45 | 0.54 | $(Na_{0.92}K_{0.84}Ca_{1.05}Mg_{1.12})[Al_{8.53}Si_{27.5}O_{72}]$ | 3.22 | 0.94 |
| 2.0 | $(Na_{0.096}K_{0.5}Ca_{0.61}Mg_{0.26})[Al_{3.43}Si_{32.6}O_{72}]$ | 9.5 | 0.44 | $(Na_{0.72}K_{0.90}Ca_{0.84}Mg_{0.76})[Al_{7.88}Si_{28.1}O_{72}]$ | 3.57 | 0.87 |

[1] [HCl]—concentration of the hydrochloric acid solution.

An increase in the Si/Al molar ratio from 3.6 to 9.5, and a decrease in the aluminum content to 44% of the original, indicates deep acid-mediated dealumination in the Georgian heulandite, while in the Kazakhstani zeolite, aluminum is not so significantly leached. A significant increase in the Si/Al ratio was observed by Akimkhan [30] for clinoptilolite treated with HCl solutions, but at higher a temperature (96–98 °C) and longer treatment time. Based on their data from FTIR analysis, Silva and coauthors [32] also concluded that clinoptilolite treated with a HCl solution leads to an increase in the Si/Al ratio.

### 3.1.2. Decationization

The acid treatment also results in strong decationization of the aluminum-rich Georgian and Kazakhstani samples, as shown in Figure 1, by a decrease in the total charge of the compensating ions $Na^+$, K+, $Ca^{2+}$, and $Mg^{2+}$, as well as the share of each of these cations in compensating the negative charge of the zeolite framework as they are leached and replaced in the framework by $H^+$.

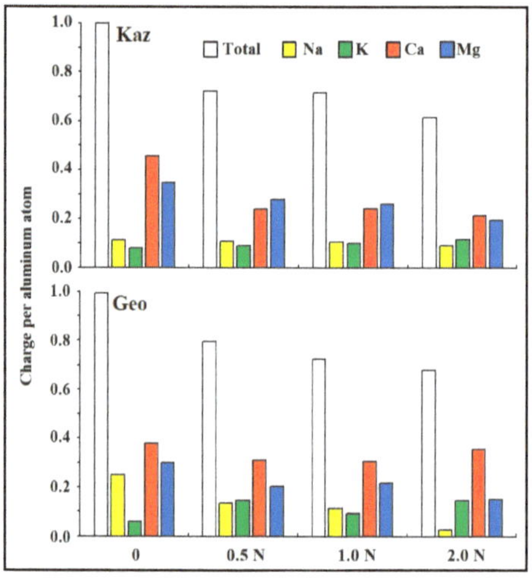


**Figure 1.** Cationic charge per Al atom of native (0) and acid-treated Georgian (Geo) and Kazakhstani (Kaz) zeolites.

In Georgian heulandite, the total charge monotonically decreases from ≈1 to ≈0.68 with increasing acid concentration. The contribution of $Na^+$ ions to compensate for the negative charge on aluminum atoms decreases, so that sodium is leached to the greatest extent (the sodium content decreased ≈9 times). The contribution of $Mg^{2+}$ ions decreases more slowly (≈2 times), the contribution of $Ca^{2+}$ ions changes insignificantly (−7%), and the contribution of $K^+$ ions increases nonmonotonically.

The degree of decationization of the Kazakhstani zeolite is also very high (from ≈1 to ≈0.62), but this process proceeds qualitatively differently. So, although the contribution of sodium decreases, and the contribution of potassium increases, these changes are small (−20% and +30%, respectively), while the contribution of divalent cations, especially $Ca^{2+}$, decreases to the greatest extent (≈2 and ≈1.8 times for $Ca^{2+}$ and $Mg^{2+}$, respectively).

Thus, in Georgian heulandite, the decationization mainly proceeds due to sodium ions, while calcium and, in particular, potassium are practically not involved in this process. This conclusion is consistent with the results of the study of decationization and dealumination of clinoptilolite tuff [29], where it was also found that natural clinoptilolite samples treated with HCl solutions exchanged mainly $Na^+$ ions, followed by $Mg^{2+}$ and $Ca^{2+}$ ions, whereas K+ ions from the clinoptilolite practically did not participate in an exchange process. However, our results do not correspond to the data of reference [40], which show that the removal of monovalent cations, such as $Na^+$ and $K^+$ ions, is insignificant for the temperatures of 25–100 °C, and changed little with acid concentration. This conclusion is more consistent with the decationization process in the Kazakhstani mixture of heulandite and chabazite, but the results are difficult to compare, since clinoptilolite with a high Si/Al ratio and a different cationic composition (Si/Al = 6.22, (Na+K) > Ca) was used in reference [40].

### 3.2. Structure of the Acid-Treated Samples

Structural changes of the studied materials are fixed by powder XRD patterns. The powder XRD pattern of Georgian heulandite shows no changes after treatment with a dilute HCl solution (0.5 N), but treatment with concentrated solutions leads to a change in the intensity of some peaks (see Figures 2 and 3). Thus, the intensity of the low angle peak at 2Θ = 9.85° (Miller indices *hkl* 020; d-spacing 8.98 Å) decreases with increasing acid concentration. The peak intensities at ≈22° (*hkl* 131, 400, 330; ≈4 Å) first increase (at 1.0 N), and then decrease (2.0 N), while the intensity of the weak peak at ≈28° (*hkl* −422 and/or −441; ≈3.15 Å) initially decreases, and then sharply increases with increasing the acid concentration to 2N.

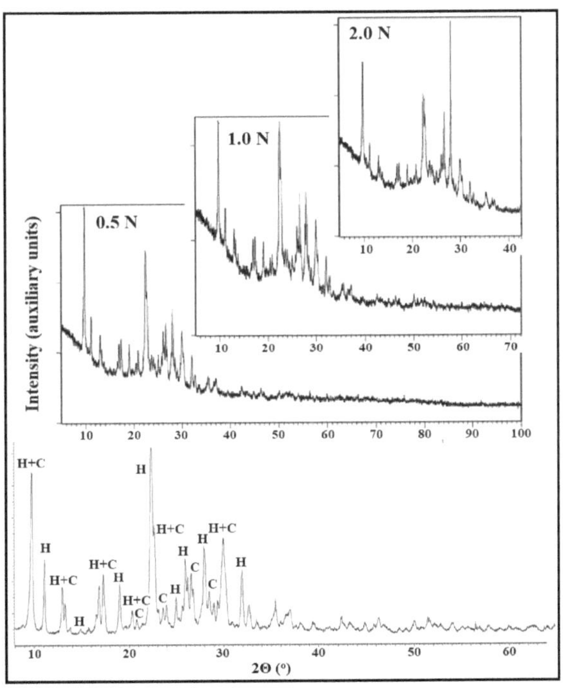


**Figure 2.** Powder XRD patterns of original (bottom left, H—peaks of heulandite, C—peaks of chabazite) and acid treated Georgian heulandite.

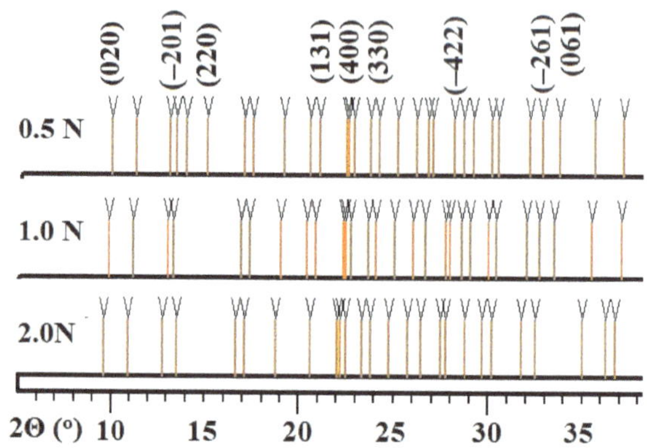


**Figure 3.** Schematic representation of powder XRD patterns of acid treated Georgian heulandite (numbers in parenthesis are Miller indices *hkl* of the most varying peaks).

In addition, some peaks decrease and disappear from the X-ray diffraction patterns with increasing acid concentration; this applies to peaks at 2Θ = 13° (*hkl* −201; 6.8 Å), 15° (*hkl* 220; ≈6 Å), and ≈33° (*hkl* −261 and/or 061; ≈2.8 Å), as schematically shown in Figure 3.

No line broadening is observed, so that the acid treatment does not cause amorphization of the sample. The overall intensity of the XRD pattern decreases slightly with increasing acid concentration, not correlating with the much greater weight loss resulting from the treatment, as shown in Table 3. Although the chemical compositions of the zeolites do not change in the third step of the acid treatment, there is still a slight weight loss, which indicates the gradual dissolution of the samples by the acid.

**Table 3.** Weight loss (%) from the acid treatment steps (1-2-3) of Georgian and Kazakhstani zeolites.

| Zeolite Origin | From 0.5 N | From 1.0 N | From 2.0 N |
|---|---|---|---|
| Georgian | 9.75–7.05–4.5 | 12.8–7.8–6.1 | 15.4–7.95–5.25 |
| Kazakhstani | 6.3–3.5–1.4 | 7.4–5.1–2.5 | 9.2–5.4–3.5 |

In a recent work [41], the amorphization of clinoptilolite recorded using XRD patterns was noted only after treatment of the zeolite with solutions with a high concentration of hydrochloric acid (5 and 10 N).

The same conclusions could be expected at high concentration of hydrochloric acid, based on a comparison of powder XRD patterns of the original and acid-treated samples of Kazakhstani zeolite (see Figure 4); since this zeolite is a mixture of heulandite and chabazite, the patterns are a superposition, and their detailed interpretation is quite difficult.

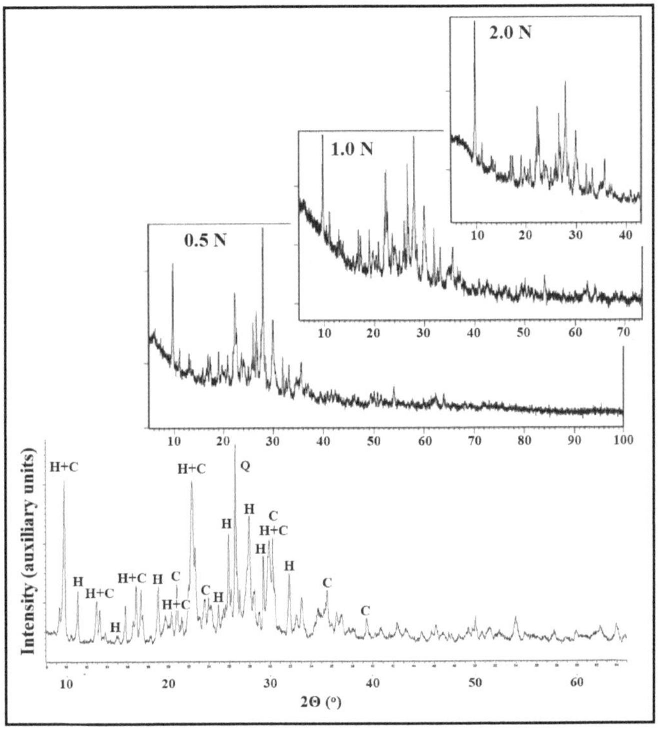

**Figure 4.** Powder XRD patterns of original (bottom left, H—peaks of heulandite, C—peaks of chabazite, Q—peak of quartz) and acid treated Kazakhstani zeolite.

### 3.3. Sorption Properties of the Acid-Treated Samples

Sorption properties carry information about micro- and mesoporous systems. The adsorption of small water molecules, with a kinetic molecular diameter of 0.266 nm, is a measure of the pore volume of high-aluminum zeolites due to their hydrophilicity. The adsorption capacity for water vapor at a relative pressure $p/p_o = 0.4$ reflects the micropore volume, and the adsorption capacity at saturated water vapor pressure ($p/p_o = 1.0$) reflects the total volume of micro-, meso-, and macropores [42].

The kinetic diameter of the benzene molecule is 0.585 nm, which significantly exceeds the sizes of micropores and channels in the studied zeolites. This non-polar molecule can be adsorbed only on the zeolite surface due to the presence of meso- and macropores; benzene adsorption capacity is a relative measure of surface area and its hydrophobicity.

The kinetic diameter of a nitrogen molecule $N_2$ (0.364 nm) is comparable to the size of the entrance windows in the chabazite crystal structure (eight-membered ring 0.38 × 0.38 nm [26]). In heulandite, a 10-membered ring (0.75 × 0.31 nm) and one of the 8-membered rings (0.47 × 0.28 nm) cannot accommodate a nitrogen molecule, which can pass only into one 8-membered ring (0.46 × 0.36 nm), which is due to the considerable flexibility of the framework [26].

#### 3.3.1. Water and Benzene Sorption

The adsorption capacity of the studied zeolites, for water vapor at relative pressures of 0.4 and 1.0, as well as for benzene at a relative pressure of 1.0, is shown in Figure 5.

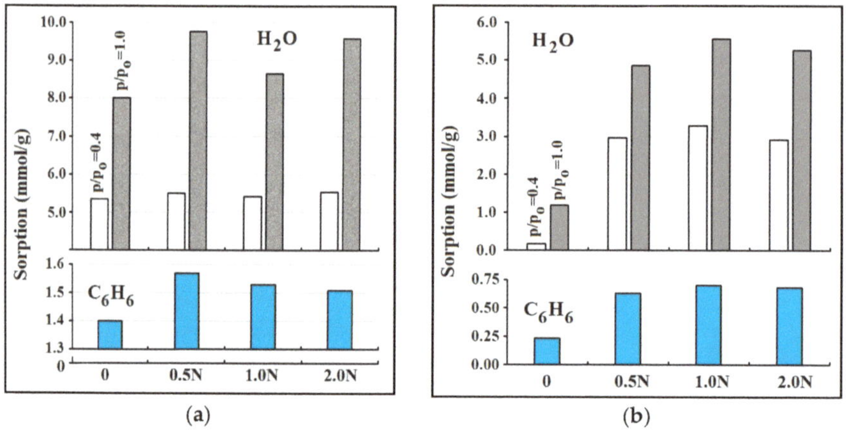


**Figure 5.** Water vapor ($H_2O$) and benzene ($C_6H_6$) adsorption capacity on original (0) and acid-treated samples of: (**a**) Georgian heulandite; (**b**) Kazakhstani zeolite.

According to the results obtained, for untreated Georgian heulandite, the volume of micropores accessible for water molecules is about 60% of the total pore volume; adsorption in the micropores practically does not change as a result of acid treatment, and the total adsorption in all the pores slightly increases. For untreated Kazakhstani zeolite, the volume of the micropores is only 10% of the total pore volume; adsorption in the micropores, as well as adsorption in all the pores, increases significantly after acid treatment; the volume of the micropores is increased $\approx$18 times and exceeds 60% of the total pore volume, as a result of treatment with a 1.0 N hydrochloric acid solution.

As a result of the study of high-silica zeolites [43–45], it was found that the strongest interaction of water molecules occurs with acidic protons bound to framework aluminum atoms H(Al), and the amount of adsorbed water is proportional to the content of framework aluminum. The zeolites studied by us, including acid-treated ones, have much lower values of the Si/Al ratio (3–9.5) than the samples studied in the indicated works (for example, Si/Al—35 [43] or Si/Al > 42 [44]), therefore, the effects described in [43–45] are not observed. Previously, Sano et al. [44] described the formation of mesopores in an acid-treated dealuminated synthetic zeolite, and this process apparently causes the observed increase in the total pore volume, which is especially characteristic of the Kazakhstani zeolite, but also occurs in Georgian heulandite.

Benzene adsorption changes insignificantly, but the results obtained indicate an increase in the hydrophobicity of the surfaces of both zeolites after acid treatment. More detailed information about the surface and mesopores was obtained by analyzing nitrogen adsorption-desorption isotherms.

#### 3.3.2. Nitrogen Sorption

The porosity parameters of natural Georgian heulandite and its acid-treated forms (calculated on the basis of nitrogen adsorption-desorption isotherms—Figure S1) are shown in Table 4.

**Table 4.** Porosity parameters of Georgian heulandite and its acid-treated forms.

| Concentration of HCl Solution (N) | 0 | 0.5 | 1.0 | 2.0 |
|---|---|---|---|---|
| Total volume of pores ($cm^3/g$) * | 0.0895 | 0.109 | 0.1129 | 0.1256 |
| Volume of micropores ($cm^3/g$) | 0.00647 | 0.0782 | 0.0856 | 0.0889 |
| BET surface area ($m^2/g$) | 12.8 | 126.7 | 154.7 | 175.0 |
| BJH desorption average pore diameter (nm) | 17.2 | 13.1 | 11.6 | 11.1 |

* The total volume of pores with a diameter less than 200 nm.

Both the volume of micropores accessible to nitrogen molecules at $p/p_0 = 0.4$ and the surface area (calculated using the BET model) increase sharply for samples treated with acid, and continue to increase with increasing acid concentration, while the total pore volume, according to the number of molecules adsorbed at saturated vapor pressure, varies nonmonotonically with changes in acid concentration. The total pore volume increases monotonically with increasing acid concentration, while the diameter of nanosized pores, calculated using the BJH model, decreases abruptly. However, little depends on the acid concentration, indicating that acid treatment affects not only the heulandite micropore system, but also the mesopores existing in the sample.

The pore size distribution curve volume vs. pore diameter V(D) and differential dV/dD curve calculated by the BJH model for Georgian heulandite and its acid-treated forms are shown in Figure 6.

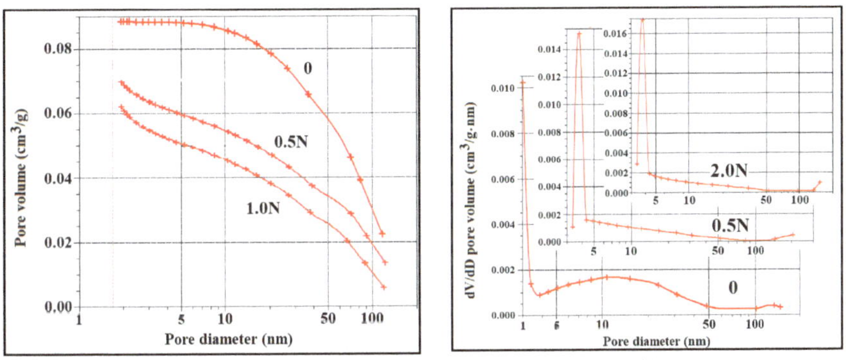


(**a**) (**b**)

**Figure 6.** Pore size distribution curves calculated by the BJH model for Georgian heulandite and its acid-treated forms: (**a**) volume V vs. pore diameter D; (**b**) differential curve dV/dD.

As the pore size distribution curve V(D) shows, the volume of mesopores decreases as a result of acid treatment; according to the differential curve dV/dD, the maximum at ≈12 nm disappears, but a sharp maximum appears at ≈4 nm. Taking into account the, albeit small, increase in the total pore volume with increasing acid concentration, it can be concluded that after acid treatment, small in size, up to 4 nm, pores become predominant in the mesopore system of Georgian heulandite.

Akimkhan [30] observed no increase in the total pore volume when clinoptilolite was treated at room temperature with a 3 M HCl solution for 24 h; the increase was about two times when the acid was 12 M, and over three times when the acid treatment was at 96–98 °C. Silva and coworkers [32] found that HCl treatment decreased the zeolite pore diameter about 2.5 times, and increased the surface area approximately 16 times.

In a recent work by Çakicioğlu-Özkan and Becer [41], it was noted that in the studied original clinoptilolite, the volume of micropores accessible to nitrogen was only 0.005 $cm^3/g$, which the authors attributed to the blocking of most of the pores by cations, considered as impurities. With an increase in the concentration of HCl in the solution, the volume of micropores increased by more than 12 times, up to 0.064 $cm^3/g$ for a 3 N HCl solution, but with further increase in the acid concentration, it decreased. The outer surface area of the HCl-treated samples gradually increased with increasing the Si/Al ratio from 3 $m^2/g$, and reached a maximum of 33 $m^2/g$ after soaking in a 10 N solution [41].

The values of micropore volume and surface area obtained by us for original and acid-treated heulandites are much higher than reported in [41], but the general tendencies of their change with increasing acid concentration are the same. The authors of [41] also processed data in the region of low relative pressures ($p/p_0 < 0.1$) using the Dubinin-Astakhov equation ($W = W_0\exp\{-[RT\ln(p_0/p)/E]^n\}$, where $W_0$ is total weight adsorbed, E is the characteristic energy, R—gas constant, T—temperature) and obtained fractal values of the heterogeneity parameter n for acid-treated samples, while the experimental data obtained by us are ideally described by the simple BET equation ($W = W_0[1 - (p/p_0)]$), at least up to pressures $p/p_0 = 0.08$. The pore size distributions were not calculated in [41].

The porosity parameters of natural Kazakhstani zeolite and its acid-treated forms (calculated on the basis of nitrogen adsorption-desorption isotherms—Figure S2), are shown in Table 5. As in Georgian heulandite, the quantity of nitrogen molecules adsorbed in micropores and the BET surface area increase sharply for samples treated with acid, and continue to increase with increasing acid concentration. Total pore volume, on the contrary, first increases and then decreases with increasing acid concentration; the diameters of nanosized pores calculated by the BJH model decrease monotonically.

**Table 5.** Porosity parameters of Kazakhstani zeolite and its acid-treated forms.

| Concentration of HCl Solution (N) | 0 | 0.5 | 1.0 | 2.0 |
|---|---|---|---|---|
| Total volume of pores ($cm^3/g$) * | 0.0732 | 0.0881 | 0.0735 | 0.0686 |
| Volume of micropores ($cm^3/g$) | 0.00152 | 0.0098 | 0.0189 | 0.0209 |
| BET surface area ($m^2/g$) | 2.75 | 20.0 | 38.0 | 44.9 |
| BJH desorption average pore diameter (nm) | 24.3 | 15.4 | 12.2 | 13.8 |

* The total volume of pores with a diameter less than 200 nm.

The differences in the systems of mesopores of Georgian and Kazakhstani zeolites should be noted: according to the pore size distribution V(D) and dV/dD curves calculated by the BJH model, using adsorption isotherms measured on original and acid-treated Kazakhstani zeolite and shown in Figure 7, the dV/dD curve for native zeolite from the Chankanay deposit has two maxima, at ≈6 and 60 nm.

The pore size distribution V(D) curves show that the volume of mesopores increases after treatment with a diluted (0.5 N) HCl solution, but decreases after treatment with concentrated solutions. The maximum of the differential curve dV/dD at ≈6 nm shifts to ≈4 nm, and the maximum at ≈60 nm disappears.

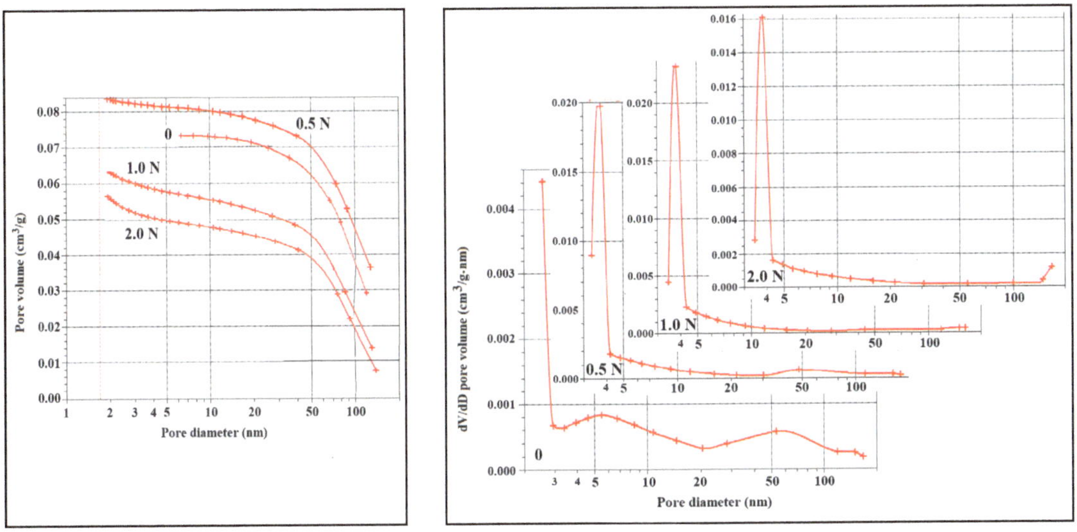


(**a**) (**b**)

**Figure 7.** Pore size distribution curves calculated by the BJH model for Kazakhstani zeolite and its acid-treated forms: (**a**) volume V vs. pore diameter D; (**b**) differential curve dV/dD.

### *3.4. Effect of Dilute Solutions on the Porosity and Ion-Exchange Properties of Georgian Heulandite*

The chemical composition of the ion-exchanged samples determines their cation exchange capacity. The cation exchange capacity (CEC) of the original and acid-treated samples was calculated from XR-EDS data, as the number of ionogenic groups of the ion exchanger entirely converted to the $H^+$ form, and expressed in milliequivalents per gram; the results are shown in Table 6, and the dependence of the obtained CEC values on the molar ratio Si/Al is shown in Figure 8.

**Table 6.** Cation exchange capacity (CEC, mEq/g) of the original and acid-treated samples.

| Concentration of HCl Solution (N) | 0 | 0.5 | 1.0 | 2.0 |
|---|---|---|---|---|
| Georgian heulandite | 3.03 | 1.90 | 1.78 | 1.46 |
| Kazakhstani zeolite | 3.43 | 3.38 | 3.25 | 3.05 |

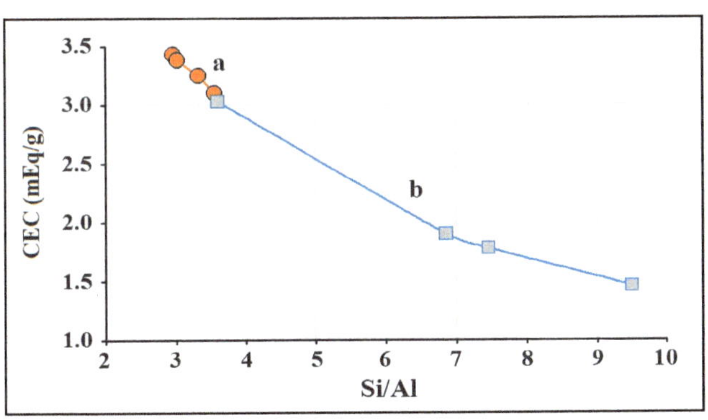


**Figure 8.** Dependence of the CEC on the Si/Al molar ratio, calculated for Kazakhstani zeolite (a) and Georgian heulandite (b).

The ion-exchange capacity of Kazakhstani zeolite remains at a high level of ≈3.4 mEq/g after treatment with a 0.5 N hydrochloric acid solution, which provides the access of large metal ions to the cages and channels of the zeolite, as well as a sufficiently large surface area (up to 20 $m^2/g$). For heulandite-bearing tuff, the ion-exchange capacity decreases significantly, which is associated with a higher degree of dealumination of this sample. In this regard, acid treatment was carried out under milder conditions (room temperature, 2 h), using more dilute solutions in the concentration range of 0.032–0.32 N.

The samples thus treated were subjected to ion exchange in solutions of silver nitrate and chlorides of copper and zinc. X-ray diffraction patterns of ion-exchanged samples confirm the preservation of the zeolite structure. The characteristics of the micro- and mesoporous systems of the acid-treated heulandites, and the chemical compositions of the exchanged samples, are shown in Table 7.

**Table 7.** Porosity parameters of acid-treated Georgian heulandite and chemical compositions of silver-, copper-, and zinc-exchanged forms.

| Concentration of HCl Solution (N) | 0.032 | 0.16 | 0.32 |
|---|---|---|---|
| Total volume of pores ($cm^3/g$) | 0.090 | 0.094 | 0.104 |
| Volume of micropores ($cm^3/g$) | 0.010 | 0.025 | 0.049 |
| BET surface area ($m^2/g$) | 19.6 | 49.8 | 87.5 |
| BJH desorption pore diameter (nm) | 17.0 | 16.2 | 15.4 |
| Empirical formula of Ag-exchanged samples | $Ag_{0.46}(Na,K,Ca,Mg)_{0.54}[AlSi_{3.75}O_{9.5}]$ | $Ag_{0.5}(Na,K,Ca,Mg)_{0.5}[AlSi_{4.6}O_{11.2}]$ | $Ag_{0.42}(Na,K,Ca,Mg)_{0.57}[AlSi_{5.5}O_{13}]$ |
| Silver content (mg/g, mmol/g) | 122 | 117 | 90 |
| | 1.1 | 1.1 | 0.82 |
| Empirical formula of Cu-exchanged samples | $Cu_{0.36}(Na,K,Ca,Mg)_{0.27}[AlSi_{3.9}O_{9.8}]$ | $Cu_{0.4}(Na,K,Ca,Mg)_{0.18}[AlSi_{4.8}O_{11.6}]$ | $Cu_{0.35}(Na,K,Ca,Mg)_{0.27}[AlSi_{5.6}O_{13.2}]$ |
| Copper content (mg/g, mmol/g) | 60 | 58.5 | 46 |
| | 0.95 | 0.92 | 0.72 |
| Empirical formula of Zn-exchanged samples | $Zn_{0.16}(Na,K,Ca,Mg)_{0.67}[AlSi_{3.8}O_{9.6}]$ | $Zn_{0.18}(Na,K,Ca,Mg)_{0.62}[AlSi_{4.7}O_{11.4}]$ | $Zn_{0.15}(Na,K,Ca,Mg)_{0.67}[AlSi_{5.6}O_{13.2}]$ |
| Zinc content (mg/g, mmol/g) | 28 | 27 | 20 |
| | 0.43 | 0.42 | 0.31 |

Compared to the untreated sample, which retains only 0.62, 0.54, and 0.22 mmol/g of silver, copper, and zinc, respectively, the largest amount of metals is immobilized by heulandites treated with dilute solutions of hydrochloric acid (0.032–0.16 N); with an increase in acid concentration, the specific amount of embedded metals in the samples decreases (for example, to only 0.24 mmol/g for silver in zeolite treated with a 2N HCl solution), due to dealumination occurring as a result of the acid treatment. Table 7 presents the data on zeolites, which is interesting from an application point of view.

The incorporation of silver ions $Ag^+$ into the microporous structure of heulandite occurs rather easily; the introduction of bivalent copper ions—$Cu^{2+}$ and zinc ions—$Zn^{2+}$, despite their smaller diameter, is difficult.

This phenomenon is likely associated with different geometry and energy characteristics of the hydration of ions entering the micropores of the zeolite. The diameter of the hydrated studied ions, in any case, exceeds the size of the entrance window, and ions cannot pass into the micropores. The dehydration energy depends on the number of water molecules in the complex, and it should be taken into account that the hydrated silver(I) ion ($Ag(H_2O)_4{}^+$) includes four water molecules arranged in a linearly distorted tetrahedron, whereas the hydrated copper(II) and zinc(II) ions ($M(H_2O)_6^{2+}$) include six water molecules located at the vertices of the regular octahedron [46].

The content of immobilized metals is ≈60, ≈75, and ≈40% of the maximum values calculated from the ion exchange isotherms obtained on Turkish clinoptilolite [47] for silver, copper, and zinc, respectively. In [48], it is shown that acid-treated clinoptilolite adsorbs more zinc ions than untreated clinoptilolite, but the ion-exchange capacity of these samples, with respect to other metals, has not been studied.

Acid-treated Georgian heulandite exhibits a higher uptake of copper and zinc than sedimentary clinoptilolite from the Serbian field Zlatokop (0.41 and 0.225 mmol/g of Cu and Zn, respectively [49]); silver uptake is higher than that of natural clinoptilolite (0.84 mmol/g [50]), which is used for the combined disinfection and heavy metal purification of water streams. The achieved uptake of silver, copper, and zinc by acid-treated Georgian heulandite is much higher than that described for the clinoptilolite-bearing tuff of the Iranian Semnan deposit (0.24, 0.28, and 0.24 mmol/g of Ag, Cu, and Zn, respectively [51]), which improves the prospect to remove and concentrate these valuable metals from wastewater.

## 4. Conclusions

Taking into account the results of studying the effect of acid treatment on the structure and properties of two natural mixtures of heulandite and chabazite, the following conclusions can be drawn:

1. Treatment of the Georgian heulandite-bearing tuff of the Rkoni plot of the Dzegvi-Tedzami deposit with hydrochloric acid solutions leads to significant dealumination and decationization with sodium washout, while the Kazakhstani zeolitic rock of the Chankanay deposit, containing heulandite and chabazite in equal amounts, is much more acid-resistant, but it also undergoes significant decationization with calcium and magnesium washout;
2. Solutions of hydrochloric acid gradually dissolve the zeolite microporous crystal structure without its amorphization; the dissolution rate of Kazakhstani zeolite is much lower, especially in dilute solutions, than that of Georgian heulandite;
3. Heulandite micropores are, in any case, accessible to small water molecules; in Kazakhstani zeolite, they become accessible only after acid treatment. The acid treatment also leads to a slight increase in surface hydrophobicity for both samples;
4. Low-temperature nitrogen adsorption-desorption isotherms show an acid-mediated sharp increase of adsorption in the micropores and of the BET surface area, as well as a decrease in the specific volume and average diameter of the mesopores;
5. Treatment of Kazakhstani zeolite with a dilute solution of hydrochloric acid (up to 0.5 N) provides the access of large metal ions to the micropores of the zeolite, while maintaining a high ion-exchange capacity; Georgian heulandite treated with more dilute solutions (0.032–0.16 N) retains its ion-exchange capacity and exhibits a high ability to uptake valuable silver, copper, and zinc ions (up to 122, 60, and 28 mg/g, respectively), making increased acidity of the zeolite processing solution impractical.

**Supplementary Materials:** The following supporting information can be downloaded at: https://www.mdpi.com/article/10.3390/min13030364/s1, Figure S1: N2 adsorption-desorption isotherms on Georgian heulandite and its acid-treated forms. Figure S2: N2 adsorption-desorption isotherms on Kazakhstani zeolite and its acid-treated forms.

**Author Contributions:** Conceptualization, V.T.; investigation, N.M. and N.D. (Nagima Dzhakipbekova) and M.N.; methodology, M.P., V.P. and N.D. (Nanuli Dolaberidze); project administration, N.M. All authors have read and agreed to the published version of the manuscript.

**Funding:** This research was funded by the International Science and Technology Center (ISTC), grant number GE-2506 "Scientific substantiation of the possibility of creating new bactericidal zeolite filter materials for purification-decontamination of water from various sources".

**Data Availability Statement:** Data is contained within the article and Supplementary Material.

**Conflicts of Interest:** The authors declare no conflict of interest.

## References

1. Yuna, Z. Review of the natural, modified, and synthetic zeolites for heavy metals removal from wastewater. *Environ. Eng. Sci.* **2016**, *33*, 443–454. [CrossRef]
2. de Magalhães, L.F.; da Silva, G.R.; Peres, A.E.C. Zeolite Application in Wastewater Treatment. *Adsorpt. Sci. Technol.* **2022**, *2022*, 4544104. [CrossRef]
3. Alvarez-Ayuso, E.; Garcia-Sanchez, A.; Querol, X. Purification of metal electroplating waste waters using zeolites. *Water Res.* **2003**, *37*, 4855–4862. [CrossRef] [PubMed]
4. Dal Bosco, S.M.; Jimenez, R.S.; Carvalho, W.A. Removal of toxic metals from wastewater by Brazilian natural scolecite. *J. Colloid Interf. Sci.* **2005**, *281*, 424–431. [CrossRef]
5. Ibrahim, H.S.; Jamil, T.S.; Hegazy, E.Z. Application of zeolite prepared from Egyptian kaolin for the removal of heavy metals: II. Isotherm models. *J. Hazard. Mat.* **2010**, *182*, 842–847. [CrossRef] [PubMed]
6. Batjargal, T.; Yang, J.-S.; Kim, D.-H.; Baek, K. Removal characteristics of Cd(II), Cu(II), Pb(II), and Zn(II) by natural Mongolian zeolite through batch and column experiments. *Separ. Sci. Technol.* **2011**, *46*, 1313–1320. [CrossRef]
7. Xie, W.-M.; Zhou, F.-P.; Bi, X.-L.; Chen, D.-D.; Li, J.; Sun, S.-Y.; Liu, J.-Y.; Chen, X.-Q. Accelerated crystallization of magnetic 4A-zeolite synthesized from red mud for application in removal of mixed heavy metal ions. *J. Hazard. Mat.* **2018**, *358*, 441–449. [CrossRef]
8. Bai, S.; Chu, M.; Zhou, L.; Chang, Z.; Zhang, C.; Liu, B. Removal of heavy metals from aqueous solutions by X type zeolite prepared from combination of oil shale ash and coal fly ash. *Energ. Sourc. Part A* **2022**, *44*, 5113–5123. [CrossRef]
9. Joseph, I.V.; Tosheva, L.; Doyle, A.M. Simultaneous removal of Cd(II), Co(II), Cu(II), Pb(II), and Zn(II) ions from aqueous solutions via adsorption on FAU-type zeolites prepared from coal fly ash. *J. Environ. Chem. Eng.* **2020**, *8*, 103895. [CrossRef]
10. Dosa, M.; Grifasi, N.; Galletti, C.; Fino, D.; Piumetti, M. Natural zeolite clinoptilolite application in wastewater treatment: Methylene blue, zinc and cadmium abatement tests and kinetic studies. *Materials* **2022**, *15*, 8191. [CrossRef]
11. Hui, K.S.; Chao, C.Y.; Kot, S.C. Removal of mixed heavy metal ions in wastewater by zeolite 4A and residual products from recycled coal fly ash. *J. Hazard. Mat.* **2005**, *127*, 89–101. [CrossRef] [PubMed]
12. Covarrubias, C.; García, R.; Arriagada, R.; Yánez, J.; Garland, M.T. Cr(III) exchange on zeolites obtained from kaolin and natural mordenite. *Micropor. Mesopor. Mat.* **2006**, *88*, 220–231. [CrossRef]
13. Zeng, Y.; Woo, H.; Lee, G.; Park, J. Adsorption of Cr (VI) on hexadecylpyridinium bromide (HDPB) modified natural zeolites. *Micropor. Mesopor. Mat.* **2010**, *130*, 83–91. [CrossRef]
14. Zanin, E.; Scapinello, J.; de Oliveira, M.; Rambo, C.L.; Franscescon, F.; Freitas, L.; de Mello, J.M.M.; Fiori, M.A.; Oliveira, V.; Magro, J.D. Adsorption of heavy metals from wastewater graphic industry using clinoptilolite zeolite as adsorbent. *Process Saf. Environ.* **2017**, *105*, 194–200. [CrossRef]
15. Huang, F.-C.; Han, Y.-L.; Lee, C.-K.; Chao, H.-P. Removal of cationic and oxyanionic heavy metals from water using hexadecyltrimethylammonium-bromide-modified zeolite. *Desalin. Water Treat.* **2016**, *57*, 17870–17879. [CrossRef]
16. Shariatinia, Z.; Bagherpour, A. Synthesis of zeolite NaY and its nanocomposites with chitosan as adsorbents for lead(II) removal from aqueous solution. *Powder Technol.* **2018**, *338*, 744–763. [CrossRef]
17. Xing, P.; Wang, C.; Ma, B.; Chen, Y. Removal of Pb(II) from aqueous solution using a new zeolite-type absorbent: Potassium ore leaching residue. *J. Environ. Chem. Eng.* **2018**, *6*, 7138–7143. [CrossRef]
18. Budianta, W.; Ardiana, A.; Andriyani, N. The removal of lead by natural zeolite. *E3S Web Conf.* **2020**, *200*, 06012. [CrossRef]
19. Jangkorn, S.; Youngme, S.; Praipipat, P. Comparative lead adsorptions in synthetic wastewater by synthesized zeolite A of recycled industrial wastes from sugar factory and power plant. *Heliyon* **2022**, *8*, e09323. [CrossRef]
20. Taffarel, S.R.; Rubio, J. On the removal of $Mn^{2+}$ ions by adsorption onto natural and activated Chilean zeolites. *Miner. Eng.* **2009**, *22*, 336–343. [CrossRef]
21. Taffarel, S.R.; Rubio, J. Removal of $Mn^{2+}$ from aqueous solution by manganese oxide coated zeolite. *Miner. Eng.* **2010**, *23*, 1131–1138. [CrossRef]
22. Bahaz, H.; Hadj Seyd, A.; Moulai, K.; Aggoun, M.S. Removal of heavy metals from an industrial effluent by synthesized zeolite: Case of Bounoura industrial zone. *Leban. Sci. J.* **2020**, *21*, 80–94. Available online: https://applications.emro.who.int/imemrf/32/2/Lebanese-Sci-J-2020-21-1-80-94-eng.pdf (accessed on 1 March 2023).
23. Eremin, O.V.; Epova, E.S.; Filenko, R.; Rusal, O.S.; Bychinsky, V.A. Use of zeolite rocks in metal recovery from mine water. *J. Min. Sci.* **2017**, *53*, 915–924. [CrossRef]
24. Ćurković, L.; Cerjan-Stefanović, Š.; Filipan, T. Metal ion exchange by natural and modified zeolites. *Water Res.* **1997**, *31*, 1379–1382. [CrossRef]
25. Iqbal, A.; Sattar, H.; Haider, R.; Munir, S. Synthesis and characterization of pure phase zeolite 4A from coal fly ash. *J. Clean. Prod.* **2019**, *219*, 258–267. [CrossRef]
26. Baerlocher, C.; McCusker, L.B. Database of Zeolite Structures. Available online: http://www.iza-structure.org/databases/ (accessed on 1 March 2023).

27. Szostak, R. Secondary synthesis methods. In *Introduction to Zeolite Science and Practice*; van Bekkum, H., Flanigen, E.M., Jacobs, P.A., Jansen, J.C., Eds.; Elsevier: Amsterdam, The Netherlands, 2001; pp. 261–297. [CrossRef]
28. Wang, S.; Peng, Y. Natural zeolites as effective adsorbents in water and wastewater treatment. *Chem. Eng. J.* **2010**, *156*, 11–24. [CrossRef]
29. Rozić, M.; Cerjan-Stefanović, S.; Kurajica, S.; Maeefat, M.R.; Margeta, K.; Farkas, A. Decationization and dealumination of clinoptilolite tuff and ammonium exchange on acid-modified tuff. *J. Colloid Interface Sci.* **2005**, *284*, 48–56. [CrossRef]
30. Akimkhan, A.M. Structural and ion-exchange properties of natural zeolite. In *Ion Exchange Technologies*; Kilislioglu, A., Ed.; IntechOpen: London, UK, 2012; pp. 261–282. [CrossRef]
31. Wang, C.; Cao, L.; Huang, I. Influences of acid and heat treatments on the structure and water vapor adsorption property of natural zeolite. *Surf. Interface Anal.* **2017**, *49*, 1249–1255. [CrossRef]
32. Silva, M.; Lecus, A.; Lin, Y.T.; Corrao, J. Tailoring natural zeolites by acid treatments. *J. Mater. Sci. Chem. Eng.* **2019**, *7*, 26–37. [CrossRef]
33. Pansini, M.; Colella, C. Dynamic data on lead uptake from water by chabazite. *Desalination* **1990**, *78*, 287–295. [CrossRef]
34. Torracca, E.; Galli, P.; Pansini, M.; Colella, C. Cation exchange reactions of a sedimentary chabazite. *Micropor. Mesopor. Mat.* **1998**, *20*, 119–127. [CrossRef]
35. Rivera-Garza, M.; Olguín, M.T.; García-Sosa, I.; Alcántara, D.; Rodríguez-Fuente, G. Silver supported on natural Mexican zeolite as an antibacterial material. *Microp. Mesopor. Mat.* **2000**, *39*, 431–444. [CrossRef]
36. Demirci, S.; Ustaoğlu, Z.; Yılmazer, G.A.; Sahin, F.; Baç, N. Antimicrobial properties of zeolite-X and zeolite-A ion-exchanged with silver, copper, and zinc against a broad range of microorganisms. *Appl. Biochem. Biotechnol.* **2014**, *172*, 1652–1662. [CrossRef] [PubMed]
37. Lalueza, P.; Monzón, M.; Arruebo, M.; Santamaría, J. Bactericidal effects of different silver-containing materials. *Mater. Res. Bull.* **2011**, *46*, 2070–2076. [CrossRef]
38. Yao, G.; Lei, J.; Zhang, W.; Yu, C.; Sun, Z.; Zheng, S.; Komarneni, S. Antimicrobial activity of X zeolite exchanged with $Cu^{2+}$ and $Zn^{2+}$ on *Escherichia coli* and *Staphylococcus aureus*. *Environ. Sci. Pollut. Res. Int.* **2019**, *26*, 2782–2793. [CrossRef]
39. Tsitsishvili, V.; Panayotova, M.; Miyamoto, M.; Dolaberidze, N.; Mirdzveli, N.; Nijaradze, M.; Amiridze, Z.; Klarjeishvili, N.; Khutsishvili, B.; Dzhakipbekova, N.; et al. Characterization of Georgian, Kazakh and Armenian natural heulandite-clinoptilolites. *Bull. Georg. Natl. Acad. Sci.* **2022**, *16*, 39–46. Available online: http://science.org.ge/bnas/vol-16-4.html (accessed on 1 March 2023).
40. Cakicioglu-Ozkan, F.; Ulku, S. The effect of HCl treatment on water vapor adsorption characteristics of clinoptilolite rich natural zeolite. *Micropor. Mesopor. Mat.* **2005**, *77*, 47–53. [CrossRef]
41. Çakicioğlu-Özkan, F.; Becer, M. Effect of the acid type on the natural zeolite structure. *J. Turk. Chem. Soc. Sect. B* **2019**, *2*, 69–74. Available online: https://dergipark.org.tr/tr/download/article-file/738982 (accessed on 1 March 2023).
42. Yamaka, S.; Malla, P.B.; Komarnani, S. Water sorption and desorption isotherms of some naturally occurring zeolites'. *Zeolites* **1989**, *9*, 18–22. [CrossRef]
43. Simonot-Grange, M.; Elm'Chaouri, A.; Weber, G.; Dufresne, P.; Raatz, F.; Joly, J. Characterization of the dealumination effect into H faujasites by adsorption: Part 1. The water molecule as a structural aluminum ion selective probe. *Zeolites* **1992**, *12*, 155–159. [CrossRef]
44. Sano, T.; Yamanhita, N.; Iwami, Y.; Takeda, K.; Kawakami, Y. Estimation of dealumination rate of ZSM-5 zeolite by adsorption of water vapor. *Zeolites* **1996**, *16*, 258–264. [CrossRef]
45. Olson, D.H.; Haag, W.O.; Borghard, W.S. Use of water as a probe of zeolitic properties: Interaction of water with HZSM-5'. *Micropor. Mesopor. Mater.* **2000**, *35/36*, 435–446. [CrossRef]
46. Persson, I. Hydrated metal ions in aqueous solution: How regular are their structures? *Pure Appl. Chem.* **2010**, *82*, 1901–1917. [CrossRef]
47. Top, A.; Ülkü, S. Silver, zinc, and copper exchange in Na-clinoptilolite and resulting effect on antibacterial activity. *App. Clay Sci.* **2004**, *27*, 13–19. [CrossRef]
48. Adinehvand, J.; Shokuhi Rad, A.; Tehrani, A.S. Acid-treated zeolite (clinoptilolite) and its potential to zinc removal from water sample. *Int. J. Environ. Sci. Technol.* **2016**, *13*, 2705–2712. [CrossRef]
49. Hrenovic, J.; Milenkovic, J.; Ivankovic, T.; Rajic, N. Antibacterial activity of heavy metal-loaded natural zeolite. *J. Hazard. Mater.* **2012**, *201–202*, 260–264. [CrossRef]
50. Akhigbe, L.; Ouki, S.; Saroj, D.; Min Lim, X. Silver-modified clinoptilolite for the removal of Escherichia coli and heavy metals from aqueous solutions. *Env. Sci. Pollut. Res.* **2014**, *21*, 10940–10948. [CrossRef]
51. Milenkovic, J.; Hrenovic, J.; Matijasevic, D.; Niksic, D.; Rajic, N. Bactericidal activity of Cu-, Zn-, and Ag-containing zeolites toward *Escherichia coli* isolates. *Environ. Sci. Pollut. Res.* **2017**, *24*, 20273–20281. [CrossRef]

minerals

MDPI

Article

# Gold Extraction from a Refractory Sulfide Concentrate by Simultaneous Pressure Leaching/Oxidation

**Juan Carlos Soto-Uribe [1,*], Jesus Leobardo Valenzuela-Garcia [1,*], Maria Mercedes Salazar-Campoy [1], Jose Refugio Parga-Torres [2], Guillermo Tiburcio-Munive [1], Martin Antonio Encinas-Romero [1] and Victor Manuel Vazquez-Vazquez [1]**

[1] Department of Chemical Engineering and Metallurgy, Universidad de Sonora, Blvd. Luis Encinas y Rosales S/N, Col. Centro, Hermosillo 83000, Mexico
[2] Institute Technology of Saltillo, Department of Metallurgy and Materials Science, Saltillo 25280, Mexico
* Correspondence: carlos.soto@unison.mx (J.C.S.-U.); jesusleobardo.valenzuela@unison.mx (J.L.V.-G.)

**Abstract:** Most gold deposits occur associated with sulphides like pyrite and arsenopyrite; thus, precious metal dissolution is possible by oxidizing auriferous sulfide concentrate using simultaneous pressure oxidation and cyanidation. The effectiveness of this process of extraction can be influenced by the temperature, cyanide (NaCN) concentration, and oxygen pressure. In this study, we conducted experiments to characterize the effects on gold extraction of ores using a range of sodium cyanide concentrations (1–8 g/L), temperatures (40–75 °C), and oxygen pressures (0.5–1.1 MPa). Characterization of the ores showed that pyrite and quartz were the main minerals present in the concentrate. The best results in terms of the highest extraction of Au were obtained with an oxygen pressure of 0.5 MPa, 6 g/L sodium cyanide, and a temperature of 75 °C, along with a constant stirring speed of 600 rpm. These conditions allowed for approximately 95% gold extraction in 90 min.

**Keywords:** cyanidation; gold-bearing; pressure leaching; pressure oxidation

**Citation:** Soto-Uribe, J.C.; Valenzuela-Garcia, J.L.; Salazar-Campoy, M.M.; Parga-Torres, J.R.; Tiburcio-Munive, G.; Encinas-Romero, M.A.; Vazquez-Vazquez, V.M. Gold Extraction from a Refractory Sulfide Concentrate by Simultaneous Pressure Leaching/Oxidation. *Minerals* **2023**, *13*, 116. https://doi.org/10.3390/min13010116

Academic Editors: Marinela Ivanova Panayotova and Vladko Panayotov

Received: 25 November 2022
Revised: 10 January 2023
Accepted: 10 January 2023
Published: 12 January 2023



## 1. Introduction

Many gold deposits contain finely disseminated gold particles in iron sulfide minerals such as pyrite. These minerals are called refractory gold ores because of the presence of gold, which occurs either as disseminated fine gold particles—typically <1 μm [1] or as fine particles of gold encapsulated in the ore host [2,3]. In certain ores, extraction rates corresponding to lower than 60% gold recovery occur when using direct cyanide leaching, even after grinding, hence indicating that the concentrate is refractory. Thus, a suitable pretreatment process is often required to make cyanidation more effective. The refractory ore must be destroyed by chemical means using oxidative processes such as oxidation by roasting [4,5], bio-oxidation [6–9], pressure oxidation [10,11], or ultrafine grinding [12,13].

The Velardeña mine is in Durango State, wherein the gold concentrate is produced by sulfide flotation and sold to external processing companies. Processing of this gold concentrate is generally conducted using roasting and leaching methods. However, pressure oxidation has been used to treat refractory gold ores or concentrates as pretreatment and subsequent cyanidation. Alkaline pressure oxidation is an effective pretreatment for releasing refractory gold from pyrite and arsenical concentrates before cyanidation [11,14]. Simultaneous pressure leaching oxidation involves the oxidation of pyrite and Au cyanidation in alkaline media, where oxidation liberates or exposes gold particles that are otherwise occluded or finely disseminated. In addition, the dissolution of gold is also dependent on oxidation due to its electrochemical behavior, which can be represented by the Elsner equation [1]. The equation proposed for gold is:

$$4Au + 8CN^- + O_2 + H_2O \rightarrow 4\left[Au(CN)_2^-\right] + 2OH^- \quad (1)$$

The pressure oxidation of pyrite involves reactions yielding ferrous ions, sulfate ions, and elemental sulfur as products [11,15–18]. Pyrite oxidation in alkaline media is described by Equation (2), where the reaction is similar to acid oxidation (Equations (2)–(6)) [1]. The reactions are shown as follows:

$$2FeS_2 + 7.5O_2 + 7H_2O \leftrightarrow 2Fe(OH)_3 + 8H^+ + 4SO_4^{2-} \quad (2)$$

$$2FeS_2 + 7O_2 + 2H_2O \rightarrow 2FeSO_4 + 2H_2SO_4 \quad (3)$$

$$2FeSO_4 + H_2SO_4 + 0.5O_2 \rightarrow Fe_2(SO_4)_3 + H_2O \quad (4)$$

$$Fe_2(SO_4)_3 + 3H_2O \rightarrow Fe_2O_3(\downarrow) + 3H_2SO_4 \quad (5)$$

$$4FeS_2 + 15O_2 + 8H_2O \rightarrow 2Fe_2O_3 + 8H_2SO_4 \quad (6)$$

Several authors have published their results from gold extraction via oxidative pretreatment following the cyanidation process [14,19–21]. In our study, we evaluate the effect of the simultaneous application of pressure leaching and cyanidation on the rate of gold dissolution from gold sulfide concentrate.

## 2. Materials and Methods

A pyrite gold concentrate was provided by La Velardeña Mine. The sample was homogenized and screened using 100 Tyler mesh (100% passing size). X-ray diffraction (XRD), scanning electron microscopy (SEM), and energy-dispersive X-ray spectroscopy (EDX) were used to obtain characteristics of the concentrate for application in combination with physical characterization. Chemical analysis (CA) was performed using the fire assay technique to determine the gold grade, and atomic absorption spectroscopy (AAS) was conducted to determine the presence of copper, zinc, and iron.

The sample was then subjected to metallurgical tests of simultaneous pressure leaching/oxidation in media alkaline. The tests were conducted in a 1 L Parr pressure reactor with a heating jacket, stirrer, and inlet and outlet gas control valves (Figure 1) under the following conditions: a solid ratio of 20%, pressure range of 0.5–1.1 MPa, temperature range of 45–75 °C, sodium cyanide concentration of 4–8 g/L, pH of 11–12 using CaO, and a reaction time of 90 min. The experimental variables and parameters for pressure leaching/oxidation are given in Table 1.

**Figure 1.** Experimental setup for simultaneous pressure leaching and oxidation.

**Table 1.** Process variables and parameters for simultaneous pressure leaching and oxidation.

| Parameter | Value |
|---|---|
| Oxygen pressure, MPa | 0.5, 0.8, and 1.1 |
| Temperature, °C | 45–75 |
| NaCN, g/L | 4–8 |
| Solids, wt.% | 20 |
| Stirring speed, rpm | 600 |

For conventional cyanidation, the tests were conducted at room temperature and under atmospheric pressure conditions: 2 g/L NaCN, 3 kg/ton CaO for pH control, a solid ratio of 20%, 72 h leaching time, and particle size of −100 mesh (149 μm).

## 3. Results and Discussion

### 3.1. Mineralogy Characterization

The scanning electron microscopy (SEM) results in Figure 2a) indicate the existence of acanthite resulting from the high percentage of Ag and S, and it is also possible to observe encrusted quartz impurities and galena in a pyrite matrix. In the X-ray diffraction (XRD) pattern in Figure 2b), the spectrum exhibits higher crystallinity for pyrite ($FeS_2$), 81.3%, and quartz ($SiO_2$), 11.6%, with both being the main species, thus confirming the concentrate's refractory nature. Other species such as calcite ($CaCO_3$) and galena (PbS) are present in minor quantities.

The Au and Ag grades in the head concentrate and other metals were determined by CA (Table 2). Table 3 shows the mineralogical reconstruction via XRD, and CA is expressed in terms of the weight percentage (wt.%).

**Table 2.** Chemical head grade analysis through fire assay (FA) and 4-acid digestion with atomic adsorption finish.

| Mineral Concentrate Sample | | | | | | | | |
|---|---|---|---|---|---|---|---|---|
| Au, g/t | Ag, g/t | Cu, % | Zn, % | Fe, % | Pb, % | As, % | S, % | Insol., % |
| 42.2 | 3932 | 2.1 | 4.1 | 32.7 | 3.3 | 0.17 | 31.2 | 23.3 |

**Table 3.** Mineralogical reconstruction of the gold concentrate.

| Compounds | | Weight % |
|---|---|---|
| Chalcopyrite | $CuFeS_2$ | 4.4 |
| Galena | PbS | 3.03 |
| Sphalerite | ZnS | 4.8 |
| Pyrite | $FeS_2$ | 64.2 |
| Arsenopyrite | FeAsS | 0.38 |
| Quartz | $SiO_2$ | 19.8 |
| Calcite | $CaCO_3$ | 3.3 |

### 3.2. Conventional Cyanidation Test

Low extraction rates were observed for the cyanidation test at 30 °C and atmospheric pressure. A conventional cyanidation test was carried out under the following conditions: leaching time of 72 h and particle size of −100 mesh (149 μm), 2 g/L NaCN concentration, the addition of 3 kg/t CaO to control pH, and 20% solids.

As shown in Figure 3, conventional cyanide leaching results in a low extraction of 46%, which may be associated with the use of a sulfide refractory gold ore. Salazar-Campoy et al. [22] obtained 37.9% Au extraction in 72 h from sulfide refractory ore, whereas Elorza-Rodriguez et al. [23]

recovered 61.2% Au following standard cyanidation for refractory ores. Figure 3 shows that the Au extraction using conventional cyanidation was lower than 50%, in contrast to reports involving simultaneous pressure leaching and oxidation in [18,22], where >90% gold was recovered.

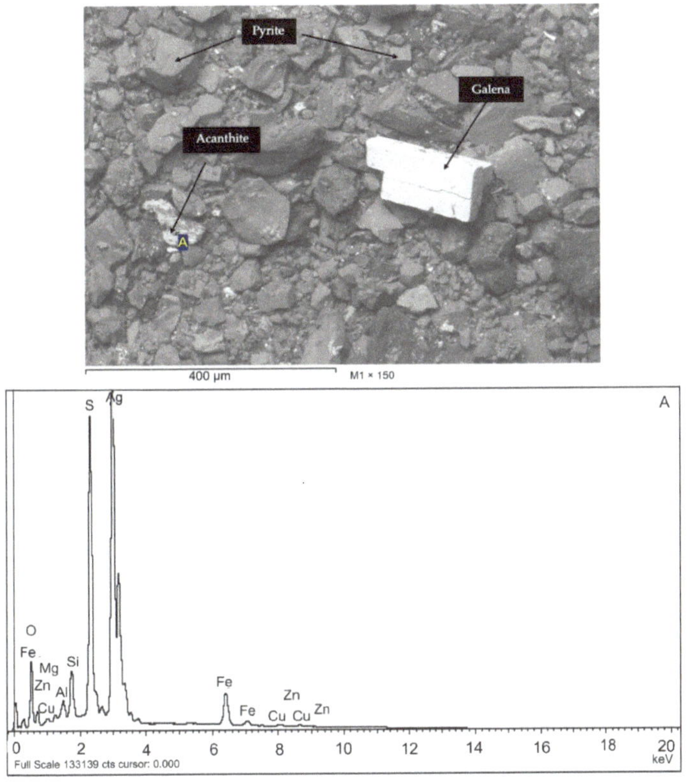


(**a**)

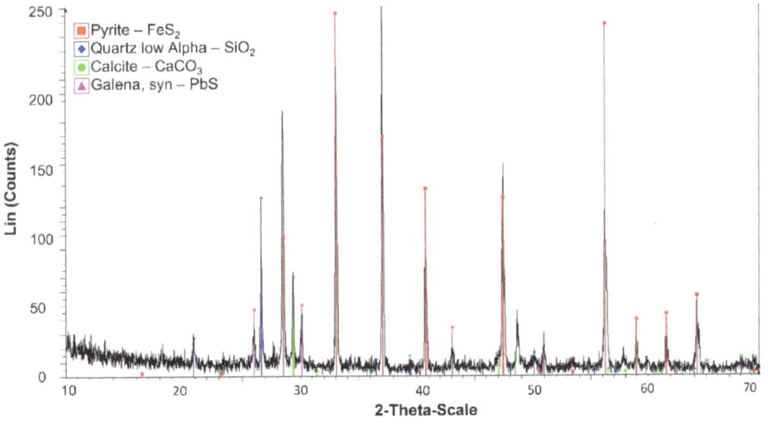


(**b**)

**Figure 2.** (**a**) Scanning electron microscopy of the sample at 400 µm showing pyrite, argentite, and galena particles, and energy dispersive X-ray spectroscopy images of particles. (**b**) X-ray diffraction pattern of concentrate sample.

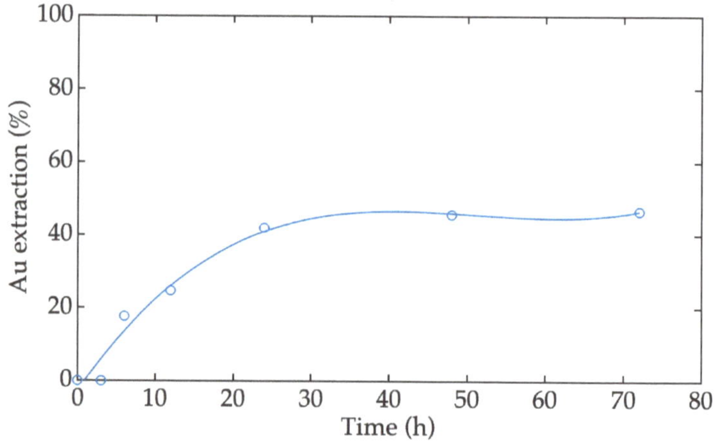


**Figure 3.** Au recovery (%) resulting from cyanidation at room temperature and atmospheric pressure as a function of time.

## 3.3. *Simultaneous Pressure Leaching and Oxidation*

Oxidation and pressure leaching tests were performed to determine temperature, oxygen pressure, and cyanide concentration effects on gold extraction. The tested variations were temperatures ranging from 45–75 °C, oxygen pressure from 0.5 to 1.1 MPa, and sodium cyanide concentrations in the range of 4 to 8 g/L.

### 3.3.1. Effects of Temperature and Pressure on Gold Extraction

Results for gold extraction from oxidation in the leaching test at a temperature of 40–75 °C and with simultaneous pressure increases are shown in Figure 4. For the range of tested pressures and 6 g/L of NaCN, the Au (%) extracted increased due to increasing oxygen pressure, whereas the temperature raise, recovery of Au (%) decreased at low pressures due to the effects on dissolved oxygen, which decreases according to Henry's law. Rusanen et al. [11] argue that oxygen pressure is the most important factor influencing the oxidation of refractory minerals.

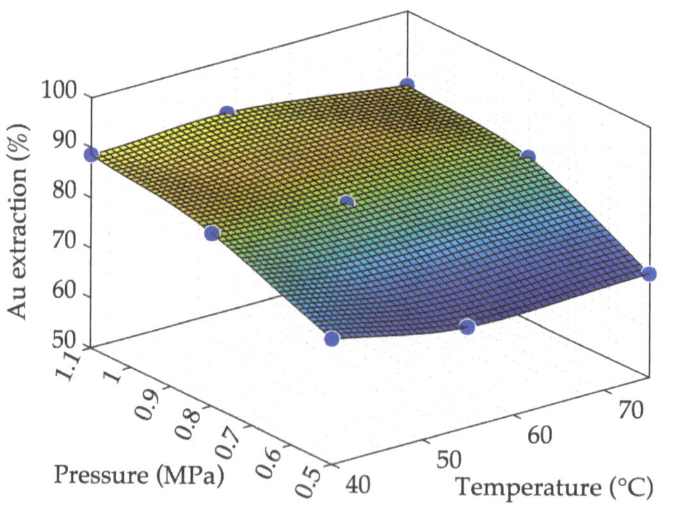


**Figure 4.** Effects of temperature and pressure on Au extraction at 6 g/L cyanide concentration.

From Figure 4, we can see that gold dissolution of 89.51% was obtained at pressure leaching and oxidation conditions of 1.1 MPa and 55 °C. Leaching and oxidation under increased pressure enhance gold extraction compared with tests conducted under atmospheric conditions.

The effects of temperature on Au extraction were studied, as shown in Figure 5. Generally, the increase in temperature also results in increased amounts of extracted gold. Nevertheless, compared with conventional cyanidation, simultaneous pressure leaching oxidation results in higher Au extraction within a short time, obtaining 41%, 54%, and 82% for each temperature in 10 min. Azizi et al. [24] indicated that the presence of galena during the cyanidation of pyrite ore increases the recovery of gold from 2% to 58%.

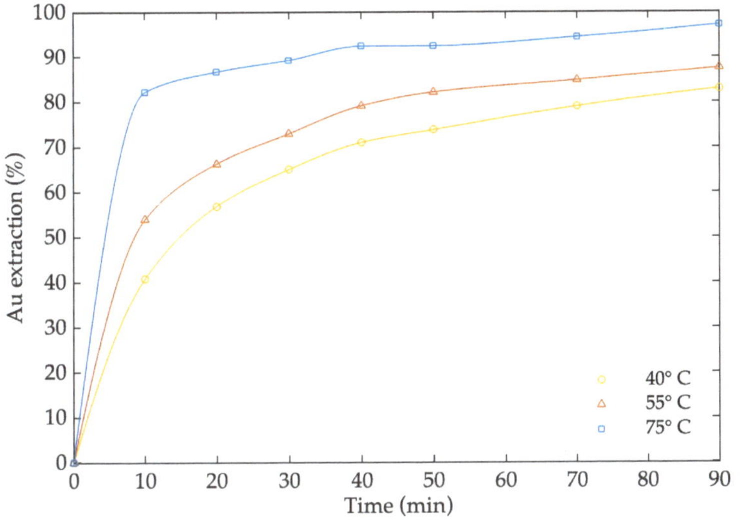


**Figure 5.** Au extraction (%) from simultaneous pressure leaching and oxidation at a pressure of 0.5 MPa and 6 g/L cyanide concentration.

According to Thomas [16], an increase in temperature will, in nearly all cases, increase the rate of a chemical reaction to a significant extent: for every 10 °C rises in temperature, the specific rate is increased by a factor of two or three. In a process similar to the chlorination leaching of pressure oxidized for refractory gold concentrate, the maximum temperature of 50 °C decreases the extraction due to the decomposition of agent leaching [25].

### 3.3.2. Effect of Cyanide Concentration

Sodium cyanide concentration (4–8 g/L) has an insignificant effect on gold extraction, as shown in Figure 6. According to Fleming [2], the cyanide concentration is less important than the oxygen concentration in gold extraction.

Results show that the use of excessive cyanide (above the optimum level) usually results in unnecessary cyanide consumption and has no beneficial effect on gold extraction. However, a high concentration of free cyanide is required for effective gold leaching because of the nature of the ore, which contains soluble sulfides. Cyanide consumption occurs because of the presence of other metal ions, such as copper and other metal ions. In this sulfide concentrate, the amount of copper is low (2.1%), whereas copper is very slightly dissolved [26]. In addition, at the temperature range of 45–75 °C, cyanide is not degraded. Parga et al. [18] have reported that gold recovery decreases at temperatures above 80 °C due to the oxidation of cyanide ions.

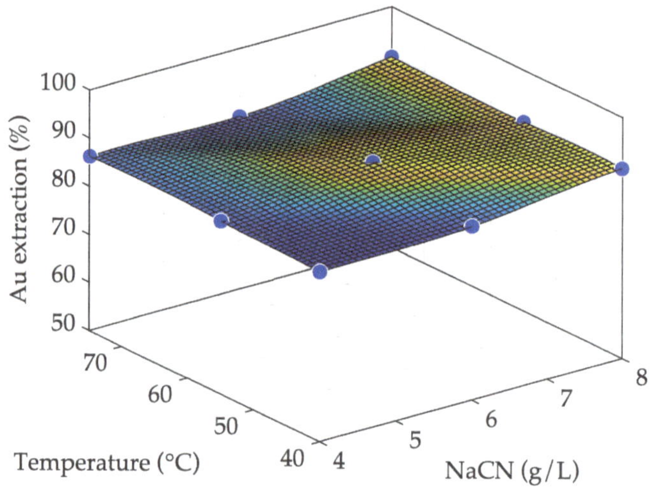


**Figure 6.** Effects of cyanide concentration on Au extraction (%) during simultaneous pressure leaching and oxidation at a pressure of 0.8 MPa.

#### 3.3.3. Effect of Particle Size on Gold Extraction

The gold concentrate was ground and screened at particle sizes of +173, −173/+147, −147/+104, −104/+74 μm to carry out tests using 6 g/L NaCN concentration, 0.5 MPa oxygen pressure, and 40 °C for 90 min of simultaneous pressure leaching and oxidation. Figure 7 shows that the particle size has an important effect on gold extraction, wherein the extraction of Au increases for small particle sizes. This effect was described by Chaidez et al. [27] for Cu extraction from chalcopyrite concentrate using acid leaching at low pressure.

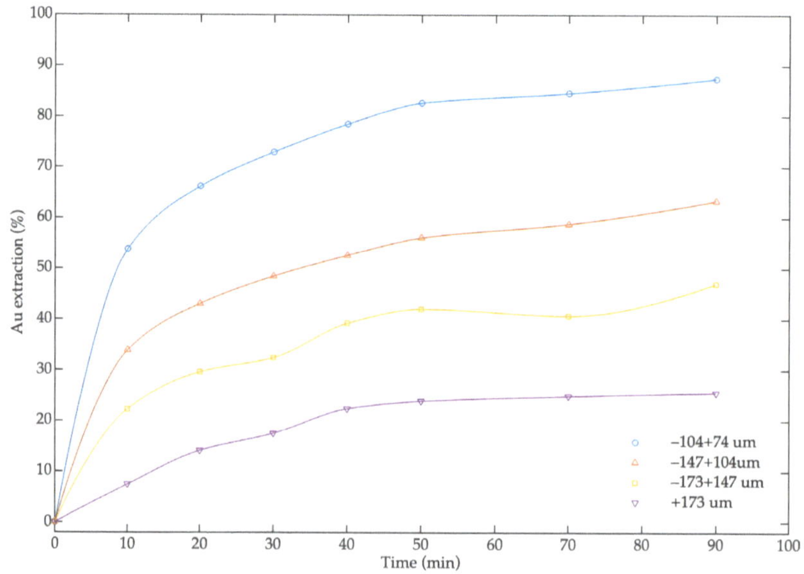


**Figure 7.** Effect of particle size (micron, μm) on Au extraction at 6 g/L cyanide concentration, 0.5 MPa pressure, and 40 °C.

### 3.3.4. Simultaneous Pressure Leaching and Oxidation

During simultaneous pressure leaching and oxidation, the gold concentrate is oxidized and leached in a single stage. Pyrite was present in a high amount in the concentrate in this study; it was oxidized in alkaline media due to NaCN working at a pH range of 11 to 12, Shown in Figure 8a) are XRD results for a sample of gold concentrate produced by alkaline oxidation via simultaneous leaching and oxidation at 75 °C and 1.0 MPa oxygen pressure. X-ray mapping of the pyrite particles is shown in Figure 8b).

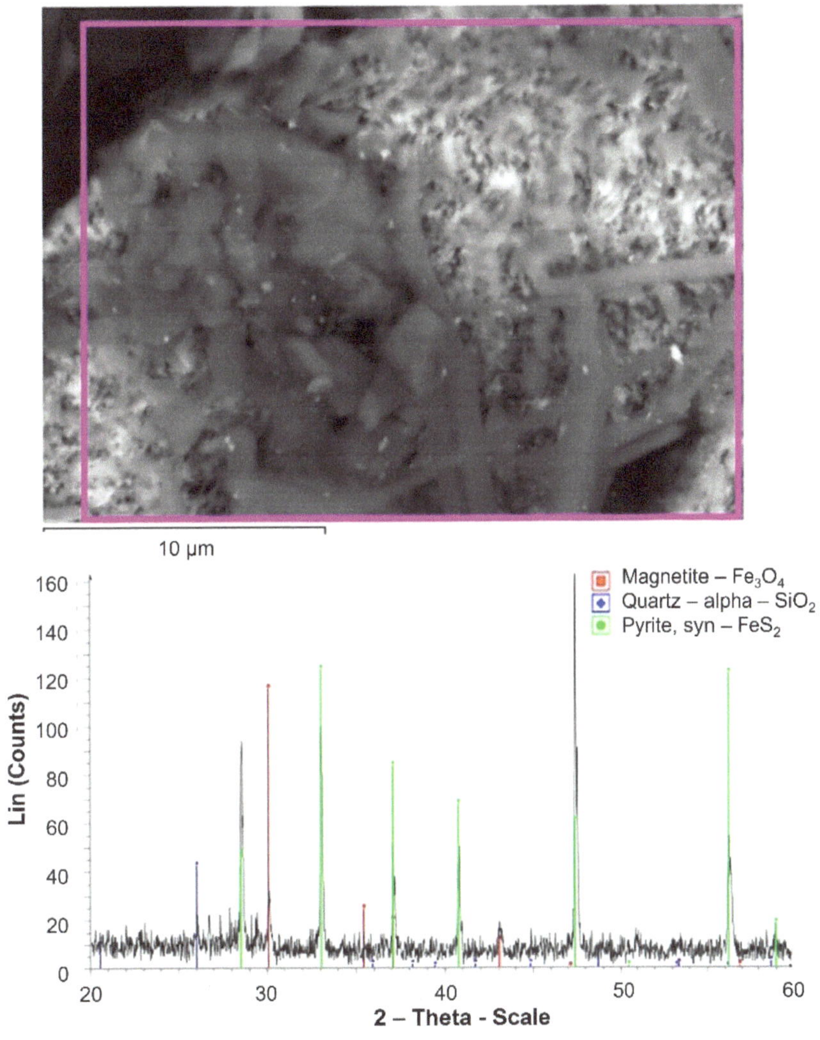


(**a**)

**Figure 8.** *Cont.*

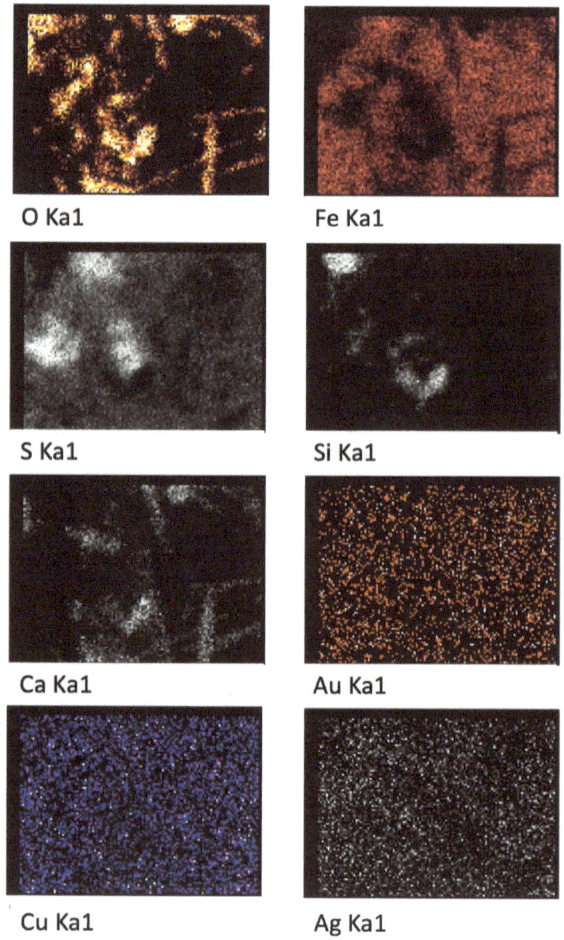


(b)

**Figure 8.** (**a**) Punctual microanalysis of particles by SEM–EDS micrograph showing a pyrite particle at 10 μm, and X-ray diffraction pattern of oxidized gold concentrate. The spectrum indicates the presence of pyrite, quartz, and magnetite. (**b**) X-ray mapping of the pyrite particles tests by SEM-EDS.

Pyrite alkaline oxidation results in hematite production. Caldeira et al. [15] indicated that the oxidative leaching of pyrite produces an insoluble iron oxide precipitate, and hematite was found in the precipitate, similar to in the main phase. Hematite is porous, and this improves the gold recovery via leaching. The magnetite in Figure 8 indicates partial oxidation of pyrite by low oxygen concentration, due the low oxygen pressure used on test.

## 4. Conclusions

The low amount of gold extracted by conventional cyanidation (46%) occurs when the mineral has refractoriness properties due to occlusion by the pyrite matrix. When the pressure leaching and oxidation during processing were 1.1 MPa and 55 °C, a gold dissolution of 89.51% was achieved. Pressure leaching and oxidation enhance gold extraction compared with tests conducted under atmospheric conditions. The effect of particle size on gold dissolution indicates that Au(%) extraction was increased with decreased particle size, 87–88% Au extraction was achieved at a size fraction of −104/+74 μm.

**Author Contributions:** Conceptualization, J.C.S.-U. and J.L.V.-G.; formal analysis, J.C.S.-U. and M.M.S.-C.; investigation, J.C.S.-U., J.L.V.-G. and G.T.-M.; methodology, J.C.S.-U., J.L.V.-G. and J.R.P.-T.; project administration, J.L.V.-G. and G.T.-M.; resources, V.M.V.-V., M.A.E.-R. and J.L.V.-G.; data curation, G.T.-M., M.A.E.-R. and M.M.S.-C.; formal analysis, J.C.S.-U. and J.L.V.-G.; supervision, J.L.V.-G. and M.M.S.-C.; validation, J.C.S.-U., M.A.E.-R. and V.M.V.-V.; writing—original draft, J.C.S.-U.; writing—review and editing, J.L.V.-G., J.R.P.-T., M.M.S.-C. and V.M.V.-V. All authors have read and agreed to the published version of the manuscript.

**Funding:** This Research received no external funding.

**Acknowledgments:** The authors gratefully acknowledge CONACYT for graduate scholarship to one author (J.C.S.U.), National Laboratory of Geochemistry and Mineralogy, Engineering Division and the Department of Chemical Engineering and Metallurgy of the University of Sonora and TNM—Institute of Technology of Saltillo.

**Conflicts of Interest:** The authors declare no conflict of interest.

## References

1. Marsden, J.; House, I. *The Chemistry of Gold Extraction*; SME: Littleton, CO, USA, 2006; ISBN 978-0-87335-240-6.
2. Fleming, C.A. Hydrometallurgy of Precious Metals Recovery. *Hydrometallurgy* **1992**, *30*, 127–162. [CrossRef]
3. Lorenzen, L.; Van Deventer, J.S.J. The Mechanism of Leaching of Gold from Refractory Ores. *Miner. Eng.* **1992**, *5*, 1377–1387. [CrossRef]
4. Angelidis, T.N.; Kydros, K.A. Selective Gold Dissolution from a Roasted Auriferous Pyrite-Arsenopyrite Concentrate. *Hydrometallurgy* **1995**, *37*, 75–88. [CrossRef]
5. Fernández, R.R.; Sohn, H.Y.; LeVier, K.M. Process for Treating Refractory Gold Ores by Roasting under Oxidizing Conditions. *Min. Metall. Explor.* **2000**, *17*, 1–6. [CrossRef]
6. Karthikeyan, O.P.; Rajasekar, A.; Balasubramanian, R. Bio-Oxidation and Biocyanidation of Refractory Mineral Ores for Gold Extraction: A Review. *Crit. Rev. Environ. Sci. Technol.* **2015**, *45*, 1611–1643. [CrossRef]
7. Deng, T.L.; Liao, M.X.; Wang, M.H.; Chen, Y.-W.; Belzile, N. Investigations of Accelerating Parameters for the Bio oxidation of Low-Grade Refractory Gold Ores. *Miner. Eng.* **2000**, *13*, 1543–1553. [CrossRef]
8. Ubaldini, S.; Vegliò, F.; Beolchini, F.; Toro, L.; Abbruzzese, C. Gold Recovery from a Refractory Pyrrhotite Ore by Biooxidation. *Int. J. Miner. Process.* **2000**, *60*, 247–262. [CrossRef]
9. Sun, L.-X.; Zhang, X.; Tan, W.-S.; Zhu, M.-L. Effects of Dissolved Oxygen on the Biooxidation Process of Refractory Gold Ores. *J. Biosci. Bioeng.* **2012**, *114*, 531–536. [CrossRef] [PubMed]
10. Pangum, L.S.; Browner, R.E. Pressure Chloride Leaching of a Refractory Gold Ore. *Miner. Eng.* **1996**, *9*, 547–556. [CrossRef]
11. Rusanen, L.; Aromaa, J.; Forsen, O. Pressure Oxidation of Pyrite-Arsenopyrite Refractory Gold Concentrate. *Physicochem. Probl. Miner. Process.* **2013**, *49*, 101–109. [CrossRef]
12. González-Anaya, J.A.; Nava-Alonso, F.; Pecina-Treviño, E.T. Gold Recovery Optimization of a Refractory Concentrate by Ultrafine Grinding—A Laboratory Study. *Min. Metall. Explor.* **2011**, *28*, 94–101. [CrossRef]
13. Birol, O.; Uçurum, M. Investigation of the Cyanide Leaching Optimization for Ultrafine Grinding Gold–Silver Ore. *Part. Sci. Technol.* **2016**, *34*, 633–638. [CrossRef]
14. Valenzuela, A.; Valenzuela, J.L.; Parga, J.R. Effect of Pretreatment of Sulfide Refractory Concentrate with Sodium Hypochlorite, Followed by Extraction of Gold by Pressure Cyanidation, on Gold Removal. *Adv. Chem. Eng. Sci.* **2013**, *3*, 171–177. [CrossRef]
15. Caldeira, C.L.; Ciminelli, V.S.T.; Dias, A.; Osseo-Asare, K. Pyrite Oxidation in Alkaline Solutions: Nature of the Product Layer. *Int. J. Miner. Process.* **2003**, *72*, 373–386. [CrossRef]
16. Thomas, K.G. Pressure Oxidation Overview. In *Developments in Mineral Processing*; Adams, M.D., Wills, B.A., Eds.; Advances in Gold Ore Processing; Elsevier: Amsterdam, The Netherlands, 2005; Volume 15, pp. 346–369.
17. Li, J.; Dabrowski, B.; Miller, J.D.; Acar, S.; Dietrich, M.; LeVier, K.M.; Wan, R.Y. The Influence of Pyrite Pre-Oxidation on Gold Recovery by Cyanidation. *Miner. Eng.* **2006**, *19*, 883–895. [CrossRef]
18. Parga, J.R.; Valenzuela, J.L.; Francisco, C.T. Pressure Cyanide Leaching for Precious Metals Recovery. *JOM* **2007**, *59*, 43–47. [CrossRef]
19. Qin, H.; Guo, X.; Tian, Q.; Yu, D.; Zhang, L. Recovery of Gold from Sulfide Refractory Gold Ore: Oxidation Roasting Pretreatment and Gold Extraction. *Miner. Eng.* **2021**, *164*, 106822. [CrossRef]
20. Lee, S.; Sadri, F.; Ghahreman, A. Enhanced Gold Recovery from Alkaline Pressure Oxidized Refractory Gold Ore After Its Mechanical Activation Followed by Thiosulfate Leaching. *J. Sustain. Metall.* **2022**, *8*, 186–196. [CrossRef]
21. Mahmoodi, F.; Azizi, A. Optimizing the Alkaline Oxidation Pretreatment of a Refractory Gold Ore Using Taguchi Orthogonal Array Method. *Mater. Res. Express* **2018**, *5*, 126516. [CrossRef]
22. Salazar-Campoy, M.M.; Valenzuela-García, J.L.; Quiróz-Castillo, L.S.; Encinas-Romero, M.A.; Tiburcio-Munive, G.; Guerrero-Germán, P.; Parga-Torres, J.R. Comparative Study of Gold Extraction from Refractory Pyritic Ores through Conventional Leaching and Simultaneous Pressure Leaching/Oxidation. *Min. Metall. Explor.* **2020**, *37*, 1279–1284. [CrossRef]

23. Elorza-Rodríguez, E.; Nava-Alonso, F.; Jara, J.; Lara-Valenzuela, C. Treatment of Pyritic Matrix Gold–Silver Refractory Ores by Ozonization–Cyanidation. *Miner. Eng.* **2006**, *19*, 56–61. [CrossRef]
24. Azizi, A.; Petre, C.F.; Larachi, F. Leveraging Strategies to Increase Gold Cyanidation in the Presence of Sulfide Minerals—Packed-Bed Electrochemical Reactor Approach. *Hydrometallurgy* **2012**, *111–112*, 73–81. [CrossRef]
25. Wen, Q.; Wu, Y.; Wang, X.; Zhuang, Z.; Yu, Y. Researches on Preparation and Properties of Sodium Polysulphide as Gold Leaching Agent. *Hydrometallurgy* **2017**, *171*, 77–85. [CrossRef]
26. Deschênes, G.; Prud'homme, P.J.H. Cyanidation of a Copper-Gold Ore. *Int. J. Miner. Process.* **1997**, *50*, 127–141. [CrossRef]
27. Cháidez, J.; Parga, J.; Valenzuela, J.; Carrillo, R.; Almaguer, I. Leaching Chalcopyrite Concentrate with Oxygen and Sulfuric Acid Using a Low-Pressure Reactor. *Metals* **2019**, *9*, 189. [CrossRef]

minerals

MDPI

*Article*

# Effects of Different Grinding Media and Milling Conditions on the Flotation Performance of a Copper-Pyrite Ore

**N. Metin Can [1,*] and Çağrı Başaran [2]**

[1] Department of Mining Engineering, Hacettepe University, Beytepe, Ankara 06800, Turkey
[2] Konutkent Mah., 2944 Sk., No. 10, Ankara 06810, Turkey
* Correspondence: metin.can@hacettepe.edu.tr

**Abstract:** Different milling conditions, such as wet or dry, and use of different grinding media have a great impact on the flotation performance of sulphide minerals. In the present study, the effects of wet and dry grinding and the use of different grinding media, such as mild steel (MS) and stainless steel (SS), were investigated on a Cu-sulphide ore. The samples were ground as dry and wet with both grinding media, to a $P_{80}$ value of $-75$ µm, and then flotation was carried out under the same conditions. The obtained data from flotation were evaluated in terms of solid/water recovery, chalcopyrite/pyrite recovery and separation efficiency. The effects of different milling conditions were discussed with the measured chemical parameters such as redox potential and dissolved oxygen level together with the flotation rate of chalcopyrite. The redox potential of the dry ground ore, irrespective of the type of milling media, was measured considerably higher than the wet grinding conditions. With SS media flotation, the rate of Cu was high for dry grinding, resulting in a higher selective concentrate in terms of grade. However, Cu recovery was lower due to the instability of the froth structure. Separation efficiency pointed out that the best flotation performance could be obtained using a wet grinding condition with MS balls.

**Keywords:** dry/wet grinding; stainless/mild steel ball; flotation; sulphide ore

**Citation:** Can, N.M.; Başaran, Ç. Effects of Different Grinding Media and Milling Conditions on the Flotation Performance of a Copper-Pyrite Ore. *Minerals* **2023**, *13*, 85. https://doi.org/10.3390/min13010085

Academic Editors: Marinela Ivanova Panayotova and Vladko Panayotov

Received: 11 December 2022
Revised: 1 January 2023
Accepted: 3 January 2023
Published: 5 January 2023



## 1. Introduction

Dissolution and electrochemical reactions, which have a great impact on flotation recovery, are influenced considerably by the grinding process. As the grinding operation is admitted as the last stage prior to concentration, the grinding environment has a significant effect on the behaviour of solids during subsequent steps, particularly in flotation [1]. The grinding process leads to the formation of highly active surface areas, some changes in the physical behaviour of solids which are attributed to permanent re-arrangement in crystal lattices and the structural alteration of particles [2,3]. Different milling procedures affect liberation, surface properties, particle morphology and surface roughness which can directly determine the hydrophobicity of the minerals and their flotation response [4,5]. The floatability of the ore, separation efficiency of minerals and selectivity are determined by surface properties which are controlled by the grinding process and its conditions [6]. Some defects such as cracks and pores may form on the surface of the particles depending on the grinding method, time and/or grindability of a mineral. These defects may have quite significant effects on the surface adsorption of collectors. Wet grinding generally produces particles of large specific surface and high surface oxidation, while dry grinding accounts for a large amount of surface defects and low surface oxidation. The results of a limited number of comparative studies performed on wet and dry grinding on flotation showed that dry ground materials are generally poorly wettable, while wet ground materials react quickly with water so that active charge centres and radicals are neutralized or disappear as a consequence of dissolving reactions. Gas molecules (especially oxygen) are physically or chemically adsorbed after dry grinding [7]. These films of gas will retard the wetting process to a greater or lesser extent. This behaviour can also be observed less prominently in many sulphides. Lepetic [8] has observed an increase in the collectorless floatability of chalcopyrite after dry autogenous grinding. This was attributed to both oxygen satisfying bonds and the formation of elemental sulphur or metal-deficient sulphur at the surface.

In industry, wet grinding is more commonly used because of the downstream processing requirements, such as target liberation of particle size, and the nature of the concentration processes, such as flotation. Feng and Aldrich [9] have compared the effects of dry and wet grinding on the flotation performance of a complex sulphide ore. They stated that flotation with dry milled ore showed more stable, highly loaded froths, and faster flotation kinetics, which were attributed to the activation of the highly formed new surface area of the particles. However, wet milled ore showed higher selectivity [10,11] and fine sulphide particles were found to have a tendency to oxidize in the air medium of dry grinding.

In addition to wet or dry grinding, the difference in the grinding media is a parameter that highly affects the flotation of sulphide minerals. The shape of the grinding media was found to have no significant effect on flotation [12]. Peng et.al [13] stated that a 30 wt.% chromium grinding media produced higher chalcopyrite selectivity against pyrite than mild steel media, while gas (oxygen and nitrogen) purging had no significant effect on chalcopyrite flotation. Other studies have been conducted on the use of balls of different types and chromium contents [14–19]. Besides grinding media, mill construction materials also have influence on the flotation performance of Cu-sulphide ores. It has been shown that ferrous ions and sulphide minerals cause galvanic interaction which reduces the redox potential of the pulp and, accordingly, results in a poor flotation response when carbon steel balls/rods are used. Utilisation of lined mills and non-ferrous or stainless steel grinding media were suggested to avoid these negative effects [1]. It has been stated that it is possible to perform pyrite flotation at alkaline pH depending on the changes in pulp potential when forged steel grinding media are used instead of chrome steel with different content [8].

In recent years, dry processing or water-free treatments have become more important for mineral processing applications in arid and semi-arid climates. In fact, water-free treatments or treatments using the minimum amount of fresh water should be encouraged for every part of the world due to changing climatic conditions and a shortage of consumable water resources. However, many enrichment methods in ore preparation, particularly flotation, are water-driven unit processes. In most of the cases, the mineralogy of complex ores does not allow for the use of separation processes other than flotation due to the necessity of fine grinding. Accordingly, the effects of dry grinding on flotation should be examined in order to be an alternative to the effective use of water. It has been well documented in the literature how dry or wet grinding and the use of different grinding media will affect the flotation of especially sulphide minerals. However, there are few studies in which all of these are combined and compared. This work thoroughly discusses how different grinding conditions and grinding media such as wet/dry and mild/stainless steel can affect flotation performance of a Cu-sulphide ore in terms of pulp potential, dissolved oxygen, recovery and separation efficiency.

## 2. Material and Methods

### 2.1. Ore Characterization

An ore sample taken from the rod mill feed of a Cu-sulphide ore processing plant was used in the experiments. The mineralogical composition of the ore was quite simple, consisting of 6.2% chalcopyrite, about 63% pyrite and non-sulphide gangue minerals, such as mostly quartz, gypsum, calcite, dolomite, chlorite, siderite/ankerite which accounted for the remainder. The general chemical composition of the ore is given in Table 1.

**Table 1.** Chemical composition of the ore.

| Element | % | Element | % |
|---|---|---|---|
| Al | 2.14 | Mn | 0.03 |
| Ca | 1.03 | Na | 0.81 |
| Co | 0.09 | P | 0.41 |
| Cu | 2.14 | Pb | 0.05 |
| Fe | 30.24 | Si | 7.83 |
| K | 0.83 | Zn | 0.33 |
| Mg | 0.87 | | |

### 2.2. Grinding Tests

Wet and dry grinding were conducted with mild steel (MS) and stainless steel (SS) balls to determine the effects of the grinding medium on flotation. The MS balls contained 20% of chromium. The ore samples, each 1150 g, were ground as dry and wet with both grinding media, 83% of which was −75 μm, based on the plant operation information, and then flotation was carried out under the same conditions. The mill was 30.5 cm × 30.5 cm in size and was used with 25 mm (161 pcs, 49% *w*/*w*), 20 mm (192 pcs, 31% *w*/*w*) and 15 mm (284 pcs, 20% *w*/*w*) balls. Grinding in wet conditions was performed at 60% solid by weight. As a result of the grinding tests, it was determined that wet and dry grinding should be 11 and 25 min, respectively, to achieve the flotation feed target size.

### 2.3. Flotation Tests

The flotation tests were carried out with freshly ground samples in a volume of 3 lt Denver DR-12 conventional flotation cell with the conditions of 4 lt/min of air flow rate, 1200 rev/min of agitator speed and 30% solids by weight. The pulp pH was adjusted to 11.5 after grinding in both conditions. A TPS meter, 90FL-T model, was used to continuously record the Eh, pH and dissolved oxygen (DO) values during the flotation tests. A mixture (50:50) of a thionocarbamate (Hostaflot X231) and dithiophosphate (Aerofin 3477) was used as the collector after pH adjustment in 10 g/t, 20 g/t and 30 g/t dosages in the flotation and conditioning for 3 min for each collector. Then, Dowfroth 250 was added as a frother at 10 g/t dosage and conditioned for 1 min. Reagents were selected based on plant operating information. Four concentrates were obtained after a total of 8 min of flotation time by collecting the top-layer froth once every 15 s and analysing with a Varian Spectra AA5 Atomic Absorbance Spectrometer. The average data of the repeated flotation tests are presented in this paper. To examine the froth phase, the images captured during the flotation by a digital video camera, which was placed over the flotation cell, were analysed by the SmartFroth software. This software was used to measure the size and the speed of the bubbles.

## 3. Results and Discussion

### 3.1. Variations in Pulp Redox Potential

Variations in redox potential, dissolved oxygen and pH of the pulp were measured continuously during the flotation tests. Figure 1 illustrates the variations in redox potential of the pulp with the time for different grinding conditions. This graph represents the change in pulp potential in the system during flotation from the moment air is introduced to initiate flotation following conditioning. It is obvious that the wet-MS grinding process produced the lowest redox potential, as low as −400 mV. MS media decreased the pulp potential drastically due to the release of ferrous ions which in turn consumed all of the dissolved oxygen in the pulp. It is known that the optimum Eh value for collector adsorption ranges is between 0 to 200 mV for this sulphide type of ores. Increasing oxygen content in the pulp consumes the excess electrons and therefore increases the pulp potential. As a result of that, adsorption of thiol collector on sulphide minerals takes place at suitable pulp potential values. Therefore, it is not likely for the collectors to adsorb on the mineral surfaces until the potential reaches the suitable value. It is known that particularly in wet grinding conditions, oxygenation during conditioning and flotation is a significant parameter in the selective flotation of chalcopyrite from pyrite [20,21]. Introduction of air to the pulp at the beginning gradually increased the potential after 2–3 min of flotation, which affected the rate of flotation of copper minerals significantly. In wet grinding, the pulp potential increased from −400 mV to −200 mV by changing the medium from MS to SS, but it was observed that the potential still remained in the reducing condition, which was attributed to the galvanic interaction between chalcopyrite and pyrite. The dissolved oxygen was almost completely consumed in wet grinding by both grinding media. However, the presence of dissolved ferrous ions due to the galvanic interaction between the MS media and the sulphide minerals resulted in lower Eh. The redox potential of the dry ground ore, irrespective of the type of milling media, was measured considerably higher than the wet grinding conditions. In parallel to that, the DO concentrations were about 3–4 ppm in the beginning of the flotation. The absence of galvanic interaction in dry grinding affected the pulp chemistry considerably with both grinding media, and resulted in oxidizing pulp conditions. The higher redox potential and DO concentration enhanced the adsorption of collectors and flotation kinetics accordingly, as expected.

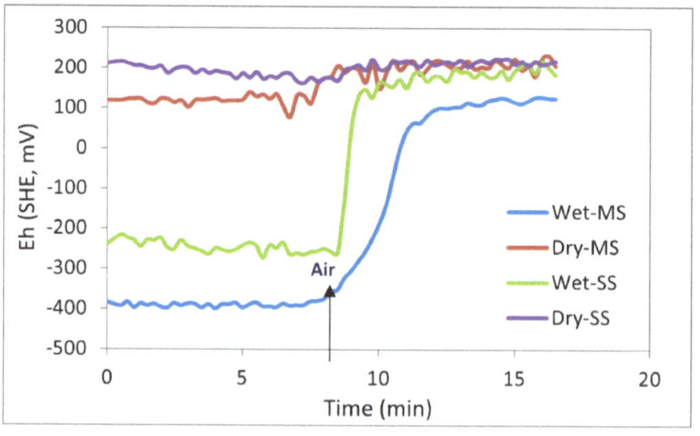


**Figure 1.** Redox potential of the pulp during at different milling conditions.

Figure 2 shows the relationship between the flotation rate constant of the chalcopyrite and redox potential of the pulp as measured in the conditioning stage. As the pulp potential was low in wet grinding conditions, the flotation rate of Cu was reduced. However, the dry grinding condition, particularly with SS media, resulted in the highest flotation rate.

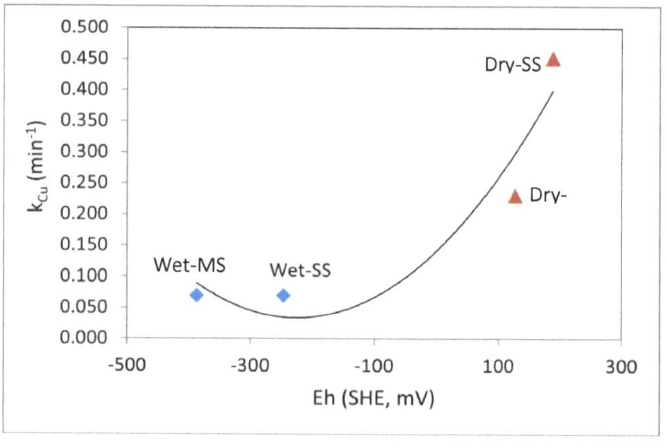


**Figure 2.** The relationship between pulp redox potential and flotation rate of Cu after different milling conditions.

### 3.2. Variations in DO Concentrations

There is a close relationship between the rate of DO consumption, pulp redox potential and therefore flotation selectivity [7,10]. Figure 3 shows the DO measurements taken after introducing air at a 4 lt/min rate at every 15 s interval. The curves indicate a clear difference between the rate of increase in DO concentration and the grinding conditions. The lowest rate was obtained with wet-MS grinding and the highest rate with dry-SS milling. Under wet milling conditions, the DO concentration appeared to increase gradually in the pulp and stabilized at around 7–8 ppm. However, in dry grinding conditions, DO concentration increased with an instant increase and remained stable around 9 ppm, especially in the use of SS media. It can be seen that the shapes of the curves in Figure 3 are quite similar to the metal recovery curves. Therefore, the rate of increase in DO concentration was calculated using the following equation:

$$DO = DO_{max}\left(1 - e^{-k_{DO}*t}\right) \quad (1)$$

where:

$DO$: Dissolved Oxygen

$DO_{max}$: Maximum dissolved oxygen

$k_{DO}$: Rate of dissolved oxygen concentration

$t$: time

$DO_{max}$ was approximately 9 ppm for this condition, and it was the maximum attainable DO concentration for fully aerated pulp. In this case, the rate of increase of DO concentration was determined by the rate of galvanic interaction between the grinding media and sulphide minerals, and between the sulphide minerals.

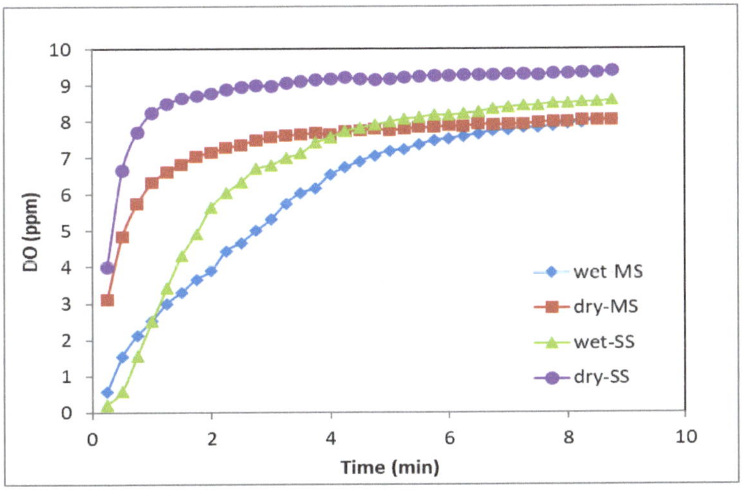


**Figure 3.** DO concentrations in the pulp during flotation for different milling conditions.

Figure 4 shows the relationship between the flotation rate constant of copper ($k_{Cu}$) and DO rate constant. There is a linear relationship between the two parameters, showing the dependence of flotation on the DO consumption of the ore.

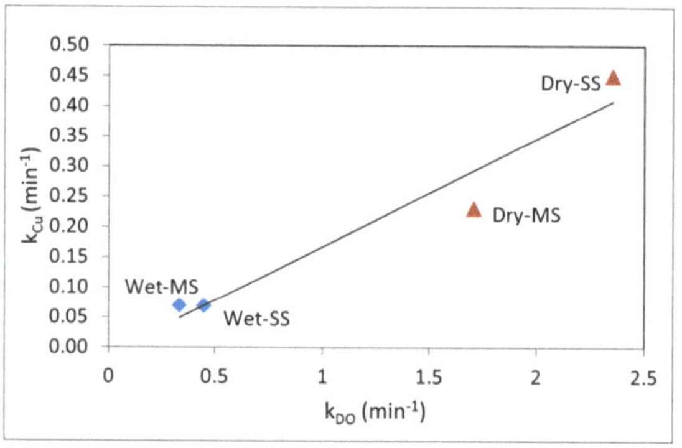


**Figure 4.** The relationship between DO concentration and flotation rate of Cu after different milling conditions.

### 3.3. Variations in Flotation Performance

Flotation tests performed after grinding processes with both MS and SS balls in wet and dry conditions were evaluated briefly below for three different dosages of collector mixture in terms of solid-water recovery (Rsolid and Rwater), copper-pyrite recovery and pyrite recovery mechanism. The water recovery can be used as an indicator of changes in the froth phase. High water recovery indicates a fast and watery froth structure, while low water recovery indicates a dry froth. Figure 5 shows that there is a linear relationship between the solid and the water recovery values. It is seen that after grinding with MS balls, both solid and water recoveries were much higher than the grinding tests with SS balls. This pointed out that froth fluidity and stability were higher in MS-ball milling conditions. Compared to the tests with MS balls, it was observed that the froth obtained in flotation after wet grinding with SS balls was very dry and a much smaller amount of solid recovered to the concentrate (Figure 5a). This situation can be attributed to the presence of metallic iron mixing with the pulp via MS balls which results in the formation of iron oxidation species. When the metallic iron does not exist, the froth is very brittle and the air bubbles may burst immediately. Higher water recovery can be obtained because of the more stable froth structure [22].

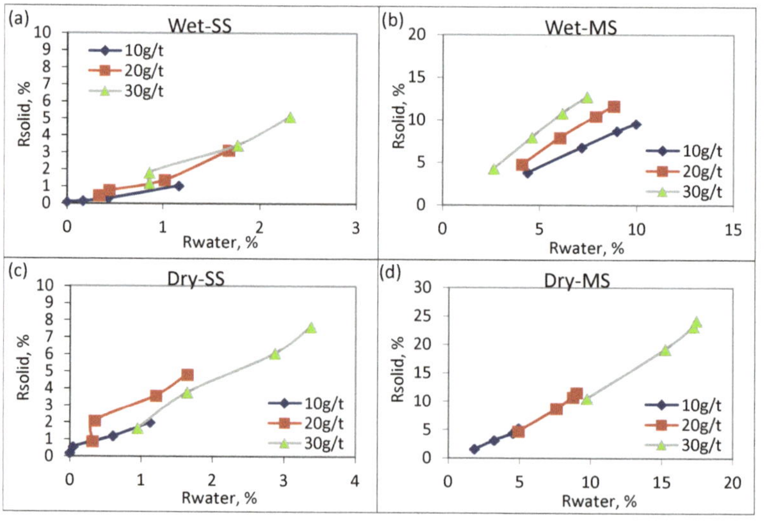


**Figure 5.** Solid vs. water recovery in flotation after milling conditions of (**a**) Wet-SS (**b**) Wet-MS (**c**) Dry-SS (**d**) Dry-MS.

Figure 6 shows the relationship between cumulative pyrite and copper recovery values after flotation for different milling conditions. Obtained data indicated that cumulative copper and pyrite recoveries increased with the collector dosages. Metallic iron mixed from the MS balls to the pulp during grinding caused an increase in the pyrite and copper recoveries by making the froth more stable in the flotation experiments. However, it is seen that the water recovery was very low in grinding conditions with SS balls (Figure 5a,c). The excessive viscous nature of the froth caused bubble coalescence and then easily burst (Figure 7). As a result, a significant decrease in copper recovery was noticed in the case of using SS balls compared to grinding conditions with MS balls. A similar observation was found by [23]. Chalcopyrite selectivity against pyrite was found to be reduced with SS media, while it was restored when grinding was performed using MS media. Ekmekçi [24] stated that the flotation behaviour of chalcopyrite and pyrite in mixtures were different from experiments with single minerals, and due to the galvanic interaction between the two minerals, the surface properties of the minerals were affected by each other. For example, it has been determined that copper ions dissolved from chalcopyrite were replaced by $Fe^{2+}$ ions on the pyrite surface and caused hydrophobic sulphur rich CuS layer formation [25]. It is also reported by Peng et al. [13] that MS ball media generate reducing grinding conditions which favoured the formation of a copper(I) sulphide phase on pyrite surfaces, which results in high pyrite activation. However, in this case, besides pyrite activation, it is thought that the significant increase in pyrite recovery, particularly in the milling condition with MS balls, was most likely due to the increase in water recovery. The highest cumulative copper recovery for dry MS condition was 60.25% and cumulative pyrite recovery was 25.21% at a 30 g/t collector dosage. In addition, it was determined that the pyrite recovery, corresponding to one unit of copper recovery, increased with the increase in the collector dosage. This shows that the increase in the collector dosage significantly increased the amount of floating pyrite.

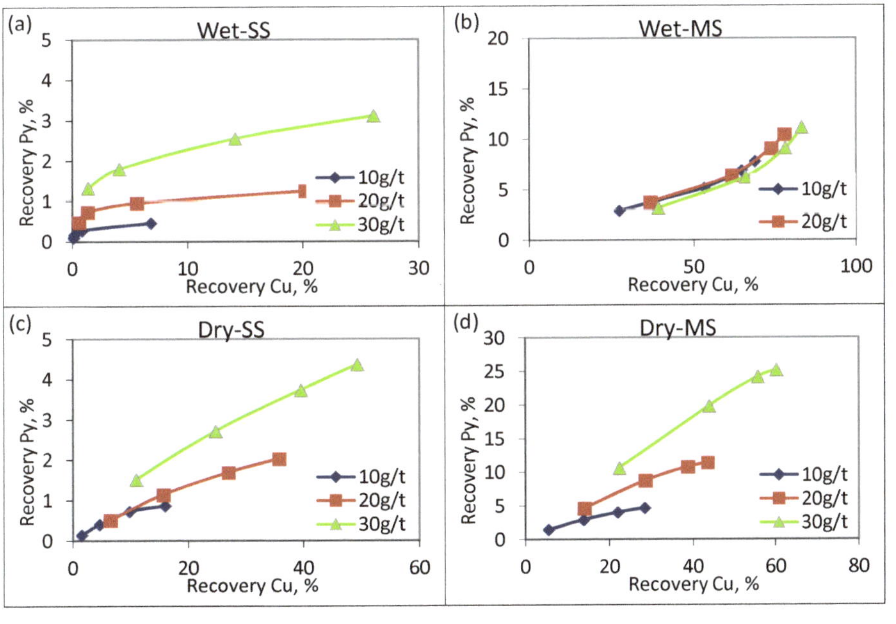


**Figure 6.** Pyrite vs. copper recovery in flotation after milling conditions of (**a**) Wet-SS (**b**) Wet-MS (**c**) Dry-SS (**d**) Dry-MS.

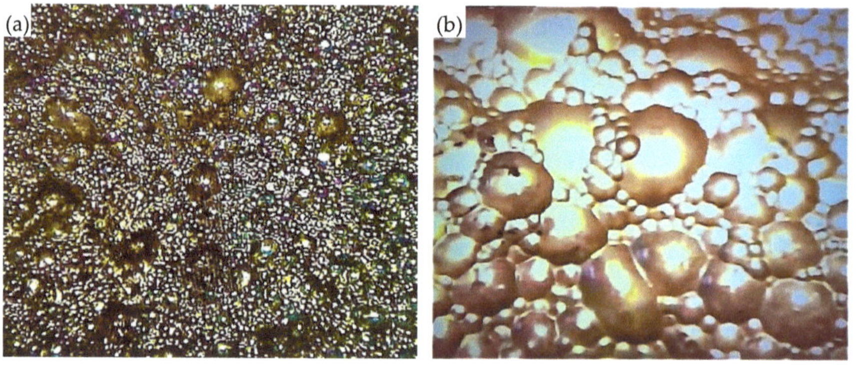


**Figure 7.** Froth images captured by Smart Froth for 30 g/t collector dosage. (**a**) Wet grinding with MS (**b**) Wet grinding with SS.

The effect of different grinding media on the flotation behaviour of pyrite is shown in Figure 8. It was observed that the least pyrite recovery to the concentrate was obtained in the experiments with SS balls grinding. However, it should be noted that the water and solid recoveries obtained under this condition were also very low. It is seen that pyrite was reported to concentrate depending on the water recovery, or, in other words, with entrainment, in the dry grinding condition with MS balls (Figure 8d). However, in the wet grinding condition with MS balls, when the collector concentration was increased to 30 g/t, it is seen that the amount of pyrite corresponding to the unit water recovery was recovered by true flotation (Figure 8b). It can be said that the similar situation is valid for the grinding condition made with SS balls, although the pyrite recovery is very low (Figure 8a,c). The increase in the collector dosage activated the pyrite, and hence increased its recovery.

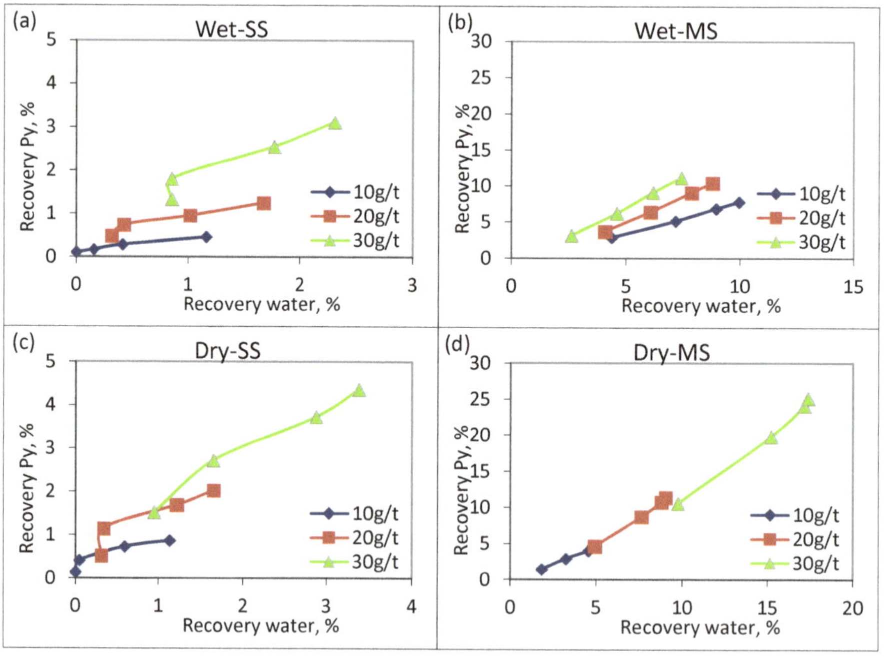


**Figure 8.** Pyrite vs. water recovery in flotation after milling conditions of (**a**) Wet-SS (**b**) Wet-MS (**c**) Dry-SS (**d**) Dry-MS.

The wet grinding condition with MS balls provided higher Cu recoveries in copper flotation (Figure 6). Although the copper flotation efficiency was very low in the experiments with SS balls, it was observed that the highest Cu concentrate grade was obtained under these conditions (Figure 9). After grinding with SS balls, the stability of the froth was very low and the chemical conditions were suitable for chalcopyrite flotation. However, the flotation efficiency was very low due to the very high hydrophobic particles making the froth extremely brittle and unstable. Therefore, in copper flotation after grinding with SS balls, a more selective concentrate in terms of Cu grade could be obtained.

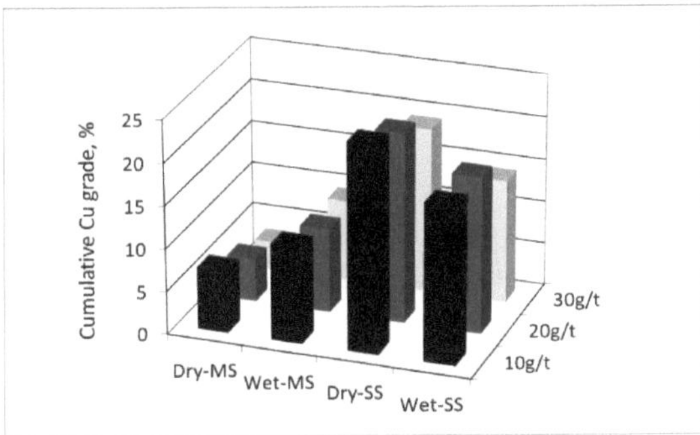


**Figure 9.** Cumulative Cu grade of the concentrates in flotation after different milling conditions.

The following separation efficiency equation [26] was used in order to evaluate the overall flotation performance in terms of both recovery and grade depending on the difference in the grinding medium and the concentration of the collector dosage:

$$E = R(1 - (f(c_m - c)/c(c_m - f))) \tag{2}$$

where:

$E$: Separation efficiency
$R$: Recovery of valuable mineral (%)
$f$: Grade of feed (%)
$c$: Grade of concentrate (%)
$c_m$: Maximum grade of concentrate (%) (Maximum Cu grade can be obtained from chalcopyrite in theory)

The calculated separation efficiency values for different milling conditions and the collector dosages are plotted in Figure 10. It is seen that the highest separation efficiency in terms of recovery and grade value was obtained in the wet grinding condition made with MS balls. In any case, the separation efficiency increased with the increase in the collector dosage.

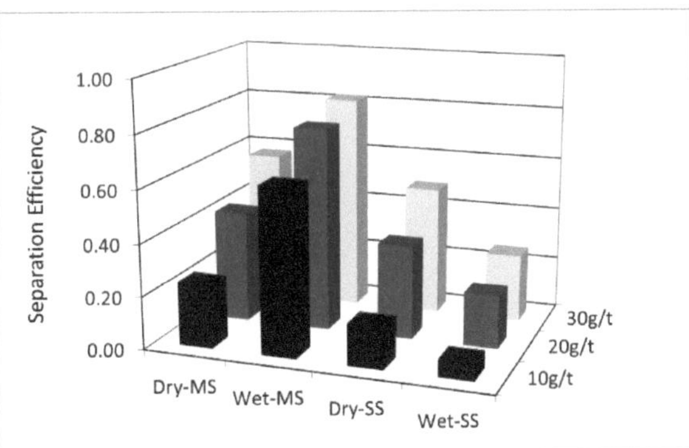


**Figure 10.** Separation efficiencies in flotation after different milling conditions.

## 4. Conclusions

In view of investigating different grinding conditions with wet/dry operation and different grinding media, in terms of grade and recovery in sulphide mineral flotation, the following conclusions can be deduced:

1. Dry grinding eliminates the galvanic interaction between sulphide minerals and the grinding media. Therefore, the redox potential of the dry ground ore, irrespective of type of milling media, was measured considerably higher than the wet grinding conditions. This situation resulted in a high flotation rate of Cu, particularly in the dry grinding condition with SS media.
2. In the dry grinding condition in which SS media were used, the dissolved oxygen amount was also measured as the highest value, around 9 ppm, and showed a linear relationship with the Cu flotation rate.
3. The dry grinding condition, particularly with SS balls, resulted in a higher selective concentrate in terms of Cu grade due to it serving a better electrochemical condition for sulphide minerals. This could be explained with the high amount of dissolved oxygen which enhanced the adsorption of collectors and flotation kinetics, accordingly. However, the Cu recoveries of the concentrates were lower due to the instability of the froth structure and the coalescence of air bubbles in flotation after grinding with SS balls. The most stable froth structure was observed in the tests in which there was milling with MS balls, and it was concluded that it was due to the metallic iron mixed into the pulp from the MS balls.
4. It was determined that the pyrite dilution was higher in Cu concentrate after milling with MS balls, and the pyrite recovery mechanism was by entrainment, especially in the dry milling condition. The pyrite dilution in the concentrate was lower in the case of milling with SS balls, but another possible reason was that the solid and water recoveries were already low under this condition.
5. Separation efficiency pointed out that the best flotation performance could be obtained using the wet grinding condition with MS balls.
6. Since dry grinding eliminates galvanic interaction, it has potential for the future in reducing the contamination of recycling plant water, using less fresh water and providing higher flotation kinetics. It is known that the flotation of sulphide minerals is strongly dependent on the redox potential and dissolved oxygen content of the pulp. After dry grinding, the surface of sulphide minerals will not be oxidized when they are introduced into the flotation pulp. Therefore, very high flotation rates and even collectorless flotation in some cases can be obtained after the dry grinding of sulphide ores. However, after dry grinding, the instability of the froth structure needs to be improved, especially in case SS balls are used.

**Author Contributions:** Conceptualization, N.M.C.; methodology, N.M.C. and Ç.B.; validation, Ç.B.; formal analysis, N.M.C. and Ç.B.; investigation, N.M.C. and Ç.B.; data curation, Ç.B.; writing—original draft preparation, Ç.B.; writing—review and editing, N.M.C.; supervision, N.M.C. All authors have read and agreed to the published version of the manuscript.

**Funding:** This research received no external funding.

**Data Availability Statement:** Not applicable.

**Conflicts of Interest:** The authors declare no conflict of interest.

## References

1. Gonçalves, K.L.C.; Andrade, V.L.L.; Peres, A.E.C. The effect of grinding conditions on the flotation of a sulphide copper ore. *Miner. Eng.* **2003**, *16*, 1213–1216. [CrossRef]
2. Boldyrev, V.V.; Avvakumov, E.G. Mechanochemistry of inorganic solids. *Russ. Chem. Rev.* **1971**, *40*, 847–859. [CrossRef]
3. Butyagin, P.Y. Kinetics and nature of mechanochemical reactions. *Russ. Chem. Rev.* **1971**, *40*, 901–915. [CrossRef]
4. Wang, X.H.; Xie, Y. The effect of grinding media and environment on the surface properties and flotation behaviour of sulfide minerals. *Miner. Process. Extr. Metall. Rev.* **1990**, *7*, 49–79. [CrossRef]

5. Chapman, N.A.; Shackleton, N.J.; Malysiak, V.; O'Connor, C.T. Comparative study of the use of HPGR and conventional wet and dry grinding methods on the flotation of base metal sulphides and PGMs. *J. South. Afr. Inst. Min. Metall.* **2013**, *113*, 407–413.
6. Zang, S.R.; Moon, K.S.; Leja, J. Effect of grinding media on the the surface reactions and flotation of heavy metal sulphides. In *Flotation*; Furstenau, M.C., Ed.; AIEM: New York, NY, USA, 1976; pp. 509–527.
7. Chelgani, S.C.; Parian, M.; Parapari, P.S.; Ghorbani, Y.; Rosenkranz, J. A comparative study on the effects of dry and wet grinding on mineral flotation separation—A review. *J. Mater. Res. Technol.* **2019**, *8*, 5004–5011. [CrossRef]
8. Lepetic, V.M. *Flotation of Chalcopyrite without Collector after Dry Autogenous Grinding*; C.I.M. Bull.: Westmount, QC, Canada, 1974; pp. 71–77.
9. Feng, D.; Aldrich, C. A comparison of the flotation of ore from the Merensky Reef after wet and dry grinding. *Int. J. Miner. Process.* **2000**, *60*, 115–129. [CrossRef]
10. Seke, M.D.; Pistorius, P.C. Effect of cuprous cyanide, dry and wet milling on the selective flotation of galena and sphalerite. *Miner. Eng.* **2006**, *19*, 1–11. [CrossRef]
11. Peltoniemi, M.; Kallio, R.; Tanhua, A.; Luukkanen, S.; Perämäki, P. Mineralogical and surface chemical characterization of flotation feed and products after wet and dry grinding. *Miner. Eng.* **2020**, *156*, 106500. [CrossRef]
12. Corin, K.C.; Song, Z.G.; Wiese, J.G.; O'Connor, C.T. Effect of using different grinding media on the flotation of a base metal sulphide ore. *Miner. Eng.* **2018**, *126*, 24–27. [CrossRef]
13. Peng, Y.; Grano, S.; Fornasiero, D.; Ralston, J. Control of grinding conditions in the flotation of chalcopyrite and its separation from pyrite. *Int. J. Miner. Process.* **2003**, *69*, 87–110. [CrossRef]
14. Huang, G.; Grano, S.; Skinner, W. Galvanic interaction between grinding media and arsenopyrite and its effect on flotation: Part II. Effect of grinding on flotation. *Int. J. Miner. Process.* **2006**, *78*, 198–213. [CrossRef]
15. Zhang, X.; Qina, Y.; Han, Y.; Li, Y.; Gao, P.; Li, G.; Wang, S. A potential ceramic ball grinding medium for optimizing flotation separation of chalcopyrite and pyrite. *Powder Technol.* **2021**, *392*, 167–178. [CrossRef]
16. Song, Z.G.; Corin, K.C.; Wiese, J.G.; O'Connor, C.T. Effect of different grinding media composition on the flotation of a PGM ore. *Miner. Eng.* **2018**, *124*, 74–76. [CrossRef]
17. Rabieh, A.; Eksteen, J.J.; Albijanic, B. Galvanic interaction of grinding media with arsenopyrite and pyrite and its effect on gold cyanide leaching. *Miner. Eng.* **2018**, *116*, 46–55. [CrossRef]
18. Mu, Y.; Cheng, Y.; Peng, Y. The interaction of grinding media and collector in pyrite flotation at alkaline pH. *Miner. Eng.* **2020**, *152*, 106344. [CrossRef]
19. Ke, B.; Chen, J.; Cheng, W. Galvanic interaction between different grinding media and galena (100) surface and its influence on galena flotation behavior: A DFT study. *Appl. Surf. Sci.* **2022**, *571*, 151379. [CrossRef]
20. Houot, R.; Duhamet, D. Importance of oxygenation of pulps in the flotation of sulphide ores. *Int. J. Miner. Process.* **1990**, *29*, 77–87. [CrossRef]
21. Kuopanportti, H.; Suorsa, T.; Dahl, O.; Niinimäki, J. A model of conditioning in the flotation of a mixture of pyrite and chalcopyrite ores. *Int. J. Miner. Process.* **2000**, *59*, 327–338. [CrossRef]
22. Van Deventer, J.S.J. Dependence of froth behaviour on galvanic interactions. In *Frothing in Flotation II*; Laskowski, J.S., Woodburn, E.T., Eds.; OPA: Amsterdam, The Netherlands, 1998; pp. 337–365.
23. Yuan, X.M.; Pålsson, B.I.; Forssberg, K.E. Flotation of a complex sulphide ore: II. Influence of grinding environments on Cu/Fe sulphide selectivity pulp chemistry. *Int. J. Miner. Process.* **1996**, *46*, 181–204. [CrossRef]
24. Ekmekçi, Z. Role of Galvanic Interaction on Collectorless Flotation of Chalcopyrite and Pyrite. PhD. Thesis, Hacettepe University, Ankara, Turkey, 1995; p. 225.
25. Weisener, C.; Gerson, A. Cu(II) adsorption mechanism on pyrite: An XAFS and XPS study. *Surf. Interface Anal.* **2000**, *30*, 454–458. [CrossRef]
26. Schulz, N.F. Separation efficiency. *Trans. AIEM* **1970**, *247*, 81–87.

*minerals*

MDPI

*Article*

# Cost-Effective and High Purity Valuable Metals Extraction from Water Leaching Solid Residues Obtained as a By-Product from Processing the Egyptian Boiler Ash

**Ahmed H. Ibrahim [1,2], Xianjun Lyu [1,*], Bahig M. Atia [3], Mohamed A. Gado [3] and Amr B. ElDeeb [2]**

1 College of Chemical and Biological Engineering, Shandong University of Science and Technology, Qingdao 266590, China
2 Mining and Petroleum Department, Faculty of Engineering, Al-Azhar University, Cairo 11884, Egypt
3 Nuclear Materials Authority, Cairo 530, Egypt
* Correspondence: lyuxianjun@163.com; Tel.: +86-532-86057101

**Abstract:** The water leaching solid residues (WLSR) obtained from salt-roasting Egyptian boiler ash are considered an essential secondary resource for (13%) nickel and (5.6%) zinc extraction. Hence, the current study aims for the cost-effective and high purity Ni, Zn, Fe and Mg metal ion extraction from (WLSR) using a sulfuric acid leaching process. The factors affecting the percentage recovery of Ni, Zn, Fe and Mg from WLSR, including leaching temperature, time, acid concentration and solid/liquid ratio, have been investigated. The obtained leaching solutions were analyzed chemically using ICP, and the different precipitates were analyzed mineralogically using XRD and EDX analysis and chemically using XRF. The maximum percentage recovery of Ni, Zn, Fe and Mg was 95.02%, 90.13%, 66.29% and 75.73%, which was obtained under the optimum leaching conditions of 8% $H_2SO_4$ concentration and 1/15 solid/liquid ratio at 85 °C for 240 min. The effect of pH, $Fe_2O_3$ dosage as nucleating agent and the precipitation duration on iron removal and Ni and Zn loss have been thoroughly studied. It has been found that >95% of the contained iron impurity can be removed, while nickel and zinc losses are around 4.2% and 3.8%, respectively. Additionally, a pH of 6 and 0.45 mol/L concentration of $H_2C_2O_4$ was utilized to precipitate Mg as $MgC_2O_4.2\,h_2O$, demonstrating that the precipitation efficiency of Mg reaches 96.9%. Nickel and zinc precipitation efficiency was 92.25% and 85.51%, respectively, by raising the solution pH to approximately 9. The kinetic of Ni and Zn dissolution has been investigated to explain the mechanism prevalent and the factors influencing the leaching process. It has been found that the nickel leaching kinetic is controlled by both diffusion through an inert porous layer and by chemical reaction with an activation energy of 20.25 kJ.mol$^{-1}$. Meanwhile, the kinetic of zinc leaching is controlled by solid product layer diffusion with an activation energy of 11.67 kJ mol$^{-1}$.

**Keywords:** extractive metallurgy; nickel; zinc; Egyptian boiler ash; waste recycling; sulfuric acid leaching; precipitation; leaching kinetics; activation energy

**Citation:** Ibrahim, A.H.; Lyu, X.; Atia, B.M.; Gado, M.A.; ElDeeb, A.B. Cost-Effective and High Purity Valuable Metals Extraction from Water Leaching Solid Residues Obtained as a By-Product from Processing the Egyptian Boiler Ash. *Minerals* **2022**, *12*, 1084. https://doi.org/10.3390/min12091084

Academic Editors: Kyoungkeun Yoo, Marinela Ivanova Panayotova and Vladko Panayotov

Received: 31 July 2022
Accepted: 22 August 2022
Published: 27 August 2022



## 1. Introduction

Nickel has unique physical and chemical properties that enhance its utilization in a wide range of industrial and everyday applications including alloy production; food, military, electroplating and energy industries; and the production of nickel-cadmium batteries and automobiles [1–3]. Until now, stainless steel has been the primary use of nickel production by about 65% due to the mechanical and anticorrosive qualities of nickel when mixed with other metals [4–6]. On the other hand, there are two main categories of users in industrial applications of zinc first and end-users. Brass, die-casters, galvanizers manufacturers and comparatively modest applications such as chemical processing are among the first users. End-users of Zn, such as cathodes in batteries and those in the hardware, automotive, building, furniture, electronics, medical, toy and textile sectors, utilize the product of the initial users as an input in their manufacturing processes [7]. The International Zinc Association has confirmed that over 70% of Zn production globally comes from mined ores, with the remaining 30% coming from recycled or secondary Zn resources. The production of Zn from natural sources is both expensive and complicated. Currently, the firstly mined Zn ores only contain 5%–15% Zn, and sphalerite mineral is the most abundant source of Zn. Oxidized ores containing Zn in various carbonate and silicate minerals have long been a significant source of Zn [8,9]. Ni-sulphide ore contains 0.2%–2% Ni and contributes approximately 60% of global Ni output, and it represents 30% of all known reserves. On the other hand, Ni laterites contain 1%–4% Ni, which accounts for 70% of all known resources, and are responsible for the remaining 40% of worldwide Ni output [10–13].

The depletion of Ni/Zn ores and the increasing global demand for these metals has necessitated extensive research on their extraction from secondary resources and/or low-grade ores [14]. Hence, its availability is a vital criterion for maintaining the output levels required to meet the demands of end customers [6,15]. Compared to the low content of Ni and Zn in the Ni/Zn-containing ores, Ni/Zn-containing wastes and scrap are considered a promising alternative to produce Ni and Zn due to their high Ni and Zn contents and easy extraction. Only around 40% of accessible nickel-bearing scrap is now recycled, indicating that nickel production from secondary materials containing (30%–72% Ni) has a lot of potential. The other 60%, in the form of solid wastes such as batteries, spent nickel-based catalysts, scrap metal, spent solutions, superalloys and non-land resources such as sea nodules, can be found and treated [16]. On the other hand, some industrial processes generate solid wastes in the form of ashes with relatively high amounts of vanadium (4.4%–19.2%), nickel (2.7%–8.5%) and zinc (1.3%–4.2%) [17–20], which makes it a beneficial and promising resource for the recovery of these important elements [14,21,22]. This material has no mining or grinding expenses, unlike other resources that are restricted and/or of low quality, because the material is generated on a regular basis as fine and agglomerate wastes from electricity stations [23].

Boiler ash is a solid waste formed by the combustion of heavy fuel oil (HFO) in power plants, from which several million tons of waste are produced annually. The generation of this ash has negative effects on the environment [24]. In order to reduce the environmental negative effects of this ash, the extraction of metals including nickel and zinc from boiler ash is of critical importance because: (1) nickel and zinc are two of the most hazardous metals and need to be reclaimed; and (2) their existence in this ash poses a serious hydrometallurgical and environmental conservation challenge [17,19,25]. Currently, there are many methods and techniques used for the processing of boiler ash, including salt roasting-water leaching and direct acidic leaching. One of the previously stated technologies, direct acidic leaching, has numerous drawbacks, such as the lack of selectivity in leaching the target metals, such as vanadium, nickel, zinc, iron, and magnesium, that are all dissolved in the solution. Additionally, the process uses large quantities of high acid concentration and generates leachate containing a lot of hazardous elements that cause some problems in the later precipitation and purification processes; the process also uses an oxidizing agent, which makes it costly [26–29]. On the other hand, salt roasting–water/acid leaching of boiler ash is a traditional way to recover vanadium and concentrate the ashes to increase its nickel and zinc content [30–35]. After the extraction of high-purity vanadium, the water leaching residue (WLSR) still contains a significant level of heavy metals and other important minerals, such as Ni, Zn and Pb; therefore, they are regarded as prospective resources for the extraction of Ni and Zn [36–38].

There are two principal methods for processing the solid waste to extract the valuable metals from it, including the following. (1) Hydrometallurgy, in which the valuable metal is leached from solid waste into a solution and then recovered from this solution. The main hydrometallurgical methods are classified according to the type of used leaching agent into three main groups of acidic leaching (i.e., HCl, $H_2SO_4$ and $HNO_3$), alkaline leaching (i.e., NaOH and $Na_2CO_3$) and a combination of acidic and alkaline leaching.

(2) Pyrometallurgy, in which the solid waste is heated to a high temperature to allow the separation of valuable metal [16,39,40]. Due to the cheap operational and energy costs, hydrometallurgical processing is frequently chosen as the preferred method for recovering nickel and zinc from fly ash [25]. Direct acidic leaching is used to dissolve almost all valuable metals into the solution from various types of solid wastes and then extract these metals separately by appropriate methods [41–56].

Due to the previous mentioned advantages, extracting nickel and zinc from boiler ash is a compelling alternative, as it allows the recovery and recycling of the metal values and, at the same time, avoids potential environmental damage or risks of these wastes. The main objective of the current study is the extraction of nickel, zinc, iron and magnesium with high purity from water leaching solid residue (WLSR) using a cost-effective and selective dilute sulfuric acid leaching process. The parameters affecting the leaching process, in addition to the mechanism of metal leaching in solutions and its subsequent precipitation, have been investigated. The kinetics of metal ion leaching process have been also studied.

## 2. Materials and Method

### 2.1. Materials

Water leaching solid residue (WLSR) containing high-grade nickel and zinc was obtained from the processing of Egyptian boiler ash (collected from one of the biggest Egyptian thermal power plants, which uses natural gas and heavy oil as a fuel, located in kuriemat, Egypt, supplied by Nuclear Material Authority) using salt roasting–water leaching processes for Vanadium extraction. The optimum conditions for roasting and water leaching processes have been optimized in the previous study as follows: 850 °C, 20 wt.% NaCl dosage, 2.5 h and 1/10 solid/liquid ratio for 90 min. The used water leaching solid residue (WLSR) sample was dried and ground to 63 μm and then mixed thoroughly to be processed in the subsequent leaching processes for the extraction of valuable metals including Ni, Zn and others.

### 2.2. Methods

#### 2.2.1. Acidic Leaching of Nickel and Zinc

In the leaching process, the WLSR sample was digested using a sulfuric acid solution in a 250 mL three-neck flask. The sulfuric acid solution with different concentration was prepared by adding the predetermined specific amount of $H_2SO_4$ to the distilled water at atmospheric pressure, and all experiments were carried out using a constant weight of the solid sample of 20 g. The three-neck flask was designed so that one neck was fitted with a reflux condenser to maintain the concentration of the species in the solution, the second for a thermometer, and the last was used either for a mechanical starrier or inlet/withdrawal of the samples or for pH measurements. The reaction suspension was agitated with a mechanical stirrer at a rate of 500 rpm, then heated directly to the specified temperature using a hot plate. After the completion of the leaching experiment, the obtained slurry was cooled and then filtered with a vacuum pump. The filter cake was washed with a small amount of warm distilled water to remove and separate the released Zn and Ni ion and then made up to volume; then, the washed spent cake was dried at 110 °C in a drying oven. The metals (Ni, Zn, Mg and Fe) content in the pregnant solution was analyzed by inductively coupled plasma–atomic mass spectrometry (ICP-MS, a Perkin Elmer ELAN model 9000, Waltham, MA, USA). To confirm the results obtained by (ICP-MS), the differences in the chemical compositions of the WLSR before leaching and their sulfuric acid solid residue (SASR) after leaching were measured using the X-ray fluorescence (XRF) data, which represent the Ni and Zn percentage recovery. The extraction of these metals was calculated according to Equation (1).

$$\eta\ (\text{Ni, Zn}) = \left(\frac{m_1 \cdot v_1}{m_0 \cdot v_0}\right) \times 100\% \tag{1}$$

where $m_0$ and $m_1$ are the mass of WLSR used in leaching experiments (g) and the concentration of metal ions (Ni, Zn, Mg and Fe) in the filtrate (g/L), respectively. $v_0$ and $v_1$ are the mass percentage age of metal ions (Ni, Zn, Mg and Fe) in WLSR (%) and the volume of filtrate (L), respectively.

#### 2.2.2. Removal of Iron and Magnesium from Pregnant Solution

During sulfuric acid leaching, some impurities of Mg, Si, Fe and Al are transferred into the pregnant solution accompanied with Ni and Zn, which, in turn, affects the Ni and Zn precipitation rate and hence the purity of the final product, so that they need firstly to be removed from the pregnant solution.

Firstly, the iron contained in the leachate mainly exists in the form of $Fe^{2+}$, but $Fe^{3+}$ ions are inevitable. To maximally improve the purity of the finally produced Ni-Zn, they had to be eliminated to the minimum possible level. Consequently, the leachate needed to be pre-treated with an excess of hydrogen peroxide to ensure that all the low valence $Fe^{2+}$ were oxidized completely to $Fe^{3+}$ and reduce the loss of nickel and zinc due to pH changes in the solution. Then, the iron ions in the pregnant solution were removed by the method of chemical precipitation method at room temperature. The pregnant solution pH value was adjusted at pH = 3 by using sodium hydroxide. Additionally, $Fe_2O_3$ powder was added in amounts from 3 g/L to 15 g/L, which was used as nucleating agent. Then, this mixture was agitated for 10 min to precipitate the iron ions [57]. The obtained precipitate was filtered from the purified solution by vacuum filtration. Finally, the contents of Ni, Zn and iron in the purified solution were determined by using ICP-MS instruments. Then, the concentration of metal ion and the Ni and Zn loss rate were calculated by generalized Equation (2).

$$\rho = \left(1 - \frac{C_a \cdot \beta_a}{C_b \cdot \beta_b}\right) \times 100\% \tag{2}$$

where $C_b$ and $C_a$ refer to the concentration of metal ions (Ni, Zn, Mg and Fe) in the liquor before and after the precipitation experiment (g/L), respectively. $\beta_b$ and $\beta_a$ are volumes of the liquor before and after the precipitation experiment (L), respectively.

Additionally, the possibility of recovering magnesium from a solution using the oxalic acid precipitation method to obtain magnesium oxalate has been investigated, and the optimal conditions of precipitation have been investigated [57].

#### 2.2.3. Nickle and Zinc Precipitation

After the iron and magnesium were precipitated and removed from the pregnant solution, the Ni and Zn concentrations in the purified solution increased. The recovery of Ni and Zn from the leachate was conducted by a direct precipitation method at pH 8.0 for 10 min at 25 °C. The Ni and Zn in the solution were subsequentially precipitated as Ni $(OH)_2$ and Zn $(OH)_2$ until the solution settled. Then, the precipitate was filtered out, washed, dried at 60 °C and calcined at 450 °C for 2 h to obtain NiO and ZnO products.

### 2.3. Sample Characterization

The chemical compositions of the original water leaching solid residue (WLSR) sample and the chemical changes that resulted from the acidic leaching process were determined and measured by the method of X-ray fluorescence (XRF), using a Shimadzu XRF-1800 analyzer (XRF-1800, 90 mA, 40 kV, Re anode, Kyoto, Japan). The changes in the mineralogical composition of WLSR, microstructural compositions during the phase transformations of the WLSR as well as sulfuric acid solid residue (SASR) and the final precipitate of nickel and zinc were determined by the method of X-ray diffraction (XRD) (Analytical X-Ray Diffraction equipment model X″ Pert PRO with Monochromator, Cu-Kα radiation ($\lambda$ = 1.542 A) at 50 KV, 40 mA and scanning speed 0.02/s). In order to determine the phase transformation during the leaching process, different products were also prepared and carefully investigated using a scanning electron microscope (SEM) Tescan TS 5130MM equipped with an energy dispersive X-ray (EDX) detector (prepared by Oxford Instruments,

Abingdon, UK, active crystal area—50 mm$^2$) in addition to a microanalysis system and YAG crystal as a backscattered electron (BSE) detector. Inductively coupled plasma atomic mass spectrometry (ICP-MS, a Perkin Elmer ELAN model 9000, Waltham, MA, USA) was also used to determine metal content accurately on the obtained pregnant solutions after the acidic leaching process and chemical precipitation.

## 3. Results and Discussion

### *3.1. Chemical and Mineralogical Composition of the Used WLSR*

The chemical composition of the water leaching solid residue (WLSR) using X-ray fluorescence analysis (XRF) tabulated in Table 1.

**Table 1.** Overall chemical analysis of water leaching solid residue (WLSR).

| Compound wt.% | $Fe_2O_3$ | NiO | $SiO_2$ | ZnO | $V_2O_5$ | CaO | $Na_2O$ | $Al_2O_3$ | $TiO_2$ | MnO | MgO | $K_2O$ | $P_2O_5$ | $SO_3$ | PbO | L.O.I | Others |
|---|---|---|---|---|---|---|---|---|---|---|---|---|---|---|---|---|---|
| Water Leaching Solid Residue | 24.83 | 13.23 | 15.04 | 5.67 | 1.25 | 4.54 | 1.04 | 4.67 | 0.552 | 0.628 | 14.21 | 0.106 | 0.302 | 1.38 | 1.61 | 1.09 | 9.85 |

The mineralogical phase composition of the WLSR analyzed by X-Ray Diffraction analysis is presented in Figure 1a,b. The results revealed that the WLSR was composed mainly of the following phases: hematite $Fe_2O_3$, magnesioferrite $MgFe_2O_4$, nickel magnesium silicate (Ni, Mg)$_2$($SiO_4$), zinc aluminum iron oxide Zn(Al, Fe)$O_4$, calcium silicate ($Ca_2SiO_4$) and jadeite ($NaAlSi_2O_6$). Meanwhile, the minor phases were nickel sulfide ($NiS_2$), lead silicate ($Pb_2SiO_4$) and nickel sulfate oxide ($NiSO_4$).

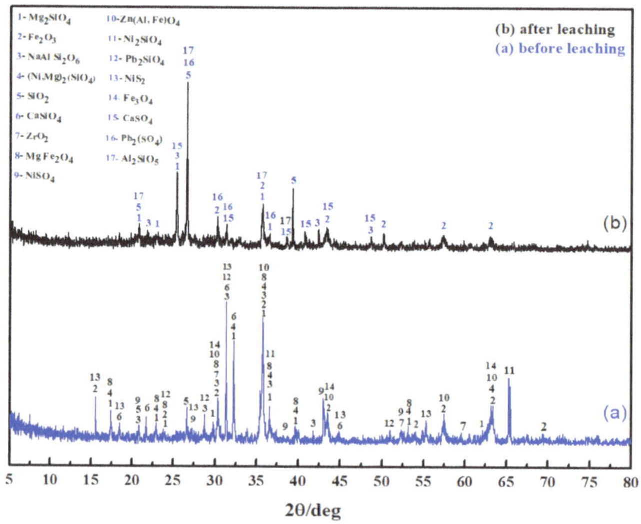


**Figure 1.** XRD patterns of water leaching solid residue (WLSR) of boiler ash (a) before and (b) after leaching with $H_2SO_4$ at the optimum operating conditions.

Further investigation of the water leaching solid residue (WLSR) was carried out using the backscattered electron image and the corresponding EDX microanalysis results of the whole area, as shown in Figure 2. The leaching residue had increased roughness and degradation with irregular semi-circular and cube surface and exhibited very different morphology from those of the ash roasting process obtained at the optimum operating conditions optimized in a previous study by the research group, as shown by Figure 2A. This confirms that most of the vanadium was leached, leaving nickel and zinc in WLSR. The EDX results in Figure 2B show that the particles mainly consisted of Ni, 10.47 wt.%; Fe, 19.48 wt.%; Si, 12.49 wt.%; Zn, 5.11 wt.%; and O, 35.15 wt.%, in good agreement with the data obtained by XRF analysis.

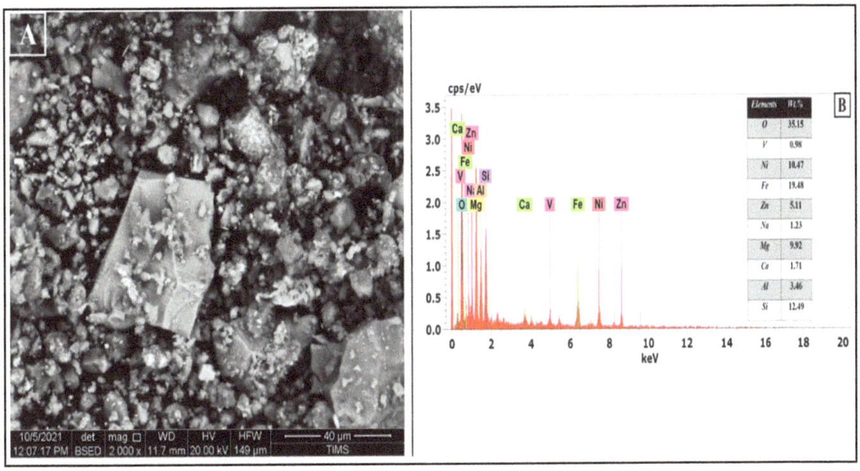


**Figure 2.** SEM images of used water leaching solid residue (WLSR) obtained under optimum leaching conditions of the Egyptian boiler ash (**A**) and EDX element analysis (**B**).

### *3.2. Thermodynamics Analysis of Leaching Process*

The reactions of the components in WLSR with $H_2SO_4$ according to Equations (3)–(14) are presented in the thermodynamic diagrams (Figure 3). The original crystal structure of the residue has been totally changed by means of sulfuric acid leaching. The metallic components such as iron, zinc, nickel and magnesium in the WLSR were converted into the corresponding sulfate and leached by the sulfuric acid solution; on the other hand, calcium, lead and silicon were converted into lead sulfate ($PbSO_4$), calcium sulfate ($CaSO_4$) precipitate and metasilicic acid ($H_2SiO_3$) as colloidal precipitate, respectively. The thermodynamic results shown in Figure 3 illustrate the feasibility of this method for leaching water leaching solid residue (WLSR) with sulfuric acid to obtain dissolved metal sulfate because the Gibbs free energy change (ΔG) of the above Equations (3)–(14) had negative values in the range from 0 to 100 °C [57,58]. The free energy change of the metals varies in the sequence of dissolution of Ni > Fe > Mg > Zn.

$$NiS_2 + 2.5O_2 + H_2SO_4 = NiSO_4 + 2SO_2\uparrow + H_2O \quad (3)$$

$$NiSO_4 + 2\,H^+ + 1/2O_2 = NiO + SO_3\uparrow + H_2O \quad (4)$$

$$NiO + H_2SO_4 = NiSO_4 + H_2O \quad (5)$$

$$Ca_2SiO_4 + 2\,H_2SO_4 = 2CaSO_4\downarrow + H_2SiO_3\downarrow + H_2O \quad (6)$$

$$MgFe_2O_4 + H_2SO_4 = MgSO_4 + Fe_2O_3 + H_2O \quad (7)$$

$$Pb_2SiO_4 + 2\,H_2SO_4 = 2PbSO_4\downarrow + H_2SiO_3\downarrow + H_2O \quad (8)$$

$$ZnO + H_2SO_4 = ZnSO_4 + H_2O \quad (9)$$

$$ZnFe_2O_4 + H_2SO_4 = ZnSO_4 + Fe_2O_3 + H_2O \quad (10)$$

$$Mg_2SiO_4 + 2\,H_2SO_4 = 2MgSO_4 + H_2SiO_3\downarrow + H_2O \quad (11)$$

$$Fe_2O_3 + 3H_2SO_4 = Fe_2(SO_4)_3 + 3H_2O \quad (12)$$

$$Ni_2SiO_4 + 2\ H_2SO_4 = 2NiSO_4 + H_2SiO_3\downarrow + H_2O \quad (13)$$

$$Fe_3O_4 + 4H_2SO_4 = FeSO_4 + Fe_2(SO_4)_3 + 4H_2O \quad (14)$$

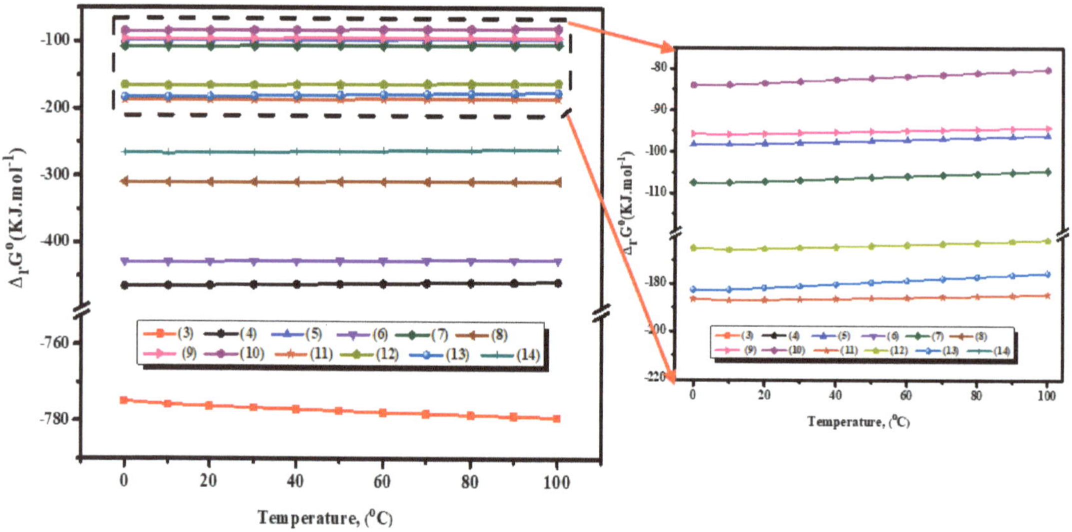

**Figure 3.** Relationship between $\Delta rG^0$-T for main reactions in $H_2SO_4$ leaching process of WLSR.

### 3.3. *Optimizing the Factors Affecting the Leaching Process of WLSR*

The factors affecting the leaching process of WLSR and hence the percentage recovery of Ni, Zn and Fe have been investigated to obtain the maximum recovery. These factors include the acid concentration, the leaching temperature, solid/liquid ratio and leaching time.

The effect of sulfuric acid concentration on the extraction of Ni, Zn and Fe has been investigated by using different acid concentrations in the range 2%–12% (vol%), for 120 min leaching time, 75 °C leaching temperature, 1/10 solid/liquid ratio and 500 rpm rotation speed [17,20]. The results indicate that the WLSR sample is easily amenable to sulfuric acid leaching, as shown in Figure 4a. Hence, by gradually increasing acid concentration from 2% to 8%, the dissolution percentage of Ni, Zn and Fe increased almost linearly from 43%, 36% and 15% until they reached 79%, 74%, 42%, respectively. This means that low acid concentration is insufficient to dissolve metal ions from WLSR sample, while at 8% acid concentration, the participation of a large part of the acid in the internal reactions has been noted in Equations (10)–(13), which improves the extraction of metal ions. Otherwise, by increasing the acid concentration more than 10%, the metal leaching efficiency remained stable due to more generation of colloidal silicic acid and solid $CaSO_4$, which prevent unreacted particles and hinder the dissolution of more desirable ions [14,19,59]. In consideration of reducing the acid consumption cost, a sulfuric acid concentration of 8% was selected as the best value and used in the subsequent leaching experiments.

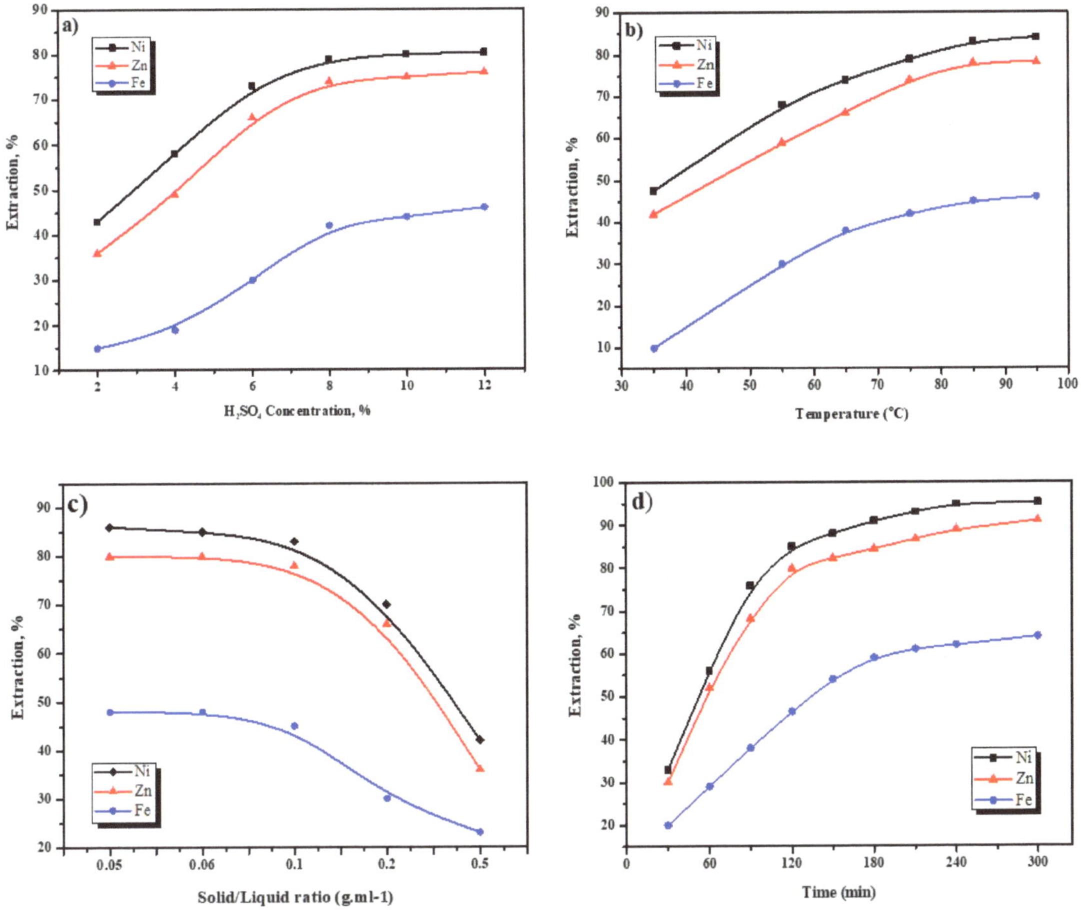


**Figure 4.** Effect of $H_2SO_4$ acid concentration (**a**), leaching temperature (**b**), solid/liquid ratio (**c**) and leaching time (**d**) on Ni, Zn and Fe extraction.

The effect of leaching temperature on the dissolution of Ni, Zn and Fe has been studied by varying the leaching temperature range from 35 °C to 95 °C, and the other leaching conditions were fixed as mentioned above. The results presented in Figure 4b reveal the important effect of leaching temperature on obtaining a reasonable dissolution of Ni, Zn and Fe from WLSR. Leaching at a low temperature (35 °C) under the above conditions gives a low recovery of Ni, Zn and Fe of no more than 47.9%, 42% and 10%, respectively. Further increasing the leaching temperature to 85 °C increases the dissolution of Ni, Zn and Fe to 83%, 78% and 45%, respectively. This can be attributed to the fact that the increased temperature leads to the breaking of the boundary founded in surface of WLSR particles by increasing the diffusion of $H^+$ ions on their surfaces, which increased the activity of the molecules, thereby increasing the collision probability between WLSR and $H_2SO_4$ molecules [14,20,22]. The further increase in the leaching temperature above 85 °C had no significant effect on the extraction of the different metal ions; therefore, 85 °C was selected to be the best leaching temperature in the subsequent leaching experiments.

The effect of the solid/liquid ratio on the dissolution percentage of Ni, Zn and Fe was studied using 1/2, 1/5, 1/10, 1/15 and 1/20 S/L ratios, while other leaching parameters have been fixed. The obtained results presented in Figure 4c show that the solid/liquid ratio has a considerable effect on the percentage recovery of Ni, Zn and Fe from WLSR.

The dissolution of the metal ions increases with decreasing the pulp density to 1/15 S/L ratio, which leads to an increase in the recovery to 85% Ni, 80% Zn and 46% Fe. This can be attributed to a sufficient amount of acid present in the liquid to attack the solid particles. However, by increasing the S/L ratio to higher values, the Ni, Zn and Fe percentage recovery decreases [9,20]. This effect could be attributed to the increased solid percentage in the solution, which decreases the interaction between the ions, thus reducing the proton ions' concentration in the solution. Therefore, the best S/L ratio has been selected to be 1/15.

The effect of leaching time on Ni, Zn and Fe percentage recovery has been investigated by varying the leaching time from 30 to 300 min with all other leaching parameters fixed. The obtained results presented in Figure 4d indicate that the dissolution percentage age of Ni, Zn and Fe increases as the leaching time increases to 240 min, to maximum values of 95.02% Ni, 90.13% Zn and 66.29% Fe. Hence, the leaching time of 240 min was selected as the best leaching time. Table 2 presents the XRF analysis results of the sulfuric leaching solid residue (SLSR) under the optimum leaching conditions (8% $H_2SO_4$, 85 °C, 1/15 S/L, 240 min), which demonstrates that the main elements in the final residue were Fe, Si, Ca and Mg.

**Table 2.** Overall chemical analysis of sulfuric acid solid residue (SASR).

| Compound wt.% | $Fe_2O_3$ | NiO | $SiO_2$ | ZnO | $V_2O_5$ | CaO | $Na_2O$ | $Al_2O_3$ | $TiO_2$ | MnO | MgO | $K_2O$ | $P_2O_5$ | $SO_3$ | Pb | L.O.I | Others |
|---|---|---|---|---|---|---|---|---|---|---|---|---|---|---|---|---|---|
| Final Residue after $H_2SO_4$ | 18.93 | 0.89 | 26.57 | 0.64 | 0.072 | 10.2 | <0.01 | 5.81 | 0.377 | 0.101 | 6.83 | 0.052 | <0.01 | 9.75 | 3.61 | 8.13 | 6.34 |

The sulfuric acid solid residue (SASR) was characterized by XRD analysis, as shown in Figure 1a,b. In comparison with the XRD analysis of the WLSR sample, it can be found that almost no peaks attributed to minerals bore nickel (e.g., $Ni_2SiO_4$, $NiS_2$ and $NiSO_4$) or zinc. This confirms the thermodynamic analysis of the chemical reactions described above in Section 3.2.

However, after leaching, the intensities of peaks corresponding to $CaSO_4$ and $PbSO_4$ were increased. In addition, the intensity of $SiO_2$ peak increased after leaching, indicating the complete dissolution of these accompanying elements. For further explanation of the phase transformation, SASR samples were investigated using scanning electron microscopy (SEM), and the obtained results are shown in Figure 5. The results indicated that the residue contained a negligible concentration of heavy metals. However, this solid residue (SASR) is recommended to be used in the next study as thermally treated adsorbents to remove undesirable ions from industrial wastewater.

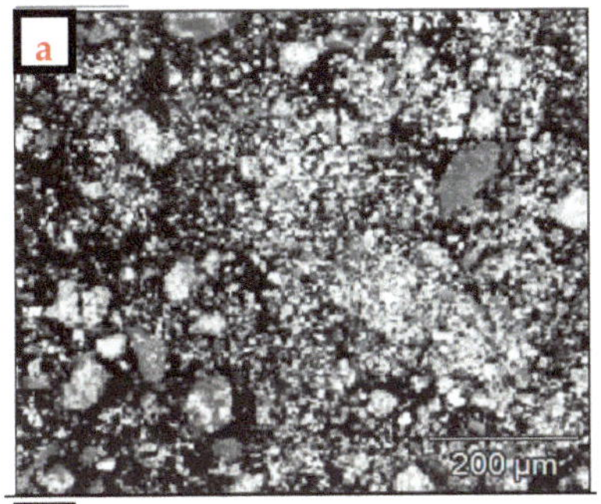


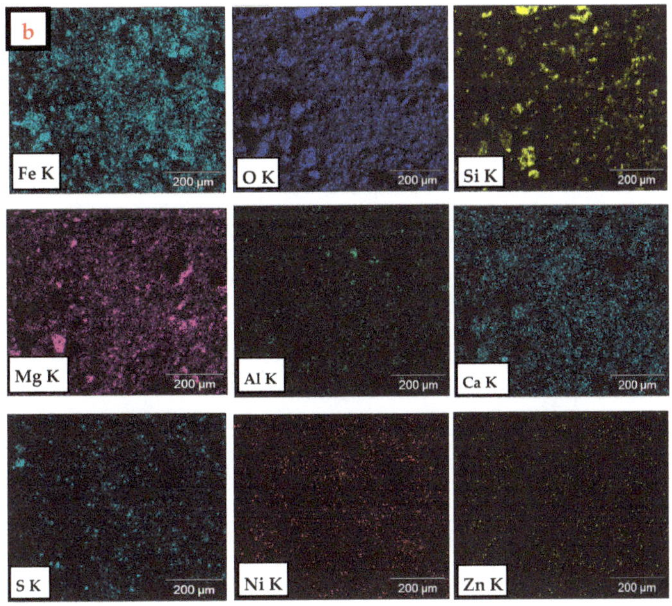


c

| Elements | Wt.% |
|---|---|
| O | 20.15 |
| Fe | 39.12 |
| Ca | 7.25 |
| Si | 21.61 |
| Mg | 5.97 |
| Al | 3.15 |
| S | 1.53 |
| Ni | 1.01 |
| Zn | 0.18 |

**Figure 5.** Backscattered electron image of SASR (**a**); distribution mapping of the different elements (**b**) and EDX element analysis (**c**).

### *3.4. Purification and Recovery of Ni and Zn from Leachate*

#### 3.4.1. Precipitation Behavior of Fe and Mg from the Leachate

Table 3 shows the complete chemical analysis of leachate using ICP-OES analysis, which contains Ni (11.417 g/L) and Zn (4.6462 g/L), as well as Fe (13.168 g/L) and Mg (9.783 g/L), which are the most inevitable dissolved impurities in solution that need to be removed to obtain a final Ni-Zn product of high purity.

**Table 3.** Chemical composition of the leachate ($g \cdot L^{-1}$).

| Components | Ni | Zn | Fe | Mg | V | Al | Ca | Ti | K | Na | Mn | Pb | Si | $SO_4^{2}$ |
|---|---|---|---|---|---|---|---|---|---|---|---|---|---|---|
| Content, (g/L) | 11.42 | 4.65 | 13.17 | 9.79 | 1.17 | 0.96 | 0.13 | 0.31 | 0.06 | 1.31 | 0.46 | 0.02 | 0.88 | 150.25 |

The Eh–pH graphs were obtained using the software HSC Chemistry 9.3 at 25 °C, and the obtained results are shown in Figure 6. The top and lower dotted lines in each graph depict the $O_2/H_2O$ and $H_2O/H_2$ stability limitations, respectively. In the water-stable zone, nickel may stably exist in the stable form of $Ni^{2+}$ when the solution pH value is less than 6, as illustrated in the Ni-$H_2O$ system in Figure 6a, and when the pH value rises above 8, $Ni^{2+}$ is transformed to $Ni(OH)_2$. As demonstrated in Figure 6b, zinc can exist in the solution in the form of $Zn^{2+}$ at a pH lower than 7; in the stable region of water, as indicated by the Zn-$H_2O$ system, $Zn^{2+}$ will be transformed to $Zn(OH)_2$ when the pH value rises over 8. The Fe-$H_2O$ system is depicted in Figure 6c. It is shown that, in the water-stable region of the lower pH solution (<1.4), iron can exist in the form of $Fe^{2+}$ and $Fe^{3+}$. The latter ($Fe^{3+}$) will be transformed to $Fe(OH)_3$ precipitation when the pH value surpasses 1.4. $Mg^{2+}$ can stably exist in acidic solutions, whereas $Mg(OH)_2$ can stably exist in alkaline solutions, as shown in the Mg-$H_2O$ system in Figure 6d.

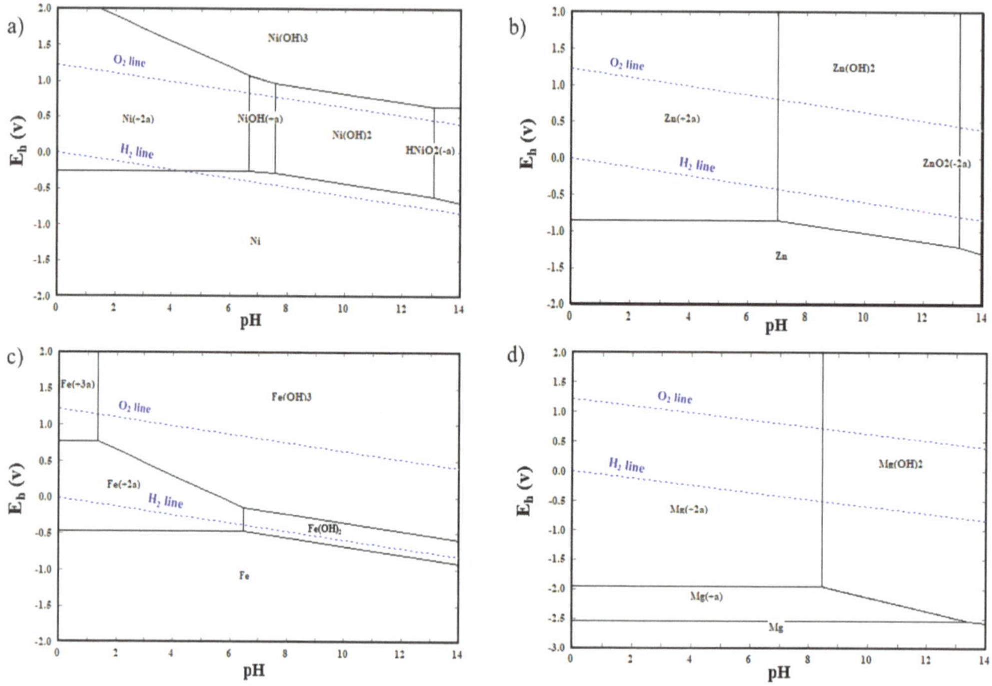


**Figure 6.** Eh–pH diagrams of Ni-$H_2O$ system (**a**), Zn-$H_2O$ system (**b**), Fe-$H_2O$ system (**c**) and Mg-$H_2O$ system (**d**) at 25 °C, where all the concentrations of Zn, Ni, Fe and Mg are 1 mol/L.

Figure 7a illustrates the species distribution of 13.168 g/L Fe in sulfuric acid solution (156.25 g/L $SO_4^{2-}$) at various pH levels. It can be observed that $Fe^{3+}$ ions react with $SO_4^{2-}$ to create $FeHSO_4^{2+}$, $FeSO_4^{+}$, and Fe $(SO_4)_2^{-}$ complex ions at pH value below 3. When the pH rises over 3.5, $Fe(OH)_3$ precipitates progressively, and beyond 4.8, virtually all iron complex ions are turned into $Fe(OH)_3$.

At different pH settings, the rate of Fe ion precipitation and the rate of Ni and Zn ion losses were studied. The precipitation time was set at 10 min, the $Fe_2O_3$ nucleating agent was introduced at a concentration of 8 g/L, and the stirring speed was set at 500 rpm. Figure 7b shows the measured findings, which indicate that the precipitation rate of Fe ions was more than 89% in the pH range of 3 to 6 but only 25.7 percent when the pH was 2. With a rise in pH from 2 to 8, the rate of nickel and zinc loss increased dramatically from 7% and 3% to 65% and 60%, respectively. As a result, a pH of about 4.8 ± 0.2 was confirmed to be the best value. The loss of nickel and zinc ions can be attributable to two factors. One is the precipitation of certain nickel and zinc ions due to the solution's high pH value, which may be avoided by lowering the pH as much as possible. The other is the adsorption loss due to iron ion precipitation, which is unavoidable but can be reduced by shortening the precipitation reaction time. Thus, in the next section, the appropriate precipitation time was determined.

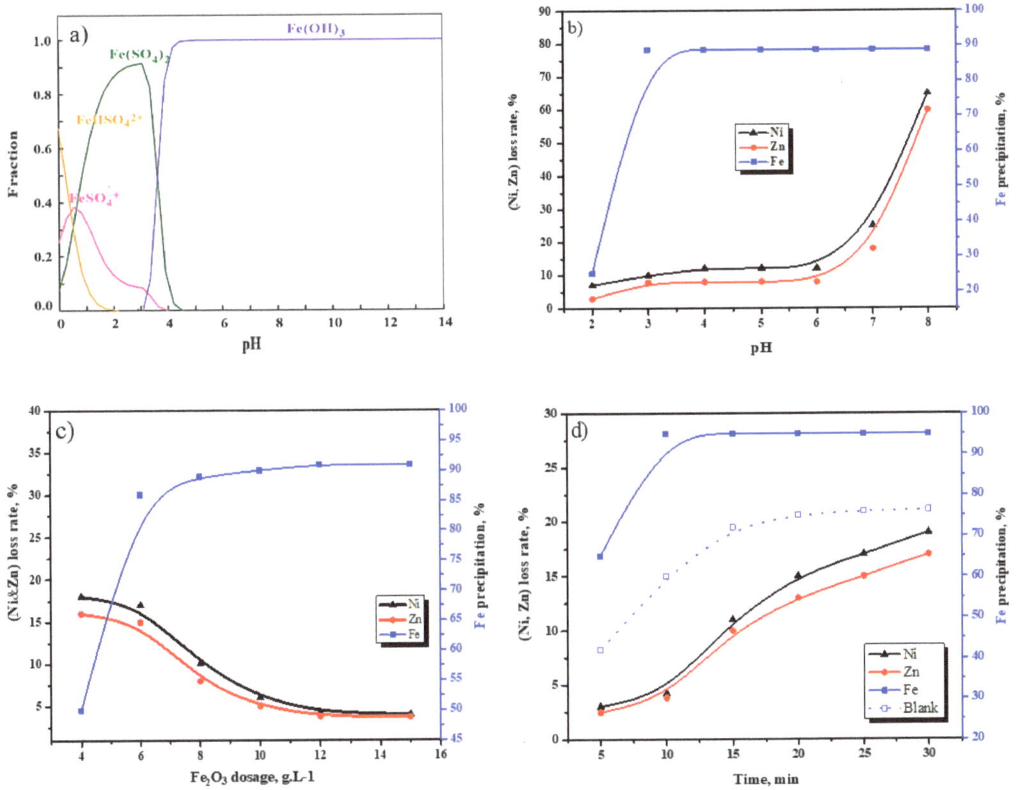


**Figure 7.** Species distribution diagrams of Fe-$H_2SO_4$ systems (**a**) containing 0.235 mol/L of Fe and 1.56 mol/L of $SO_4^{2-}$, Effect of pH value (**b**), $Fe_2O_3$ added amount (**c**) and precipitation time (**d**) on Fe precipitation rate and Ni, Zn loss rates.

The effect of $Fe_2O_3$ addition dosage (4, 6, 8, 10, 12 and 15 g/L) as a crystallizing agent on the removal of Fe and loss of Ni, Zn was examined under the following conditions: a pH value of nearly 5, for 10 min precipitation time, 500 rpm stirring speed. Figure 7c presents the Fe precipitation rate as well as the Ni and Zn loss rates. It can be observed that, as the amount of $Fe_2O_3$ addition rises, the Fe precipitation rate increases dramatically. At the same time, the adsorption loss of Ni and Zn is significantly reduced. As a result, the dose of 12 g/L $Fe_2O_3$ was selected as the best value. This confirms that $Fe_2O_3$ is an efficient crystallizing agent for eliminating Fe from leachate.

The effect of precipitation time on Fe ion precipitation rate and the rate of loss of Ni and Zn ions was investigated by varying the precipitation time in the range from 5 to 30 min with a 5 min interval. The precipitation was carried out using a dosage of 12 g/L $Fe_2O_3$ as a crystallizing agent, at 5 pH and 500 rpm stirring speed. Figure 7d shows that, with the addition of a $Fe_2O_3$ powder as a crystallizing agent, the iron precipitation rate can approach 95% after 10 min, while the iron precipitation rate does not exceed 60% without the addition of a $Fe_2O_3$ powder as a crystallizing agent, as shown by the dotted line. On the other hand, the losing rate of Ni and Zn is noticeably increased after 10 min. As a result, the precipitation time of 10 min was selected as the best magnitude to precipitate the maximum available amount of Fe ions and, at the same time, reduce the losing rate of Ni and Zn ions to the minimum value.

The obtained Fe precipitate was then separated from the leachate, calcined at 600 °C for 2 h and characterized using XRD and SEM-EDX. The XRD pattern of the calcined precipitate is shown in Figure 8, demonstrating that the calcined product was $Fe_2O_3$. Figure 9 displays a micrograph of the precipitate, with EDX analysis and distribution mapping of Ni, Zn and Fe elements. The obtained Fe precipitate was composed mainly of iron, with very small amounts of Ni and Zn ions.

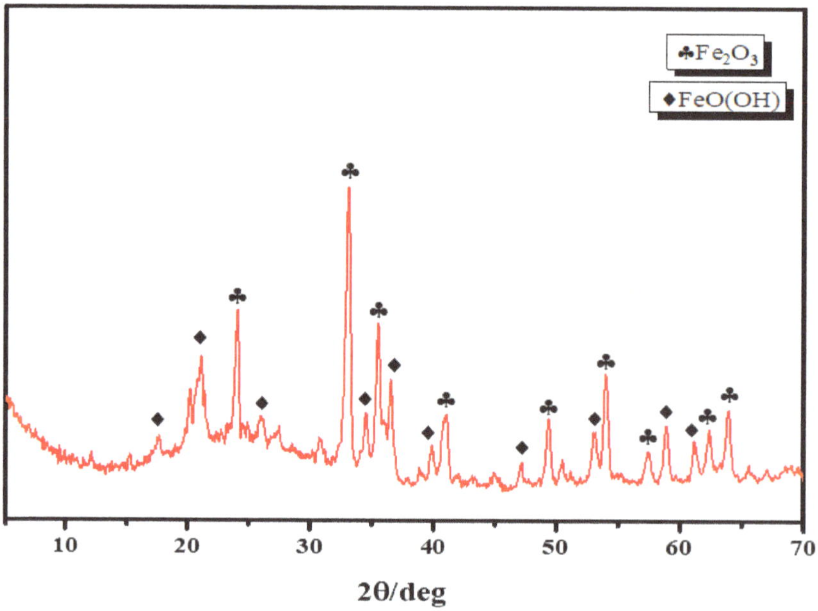


**Figure 8.** XRD patterns of $Fe_2O_3$ calcined precipitate.

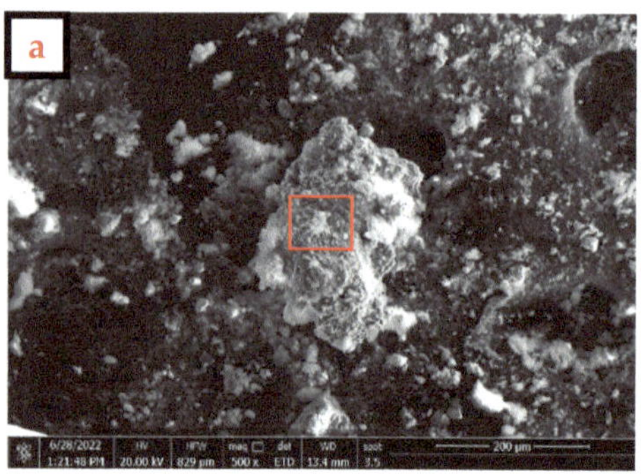


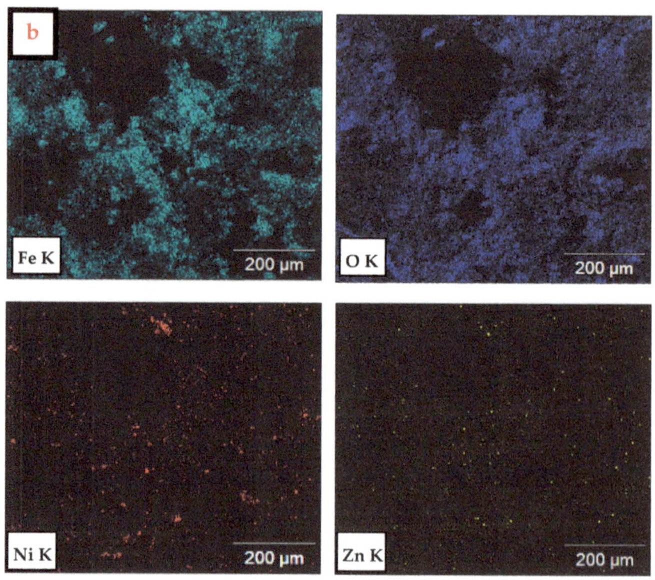


c

| Elements | Wt.% |
|---|---|
| O | 22.27 |
| Fe | 76.75 |
| Ni | 0.11 |
| Zn | 0.03 |
| Ca | 0.15 |
| Si | 0.13 |
| Al | 0.55 |

**Figure 9.** The backscattered electron image of the iron precipitate (**a**); distribution mapping of the different elements (**b**) and EDX element analysis of the red rectangle area (**c**).

According to the above-mentioned results, at 5 pH, 12 g/L added $Fe_2O_3$ dosage and 10 min precipitation time, the rate of iron precipitation reaches 95.5%, whereas Ni and Zn ions loss are around 4.2% and 3.8%, respectively. The purification steps of the Ni/Zn-bearing leachate are listed in Table 4. It was found that the predominant constituent in the purified solution after Fe precipitation was Ni and Zn ions, which have impurity components presented in an undesirable amount (e.g., Mg) and hence need extra purification.

**Table 4.** Overall composition of the purified leachate before and after Fe and Mg precipitation (g/L).

| Components | Ni | Zn | Fe | Mg | V | Al | Ca | Ti | K | Na | Mn | Pb | Si | $SO_4^{2-}$ |
|---|---|---|---|---|---|---|---|---|---|---|---|---|---|---|
| Before purification | 11.417 | 4.6462 | 13.168 | 9.783 | 0.981 | 1.016 | 0.1283 | 0.3142 | 0.060 | 1.3087 | 0.461 | 0.0228 | 0.883 | |
| After purification (Fe precipitation) | 10.901 | 4.397 | 0.305 | 9.564 | 0.875 | 0.862 | 0.1323 | 0.298 | 0.0596 | 1.288 | 0.062 | 0.0513 | 0.747 | |
| After purification (Mg precipitation) | 10.78 | 4.355 | 0.182 | 0.254 | 0.564 | 0.314 | 0.0216 | 0.215 | 0.0544 | 0.991 | 0.035 | 0.0489 | 0.605 | 150.25 |
| Composition of the residual solution (Ni–Zn precipitation) | 0.051 | 0.046 | 0.082 | 0.098 | 0.235 | 0.076 | 0.032 | 0.209 | 0.0383 | 0.087 | 0.062 | 0.0416 | 0.327 | |

As previously stated in Section 3.4.1, the magnesium content in the sulfuric acid solution can be converted to magnesium hydroxide precipitate by elevating the pH over 9.0. This procedure uses a lot of NaOH; in addition, the alkaline solution that results is harmful to the environment. Magnesium may be converted into magnesium oxalate by oxalic acid, which is an ecologically beneficial organic acid. Therefore, in our study, oxalic acid was employed as a precipitant to remove magnesium from the leachate. Figure 10 illustrates the species distribution of 9.564 g/L Mg and 40 g/L $C_2O_4^{2-}$ (equal 1.5 molar ratio $(H_2C_2O_4)/(Mg)$) in a sulfuric acid solution (150.25 g/L $SO_4^{2-}$) at various pH levels. It can be observed that magnesium mostly occurs in the form of a $MgSO_4$ complex or $Mg^{2+}$ in a highly acid solution (pH < 2). On the other hand, the soluble magnesium ions are progressively converted into a precipitate of magnesium oxalate as the pH of the solution rises, and under a pH range of 4 to 10, magnesium oxalate is the most common form of magnesium. Magnesium oxalate is gradually transformed into magnesium hydroxide precipitation when the pH of the solution is raised. Anyway, adopting the relevant process conditions as follows—0.45 mol/L oxalic acid, 6 pH solution, 2 h precipitation time and 25 °C precipitation temperature—and according to the findings, nearly all magnesium (96.97%) existed as $MgC_2O_4$ precipitate. Then, the slurry was filtered to separate the precipitate and then dried in a drying oven at 90 °C. The dried Mg containing precipitate was then characterized using XRD to detect the phases of the precipitate as shown in Figure 11. The chemical composition of the Zn and Ni containing solution after Mg precipitation is presented in Table 3, which indicates that the Mg concentration was reduced to 0.254 g/L.

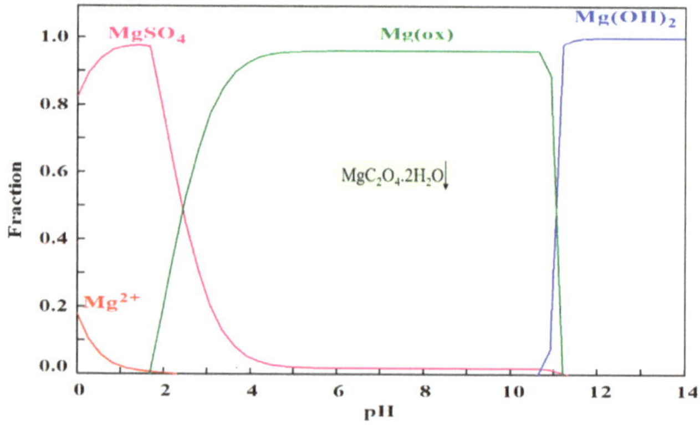


**Figure 10.** Species distribution diagrams of $C_2O_4^{2-}$/Mg-$H_2SO_4$ systems containing 0.393 mol/L $Mg^{2+}$, 0.45 mol/L $C_2O_4^{2-}$ and 1.5625 mol/L $SO_4^{2-}$.

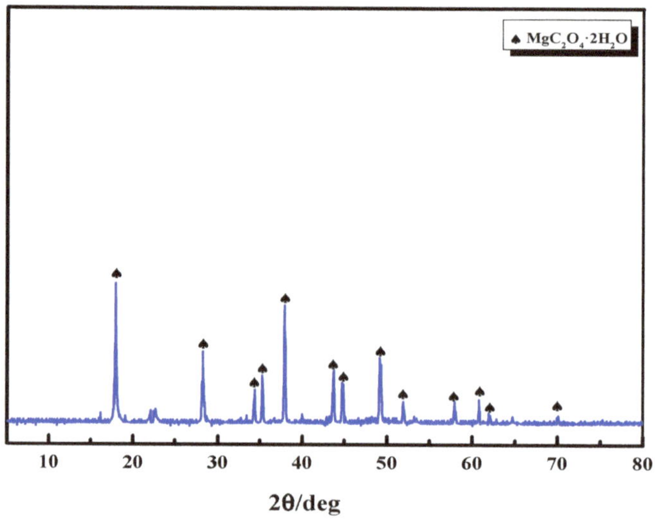


**Figure 11.** XRD patterns of $MgC_2O_4$.2 $h_2O$ precipitate.

3.4.2. Recovery of Nickel and Zinc from Raffinate

Once Fe and Mg have been removed from the solution, Ni and Zn can then be precipitated by raising the solution's pH to approximately 8.0 by adding NaOH. Figure 12a,b illustrates the species distribution of 10.78 g/L Ni and 4.335 g/L Zn in sulfuric acid solution (150.25 g/L $SO_4^{2-}$) at various pH levels. It can be observed that nickel mostly occurs in the form of Ni $(SO_4)_2^{2-}$ complex in a weak acid solution (pH < 5). When the pH rises over 6.0, $Ni(OH)_2$ precipitates progressively, and beyond 8.0, almost all nickel complex ions are turned into $Ni(OH)_2$, as shown in Figure 12a.

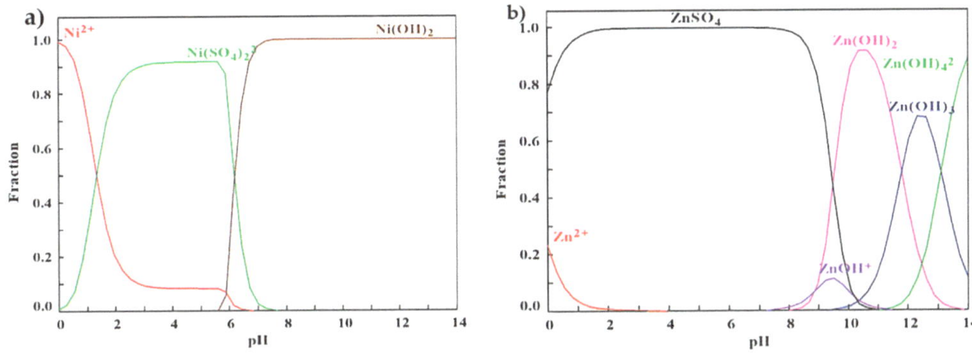


**Figure 12.** Species distribution diagrams of Ni-Zn/$H_2SO_4$ systems containing 0.1836 mol/L Ni (**a**), 0.0662 mol/L Zn (**b**) at 1.5625 mol/L $SO_4^{2-}$.

On the other hand, Figure 12b demonstrates that zinc ion mainly exists in the formed in $ZnSO_4$ at the pH value of solution less than 7, and as the pH value increases above 8.0, the soluble Zn ions are progressively converted into $ZnOH^+$ and $Zn(OH)_2$ precipitate. With the further increase in the pH value to above 10, the Zn $(OH)_2$ precipitate is re-dissolved as $Zn(OH)_3^-$ and $Zn(OH)_4^{2-}$. In any case, the recovery of Ni and Zn from purified solution was conducted by a direct precipitation method at pH 9, allowing it to stand for 1 h at 25 °C, and the precipitation efficiency was over 92.25% and 85.51% for Ni and Zn, respectively. The precipitates were then calcined at 450 °C for 2 h before being characterized using chemical analysis and SEM-EDX. The XRD pattern of the calcined precipitated is shown in Figure 13, demonstrating that the calcined product consisting mainly of mixed Ni-Zn oxide.

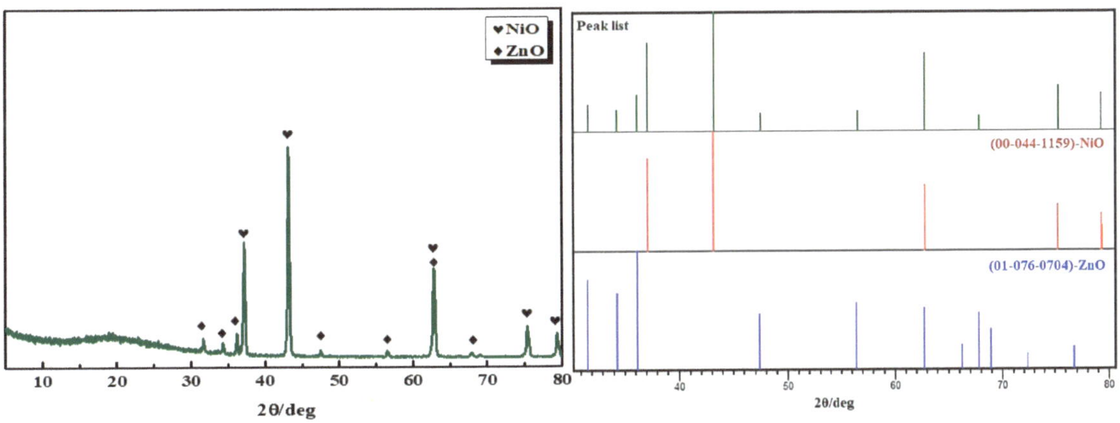


**Figure 13.** XRD patterns of NiO-ZnO precipitate.

According to the chemical analysis presented in Table 5 and EDX of Ni/Zn hydroxide precipitate shown in Figure 14, the product composed mainly of Ni (37%) and Zn (23%) would require additional processing steps (i.e., re-leaching and/or solvent extraction/electrowinning) to recover the containing metals' values. While this is a saleable product and easily integrated into any existing Ni and Zn extraction flowsheets, it would command a considerably lower price than the pure metal, and thus, this would affect the economics of any plant using this process.

**Table 5.** Overall composition of the Ni/Zn product (wt.%).

| Elements | NiO | ZnO | Fe | Al | Mn | Ca | Mg | K | pb | Na | Si | V | Ti |
|---|---|---|---|---|---|---|---|---|---|---|---|---|---|
| (wt.%) | 57.75 | 36.86 | 0.85 | 0.195 | 0.068 | 1.883 | 0.132 | 0.016 | 0.043 | 0.985 | 1.138 | 0.016 | 0.062 |

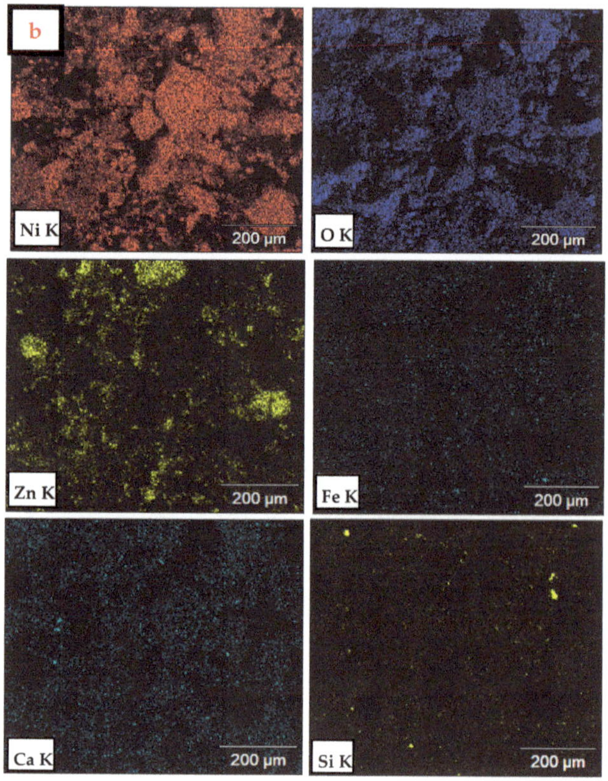


c

| Elements | wt.% |
|---|---|
| O | 22.31 |
| Ni | 60.27 |
| Zn | 15.52 |
| Fe | 0.38 |
| Ca | 0.95 |
| Si | 0.48 |

**Figure 14.** Scanning electron micrographs for Ni-Zn precipitation powder (**a**), distribution mapping of the different elements analysis (**b**) and EDX element analysis of the red rectangle area (**c**).

According to the obtained results of the current study mentioned above, the proposed and recommended flowsheet for the extraction of Ni, Zn, Fe and Mg from water leaching solid residue is shown in Figure 15.

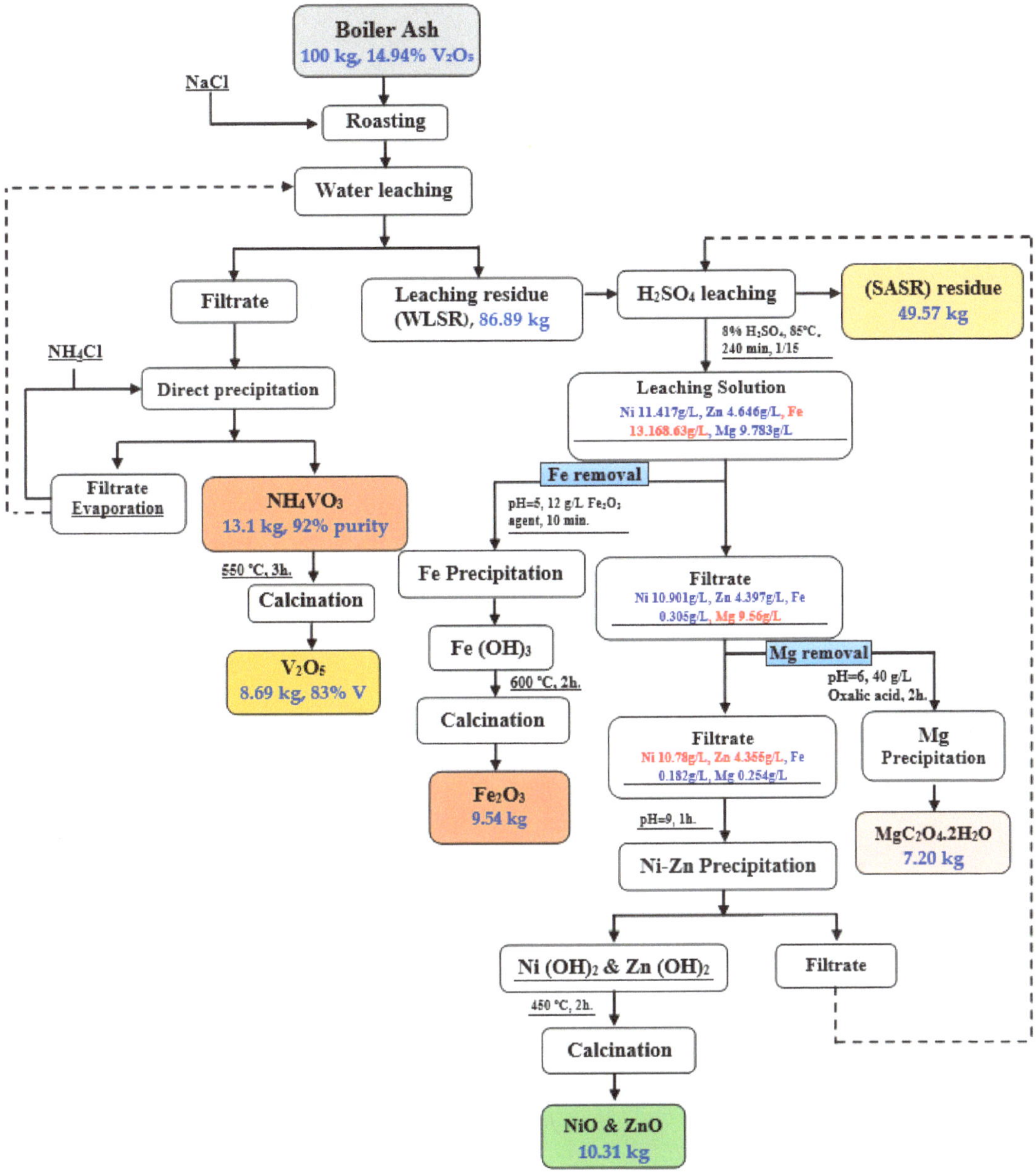


**Figure 15.** The proposed scheme for the recovery of Ni, Zn, Fe and Mg from water leaching solid residue obtained from the salt roasting of Egyptian boiler ash.

## *3.5. Kinetics Analysis*

### 3.5.1. Leaching Kinetics of Nickel

Heterogeneous fluid–solid reaction may be used in leaching processes. The reaction progress can be controlled by the following steps: (1) the mass transfer of reactants and products between the fluid molecules and the external surface of the solid particle, (2) the diffusion of reactants and products within the pores of the solid, (3) chemical reaction between the reactants in the fluid and in the solid [60,61]. One or more of these factors might control the rate of the reaction according to the slowest link step. The un-reacted shrinking core model (SCM) is the most used mathematical model to describe the heterogeneous reactions in leaching mineralogical ores. In discussing the following kinetic properties, the controlling steps in the SCM were checked for the whole reaction time.

$$1 - (1 - x)^{2/3} = K_1 t \quad (15)$$

$$1 - (1 - x)^{1/3} = K_c t \quad (16)$$

$$1 - 3 \times (1 - x)^{2/3} + 2 \times (1 - x) = K_d t \quad (17)$$

where x is the conversion fraction of Ni or Zn particles, $K_1$ is the apparent rate constant ($min^{-1}$) for the fluid film diffusion, t is the reaction time, $K_c$ is the rate constant for the chemical reaction, and $K_d$ is the rate constant for diffusion through product layer.

The relationships of nickel extraction versus leaching time at different temperatures are plotted in Figure 16a, which presents that the dissolution of nickel increases gradually by increasing time and temperature. However, to understand the type of leaching mechanism prevalent of nickel leaching in sulfuric acid process, the leaching experiment during the whole reaction time was investigated to find which kinetic Equations (15)–(17) model can fit the reaction isotherms. Amongst the three controlling mechanisms, the liquid–film diffusion resistance is eliminated or minimized by effective stirring.

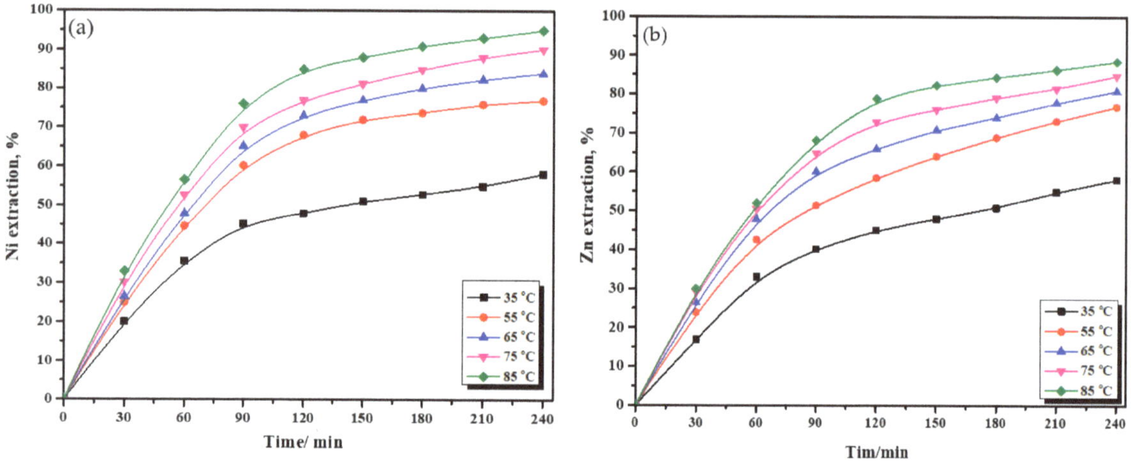


**Figure 16.** Effect of leaching time on the nickel (**a**) and zinc (**b**) dissolution at different temperature.

The obtained results are presented in Table 5 and Figure 17a,b. The left sides of Equations (15)–(17) were plotted as a function of time at different leaching temperatures and according to the obtained data from the optimal operating conditions. The values of the reaction rate constants (K) were determined from the slope of the straight line of the relation between kinetic model and time. It was found that the correlation coefficient value for internal diffusion ($k_d$) was 0.9990, while for chemical reaction ($k_c$), it was 0.9763 and 0.9363 for liquid film diffusion. Based on the $R^2$ values, it can be inferred that the predominant dissolution mechanism of nickel from WLSR is diffusion-controlled only. This due to production of a solid layer that appears to be a hindrance to the progression of the reactions and cannot be dissolved quickly [9]. Hence, the mass transfer efficiency in the generated layer was lower, which made the diffusion of reactants through it the control step.

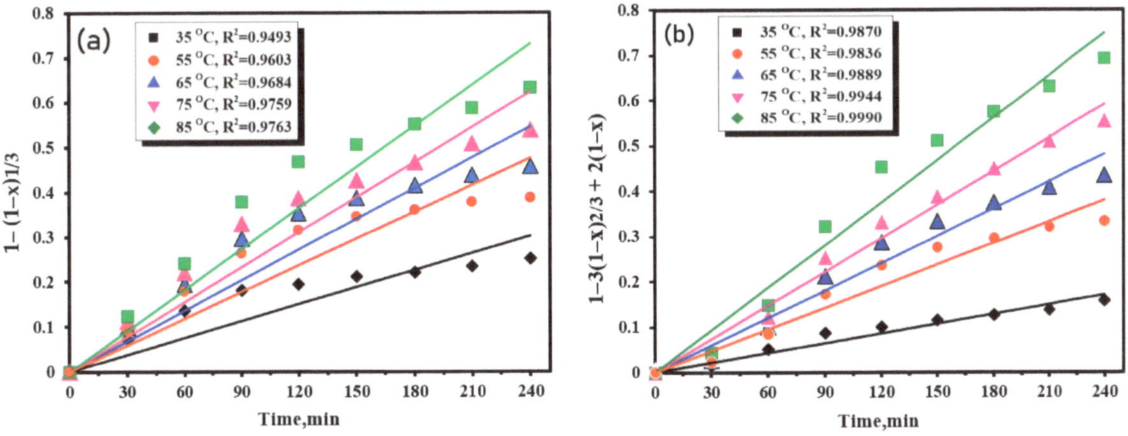


**Figure 17.** Plot of $1 - (1 - x)^{1/3}$ (**a**) and $1 - 3(1 - x)^{2/3} + 2(1 - x)$ (**b**) vs. time at different temperature.

The apparent activation energy can be calculated by the Arrhenius Equation (18):

$$\text{Lnk} = \text{LnA} - \frac{E_a}{RT} \tag{18}$$

where A is the frequency factor, Ea is the apparent activation energy of the reaction, R is the universal gas constant, and T is the reaction temperature in kelvin.

The logarithmic values of the reaction rate constants (K) of WLSR in sulfuric acid leaching were plotted against the reciprocal of the absolute reduction temperature (T), with obtained the $K_d$ value in Table 6, and the result is displayed in Figure 18. Using the Arrhenius Equation (18), apparent activation energy ($E_a$) in the whole Ni leaching process was calculated from the slope of straight line to be 20.26 KJ/mol ($R^2$ = 0.9917); this value is lower than the apparent activation energy compared to other studies [14,36]. The diffusion-controlled process has an activation energy of around <12 kJ/mol, while chemical reaction-controlled process often has an activation energy of >40 kJ/mol; when activation energy is between 12 and 40 kJ/mol, the process is controlled by both diffusion and chemical reaction [62].

**Table 6.** Nickel leaching kinetics parameters of different models during the reaction time.

| Temperature | Liquid Film Diffusion Control | | Surface Chemical Reaction Control | | Solid Product Diffusion Control | |
|---|---|---|---|---|---|---|
| | $1 - (1 - x)^{2/3} = k_1t$ | | $1 - (1 - x)^{1/3} = k_ct$ | | $1 - 3(1 - x)^{2/3} + 2(1 - x) = k_dt$ | |
| | $k_1$ | $R^2$ | $K_c$ | $R^2$ | $K_d$ | $R^2$ |
| 35 °C | 0.0022 | 0.9201 | 0.0013 | 0.9493 | 0.001 | 0.9870 |
| 55 °C | 0.0033 | 0.9282 | 0.0020 | 0.9603 | 0.0016 | 0.9836 |
| 65 °C | 0.0036 | 0.9336 | 0.0023 | 0.9684 | 0.0020 | 0.9889 |
| 75 °C | 0.0040 | 0.9369 | 0.0026 | 0.9759 | 0.0025 | 0.9944 |
| 85 °C | 0.0044 | 0.9363 | 0.0030 | 0.9763 | 0.0030 | 0.9990 |

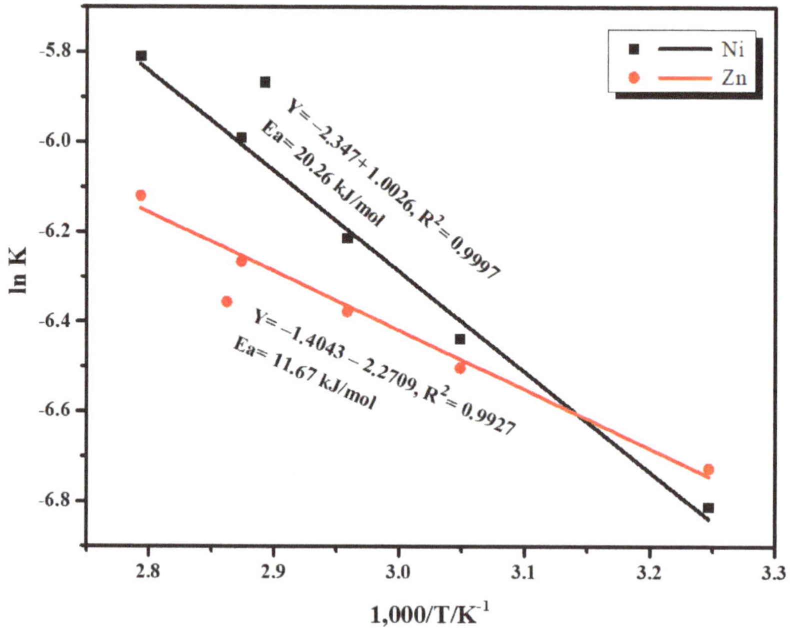


**Figure 18.** Arrhenius plot for nickel and zinc leaching from WLSR residue.

The current study shows that the apparent activation energy was between those of a typical diffusion-controlled process and a chemical reaction-controlled process, providing more evidence to prove that the sulfuric acid leaching process was controlled by both internal diffusion and interface chemical reactions. This can be attributed to the fact that the sulfuric acid concentration decreased as the leaching process proceeds, and thus the chemical reaction rate decreased, resulting in the chemical reaction becoming part of the rate-controlling step. Since the calculated value of 20.26 KJ/mol is near to the value of 12 kJ/mol significantly, the diffusion of the solid product layer contributed more than the chemical reaction to control the rate of leaching process [36]. Hence, the leaching rate improved with the increase in the leaching temperature and $H_2SO_4$ concentration and decreased the particle size of WLSR.

3.5.2. Leaching Kinetics of Zinc

The leaching kinetics of zinc were studied by investigating the leaching experiments at 35 °C, 55 °C, 65 °C, 75 °C and 100 °C; the obtained results were plotted in Figure 16b, applying the left sides of Equations (15)–(17) vs. time at different leaching temperatures and based on data acquired from optimization studies of each condition. Table 7 and Figure 19 summarize the results. It was found that Equation (17) presented a better fit of the leaching experimental result, and the correlation coefficient value for (internal diffusion) $k_d$ was 0.9566. This proved that the predominant dissolution kinetic mechanism of zinc from WLSR was controlled by the diffusion of reactants through a solid generated layer [9]. This might be because leaching occurs on the thin layer around the surface rather than on the surface itself, as the SCM suggests.

**Table 7.** Zinc leaching kinetics parameters of different models during the reaction time.

| Temperature/°C | Liquid Film Diffusion Control | | Surface Chemical Reaction Control | | Solid Product Diffusion Control | |
|---|---|---|---|---|---|---|
| | $1-(1-x)^{2/3} = k_1 t$ | | $1-(1-x)^{1/3} = k_c\ t$ | | $1-3(1-x)^{2/3}+2(1-x) = k_d t$ | |
| | $k_1$ | $R^2$ | $K_c$ | $R^2$ | $K_d$ | $R^2$ |
| 35 °C | 0.0022 | 0.9010 | 0.0012 | 0.9564 | 0.0011 | 0.9970 |
| 55 °C | 0.0030 | 0.9382 | 0.0018 | 0.9613 | 0.0014 | 09986 |
| 65 °C | 0.0033 | 0.9214 | 0.0020 | 0.9766 | 0.0016 | 0.9965 |
| 75 °C | 0.0036 | 0.9352 | 0.0023 | 0.9677 | 0.0018 | 0.9921 |
| 85 °C | 0.0039 | 0.9344 | 0.0025 | 0.9688 | 0.0021 | 0.9872 |

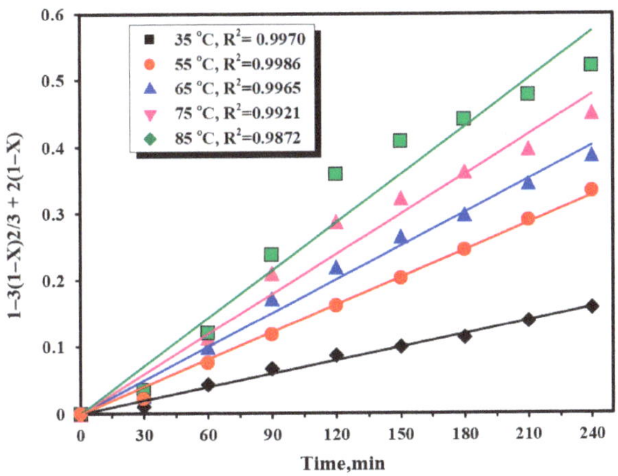


**Figure 19.** Plot of $1-3(1-x)^{2/3}+2(1-x)$ vs. time at different temperature.

The Arrhenius Equation (18) was plotted as Lnk vs. 1/T for different temperatures, as shown in Figure 18. Arrhenius Equation is used to calculate the apparent activation energy ($E_a$) of the zinc acid leaching process from the slope of the straight line, which is 11.67 kJ/mol ($R^2$ = 0.9927). Both nickel and zinc have a lower apparent activation energy, implying that they are easy to leach, which agrees with the thermodynamic results in Section 3.2.

## 4. Conclusions

In the current study, the hydrometallurgical processing of the water leaching solid residue obtained from salt-roasting of the Egyptian boiler ash for high purity and cost-effective extraction of valuable metals including Ni, Zn, Fe and Mg has been investigated. The obtained results can be concluded as follows:

1. The maximum extraction of Ni, Zn, Fe and Mg from water leaching solid residue after vanadium extraction from salt-roasting of the Egyptian boiler ash was 95.02%, 90.13%, 66.29%, and 75.73%, respectively, under the optimum leaching conditions of 8% (vol%) $H_2SO_4$ concentration, 85 °C leaching temperature and 1/15 S/L ratio for 240 min leaching time.
2. The precipitation approach using $Fe_2O_3$ as a nucleating agent was effective in removing the iron ions from the pregnant solution. The removal rate of iron ions increases to its maximum value of 95% by modifying the pH of the solution and the precipitation duration, with a minimal loss rate of Ni (4.2%) and Zn (3.8%).
3. An effective Mg precipitation efficiency of 96.9% was obtained after Fe removal, using 0.45 mol/L oxalic acid dose, 6 pH and 2 h precipitation time at ambient temperature.

Then, nickel and zinc in the purified solution was precipitated at pH of 9 as Ni-Zn hydroxide, which was subsequently transformed into $Ni(OH)_2$, $Zn(OH)_2$, and NiO-ZnO by its calcining at (450 °C). The precipitation efficiency of Ni and Zn was 92.25% and 85.51%, respectively, and the final calcined product was composed mainly of 37% Ni and 23% Zn.

4. A kinetics analysis revealed that the nickel leaching process is controlled by both diffusion through solid product layer and chemical reaction, where diffusion through solid product layer contributes more, with an activation energy of 20.26 kJ $mol^{-1}$. The kinetics of zinc dissolution are controlled by diffusion through solid product layer with an activation energy of 11.67 kJ $mol^{-1}$.
5. The chemical and physical composition and properties of the obtained sulfuric acid solid residue (SASR) enhance and recommend its utilization as adsorbent to eliminate undesirable ions from industrial effluent.

**Author Contributions:** Conceptualization, A.H.I. and X.L.; methodology, A.H.I., X.L., B.M.A., M.A.G. and A.B.E.; software, A.H.I., B.M.A., M.A.G. and A.B.E.; validation, X.L., M.A.G. and A.B.E.; formal analysis, A.H.I., B.M.A., M.A.G. and A.B.E.; investigation, A.H.I., B.M.A., M.A.G. and A.B.E.; resources, B.M.A. and M.A.G.; data curation, A.H.I. and A.B.E.; writing—original draft preparation, A.H.I.; writing—review and editing, X.L., B.M.A. and A.B.E.; visualization, A.H.I. and B.M.A.; supervision, X.L.; project administration, X.L.; funding acquisition, X.L. All authors have read and agreed to the published version of the manuscript.

**Funding:** This work was supported by Qingdao West Coast New Area 2020 Annual Science and Technology Project (3-2-2020), the National Natural Science Foundation of China (No. 51674161) and the Natural Science Foundation of Shandong Province (No. ZR2017ZC0735).

**Acknowledgments:** All the authors would like to express their sincere gratitude and appreciation for Nuclear Materials Authority, Cairo, Egypt for providing the Egyptian boiler ash sample. Also, they are grateful for Shandong University of Science and Technology, Qingdao, China for providing the required funding. Great thanks and sincere gratitude to Amr B. ElDeeb (Mining and Petroleum Department, Faculty of Engineering, Al-Azhar University in Cairo, Egypt) for his contribution by the critical revisions and improvement of the manuscript.

**Conflicts of Interest:** The authors declare no conflict of interest.

## References

1. Roskill Information Services Ltd. Nickel Market Beware: Batteries Can No Longer Be Ignored. *PR Newswire*, 21 April 2017.
2. Bide, T.; Hetherington, L.; Gunn, G. Mineral Profile: Nickel. Available online: https://www.mendeley.com/catalogue/cd4a0cf6-0bd3-3ebb-b763-b6b6bc4de59c/?utm_source=desktop&utm_medium=1.19.8&utm_campaign=open_catalog&userDocumentId=%7Bd0d00b50-22a8-42cb-900d-f51663c38740%7D (accessed on 18 August 2022).
3. British Geological Survey. British Geological Survey Briefing Note–Battery Raw Materials. Available online: http://batteryuniversity.com/learn/article/bu_311_supplementary_battery_raw_materials (accessed on 18 August 2022).
4. Aalco Stainless Steel—Introduction to Grades, Properties and Applications. 2011. Available online: https://www.aalco.co.uk/literature/files/aalco-stainless-steel.pdf (accessed on 18 August 2022).
5. Pariser, H.H.; Backeberg, N.R.; Masson, O.C.M.; Bedder, J.C.M. Changing Nickel and Chromium Stainless Steel Markets—A Review. *J. S. Afr. Inst. Min. Metall.* **2018**, *118*, 6. [CrossRef]
6. Richardson, C. Energy transition, Nickel helping to combat climate change. *Nickel Inst. Nickel Mag.* **2018**, *33*, 1–16. Available online: https://nickelinstitute.org/media/2697/nickelvol33no1spring2018_fb_en_final.pdf (accessed on 18 August 2022).
7. Kaya, M.; Hussaini, S.; Kursunoglu, S. Critical review on secondary zinc resources and their recycling technologies. *Hydrometallurgy* **2020**, *195*, 105362. [CrossRef]
8. Rabah, M.A.; El-Sayed, A.S. Recovery of zinc and some of its valuable salts from secondary resources and wastes. *Hydrometallurgy* **1995**, *37*, 23–32. [CrossRef]
9. Espiari, S.; Rashchi, F.; Sadrnezhaad, S.K. Hydrometallurgical treatment of tailings with high zinc content. *Hydrometallurgy* **2006**, *82*, 54–62. [CrossRef]
10. Warner, A.E.M.; Díaz, C.M.; Dalvi, A.D.; Mackey, P.J.; Tarasov, A.V.; Jones, R.T. JOM world nonferrous smelter survey Part IV: Nickel: Sulfide. *JOM* **2007**, *59*, 58–72. [CrossRef]
11. Crundwell, F.K.; Moats, M.S.; Ramachandran, V.; Robinson, T.G.; Davenport, W.G. Overview of the Smelting of Nickel Laterite to Ferronickel. In *Extractive Metallurgy of Nickel, Cobalt and Platinum Group Metals*; Elsevier: Amsterdam, The Netherlands, 2011; pp. 49–53. ISBN 9780080968094.

12. Meshram, P.; Abhilash; Pandey, B.D. Advanced Review on Extraction of Nickel from Primary and Secondary Sources. *Miner. Process. Extr. Metall. Rev.* **2019**, *40*, 157–193. [CrossRef]
13. Wang, F.; Liu, F.; Elliott, R.; Rezaei, S.; Khajavi, L.T.; Barati, M. Solid State Extraction of Nickel from Nickel Sulfide Concentrates. *J. Alloys Compd.* **2020**, *822*, 153582. [CrossRef]
14. Abdel-Aal, E.A.; Rashad, M.M. Kinetic study on the leaching of spent nickel oxide catalyst with sulfuric acid. *Hydrometallurgy* **2004**, *74*, 189–194. [CrossRef]
15. Henckens, M.L.C.M.; Worrell, E. Reviewing the availability of copper and nickel for future generations. The balance between production growth, sustainability and recycling rates. *J. Clean. Prod.* **2020**, *264*, 121460. [CrossRef]
16. Coman, V.; Robotin, B.; Ilea, P. Nickel recovery/removal from industrial wastes: A review. *Resour. Conserv. Recycl.* **2013**, *73*, 229–238. [CrossRef]
17. Abd El-Hamid, A.A.M.; Abu Khoziem, H.A. Physical and chemical characterization of El Kriymat boiler ash to optimize the leachability of some valuable elements. *J. Environ. Chem. Eng.* **2019**, *7*, 103362. [CrossRef]
18. Di Pietro, A.; Visalli, G.; Munaò, F.; Baluce, B.; La Maestra, S.; Primerano, P.; Corigliano, F.; De Flora, S. Oxidative damage in human epithelial alveolar cells exposed in vitro to oil fly ash transition metals. *Int. J. Hyg. Environ. Health* **2009**, *212*, 196–208. [CrossRef]
19. Amer, A.M. Processing of Egyptian boiler-ash for extraction of vanadium and nickel. *Waste Manag.* **2002**, *22*, 515–520. [CrossRef]
20. Abd El-Hamid, A.M.; Zahran, M.A.; Khalid, F.M.; Mahmoud, A.H. Leaching of hafnium, zirconium, uranium and other nuclear economic elements from petroleum ash. *RSC Adv.* **2014**, *4*, 12506. [CrossRef]
21. Masud, A. Abdel-Latif Recovery of vanadium and nickel from petroleum flyash. *Miner. Eng.* **2002**, *15*, 953–961. [CrossRef]
22. Al-Mansi, N.M.; Abdel Monem, N.M. Recovery of nickel oxide from spent catalyst. *Waste Manag.* **2002**, *22*, 85–90. [CrossRef]
23. Mohammed, H.; Sadeek, S.; Mahmoud, A.R.; Diab, H.; Zaky, D. Natural radioactivity and radiological hazard assessment of Egyptian oil ashes. *Environ. Sci. Pollut. Res.* **2016**, *23*, 15584–15592. [CrossRef]
24. Al-Ghouti, M.A.; Al-Degs, Y.S.; Ghrair, A.; Khoury, H.; Ziedan, M. Extraction and separation of vanadium and nickel from fly ash produced in heavy fuel power plants. *Chem. Eng. J.* **2011**, *173*, 191–197. [CrossRef]
25. Nazari, E.; Rashchi, F.; Saba, M.; Mirazimi, S.M.J. Simultaneous recovery of vanadium and nickel from power plant fly-ash: Optimization of parameters using response surface methodology. *Waste Manag.* **2014**, *34*, 2687–2696. [CrossRef]
26. Ettler, V.; Kvapil, J.; Šebek, O.; Johan, Z.; Mihaljevič, M.; Ratié, G.; Garnier, J.; Quantin, C. Leaching behaviour of slag and fly ash from laterite nickel ore smelting (Niquelândia, Brazil). *Appl. Geochem.* **2016**, *64*, 118–127. [CrossRef]
27. Hu, P.; Zhang, Y.; Liu, T.; Yuan, Y.; Xue, N. Source separation of vanadium over iron from roasted vanadium-bearing shale during acid leaching via ferric fluoride surface coating. *J. Clean. Prod.* **2018**, *181*, 399–407. [CrossRef]
28. Hu, P.; Zhang, Y.; Liu, T.; Huang, J.; Yuan, Y.; Zheng, Q. Highly selective separation of vanadium over iron from stone coal by oxalic acid leaching. *J. Ind. Eng. Chem.* **2017**, *45*, 241–247. [CrossRef]
29. Liu, Z.; Huang, J.; Zhang, Y.; Liu, T.; Hu, P.; Liu, H.; Zheng, Q. Separation and recovery of iron impurities from a complex oxalic acid solution containing vanadium by $K_3Fe(C_2O4)_3{\cdot}3H_2O$ crystallization. *Sep. Purif. Technol.* **2020**, *232*, 115970. [CrossRef]
30. Holloway, P.C.; Etsell, T.H. Salt Roasting of Suncor Oil Sands Fly Ash. *Can. Metall. Q.* **2004**, *43*, 535–544. [CrossRef]
31. Holloway, P.C.; Etsell, T.H. Alternative reagents for roasting Suncor oil sands fly ash. *Miner. Process. Extr. Metall.* **2004**, *113*, 153–160. [CrossRef]
32. He, D.; Feng, Q.; Zhang, G.; Ou, L.; Lu, Y. An environmentally-friendly technology of vanadium extraction from stone coal. *Miner. Eng.* **2007**, *20*, 1184–1186. [CrossRef]
33. Zhang, Y.-M.; Bao, S.-X.; Liu, T.; Chen, T.-J.; Huang, J. The technology of extracting vanadium from stone coal in China: History, current status and future prospects. *Hydrometallurgy* **2011**, *109*, 116–124. [CrossRef]
34. Tavakoli, M.R.; Dreisinger, D.B. Separation of vanadium from iron by solvent extraction using acidic and neutral organophosporus extractants. *Hydrometallurgy* **2014**, *141*, 17–23. [CrossRef]
35. Wang, C.; Yuan, Y.; Xie, B.; Li, H.-Y. Mechanism of Extraction of Vanadium from Vanadium Slag with MgO. In *Rare Metal Technology 2020*; Azimi, G., Forsberg, K., Ouchi, T., Kim, H., Alam, S., Baba, A.A., Eds.; Springer International Publishing: Cham, Switzerland, 2020; pp. 317–324. ISBN 978-3-030-36758-9.
36. Xiao, W.; Liu, X.; Zhao, Z. Kinetics of nickel leaching from low-nickel matte in sulfuric acid solution under atmospheric pressure. *Hydrometallurgy* **2020**, *194*, 105353. [CrossRef]
37. Xiao, J.; Xiong, W.; Zou, K.; Chen, T.; Li, H.; Wang, Z. Extraction of Nickel from Magnesia–Nickel Silicate Ore. *J. Sustain. Metall.* **2021**, *7*, 642–652. [CrossRef]
38. ElDeeb, A.B.; Brichkin, V.N.; Povarov, V.G.; Kurtenkov, R.V. The activating effect of carbon during sintering the limestone–kaolin mixture. *Tsvetnye Metally* **2020**, *7*, 18–25. [CrossRef]
39. Li, J.; Chen, Z.; Shen, B.; Xu, Z.; Zhang, Y. The extraction of valuable metals and phase transformation and formation mechanism in roasting-water leaching process of laterite with ammonium sulfate. *J. Clean. Prod.* **2017**, *140*, 1148–1155. [CrossRef]
40. Mu, W.; Huang, Z.; Xin, H.; Luo, S.; Zhai, Y.; Xu, Q. Extraction of Copper and Nickel from Low-Grade Nickel Sulfide Ore by Low-Temperature Roasting, Selective Decomposition and Water-Leaching Process. *JOM* **2019**, *71*, 4647–4658. [CrossRef]
41. Li, J.; Li, D.; Xu, Z.; Liao, C.; Liu, Y.; Zhong, B. Selective leaching of valuable metals from laterite nickel ore with ammonium chloride-hydrochloric acid solution. *J. Clean. Prod.* **2018**, *179*, 24–30. [CrossRef]

42. Tokuyama, H.; Nii, S.; Kawaizumi, F.; Takahashi, K. Process development for recovery of vanadium and nickel from heavy oil fly ash by leaching and ion exchange. *Sep. Sci. Technol.* **2003**, *38*, 1329–1344. [CrossRef]
43. Huaiwei, Z.; Xin, H. An Overview for the Utilization of Wastes from Stainless Steel Industries. *Resour. Conserv. Recycl.* **2011**, *55*, 745–754. [CrossRef]
44. Alex, P.; Mishra, P.; Suri, A.K. Studies on Processing of an Alnico Scrap. *Miner. Process. Extr. Metall. Rev.* **2001**, *22*, 547–565. [CrossRef]
45. Shen, Y.; Xue, W.; Niu, W. Recovery of Co(II) and Ni(II) from Hydrochloric Acid Solution of Alloy Scrap. *Trans. Nonferrous Met. Soc. China* **2008**, *18*, 1262–1268. [CrossRef]
46. Goel, S.; Pant, K.K.; Nigam, K.D.P. Extraction of Nickel from Spent Catalyst Using Fresh and Recovered EDTA. *J. Hazard. Mater.* **2009**, *171*, 253–261. [CrossRef]
47. Vuyyuru, K.R.; Pant, K.K.; Krishnan, V.V.; Nigam, K.D.P. Recovery of Nickel from Spent Industrial Catalysts Using Chelating Agents. *Ind. Eng. Chem. Res.* **2010**, *49*, 2014–2024. [CrossRef]
48. Tsai, S.-L.; Tsai, M.-S. A Study of the Extraction of Vanadium and Nickel in Oil-Fired Fly Ash. *Resour. Conserv. Recycl.* **1998**, *22*, 163–176. [CrossRef]
49. Sahu, K.K.; Agarwal, A.; Pandey, B.D. Nickel recovery from spent nickel catalyst. *Waste Manag. Res. J. A Sustain. Circ. Econ.* **2005**, *23*, 148–154. [CrossRef] [PubMed]
50. Idris, J.; Musa, M.; Yin, C.-Y.; Hamid, K.H.K. Recovery of nickel from spent catalyst from palm oil hydrogenation process using acidic solutions. *J. Ind. Eng. Chem.* **2010**, *16*, 251–255. [CrossRef]
51. Marafi, M.; Stanislaus, A. Waste Catalyst Utilization: Extraction of Valuable Metals from Spent Hydroprocessing Catalysts by Ultrasonic-Assisted Leaching with Acids. *Ind. Eng. Chem. Res.* **2011**, *50*, 9495–9501. [CrossRef]
52. Oza, R.; Shah, N.; Patel, S. Recovery of nickel from spent catalysts using ultrasonication-assisted leaching. *J. Chem. Technol. Biotechnol.* **2011**, *86*, 1276–1281. [CrossRef]
53. Ognyanova, A.; Ozturk, A.T.; De Michelis, I.; Ferella, F.; Taglieri, G.; Akcil, A.; Vegliò, F. Metal extraction from spent sulfuric acid catalyst through alkaline and acidic leaching. *Hydrometallurgy* **2009**, *100*, 20–28. [CrossRef]
54. Ferella, F.; Ognyanova, A.; De Michelis, I.; Taglieri, G.; Vegliò, F. Extraction of metals from spent hydrotreating catalysts: Physico-mechanical pre-treatments and leaching stage. *J. Hazard. Mater.* **2011**, *192*, 176–185. [CrossRef]
55. Yang, Q.Z.; Ng, R.S.; Qi, G.J.; Low, H.C.; Zhang, Y.P. Economic Viability of Nickel Recovery from Waste Catalyst. *Key Eng. Mater.* **2010**, *447–448*, 765–769. [CrossRef]
56. Yang, Q.Z.; Qi, G.J.; Low, H.C.; Song, B. Sustainable recovery of nickel from spent hydrogenation catalyst: Economics, emissions and wastes assessment. *J. Clean. Prod.* **2011**, *19*, 365–375. [CrossRef]
57. Ju, J.; Feng, Y.; Li, H.; Xu, C.; Yang, Y. Efficient Separation and Recovery of Vanadium, Titanium, Iron, Magnesium, and Synthesizing Anhydrite from Steel Slag. *Min. Metall. Explor.* **2022**, *39*, 733–748. [CrossRef]
58. Stefanova, V.P.; Iliev, P.K.; Stefanov, B.S.; Avramov, A.G. Selective Dissolution of FeCuNiCoMn Alloy Obtained after Pyrometallurgical Processing of Polymetallic Nodules. In Proceedings of the Eighth ISOPE Ocean Mining Symposium, Chennai, India, 20–24 September 2009; pp. 186–190.
59. Yang, Z.; Li, H.-Y.; Yin, X.-C.; Yan, Z.-M.; Yan, X.-M.; Xie, B. Leaching kinetics of calcification roasted vanadium slag with high CaO content by sulfuric acid. *Int. J. Miner. Process.* **2014**, *133*, 105–111. [CrossRef]
60. Levenspiel, O. *Electrochemical Reaction Engineering*, 3rd ed.; John Wiley & Sons: New York, NY, USA, 1999; Volume 35, pp. 566–588. [CrossRef]
61. Rasoulnia, P.; Mousavi, S.M. V and Ni recovery from a vanadium-rich power plant residual ash using acid producing fungi: Aspergillus niger and Penicillium simplicissimum. *RSC Adv.* **2016**, *6*, 9139–9151. [CrossRef]
62. Ashraf, M.; Zafar, Z.I.; Ansari, T.M. Selective leaching kinetics and upgrading of low-grade calcareous phosphate rock in succinic acid. *Hydrometallurgy* **2005**, *80*, 286–292. [CrossRef]

MDPI AG
Grosspeteranlage 5
4052 Basel
Switzerland
Tel.: +41 61 683 77 34

*Minerals* Editorial Office
E-mail: minerals@mdpi.com
www.mdpi.com/journal/minerals

www.ingramcontent.com/pod-product-compliance
Lightning Source LLC
Chambersburg PA
CBHW042013260926
51850CB00020B/3306